THE SURFACE OF MARS

YALE PLANETARY EXPLORATION SERIES

THE SURFACE OF MARS

MICHAEL H. CARR

NEW HAVEN AND LONDON YALE UNIVERSITY PRESS

Published with assistance from the foundation
established in memory of Calvin Chapin
of the Class of 1788, Yale College.

Designed by Christopher Harris
and set in Electra and Memphis types by
The Composing Room of Michigan, Inc.
Printed in the United States of America by
The Murray Printing Co., Westford, Mass.

Library of Congress Cataloging in Publication Data

Carr, M. H. (Michael H.)
The surface of Mars.

Bibliography: p.
Includes index
1. Mars (Planet)—Surface. I. Title.
QB641.C363 559.9'23 81-3425
ISBN 0-300-02750-8 AACR2
(pbk.) 0-300-03242-0

10 9 8 7 6 5 4 3 2

CONTENTS

PREFACE

This book was conceived early in 1977, after the tension and excitement of the early phases of the Viking mission had subsided. Most of the participating scientists, myself included, had by this time returned home after an exhilarating six months at the Jet Propulsion Laboratory in Pasadena. The six months had been packed with drama. We had seen the surface of Mars unfold in exquisite detail in the pictures from the orbiters. We had landed two spacecraft on the surface and photographed their surroundings with extraordinary clarity. We had poked the surface, analyzed it, pushed it around. We had monitored the local weather and had experienced a long and tantalizing search for living organisms and organic compounds. But by the spring of 1977 the excitement had abated considerably, and the time seemed appropriate for preparing a synthesis of the geologic results.

However, administrative duties and other scientific tasks prevented my starting the project at that time, and it was temporarily put aside. It was not until well into 1979 that I seriously began to entertain again the possibility of writing a book summarizing our knowledge of the martian surface. The time seemed particularly propitious. Interpretation of the early Viking results had by then matured significantly. But more important, three of the four Viking spacecraft were, to everyone's surprise, still operating. We had made observations of the surface and atmosphere for over a full martian year and had extended the orbiter coverage to the entire planet. Moreover, I was becoming increasingly aware of, and concerned about, a gap that was developing between the general perception of the planet's surface and the reality revealed by the continuous stream of orbiter data. By that time we had acquired 40,000–50,000 pictures of the planet. While the few of us directly involved with the imaging experiment had examined most of them, it was, and still is, difficult for the general scientific community to gain an appreciation of their content, not because of any restrictions on the distribution of the data—there are none other than cost—but because of not knowing where among the data to look for information relevant to a particular topic of interest. A summary by one who had seen most of the pictures seemed due.

Fortunately, I was at that time asked to teach a course on the geology of Mars at the California Institute of Technology in Pasadena. Seeing this as an opportunity to write the book, I requested and was granted a leave of absence from the U.S. Geological Survey, and in the fall of 1979 I went to Pasadena to teach and write. Most of the book was written there in late 1979. It was revised in early 1980 to take into account some publications that appeared that spring and some new orbiter data.

ACKNOWLEDGMENTS

Spacecraft missions are in large part engineering enterprises. Without sound engineering there is no science. My first acknowledgment is, therefore, to the thousands of engineers who made the Viking mission, upon which much of this book is based, so successful. Especially warm thanks go to the Imaging Systems Section and the various image-processing facilities at the Jet Propulsion Laboratory, whose products constitute the bulk of the book's illustrations. Thanks also go to the California Institute of Technology for providing me with the opportunity to write and to my class there who provided such stimulating discussion. F. Fanale of the Jet Propulsion Laboratory, R. Phillips of the Lunar and Planetary Science Institute, G. Briggs of NASA headquarters, L. Martin of Lowell Observatory, and P. Toulmin, C. Breed, and L. Soderblom of the U.S. Geological Survey read various chapters, while D. Wilhelms of the U.S. Geological Survey reviewed the entire manuscript. I am particularly indebted to H. Klein of NASA-Ames, who permitted me to include as a chapter one of his excellent summary articles on the search for life on Mars. N. Evans, manager of the Viking Orbiter Imaging Group at the Jet Propulsion Laboratory, helped me with many of the logistical problems of identifying and locating pictures and provided the Orbiter Imaging Team with much enthusiastic support during the Viking Extended Mission. While the writing of the book was not directly supported by NASA, my work on Mars over several years has benefited from grants from the Office of Space Sciences, NASA Headquarters. Finally, I want to thank the U.S. Geological Survey for its encouragement and my many colleagues there who provided much needed advice.

U.S. Geological Survey
Menlo Park, Calif.

INTRODUCTION

Mars, the fourth planet from the Sun, has long been an object of special interest. It was realized as far back as we have historical records that most objects in the night sky remain fixed with respect to each other in their slow rotation across the sky. These objects were grouped into constellations and given names such as the Great Bear, the Swan, and the Huntsman, which purported to describe the shapes of the groups. Five objects, however, could not be fitted into the scheme. They wandered slowly through the constellations from west to east and occasionally reversed direction for a few months. They also varied greatly in brightness, in contrast to the invariant stars. These five we now call planets, after the Greek word *planes*, meaning "wanderer." Distinctive among them because of its red color is Mars. To the ancients, the color evoked images of fire and blood, and the planet came to symbolize the carnage and destruction of war. The Greeks accordingly called it Ares and the Romans Mars, for their gods of war. We now know the original five wanderers as Mercury, Venus, Mars, Jupiter, and Saturn and have discovered three more, Uranus, Neptune, and Pluto, for a total count of nine planets, including Earth (table I.1).

In more recent times the fascination of Mars has stemmed largely from the possibility that life might exist there under conditions similar to those here on Earth. Following the invention of the telescope in the early 1600s, Mars was recognized as having many Earth-like characteristics. It had surface markings which, by analogy with Earth, were thought to be seas and termed *maria*. The early observers were able to track the markings and determine that the length of day on Mars is almost identical to that on Earth. Polar caps were also observed, as were transient brightenings interpreted as clouds. For 200 years Mars was thus perceived as being very Earth-like. A dramatic change followed the close opposition of 1877. Several maps made at that time showed numerous linear markings, which were termed canals, and these rapidly became the planet's most renowned feature. While some observers failed to detect the markings, despite access to some of the world's best telescopes, others scorned either the visual perception of the unbelievers or the optics of their instruments and, insisting that the markings were there in profusion, drew ever more elaborate maps of the canal network. The possibility of canals built by an advanced civilization struggling to survive on a slowly dessicating planet caught the public's imagination and stimulated a considerable literature, with numer-

TABLE I.1. ORBITAL AND PHYSICAL PARAMETERS OF THE PLANETS, THE MOON, AND THE MARS SATELLITES

Object	*Diameter (km)*	*Mass (gm)*	*Mean Density* (gm/cm^3)	*Visual Geometric Albedo*	*Rotation Period (days)*	*Obliquity*	*Revolution Period*	*Semimajor Axis (A.U. for planets;* 10^3 *km for satellites)*	*Orbit Inclination (with respect to ecliptic for planets; planetary equator for satellites)*	*Orbit Eccentricity*
Sun	1,391,400	1.987 (33)	1.4	—	25.4	7°.25	—	—	—	—
Mercury	4,864	3.30 (26)	5.5	0.10	58.6	<7°	0.2408 yr	0.387	7.0	0.206
Venus	12,100	4.87 (27)	5.2	0.586	243R	~179°	0.6152 yr	0.723	3.39	0.007
Earth	12,756	5.98 (27)	5.52	0.39	1.00	23°.5	1.000 yr	1.000	0.00	0.017
Mars	6,788	6.44 (26)	3.9	0.15	1.02	25°.0	1.881 yr	1.524	1.85	0.093
Jupiter	137,400	1.90 (30)	1.40	0.44	0.41	3°.1	11.86 yr	5.203	1.31	0.048
Saturn	115,100	5.69 (29)	0.71	0.46	0.43	26°.7	29.46 yr	9.54	2.49	0.056
Uranus	50,100	8.76 (28)	1.32	0.56	0.45R	97°.9	84.0 yr	19.18	0.77	0.047
Neptune	49,400	1.03 (29)	1.63	0.51	0.6	28°.8	164.8 yr	30.07	1.78	0.008
Pluto	5,800	6.6 (26)	6?	0.13	6.4	?	284.4 yr	39.44	17.17	0.249
Moon	3,476	7.35 (25)	3.34	0.115	27.3	6°.7	27.3 d	384	18–29°	0.055
Phobos	18 × 22	9.2 (18)	1.9	0.06	7.65h	—	0.319 d	9	1.1	0.021
Deimos	12 × 13	1.7 (18)	1.5	0.06	30.29h	—	1.26 d	23	1.6	0.003

SOURCE: Hartmann 1972; Veverka and Thomas 1979.
NOTE: R under rotation period indicates retrograde motion.

ous authors, such as Wells, Burroughs, and Bradbury, telling stories of invasions from Mars, of colonies on the planet, and contacts with exotic civilizations.

The facts about Mars turned out to be almost as bizarre as the fiction. The modern era of Mars exploration began on July 15, 1965, when the Mariner 4 spacecraft flew by the planet at a closest distance of 9,780 km. The first pictures revealed an apparently dead, cratered surface much like that of the Moon. There was no indication of the canals and oases of earlier maps, and the prospects for life on Mars dimmed considerably. This somewhat disappointing result was confirmed by two subsequent spacecraft, Mariners 6 and 7, which in 1969 sent back more pictures of a cratered surface and confirmed that the atmosphere was thin, less than one-hundredth that of Earth, and composed primarily of carbon dioxide. It was not until 1971 that the geological diversity of the martian surface was revealed. Following a series of misadventures, which included losing one spacecraft and arriving at the planet only to find it almost completely hidden from view by atmospheric dust, Mariner 9 started systematically to map the planet early in 1972, about 6 weeks after insertion into orbit.

The Mariner 9 pictures revealed a strange planet with huge volcanoes, vast canyons, and what were, seemingly, enormous dry river beds. They showed that the earlier spacecraft had fortuitously passed over only the older parts of the planet, where a primitive, densely cratered surface was preserved. Much of the surface was clearly younger than that seen earlier and had been subject to a wide variety of geologic processes, which had created features of surprisingly large dimensions, considering the rather modest size of the planet, a little over half that of Earth (see table I.1). The presence of what appeared to be river channels was particularly exciting. By that time it was known that liquid water is unstable on the martian surface, so, in order for the channels to have been cut by water, more clement climatic conditions must have existed in the past. The prospect of running water and temperate climates gave considerable encouragement to those concerned with detecting life on the planet. The findings were timely in that plans to land spacecraft on Mars and look for life were well advanced.

The latest episode in the exploration of Mars is the Viking mission. Planning for Viking was started in the late 1960s before the Mariner 9 results, and from the start the mission was conceived as being primarily concerned with detection of life. After numerous instrument developments and cost problems, two Viking spacecraft were successfully launched to Mars in the summer of 1975. Each consisted of an orbiter and a lander and carried an elaborate array of scientific instruments. On the landers most were related to the prime goal of life detection. The two spacecraft went into orbit around Mars in the summer of 1976, and the first successful soft landing on the planet was achieved on July 20, 1976, when the first Viking lander touched down on Chryse Planitia. A second vehicle landed shortly after, on September 3, in Utopia Planitia. Both landers operated for several years, and, while they detected no life, they have returned an enormous amount of data on the chemistry of the soil, the meteorologic conditions of the landing sites, and local landforms. Meanwhile the orbiters had been scrutinizing the rest of the planet, and by the end of the orbiter mission in August 1980, they had taken close to 55,000 pictures of the planet, from which most of the illustrations in this book have been selected.

The Viking spacecraft confirmed and added to the impression of geologic diversity. Well-integrated valley networks were observed, implying that parts of the surface have been subjected to the slow erosion of running water. Other, much larger, channels suggest episodic floods of enormous magnitude. Collapsed ground, morainelike features, and surface scour suggest the action of ice, while giant landslides and rock glaciers indicate mass wasting on a grand scale. The ejecta patterns around impact craters indicate that liquid water and ice may exist at relatively shallow depths below the surface. Vast sand seas around the poles attest to the efficacy of wind transport. Lava flows hundreds of kilometers long bear witness to repeated large eruptions of fluid lava, and the presence of sparsely cratered volcanoes suggests that volcanism has continued into the relatively recent geologic past. The Viking data also better define the conditions under which geologic processes operate at the martian surface. Of particular note are the new constraints on the stability of water near the surface, its availability, and the possibility of different conditions in the past. Integration of all these findings into a coherent story of the geologic evolution of the planet is the main intent of this book.

The book is aimed at the informed scientific reader. A working knowledge of physics and chemistry and a familiarity with scientific units are assumed. Also assumed is a basic vocabulary of geology, although the use of specialized terms is avoided where possible. Those that might be unfamiliar to most scientists are explained. This tends in places to give the text an imprecise and rather general tone, but this was thought preferable to the profuse use of technical terminology.

The main focus is on the surface geology; other subjects are discussed only insofar as they illustrate a geologic point. A possible exception is chapter 14, on the search for life, but our perception of Mars has been so colored by the prospect that life might exist there that I feel that any discussion of Mars must address the life issue. The emphasis on geology may leave dissatisfied those more interested in the atmosphere or the planet's interior, but I thought it prudent to leave detailed discussion of such subjects to others more familiar with them. Even with this narrow focus, the discussion wanders into several nongeologic topics, such as planetary motions, climate changes, and evolution of the atmosphere. I have drawn freely from Earth and the Moon for analogies, but I have deliberately avoided long asides about other planetary bodies. The book is about Mars, and most of the discussion specifically concerns that body.

One problem with writing this book is that the study of the geology of Mars is in its infancy. We have recently acquired an enormous amount of information about the planet, but much of it has not even been disseminated throughout the scientific community, let alone assimilated. Interpretation of the data is thus still tentative and controversial. The subject is also volatile, with rapid changes in perception and understanding from year to year. This inevitably leads to a lack of conclusiveness in many areas, since a general consensus is lacking. Another problem caused by the relative immaturity of the science is that some of the derivative data are mutually inconsistent. This is particularly true for crater counts, many of which differ widely according to who did the counting, even though they are reportedly of the same geologic units. On other topics there has been little systematic work, so that the basic knowledge of what exists where is lacking. Different types of plains have been recognized, for example, but their areal distribution is not known. As a result the book is somewhat uneven; the level of detail depends largely on the level of knowledge. In some cases I have tried to fill gaps from personal knowledge gained from my direct involvement in the acquisition of the imaging data. These parts of the text are easily recognized because they are only sparsely referenced.

Most books on the geology of Earth are organized according to process, and, to some extent, I have done that here. But such an organization presupposes that the processes are known, and this is commonly not the case for Mars. The book, accordingly, is arranged mainly by type of surface feature. After a historical summary and two overview chapters, one on the surface and one on the atmosphere, craters are discussed in detail. They are the dominant landform that has survived from the planet's early years; they provide a means of assessing age and offer clues concerning subsurface structure and the efficacy of such processes as volcanism and erosion in modifying the surface. Following the crater discussion in chapters 4 and 5 is a largely descriptive chapter (6) on the distribution of different types of plains and cratered uplands. These two types of terrain cover much of the martian surface, and other features are mostly superposed on them. Three chapters (7–9) follow that are concerned mainly with the internal processes of tectonism and volcanism. The canyons are included there because they appear to have been formed largely by faulting, but a tectonic origin is by no means certain. The next two chapters (10 and 11) concern modification of the surface by water and wind. Polar processes, surface chemistry, and the present distribution of volatiles are discussed next (chapters 12 and 13) to complete the overview of the surface. Additional chapters on the search for life on the planet and on the two moons, Phobos and Deimos, are not strictly about martian geology but are closely related to the central theme. The main body of the book closes with a short summary of the geologic history (chapter 16). Appendixes include a commentary on how the Viking orbiter pictures were acquired, a guide to the use of the pictures, and information on the availability of maps of Mars.

1 HISTORICAL PERSPECTIVE

EARLY OBSERVATIONS

Mars holds a special place in the history of science, for it was through study of its motions that Kepler was able to formulate his three general laws of planetary motion. In his efforts to understand the motions of the planets, Kepler made extensive use of the observations of Tycho Brahe. Although Brahe made his observations between 1580 and 1600, before the invention of the telescope, he had designed a number of special instruments to measure position accurately. Kepler started with two assumptions: the first was that Earth moves in a circular orbit with a period of 365 days, and the second was that Mars moves in an unknown orbit with a period of 687 days. The latter figure was derived by noting the position of Mars with respect to the Sun over many years. By a process of triangulation, taking advantage of the numerous positional determinations of Mars made by Brahe at times when Earth and Mars were in different parts of their orbits, Kepler was able to calculate the distance of Mars from the Sun for any specified position of the planet in the sky. After much trial and error he was forced to conclude that the orbit of Mars is an ellipse with the Sun at one focus. This, generalized to all planets, is Kepler's first law. His second law, which states that a straight line joining the sun and a planet sweeps out an equal area in an equal time, he also derived from his study of Mars. This followed from the faster motion of Mars near perihelion as compared to aphelion. The first two laws were published in 1609. Nine years later, in 1618, Kepler published his third law of planetary motions, which states that the square of the orbital period of a planet is proportional to the cube of its mean distance from the Sun. The second and third laws provide a simple explanation for the occasional retrograde motion of Mars as seen from Earth, when the planet for a short time reverses its normal motion relative to the stars (figure 1.1).

Galileo is held to be the first person to have viewed Mars through a telescope, around the year 1610. Although with Venus he was able to confirm its various phases, he was unable to make a similar determination for Mars, because the disc was too small for his crude instruments and because Mars is always almost fully illuminated at closest approach. Francisco Fontana, however, made several drawings of the planet in 1638, one of which shows a distinctly gibbous phase (between half and full illumination of the disc). By 1659 Huyghens had been able to demonstrate that Mars, like Earth, rotates around a north-south axis with a period close to 24 hours. Seven years later, in 1666, Cassini further refined the rotation period and noted the presence of polar caps. Drawings by Huyghens in 1672 also showed the polar caps and an equatorial marking that could be Syrtis Major, the planet's most prominent dark marking. By the end of the century the advance and recession of the polar caps had also been noted.

Herschel, in the late 1700s, made extensive observations of Mars and published his data in 1781 and 1784. By tracking fixed features on the surface, he estimated the length of the day to be 24 hr 39 min 22 sec, which is remarkably close to the currently accepted value of 24 hr 39 min 35 sec for the mean solar day.* He was also able to measure the inclination of the rotation axis and correctly deduced that Mars must have seasons analogous to those on Earth. He speculated on the advance and retreat of the polar caps, suggesting, again correctly, that the ice cover must be very thin. He also attributed transient brightenings to clouds. His perception of many martian phenomena was thus remarkably modern. However, one area where he and others of his day were in error is with regard to water. The dark areas on the surface were generally regarded as seas and termed maria. A dark band along the edge of the polar cap, first recorded by Maraldi, a nephew of Cassini, in the early 1700s, was interpreted as meltwater from

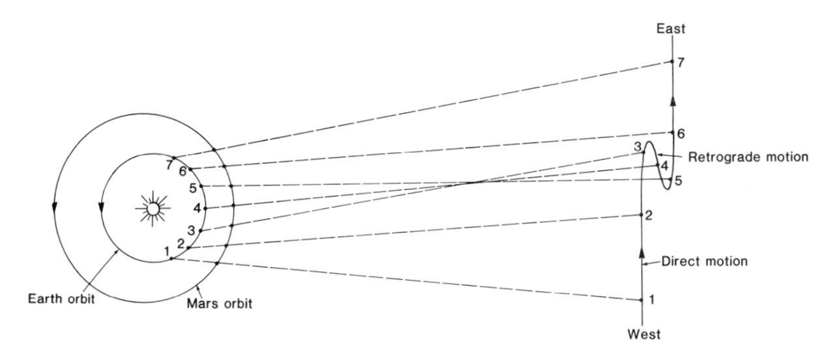

Figure 1.1. Retrograde motion of Mars as viewed from Earth. For short periods near opposition, Mars appears to reverse its direction of motion with respect to the star background. The reversal is caused by the faster angular motion of Earth, as shown.

*The sidereal day, the rotation period with respect to the stars, is 24 hr 37 min 23 sec, as compared with 23 hr 56 min 4 sec for Earth.

the cap. This perception of abundant water continued to influence observations well into the twentieth century.

During the late nineteenth century, maps of the planet were made by several observers, notably Beer and V. Mädler, Proctor, Niesten, and Flammarion, who published several maps in the 1880s. The most influential map, however, was that published by Schiaparelli (1878), based on his observations during the 1877 opposition. With his maps he introduced a new system of nomenclature (figure 1.2), which became standard by the

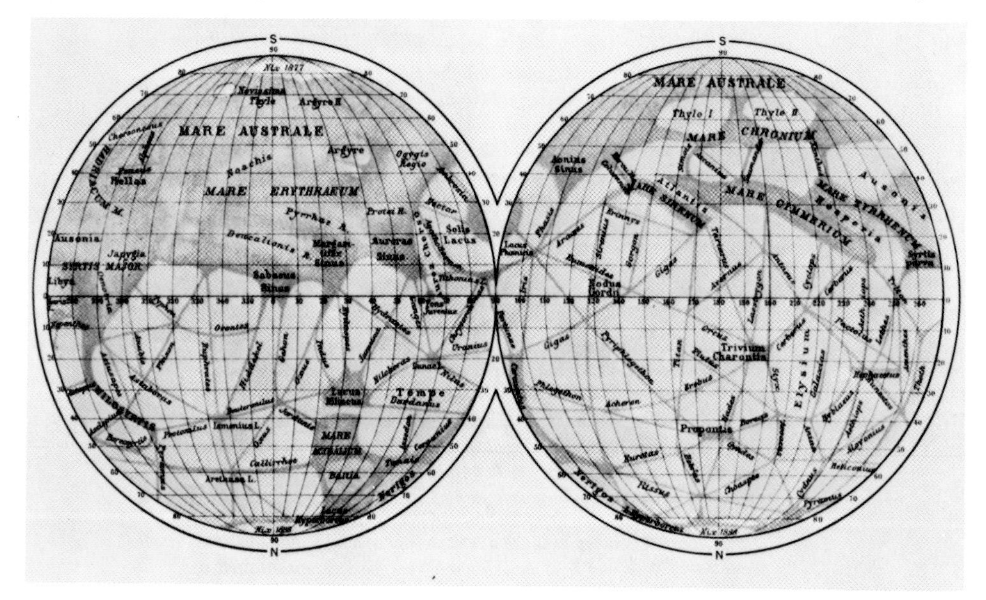

Figure 1.2. These maps of Giovanni Schiaparelli, published by Flammarion (1892), introduced the nomenclature that was to be used subsequently by most telescopic observers. The maps also depicted numerous linear features, or canals. South is up because of the inverted image at the telescope.

end of the century and which was subsequently elaborated upon by other observers, such as Antoniadi, Flammarion, and Lowell. Most of the markings were given classical or biblical names for various real or mythical geographic features of the ancient world (MacDonald 1971), and only in recent years, with the acquisition of spacecraft data, has the system been changed significantly. Schiaparelli's map is of historical importance in yet another respect. It was the first map to portray the elaborate network of linear features that was to be an essential component of so many maps that followed.

The story of canals on Mars is one of the more bizarre in the history of science. The term appears to have been introduced in the 1860s by the Italian observer Pietro Angelo Secchi, S.J., who used the Italian word *canale,* meaning "channel" or "canal," to describe some faint linear features. In describing his observations of 1877, Schiaparelli used the term to describe numerous faint linear markings that were just visible during periods of unusually good seeing. During the next opposition, in 1879, Schiaparelli was astounded to observe that many of the features formerly seen as single lines now appeared as pairs of parallel lines. This doubling became known as gemination in subsequent literature. Initially, Schiaparelli's observations of canals did not attract much attention. In the late 1880s, however, reports of canals started to proliferate as Schiaparelli's observations became more widely known and more people strained to see the linear markings.

The "canals" quickly became the focus of all discussion on Mars. Ever more elaborate maps were made showing numerous geminations and oases where canals met. The number of canals increased, from the original hundred or so identified on the map of Schiaparelli to around five hundred on the later maps of Percival Lowell. Doubts about their existence by such renowned astronomers as E. E. Barnard of Lick Observatory and G. E. Hale of Yerkes Observatory were drowned in the enthusiasm. In Lowell the canalists acquired a particularly vocal and articulate proponent. In 1894 Lowell founded an observatory in Flagstaff, Arizona, and directed much of its effort to the study of Mars. Lowell not only enthusiastically supported the existence of canals but aroused considerable public interest by speculating, in three widely read books published close to the turn of the century, that the canals had been built by an advanced civilization (Lowell 1906, 1909, 1910). He argued that the integrated planetwide network could only be sustained by a highly organized society in which war had been outlawed. He further suggested that the purpose of the network was to distribute water from the wet circumpolar regions to the dessicated equatorial deserts. He ascribed the failure of other observers to confirm the existence of the canals to inferior optics or seeing at their observatories, as compared with his at Flagstaff.

Meanwhile, skepticism about the canals became more widespread within the astronomical community, as better telescopes and photographs failed to confirm their existence. Two English astronomers, Evans and Maunder (1903), conducted an experiment with schoolchildren in which they demonstrated the tendency for an observer to connect by straight lines irregular markings which are at the threshold of visibility. This solution to the problem of canals is implicit in maps and drawings produced over a 20-year period by the French astronomer Antoniadi (1930), which showed irregular arrays of spots and streaks in areas where previous observers had shown canals. This is essentially how the matter rested until spacecraft began to explore the planet. The concept of canals, however, died hard. As late as the early 1960s, Slipher published a series of photographs that he claimed "recorded traces of so many canals and oases in the same position, form and character as drawn on the Lowell and Lick Observatory maps of the planet, that they remove all doubt . . . their existence as true markings on the planet has been clearly established" (Slipher 1962). Nevertheless, whereas most of the irregular features portrayed on the Schiaparelli and subsequent maps are clearly identifiable in spacecraft images, very few of the canals are (Sagan and Fox 1975). The irregular markings are almost certainly caused by the differential distribution of surficial debris by the wind; the canals appear to be largely an imaginary perception on the part of the observer of irregular groups of surface markings close to the limit of telescopic resolution.

PRESENT TELESCOPIC VIEW

Orbital and rotational motions

As viewed from Earth, Mars, like other planets, moves slowly through the star background and varies in brightness as the relative position of the two planets within the solar system changes. When the two planets are on opposite sides of the Sun (superior conjunction), the Earth-Mars distance is close to 400 million km, and no surface features can be distinguished at the telescope. The planet then has a brightness equivalent to a +1.5-magnitude star and subtends an angle of only 3.5 arcsec. In contrast, at closest approach, when Earth passes between Mars and the Sun (opposition), the distance between the two planets may be as small as 55 million km. At this distance Mars subtends an angle of 25 arcsec, and surface features as small as 150 km can be distinguished with the best telescopes (figure 1.3). The time interval over which Mars brightens and fades is

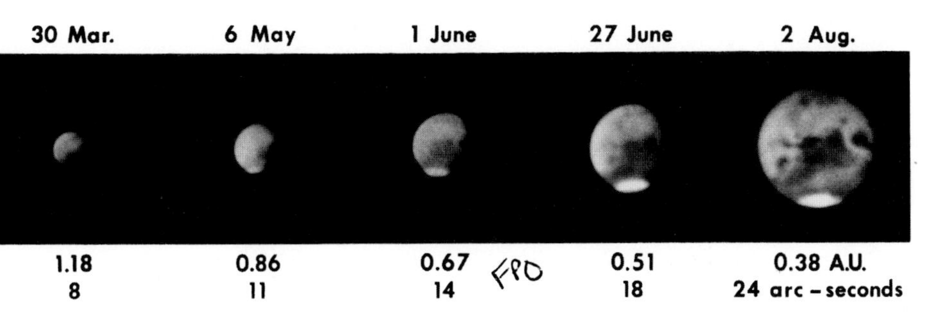

Figure 1.3. The changing size of the Mars image during opposition, as viewed from Earth. This series of images was taken during 1971 at the Lowell Observatory, Flagstaff, Arizona. It vividly demonstrates the importance of oppositions for telescopic viewing of Mars. Oppositions occur approximately every 2 years, and only during the few months on either side of opposition can the markings be seen clearly. In the August 2 image Solis Lacus is visible to the lower left of the disc, Sinus Meridiani at the middle right. (Courtesy of Lowell Observatory.)

governed by the orbital motions. Earth takes 365 days to travel around the Sun; Mars takes 687 Earth-days, or 43 days short of 2 Earth-years. The combination of the two motions causes successive close approaches to be around 780 days apart. The precise interval changes slightly because of orbital eccentricities.

The orbit of Mars (figure 1.4) is distinctly elliptical (eccentricity of 0.093), in contrast to the near-circular orbit of Earth (eccentricity of 0.017). The high eccentricity affects the planet in a number of ways. One observa-

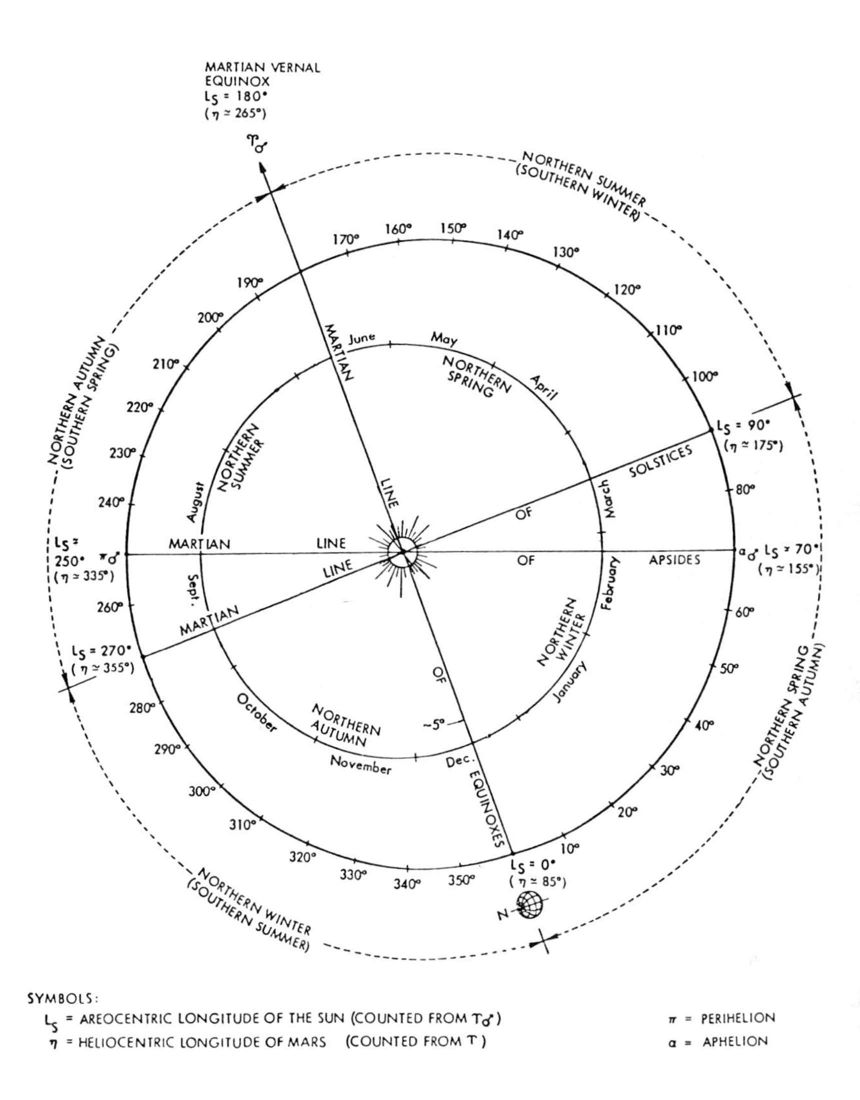

Figure 1.4. The motions of Earth and Mars around the Sun. The orbit of Mars is distinctly elliptical, while that of Earth is nearly circular. Times during the martian year are commonly referred to in terms of L_s, the areocentric longitude of the Sun. This is the angle between the Mars-Sun line and the line of equinoxes, which is measured as shown. (Adapted from Michaux and Newburn 1972, NASA/JPL.)

tional effect is the different size of the Mars image at different oppositions. At closest approach to the Sun (perihelion), the Mars-Sun distance is 1.381 A.U.,* or 206.5 million km; at its farthest distance (aphelion) the values are 1.666 A.U., or 249.1 million km. The mean distance, or semimajor axis of the orbit, is 1.524 A.U. If opposition occurs when Mars is close to perihelion, the Earth-Mars distance is close to its minimum of 55 million km. However, if opposition occurs when Mars is close to aphelion, then the Earth-Mars distance may be as large as 100 million km. Thus the

*Distances within the solar system are commonly measured in Astronomical Units (A.U.). One A.U. is the mean Earth-Sun distance, or 149.5 million km.

maximum size of the Mars image, as viewed from Earth as it passes through opposition, varies by almost a factor of two. During this century particularly favorable oppositions, when Mars passed within 60 million km of Earth, occurred in 1909, 1924, 1939, 1956, and 1971. Even at favorable oppositions Mars presents a relatively small telescopic image, being equivalent in size to a 45-km crater on the Moon, or half the size of the crater Copernicus. For most of the viewing cycle, of course, the image is much smaller.

The inclination of the rotation axis of Mars to the normal to its orbital plane is approximately 25°, close to the 23.5° tilt for Earth. Mars therefore undergoes seasonal variations in temperature analogous to those on Earth. A major difference between the two planets is caused by the high eccentricity of Mars's orbit, which results in significant differences in the lengths of their respective seasons. Because the orbital velocity changes on account of the high eccentricity, the time taken for Mars to pass through any quadrant in its orbit varies, with the times spent close to perihelion being considerably shorter than those spent close to aphelion. Since it is the southern hemisphere that tilts toward the Sun when Mars is at perihelion, the southern springs and summers are, respectively, 52 and 25 days shorter than the equivalent seasons in the north (table 1.1). A similar effect on Earth causes only a 3-day difference because of the low eccentricity. The orbital eccentricity of Mars also affects peak summer temperatures. Because Mars is about 20 percent closer to the Sun at perihelion than at aphelion, the insolation is about 45 percent higher. Southern summers are therefore not only shorter but hotter, with peak temperatures around 30°C higher than those in the north. The tilt of the southern hemisphere of Mars toward the Sun at perihelion also affects telescopic viewing of the two hemispheres. The best oppositions, those near perihelion, coincide with late southern spring and early southern summer, when the southern hemisphere is tilted not only toward the Sun, but also toward Earth. The telescopic record is therefore biased toward the southern hemisphere.

Times during the martian year are commonly referred to in terms of the

TABLE 1.1 THE LENGTH OF MARTIAN SEASONS COMPARED WITH THOSE OF EARTH

	Season		*Duration of the Seasons on* *Mars*		*Earth*
Areocentric Longitude of the Sun L_s	*Northern Hemisphere*	*Southern Hemisphere*	*Martian Days*	*Terrestrial Days*	*Terrestrial Days*
0–90°	Spring	Fall	194	199	92.9
90–180°	Summer	Winter	178	183	93.6
180–270°	Fall	Spring	143	147	89.7
270–360° or 0°	Winter	Summer	154	158	89.1
			669	687	365.3

SOURCE: Michaux and Newburn 1972.

areocentric longitude of the sun, or L_s. This is a measure of the position of Mars in its orbit, and hence of the season. It is the angle between Mars, the Sun, and the autumnal equinox, as shown in figure 1.4. Northern spring extends from $L_s = 0$ to $L_s = 90°$, northern summer from $L_s = 90°$ to $L_s = 180°$, and so forth.

Long-term alterations in the pattern of the martian seasons result from slow changes in the various rotational and orbital parameters. Two important sources of variation are the precession of the spin axis and the precession of the axis of the orbit. Precession is the term given to the slow conical motion of an axis of rotation, such as that generally observed with a spinning top. Precession of the rotational axis of Mars causes a slow rotation of the line of equinoxes—the line of intersection of the equator with the orbit plane. The time taken to complete the cycle—that is, for the axis to return to its starting orientation in inertial space—is 175,000 years (Ward 1973). Precession of the axis of the orbit causes the line of apsides—the line joining perihelion and aphelion—to rotate. Thus the position of perihelion in inertial space slowly changes. The period of this motion is 72,000 years. The net effect of the two motions is to cause the climatic regimes of the two hemispheres to change with a period of 51,000 years (Leighton and Murray 1966). The southern hemisphere is now tilted toward the Sun at perihelion, so it has the short hot summers. In about 25,000 years, however, precession of the orbital and rotational axes will have caused the northern hemisphere to be tilted toward the Sun, and this hemisphere will have the short hot summers. In another 25,000 years the orientation will have returned to approximately the present one.

Other causes of long-term climate changes are variations in orbital eccentricity, the inclination of the orbit plane, and the tilt of the spin axis. The eccentricity of the orbit ranges from 0.004 to 0.141. Short-term variations have a period of 95,000 years and long-term variations a period of approximately 2 million years (Murray et al. 1973; Ward 1974). The main effect of eccentricity is to modulate the amplitude of the hemispheric differences within the precessional cycle. We just saw how eccentricity results in different climatic conditions in the northern and southern hemispheres. If the eccentricity were zero, there would be no differences; but the larger the eccentricity, the larger the differences. Changes in the inclinations of the orbit and the spin axis cause changes in obliquity—the angle between the plane of the equator and the plane of the orbit. At present the obliquity is 25°, but it ranges from 14.9° to 35.5° with a period of 1.2×10^6 years (Ward 1974). Obliquity changes affect the latitudinal distribution of the Sun's radiation. At high obliquities more radiation falls on the poles than at low obliquities; therefore, polar temperatures increase and equatorial temperatures fall. At low obliquities the reverse holds. Obliquity variations can thus lead to significant climate changes and are important in controlling the quantity of volatiles, such as carbon dioxide and water, stored at the poles. The possibility of long-term climate changes is dealt with more fully in the discussion of the polar regions in chapter 12.

The telescopic image

Under optimum conditions, with good seeing and good telescopes, the smallest object that can be seen on Mars is around 150 km across, but for most of the time, the resolution is considerably poorer. The planet's appearance (figure 1.5) shows differences in reflectivity (albedo), either of the ground or the atmosphere; resolutions fall far short of those needed to detect topography. Although the appearance of the planet may change on time scales ranging from hours to decades, the gross pattern of surface markings has remained fairly constant over the last century. The planet can be divided into dark and light areas, which have been termed maria and deserts, respectively. Most of the dark areas are in the southern hemisphere, between 0° and 40°S, although two of the darkest regions are Mare Acidalium, centered at 50°N, 30°W, and Syrtis Major, centered at 10°N, 290°W (figure 1.6). Most markings show no obvious relation to surface topography as we now know it. Exceptions are the bright region Hellas, which coincides with an impact basin; the dark marking Coprates, which coincides with a canyon; and Mare Acidalium, which is a plains region of low elevation. The variations in brightness are now believed to be caused largely by varying amounts of fine debris on the surface: dark regions are probably relatively free of debris, whereas light regions are areas of accumulation. The pattern of markings changes as wind redistributes the debris around the planet.

Surface markings vary from year to year and over the long term. It is not known whether the gross pattern that we see is changing systematically or whether the changes are statistical variations about a relatively stable configuration, but the markings probably shift slowly as the planetwide circulation of the atmosphere is modified in response to long-term climate changes. Almost all regions undergo changes in detail from year to year, as evidenced by the maps of Mottoni (1970) and the various maps of surface markings produced by Lowell Observatory (Inge et al. 1971a,b; Inge 1974, 1976, 1978). The Syrtis Major region, for example, shows considerable variation from year to year, particularly the area just to the east of Syrtis. Other changes take place gradually over periods of decades. Solis Lacus appeared to enlarge considerably between 1909 and 1926 but then contracted, so that by 1956 the configuration was similar to that in 1909 (Slipher 1962). The region experienced additional major changes between 1971 and 1973 (Capen 1976). Other types of changes are strong darkenings in regions that are normally bright. In recent years such changes have been seen in Lunae Palus and Claritas (Baum 1974) and in Noachis and Argyre (Slipher 1962).

The planet's appearance changes seasonally as a result of dust-storm activity. Yellow clouds have long been observed on Mars and have been correctly attributed to dust storms. During years of extreme activity all markings outside the polar areas may disappear. While this is mostly due to dust suspended in the atmosphere, some loss of contrast may also be caused by dust that has settled out onto the ground. The dust may later be redistributed by the wind, allowing the former albedo markings to reappear. Such a process certainly occurs on a small scale (Sagan et al. 1973a) and when integrated over large areas may affect the view from Earth.

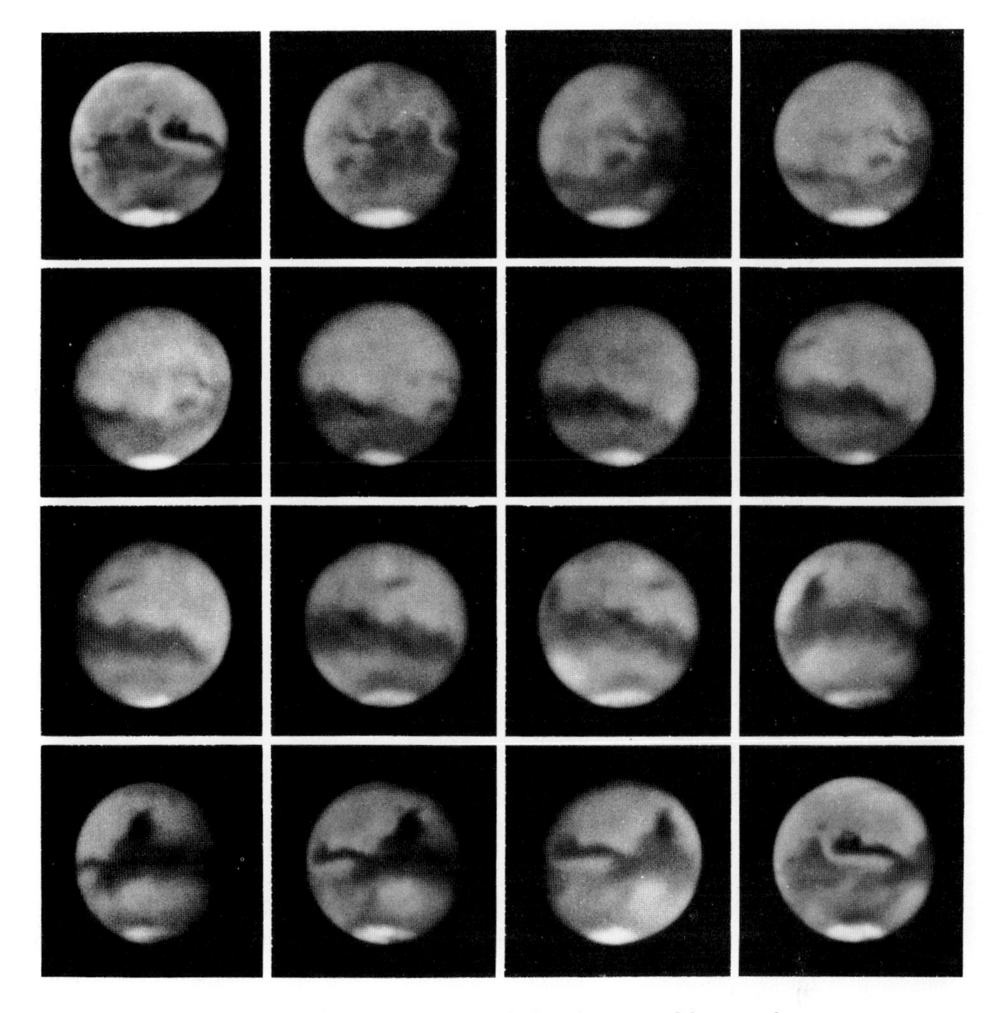

Figure 1.5. The appearance of Mars during 1971, before the onset of the great dust storm. The images show a complete rotation of the planet, starting with the zero longitude close to the middle of the disc in the top left. Syrtis Major shows prominently in the two bottom left images. (Courtesy of Lowell Observatory.)

The most obvious changes in the telescopic image occur in the polar regions. The cycle at the north pole may be conveniently considered as starting around $L_s = 180°$, the beginning of northern fall. At this time the north polar cap is at its minimum size, being a little over 20° across and arrayed symmetrically about the pole. Clouds begin to form over the polar region about this time and shortly thereafter envelop most of the region down to approximately 50°N. These clouds constitute the polar hood, which remains through fall and winter, masking the growth of the cap.

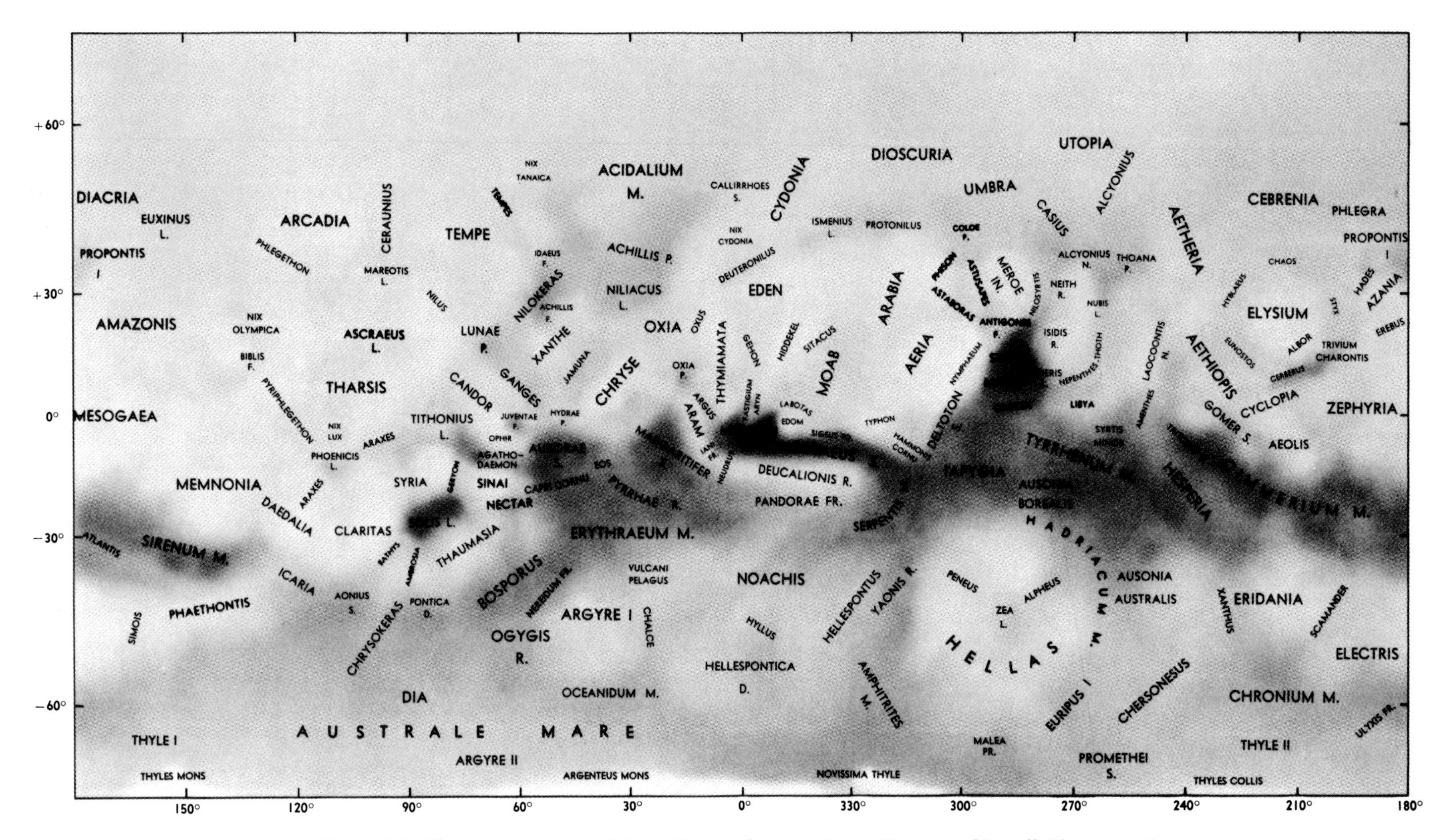

Figure 1.6. The classical names of the surface markings on Mars. (Courtesy of Lowell Observatory.)

When the hood clears at the start of northern spring (L_s = 0), the cap is revealed at its maximum, extending from the poles down to 65°N. At approximately midspring (L_s = 45°), the cap starts to retreat at a rate that averages close to 20 km per day. By the end of spring (L_s = 90°) the cap has retreated to its minimum, where it remains stable until the hood forms again in the fall.

The succession of events in the south is somewhat different. While most observers argued for the presence of a hood in the south, Tombaugh (1968) claimed that southern hoods do not form. His contention is supported by observations of the Viking orbiters, which showed only discrete clouds in the south polar regions during fall and winter. At this time the cap is at its maximum extent, reaching as far north as 40°S in the Argyre and Hellas regions. Toward the end of winter the southern cap starts its retreat, which is somewhat slower than in the north, averaging around 15 km per day. The retreat is also more irregular. Isolated outliers are left behind, the largest of which was first observed in 1845 and was named the Mountains of Mitchel after its discoverer. We now know that these "mountains" are actually low areas where ice is preferentially retained. Retreat of the cap is essentially complete by the start of southern summer (L_s = 270°), when the residual cap is just under 10° across. The cap is displaced about 5° from the south pole along the 20° longitude.

The albedo features of both hemispheres have a greater contrast in summer than in winter. The effect is real and has been confirmed photometrically (Focas 1961; Thompson 1973). It was long thought to result from the wave of darkening—a progressive darkening of the dark markings, which started at the poles and proceeded toward the equator as the polar cap receded (de Vaucouleurs 1954; Focas 1961; Dollfus 1961). The wave of darkening was believed to start in the spring with the formation of a dark collar, or fringe, around the pole, which expanded toward the equator as the cap receded, thereby producing the darkening wave. The dark collar appears real, but its relation to increases in contrast at lower latitudes is unclear. The collar was observed by many of the early workers, and much was made of it at the turn of the century, particularly by Lowell, who ascribed it to the release of water by the retreating cap and the springtime renewal of vegetation. Capen (1976) indicated that the dark collar is very evident in the spring but fades in summer as the cap retreats, rather than sustaining itself as the early workers thought. Sagan et al. (1973a) suggested that strong polar winds, caused by a steep temperature gradient between frosted and unfrosted ground, deflate a bright surface layer of dust to produce the collar. While the polar fringe and the summer increase in contrast as viewed from Earth appear real, the existence of a wave moving from the poles to the equator is highly questionable. Pollack and coworkers (1967) and Baum (1974) have suggested that the "wave" may be simply a photometric effect caused by the different illumination and viewing geometries in summer and winter. The darkening has also been questioned. Capen (1976), Thompson (1972, 1973), and Boyce and Thompson (1972) suggested that the increase in contrast in spring and early summer is actually caused by a brightening of the light areas; the phenomenon may thus be more aptly described as a wave of brightening. Such brightenings could well result from increased dust-storm activity as summer temperatures are approached.

Most surface markings are difficult to see at wavelengths short of about 4,500 A. This led to the belief that a blue haze normally envelops the planet. There are short periods, commonly lasting up to 2–3 days, when surface markings are visible in the blue, a phenomenon first observed by Slipher (1937) and termed blue clearing. It is now generally accepted that most of the lack of contrast at the blue end of the spectrum is due to the surface materials and not to haze, although hazes may certainly be present. In general the reflection spectra of the dark and light areas (see next chapter) show little contrast in the blue but substantial contrast in the red (McCord and Westphal 1971; Boyce and Thompson 1972). Markings therefore rarely show in the blue. The so-called blue clearings may be another manifestation of the seasonal brightenings in the light areas, which are probably a result of increased dust activity (Thompson 1972, 1973; Boyce and Thompson 1972; Baum 1974). The effect is most dramatic in the blue, because these images normally show no markings and any slight increase in contrast will cause the markings to emerge.

Transient brightenings are frequently observed on Mars images and are generally assumed to be caused by clouds of various types. The brightenings vary greatly in scale, from those that are close to the limiting resolution of the telescope to those that are caused by the great dust storms that may grow to encompass the whole planet. The brightenings also occur on different time scales, with durations from a few hours to several months in the case of large dust storms and the polar hoods. Two main kinds of brightenings have been identified: yellow clouds and white or blue clouds. Martin and Baum (1969) compiled data on the locations and motions of 97 brightenings, both yellow and white, that occurred between 1907 and 1958 (figure 1.7). They found that there are strongly preferred locations for

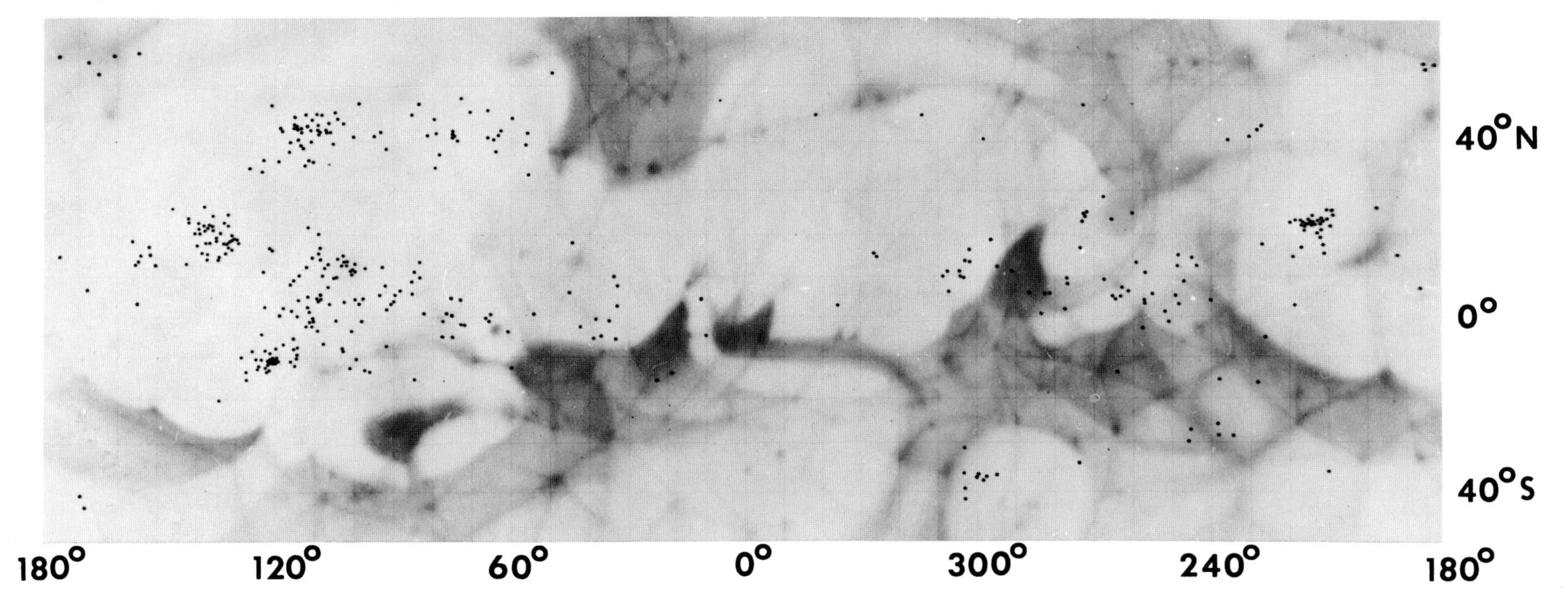

Figure 1.7. Locations of discrete transient brightenings photographed on Mars between 1907 and 1958. The brightenings are mostly white clouds and are preferentially located around the Tharsis and Elysium volcanoes and near Olympus Mons and Alba Patera, where clouds were frequently observed during the Mariner and Viking missions. (Courtesy of Lowell Observatory.)

the brightenings, most of which are in bright areas and areas of high relief. Particularly dense clusters occur around the regions known classically as Tharsis, Nix Olympica, Elysium, and Alba. Spacecraft imaging now shows that what these areas have in common is the presence of large volcanoes. Martin and Baum also showed that the white clouds are relatively stable in location, whereas the yellow clouds tend to move as fast as 15 km/hr.

The white or blue clouds are believed to be caused by condensates, either carbon dioxide or water. The most prominent are the W-clouds in the Tharsis region first photographed by Slipher in 1907. These clouds tend to form around noon and persist through the afternoon, mainly during northern spring and summer, affecting an area over 2,000 km across. We now know from spacecraft observations that the clouds occur mainly to the west of the large Tharsis volcanoes and that they are composed of water-ice (Leovy et al. 1973; Smith and Smith 1972). An orographic origin for these and other volcano-related clouds seems likely (Leovy et al. 1973). Another preferred location for white clouds is the general region south of Labyrinthus Noctis, where extensive arrays of convective clouds were photographed by Viking (Briggs et al. 1977). Brightenings at both morning and evening terminators have also been reported from telescopic observations (Dollfus 1961) and have subsequently been seen in spacecraft pictures (Leovy et al. 1973; Briggs et al. 1977).

Yellow clouds are most easily seen in yellow and red light and tend to disappear in the blue, in contrast to white clouds, which show up prominently in the blue. Even before the availability of spacecraft data, the yellow clouds were accepted as being caused by dust. Local yellow clouds are observed relatively infrequently from Earth, although Viking observations suggest that, statistically, about one hundred each year develop to the size of a few hundred kilometers (Briggs et al. 1979). Most occur within 60° of perihelion—that is, when it is late spring and early summer in the south. In a typical year, only one or two such storms grow larger than a few hundred kilometers. The intensity of these major storms varies greatly. In some years, such as 1956 and 1971, the dust storms become truly global in extent, obscuring nearly all the surface outside the polar regions.

Telescopic and spacecraft observations provide complete records of recent dust storms. Those after 1971 are particularly well documented (Martin 1974, 1976; Capen 1974; Capen and Martin 1972; Leovy et al. 1973; Briggs et al. 1979). Data on earlier storms have been reviewed by Slipher (1962), Martin and Baum (1969), Capen (1971), and Capen and Martin (1972). As perihelion is approached, storms start to appear in several preferred locations—namely, Argyre, Hellas, Hellespontus, Solis Planum, and Isidis—which, except for Isidis, are all in the southern hemisphere. The 1971 storm started in the Hellespontus-Serpentis region; within 7 days it had encompassed a region 6,000 km across, between 30°S and 50°S (Capen and Martin 1972), and within 2 weeks it had enveloped much of the planet. It was this storm that prevented the Mariner 9 spacecraft from seeing the surface when it first arrived at the planet in November 1971.

In 1977 two major dust storms developed and were closely observed by the Viking spacecraft (Briggs et al. 1979). Neither storm grew as large as the 1971 storm, but they still affected much of the planet. Their patterns were somewhat similar, in that they each developed rapidly within a few days and lifted large amounts of dust into the atmosphere, which took two to three months to settle out. The location and timing of the dust storms appear to support the classical view (Slipher 1962) that dust-storm activity is associated with intensification of winds at the time of maximum insolation near perihelion.

SPACECRAFT EXPLORATION

Other than Earth, Mars is by far the most intensively studied of the planets. As indicated in table 1.2, twelve spacecraft have either flown by, landed on, or orbited the planet. The first was Mariner 4, which on July 15, 1965, flew by the planet and took 22 close-up pictures of the surface. The best had a nominal resolution of around 3 km and gave a low-contrast, hazy view of a cratered surface that superficially resembled the lunar highlands. The planet appeared to be a dull, dead place that had remained essentially unchanged for the last 4 billion years. The results were disappointing: even the more conservative, while not expecting canals, oases, and flourishing vegetation, had expected to see a more varied landscape than we see on the Moon. The mission did, however, confirm that Mars has only a thin atmosphere, with surface pressures in the 5–10 mb range, and demonstrated that its intrinsic magnetic field is small—less than 0.00003 times that of Earth.

TABLE 1.2. SPACECRAFT MISSIONS TO MARS

Name	*Arrival Date*	*Type*
U.S. missions		
Mariner 4	July 15, 1965	Flyby
Mariner 6	July 30, 1969	Flyby
Mariner 7	August 4, 1969	Flyby
Mariner 9	November 14, 1971	Orbiter
Viking 1	June 19, 1976	Orbiter/Lander
Viking 2	August 7, 1976	Orbiter/Lander
Soviet missions		
Mars 2	November 27, 1971	Orbiter/Lander
Mars 3	December 2, 1971	Orbiter/Lander
Mars 4	March 1974	Orbiter
Mars 5	March 1974	Orbiter
Mars 6	March 1974	Lander
Mars 7	March 1974	Lander

Exploration continued with the launching of two additional spacecraft, Mariners 6 and 7, in February and March of 1969. The launches were four years later than the previous ones because of the 2-year spacing of the

Mars-Earth oppositions. After a journey of 5 months, Mariner 6 flew by the planet on July 30, 1969, and Mariner 7 on August 4. On approach to Mars both spacecraft took a series of far-encounter pictures, which clearly showed many of the classical albedo markings, although again there were no indications of canals. As the spacecraft came closer the albedo markings became less distinct and the surface topography more obvious; thus the gap was bridged between centuries of telescopic observations and close-up observations of the surface topography. At encounter Mariner 6 took 25 pictures of the surface and Mariner 7 another 33 (Leighton et al. 1969a,b). The pictures were a mix of high- and low-resolution frames, with the highest-resolution pictures having nominal resolutions as low as 300 m. The two spacecraft were also able to detect different molecular species in the atmosphere with an ultraviolet spectrometer, to measure surface temperatures with an infrared spectrometer, and to reconstruct pressure and temperature profiles of the atmosphere from attenuation of radio signals as the spacecraft were occulted by the planet.

It was left to Mariner 9 to reveal the exotic nature of the martian surface. The original plan was to send two identical spacecraft to the planet during the next opposition, in 1971. The spacecraft were to be similar to the previous Mariner spacecraft except for a larger propulsion system, to allow the spacecraft to be decelerated and inserted into orbit around Mars; a heavier and more versatile science payload; and a more flexible computer system, for increased inflight adaptability. The new computer system proved to be a crucial supplementary capability, enabling the mission to adapt to a series of unexpected events. The first was the loss of the first spacecraft soon after launch on May 9, 1971. The two spacecraft had been assigned different roles: one was to fly progressively over the entire surface of the planet and map its surface features; the other was to pass over the same parts of the planet repeatedly, in order to detect changes such as might be caused by the growth of vegetation, the response of the soil to seasonal temperature changes, or by cloud patterns. With the loss of the first spacecraft the previous plans were scrapped, and a new plan was quickly formulated, which preserved the original objectives but adapted them to a single-spacecraft mission. The second spacecraft was successfully launched on May 30, 1971, and proceeded on a relatively uneventful course to encounter Mars 161 days and over 600 million km later.

The spacecraft had on board an array of instruments similar to, but more sophisticated than, that on board the previous Mariners. Ultraviolet and infrared spectrometers were designed primarily to determine the composition of the atmosphere, although the infrared spectrometer was also expected to yield information on surface mineralogy. An infrared radiometer was included to measure surface and atmospheric temperatures. The radio transmitters on board were to be used to measure surface elevations and pressure-temperature profiles in the atmosphere during occultations, when the spacecraft passed behind the planet as viewed from Earth. The radio link also provided a means of detecting irregularities in the planet's gravity field. Of special importance for the surface geology were two cameras, a wide-angle camera to provide 1-km-resolution coverage at the nominal periapsis altitude of 1,500 km, and a narrow-angle camera to give resolutions ten times better on frames nested within the wide-angle field of view. The instruments were mounted on a scan platform, which could be pointed on command from Earth.

The second major unexpected event during the mission was the appearance of Mars at the time of arrival of the spacecraft on November 14, 1971. Much to the consternation of the Mariner 9 experimenters, the planet's surface was almost completely hidden from view by dust in the atmosphere, the aftermath of a planetwide dust storm. Similar obscurations of the planet had been previously noted by telescopic observers, who correctly attributed them to dust storms. They were known to occur during southern summers, and one was predicted for the time of arrival of Mariner 9 by C. F. Capen of Lowell Observatory. Nevertheless, the Mariner 9 experimenters were unprepared for the almost complete masking of the surface. Some features could still be seen faintly through the dust: the south polar cap was visible, as were numerous faint, bright, circular features suggestive of craters, but more intriguing were four dark spots in the Tharsis region. These appeared to be mountains poking up above the dust in the atmosphere, and close inspection showed that each contained at its center a complex crater that resembled a terrestrial volcanic caldera. This was the first clue that previous Mariner spacecraft had given a false impression of the surface. When the dust slowly settled out, the dark spots were confirmed as marking the summits of four huge volcanoes.

By January 1972 the atmosphere was judged clear enough for the spacecraft to begin its systematic photography of the planet. Actually there were strong operational arguments to start the mapping despite a still somewhat hazy atmosphere. The spacecraft had only a limited amount of attitude-control gas, which is used for stabilization; this was being slowly consumed and, once gone, would render the spacecraft unusable. Moreover, the spacecraft had been designed to operate during November, December, and January, and as time passed, the illumination conditions at the planet and the geometry with respect to Earth were changing. These factors could greatly complicate spacecraft operations. Steps were accordingly taken to start the mapping. The spacecraft had been placed in an elliptical orbit with a 12-hr period and an altitude at closest approach (periapsis) of 1,387 km. Because Mars rotates every 24 hr 37 min, this gave two periapsis passes every day and caused successive passes to be displaced slightly in longitude. Small changes in period on successive orbits and Doppler shifts of the radio signal had by this time indicated that Mars had sizeable irregularities in its gravity field. To optimize mapping and provide better overlap between successive passes, periapsis was raised to 1,650 km. Mapping was started in the high southern latitudes and moved progressively northward as each latitude band was covered. Because there was still considerable dust in the atmosphere at the start, the photography in the high southern latitudes is indistinct. The photography improved greatly, however, as the dust settled out.

During the next few months the surprising geologic diversity of the planet became apparent. Several more large volcanoes, in addition to those first seen as dark spots, were discovered, and the enormous size of those in Tharsis was realized. The first measurements of elevation were treated with disbelief, and it was not until measurements by different techniques were found to converge that the extreme heights of 27 km were accepted. As the mission progressed the now familiar canyons, flood features, fracture systems, layered sediments, etch-pitted terrains, and so forth were revealed. Changes in surface markings, the aftermath of the great dust storm, were observed, as were a wide variety of atmospheric phenomena. The mission finally ended on October 12, 1972, when the spacecraft ran out of attitude-control gas.

The Russians, however, had been less fortunate. Two Soviet spacecraft, Mars 2 and Mars 3, were launched to Mars on May 19 and May 29, 1971. Shortly before arrival at the planet, each spacecraft released a lander capsule. The first landed in Hellas and the second in the vicinity of 158°W, 45°S, but neither returned any useful scientific data. Meanwhile Mars 2 entered into orbit on November 27, and Mars 3 on December 2 (Marov and Petrov 1973). Unfortunately, the dust storm that had greeted Mariner 9 was still present, and because the sequences on the Russian spacecraft were preplanned and could not be delayed until the atmosphere cleared, the orbiter pictures showed little surface detail.

In 1973 the Soviets followed with four separate launches of two orbiters and two landers. Of the two orbiters, Mars 4 and Mars 5, only Mars 5 was successfully injected into orbit. It was placed in a 25-hr orbit with a periapsis altitude of 1,500 km and functioned for 20 revolutions, during which the cameras took 70 pictures of Mariner 9 quality. Other instruments returned data on the composition of the surface and the atmosphere, the water content of the atmosphere, and the planet's magnetic field. Only one lander reached the surface, and, like the previous lander, it failed to relay any usable data back to Earth.

The most recent stage in the exploration of Mars was initiated in the summer of 1975 with the launching of two Viking spacecraft (figure 1.8), each consisting of an orbiter and a lander. The project had started in 1968, and the array of scientific instruments had been chosen in 1969. From the first, emphasis was on life detection, which resulted in selection of two cameras, a gas chromatograph–mass spectrometer (GCMS) to determine the organic content of the soil, and a four-part life-detection experiment as the main scientific payload on the lander (table 1.3; figure 1.9). One part of the life-detection experiment was later deleted because of cost and engineering problems. Other instruments were an x-ray fluorescence device to analyze the soil, a seismometer, and a meteorological package. A simple magnet experiment to assess the type and fraction of magnetic minerals in the soil was also included (Soffen 1977). The orbiters had three scientific instruments: a pair of high-resolution cameras, an array of infrared detectors to examine the thermal properties of the atmosphere and the surface, and a spectrometer designed to measure the water-vapor content of the atmosphere (figure 1.10). In addition the radio could be used for a variety of scientific purposes, including mapping the planet's gravity field and determining atmospheric profiles and surface relief. The orbiter camera system differed from the systems flown on earlier missions in that it consisted of two identical cameras that could be shuttered alternately every 4.48 sec to take a continuous strip of pictures two frames wide. Most of the photographs in this book were taken with these cameras.

The orbiters and landers were initially coupled, and the plan was for the combined spacecraft to enter orbit and closely examine the predesignated landing site; then, if the site appeared suitable, the lander was to be deployed and set down on the surface, with the orbiter remaining above to serve as a relay link and perform its own science. If the predesignated site was considered unsuitable, the capability existed to search the planet until an appropriate site was found. This adaptive aspect, as in the case of Mariner 9, proved crucial to the success of the mission.

The first spacecraft arrived at Mars on June 19, 1976. Three days later orbiter pictures were acquired of the predesignated landing site in the southern part of Chryse Planitia. The first pictures showed a wealth of surface detail, which was unanticipated from the earlier Mariner 9 coverage. The rugged aspects of the terrain, together with radar measurements that indicated small-scale surface roughness, and geologic reasoning that argued for the presence in the area of boulders derived from some large channels to the south, suggested that it would be prudent to delay landing and look elsewhere for a safer site (Masursky and Crabill 1976). After a 3-week search, a safer-looking site was found in the western part of Chryse Planitia, at 22.5°N, 48.0°W, and on July 20, 1976, the first Viking lander made its historic soft landing on Mars and gave us our first close-up view of the surface (figure 1.11). The second spacecraft arrived at Mars on August 7, 1976. Its predesignated landing site at 44°N, 10°W was also judged unsafe, and another search was undertaken, which involved photographing much of the surface between latitudes 40°N and 50°N. A site was finally chosen at 44°N, 226°W, and a successful landing took place on September 3, 1976.

After landing, several months of intense activity followed. The biology experiments revealed a strange soil chemistry, which, in some ways, mimicked living processes. Organic compounds added to the soil were decomposed, thus simulating metabolism, and addition of water caused release of oxygen, thus simulating photosynthesis. This initially caused some excitement. However, the gas chromatograph–mass spectrometer showed that the soils contained no organic molecules despite detection sensitivities in the parts-per-billion range, and as more definitive experiments were performed, the prospects for life slowly dimmed. Disappointing also was the fact that no "marsquakes" were detected, despite the

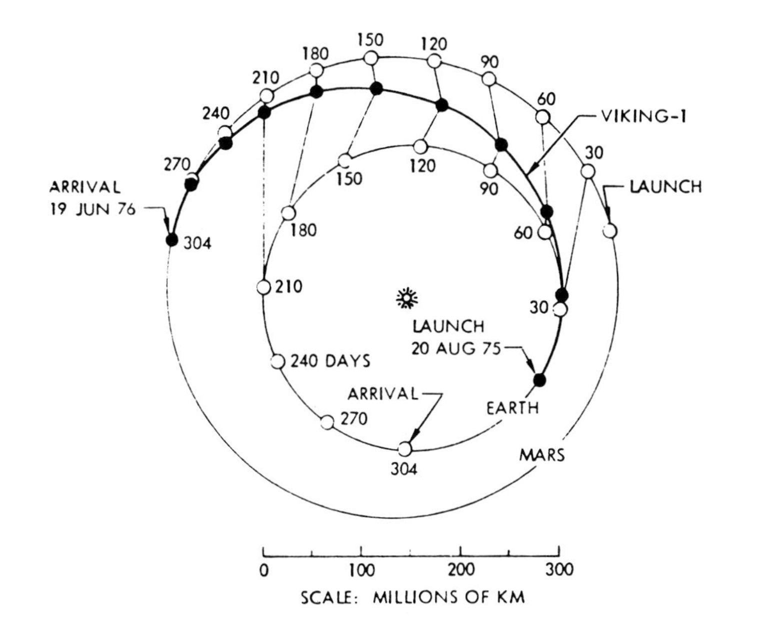

Figure 1.8. Path of the Viking 1 spacecraft to Mars. Travel to Mars took almost a year, so that on arrival at Mars, Earth was on the opposite side of the solar system and was moving away. In November of 1976, 4 months after the first Viking landing, Mars passed behind the Sun, as viewed from Earth. (NASA/JPL.)

Figure 1.9. (*above, right*) The Viking lander, showing the location of the different science instruments.

TABLE 1.3. VIKING SCIENCE INVESTIGATIONS AND INSTRUMENTS

Investigations	*Instruments*
Orbiter imaging	Two vidicon cameras
Water vapor mapping	Infrared spectrometer
Thermal mapping	Infrared radiometers
Entry science	
Ionospheric properties	Retarding potential analyzer
Atmospheric composition	Mass spectrometer
Atmospheric structure	Pressure, temperature, and acceleration sensors
Lander imaging	Two facsimile cameras
Biology	Three analyses for metabolism, growth, or photosynthesis
Molecular analysis	Gas chromatograph–mass spectrometer
Inorganic analysis	X-ray fluorescence spectrometer
Meteorology	Pressure, temperature, wind velocity sensors
Seismology	Three-axis seismometer
Magnetic properties	Magnet on sampler observed by cameras
Physical properties	Various engineering sensors
Radio science	Orbiter and lander radio and radar systems
Celestial mechanics, atmospheric properties, and test of general relativity	

SOURCE: Soffen 1976.

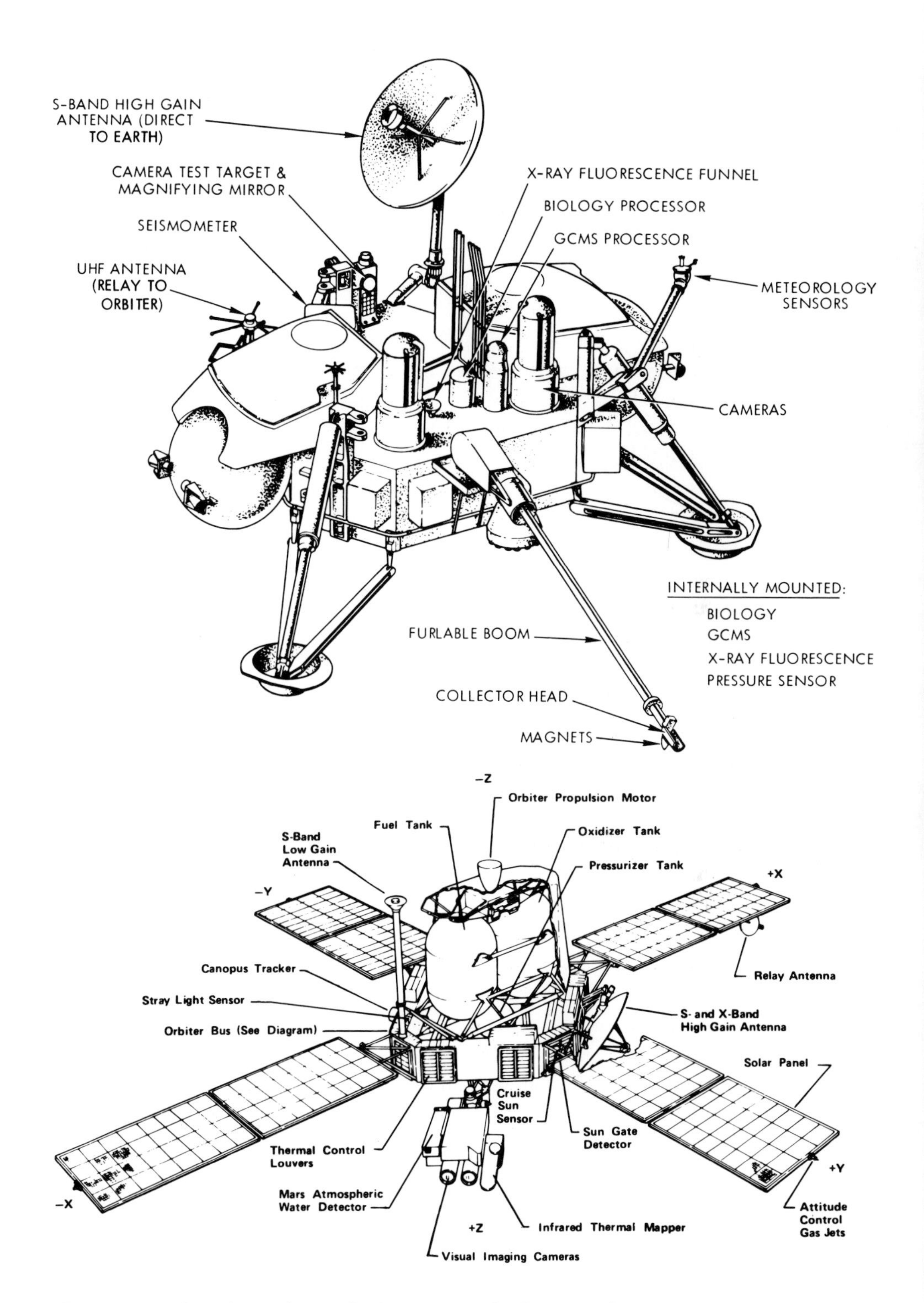

Figure 1.10. The Viking orbiter. The Sun is up in the diagram. The science instruments were mounted on a scan platform that could be rotated about the Sun line (clock) or away from the Sun line out of the plane of the solar panels (cone). See Appendix A for a description of how the orbiter was used to take pictures of the planet.

Figure 1.11. The view of the surface of Chryse Planitia from Viking lander 1. More detailed views are shown in the next chapter. The left of the scene is characterized by numerous drifts of fine-grained debris. To the right occasional outcrops of bedrock show through the rubbly surficial cover. The object in the center is the meteorology boom. (Viking lander 1 event no. 11A097.)

perfect working condition of the seismometer on the second lander. The other lander experiments, however, returned abundant information on the local geology, the chemistry of the soil, the composition of the atmosphere, meteorological conditions, variations in the dust content of the atmosphere, local precipitation of volatiles, and so forth. At the time of writing the second Viking lander had stopped functioning, but the first lander was expected to last another 10 years, an astonishing achievement for a vehicle built to specifications that required only a 90-day lifetime.

The Viking orbiters observed the planet for almost two full martian years, monitoring changing weather patterns, the waxing and waning of global dust storms, the changing water-vapor content of the atmosphere, the continual redistribution of fine-grained debris on the surface, and the advance and recession of the polar caps (Snyder 1979). More important with respect to the geology has been the systematic mapping of the surface by both the cameras and the infrared instruments. The cameras have mapped the entire surface at a resolution of 200 m and large areas at significantly higher resolutions, ranging down to 10 m. Much of the planet has been photographed in color and stereo although at somewhat lower resolutions. Late in the mission, in 1979 and 1980, large areas of Arabia and Memnonia were photographed at 30 m resolution in an effort to obtain coverage that would be useful in planning future missions to the surface of the planet. In addition, the infrared instruments have measured areal variations in the diurnal temperature fluctuations and have mapped seasonal variations in the temperatures of the surface and the atmosphere. They have also monitored the water content of the atmosphere and determined temperatures on the polar caps, from which compositions can be inferred. The only consumable on the spacecraft was the attitude-control gas, which kept the spacecraft stably oriented. The second orbiter ran out of gas in July 1978, and the first in August 1980, five years after it had been launched.

Unless indicated otherwise, the pictures in this book were taken with the Viking orbiter cameras. They are identified by a picture number, or PICNO, in which the first three digits indicate the orbit number, the middle letter indicates which orbiter, and the last two digits indicate the sequence of the picture in that orbit. For further details on picture annotation see Appendix A.

Figure 2.1. Two Lambert azimuthal equal area charts of the martian surface. The left figure is centered at 120°W; the right at 240°W. A small area around 0° longitude is not seen on either plot. The maps show the relationship between the albedo markings and the physiography and give a more accurate portrayal of the true geometry than the widely used Mercator maps. (Reproduced with permission from Inge and Baum 1973.)

2 GENERAL CHARACTERISTICS OF THE SURFACE

This chapter gives an overview of the martian surface. The intent is to provide sufficient background that when specific topics, such as channel formation or wind action, are discussed later, they can be placed in an appropriate context. The chapter includes sections on physiography, topography, albedo, color, and thermal inertia, all of which can be discussed in global terms. Included also are descriptions of the views of the surface from the two Viking landers. The degree to which the lander scenes are typical of the entire planet is not known; they are known to be atypical in one sense, in that they were chosen partly because of the lack of surface detail as seen from orbit. However, they are the only two sites that we have looked at closely. The sections vary somewhat in detail: the physiographic features, which are treated in detail later, are discussed here only superficially, whereas other topics are discussed more fully.

PHYSIOGRAPHY

The physiography of the surface is varied and has a marked north-south asymmetry (figure 2.1). Much of the southern hemisphere is heavily cratered, with a landscape that largely dates from early in the planet's history, possibly around 4 billion years ago. The density of craters larger than 20 km in diameter is particularly high and is comparable to that in the lunar highlands. However, because the larger craters are relatively shallow,

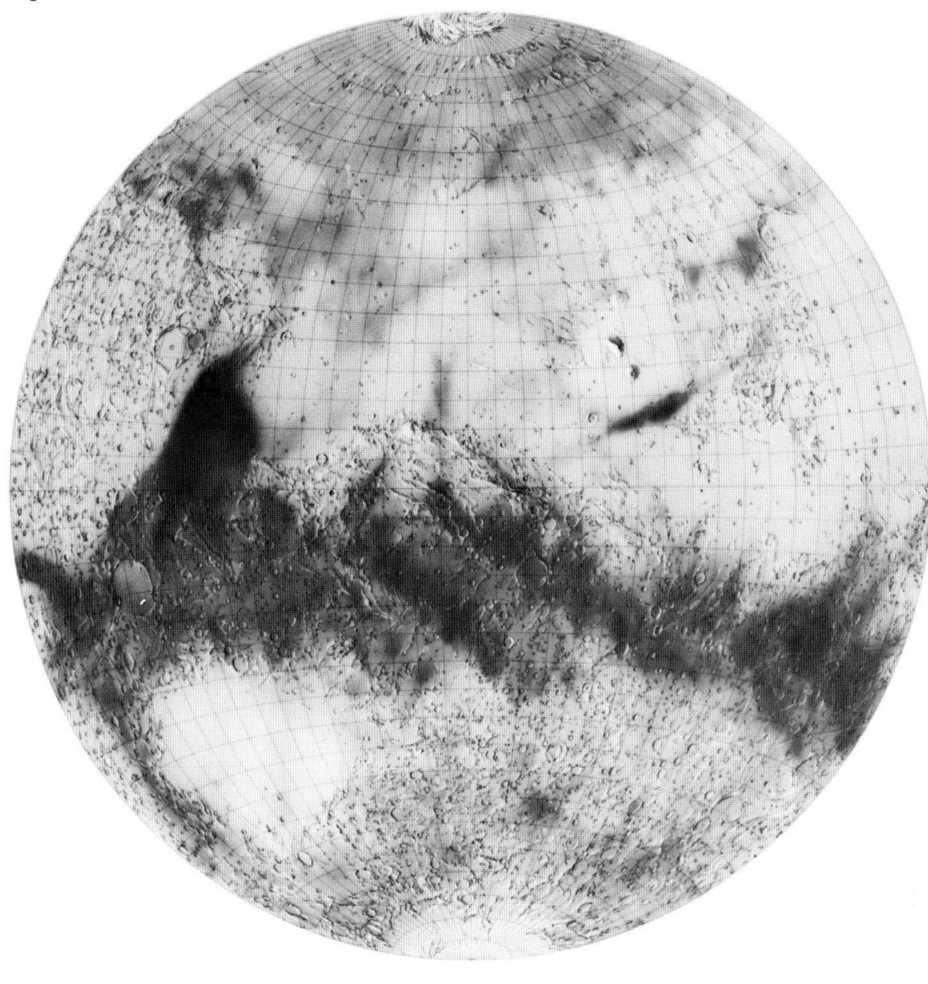

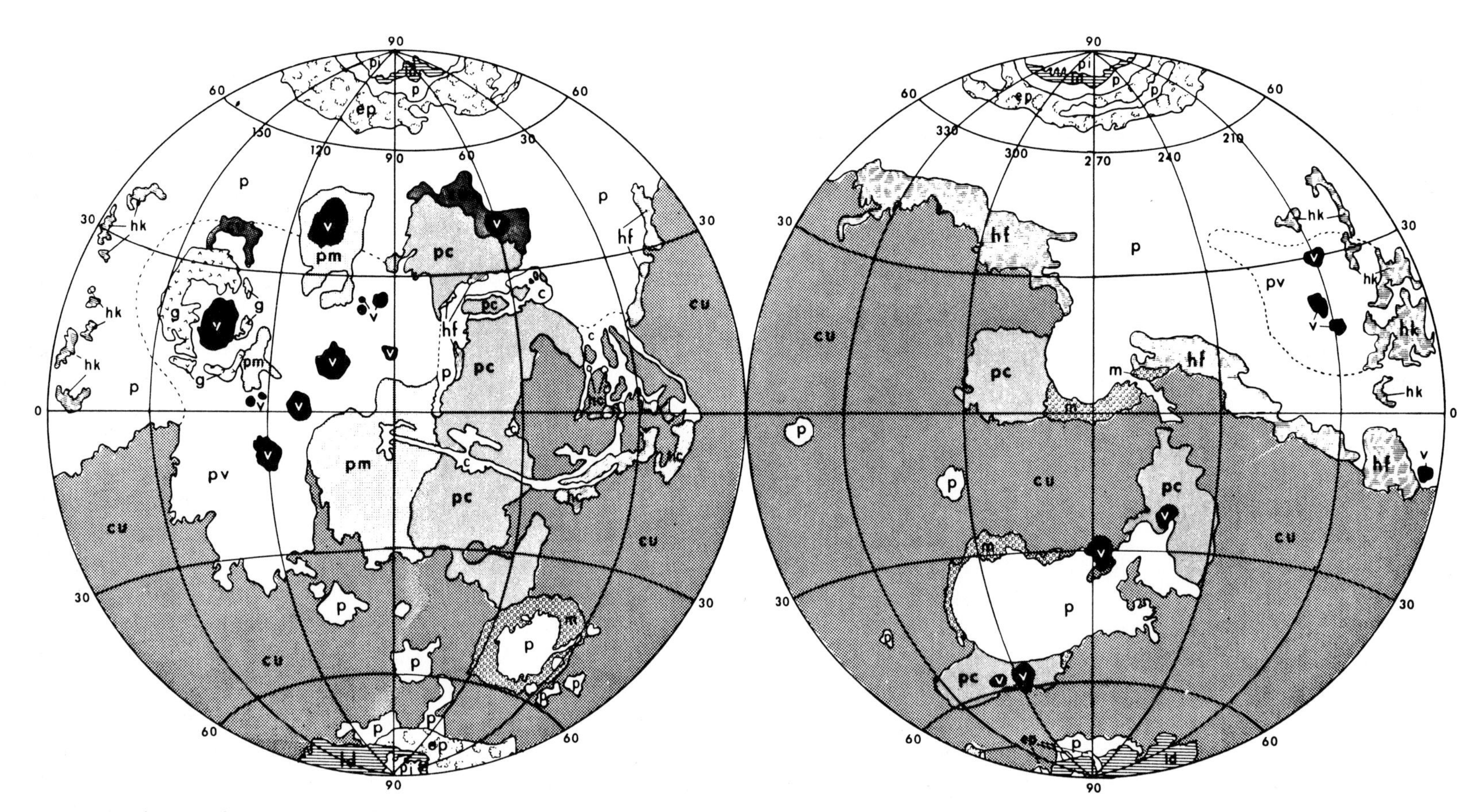

Figure 2.2. Physiographic components of the martian surface plotted on a Lambert equal area base. Polar units include pi (permanent ice), ld (layered deposits), and ep (etched plains). Volcanic units include v (volcanic constructs), pv (volcanic plains), pm (moderately cratered plains), and pc (cratered plains). Modified units include hc (hummocky terrain, chaotic), hf (hummocky terrain, fretted), hk (hummocky terrain, knobby), c (channel deposits), p (plains, undivided), and g (grooved terrain). Ancient units include cu (cratered terrain, undivided) and m (mountainous terrain). (From Mutch and Head 1975, copyrighted by the American Geophysical Union.)

the surface is more subdued than that of the heavily cratered parts of the Moon. As on the Moon, there are several large impact basins, the largest of which are Hellas, with a diameter of 1,800 km, and Argyre, with a diameter of 800 km. By comparison, the Imbrium basin on the Moon is about 1,300 km across. In much of the densely cratered terrain there are smooth plains between the large craters, and the proportion of intercrater plains to older, more rugged and more cratered surfaces varies. In many places the densely cratered terrain is cut by small gullies and elaborate channel systems, which are generally more common on the older surfaces than on the intercrater plains, although both are dissected. The densely cratered terrain is not entirely confined to the southern hemisphere. A large tongue extends into the northern hemisphere on either side of the 330° meridian, and smaller patches occur elsewhere in the north, particularly in the Tempe plateau region around 45°N, 75°W.

Most of the northern hemisphere is covered by plains that have much smaller crater densities. Extensive plains also occur in the south, to the south of Tharsis and Valles Marineris and in and around the Hellas basin. Most of the plains have crater densities that are within a factor of two or three of those for the lunar maria. The main exceptions are the lava plains of the Tharsis region, which have considerably fewer craters. At low resolution (1–2 km) the plains are relatively featureless except for the occasional crater and albedo marking. Examination at higher resolution, however, shows that they range widely in character and have numerous features indicative of volcanism, wind action, fluvial erosion, and the presence of ground-ice.

Among the most impressive features of the planet are the large volcanoes (figure 2.2), most of which occur in three regions: Tharsis, centered on the equator at 115°W; Elysium, centered at 25°N, 210°W; and around the Hellas basin, at 40°S, 290°W. The three largest Tharsis volcanoes lie along a line on the northwest flank of a broad upwarp in the martian surface, which is generally referred to as the Tharsis bulge (see next section). The tallest volcano on the planet, Olympus Mons, is 1,600 km to the northwest of the line, at the edge of the bulge. All the volcanoes are enormous by terrestrial standards. Olympus Mons is over 700 km across, with a summit 25 km above the surrounding plains; the three Tharsis volcanoes have summits 17 km above the plains. The summits of all four are concordant, at 27 km above the Mars datum (see next section). Alba Patera, a large volcano to the north of Tharsis, at 25°N, 210°W, appears to have little vertical relief, although it is over 1,500 km across. By comparison, the largest volcano on Earth (Mauna Loa) is 120 km across at its base, and its summit is 9 km above the ocean floor. Other, smaller volcanoes in the Tharsis region, are close to the line along which the larger volcanoes are located. The Elysium volcanic province centered at 25°N, 210°W, is considerably smaller than Tharsis and has only three volcanoes of any appreciable size. These volcanoes are still large compared to those on Earth, all of them being more than 150 km across. The volcanoes around the edge of Hellas tend to be older and have less relief than those in the other provinces.

The region of Tharsis is close to the center of a vast array of radial fractures that affects almost an entire hemisphere of the planet. To the north and northeast the radial fractures are especially intense. The flanks of the large volcano Alba Patera, for example, are cut by numerous north- and northeast-trending fractures, which are deflected around the volcano to form a fracture ring about its center. South of Tharsis is another intense fracture zone, Claritas Fossae, and to the southwest long persistent grabens curve away from the Tharsis volcano line to form the Sirenum and Memnonia Fossae. Fractures are less evident to the east of Tharsis, but development of the canyons and the large channel Kasei Vallis appear to have been partly controlled by the radial fractures. In the parts of the planet unaffected by the Tharsis radial fractures (180°W to 20°W), the dominant tectonic features are curvilinear scarps around large impact basins and a faint grid of northeast- and northwest-trending fractures.

To the east of Tharsis, just south of the equator, is a vast system of interconnected canyons. They extend from close to the summit of the Tharsis bulge, at 5°S, 100°W, eastward for almost 4,000 km, until they merge with what has been termed "chaotic terrain"—terrain in which the ground has seemingly collapsed, thereby producing jumbled arrays of blocks at lower elevations than the surroundings. In most places individual canyons are over 3 km deep and over 100 km across. In the central section, three parallel canyons merge to form a depression over 7 km deep and over 600 km wide. The canyon floor is generally flat, and the walls are steep and gullied. In many places, however, the walls have collapsed in gigantic landslides, whose debris covers the floor. To the north of the main canyon are several east–west-trending box canyons that are not directly connected to the main canyon system and a vague north-south canyon just west of Lunae Planum that gives rise to the large channel Kasei Vallis.

The large channels are among the most puzzling of the martian surface features. The so-called outflow channels may be several hundreds of kilometers long and tens of kilometers across and, unlike most terrestrial river valleys, generally start abruptly, without tributaries. Most occur north of the canyon-chaos region between 20°W and 80°W, where they extend northward from several box canyons and areas of chaos to converge on Chryse Planitia. They then continue north across Chryse Planitia until they disappear between 30°N and 40°N. Other large channels extend northwestward from the Elysium volcanic province; these are smaller than those around Chryse Planitia but, like the latter, also start abruptly, without tributaries. Still others occur within the old cratered terrain, mostly close to its boundary with the plains, between 150°W and 200°W.

The physiography of the poles is distinctively different from the rest of the planet. At both poles, and extending outward for a little over 10°, are some layered deposits unique to the polar regions. The layers, which range in thickness from several tens of meters down to the limiting resolution of the available photographs, are visible in the numerous escarpments and valley walls that are arranged in broad swirls around both poles. Individual layers can be traced for hundreds of kilometers. The surfaces of the deposits are almost entirely devoid of craters, indicating that the deposits are among the youngest features of the planet. In the south the layered materials lie unconformably on old cratered terrain and transect a large impact basin. In the north the polar deposits lie on plains and are surrounded by a vast array of sand dunes, which forms an almost complete collar around the polar region. Dune fields of comparable magnitude do not occur in the south, although small dune fields are common within craters in the region peripheral to the south pole.

TOPOGRAPHY

Because there is no sea level on Mars against which elevations can be referenced, an artificial datum had to be established. The reference surface was picked somewhat arbitrarily as that at which atmospheric pressure is 6.1 mb—the pressure at the triple point of water. At partial pressures of water greater than 6.1 mb liquid water can exist under appropriate temperature conditions; at pressures less than 6.1 mb liquid water is always unstable. The dimensions and shape of the surface were determined from occultation and gravity data. As the Mariner 9 spacecraft was occulted by (passed behind) Mars as viewed from Earth, its radio signal decayed and was ultimately cut off. The precise form of the decay depended on the geometry of the occultation and the temperature-pressure profile within the atmosphere (Kliore et al. 1973). The occultations thus provided a means of measuring surface pressure at a wide array of points at known distances from the center of the planet, and from these the distance from the planet's center to the 6.1-mb reference level could be readily determined. The points were then fitted to a fourth-order, spherical harmonic representation of the gravity field (Jordan and Lorell 1975) to obtain the reference level. The datum closely approaches a triaxial ellipsoid with equatorial radii of 3,394.6 and 3,393.3 km and a polar radius of 3,376.3 km (Christensen 1975).

The same occultations used to define the datum also give absolute elevations with respect to the center of mass to an accuracy of 0.25 km to 1 km (Kliore et al. 1972, 1973). The main source of information on local differences in elevation is the Mariner 9 ultraviolet spectrometer (UVS). The instrument measured the intensity of ultraviolet light scattered by the atmosphere, which depends on the amount of atmosphere between the spacecraft and the ground, which in turn depends on the surface pressure; hence, estimates can be made of elevations. The elevation differences between adjacent regions 30 km across were measured with a precision of 0.6 km (Hord et al. 1974). Similarly, the Mariner 9 infrared interferometer spectrometer (IRIS) derived topographic information from variations in absorption of carbon dioxide in the atmosphere; elevation differences can be estimated with a precision of about 0.5 km (Hanel et al. 1972). The UVS and IRIS data duplicate each other somewhat, because they are determined along the same ground track. They are both subject to large errors owing to atmospheric obscurations, such as dust and condensate clouds. Earth-based radar provides accurate measures of local elevation differences with errors in the range of 175–200 m. The elevations are determined from the delay time of radar signals reflected from the martian surface. The data are gathered along east-west tracks and are confined to latitudes below 25°. While relative elevations can be measured accurately, absolute elevations are subject to sizeable errors because of uncertainties in the geometric center of Mars with respect to Earth (as opposed to the center of mass). Photogrammetry has provided detailed information on the topography of a few specific features, such as Olympus Mons and parts of Valles Marineris, for which there is good stereo coverage. The precision of the measurements is generally in the range of hundreds of meters, depending on the resolution of the photographs and the viewing angles.

Data from all these sources were integrated to produce the planetwide topographic map, with the occultation points, being the most accurate, used as controls. The east-west radar traces were adjusted in absolute altitude to conform with nearby occultation points, so that precise elevations were available for a wide scattering of occultation points across the planet, plus a number of east-west traces within 25° of the equator. Because of the high inclination of the Mariner 9 orbit (65°), the UVS and IRIS measurements, which represent the bulk of the elevation data, are along traces which run almost north-south. Relative elevations along these traces were adjusted to conform with any radar tracks that they crossed and any nearby occultations, to give a large array of elevations. In general, the errors are around 1 km, but considerably larger errors may occur locally. Commonly tracks could not be tied closely to occultations or radar, and long extrapolations were necessary. As a consequence, errors larger than 1 km may be present in some regions where control is poor. One of the final steps in production of the map was contouring the elevations. Some discretion had to be used in this process, since considerable interpolation was necessary between the UVS and IRIS traces. This accounts for most of the differences between the contour maps of Wu (1978) and those of Christensen (1975).

The topography, like the physiography, has a marked hemispheric asymmetry (figure 2.3). The densely cratered terrain is mostly 3–5 km higher than the sparsely cratered plains. As a consequence most of the southern hemisphere is high, and most of the northern hemisphere is low. The main exceptions in the north are caused by the broad upwarp of the Tharsis bulge, centered at approximately 10°S, 110°W, and a much smaller upwarp centered on Elysium, at 25°N, 210°W. The Tharsis bulge is 5,000 km to 6,000 km across, depending on what is taken as its base, and 10 km high at its center. It straddles the equator, causing anomalously high elevations around the 110°W meridian, both to the north and the south. By comparison, the Elysium bulge is just under 2,000 km across and approximately 4 km high. High elevations also occur in the northern hemisphere around the 330°W meridian, where a large tongue of old cratered upland extends to the north. Between 50°N and 70°N, low-lying plains completely encircle the planet. Extending southward from these circumpolar plains, almost to the equator, are three low-lying plains regions that separate the high areas just described. These are Mare Acidalium—Chryse Planitia, east of the Tharsis bulge; Amazonis Planitia, between the Tharsis and Elysium bulges, and the Elysium and Isidis Planitia, to the west of Elysium.

The only exceptions to the general pattern of high elevations in the southern hemisphere are the large impact basins Hellas and Argyre, whose floors are close to, or below, the datum. The poles are also somewhat aberrant: the area within 10° of the north pole is mostly above the datum, in contrast to the surrounding plains, whereas the south pole is somewhat lower than the surrounding terrain. Superimposed on these broad regional differences are sizeable differences in local elevation. The largest positive relief is associated with the volcanoes, which reach elevations up to 27 km above the datum; the largest negative relief is in the canyons which, in places, are over 7 km deep.

ALBEDO AND COLOR

Albedo markings (brightness variations) have been observed on Mars for centuries and are the only surface features of the planet that are distinguishable from Earth. In many cases the albedo boundaries show no obvious relation to topography (see figure 2.1). They probably result largely from a thin dusting of fine-grained, light-colored debris that can be

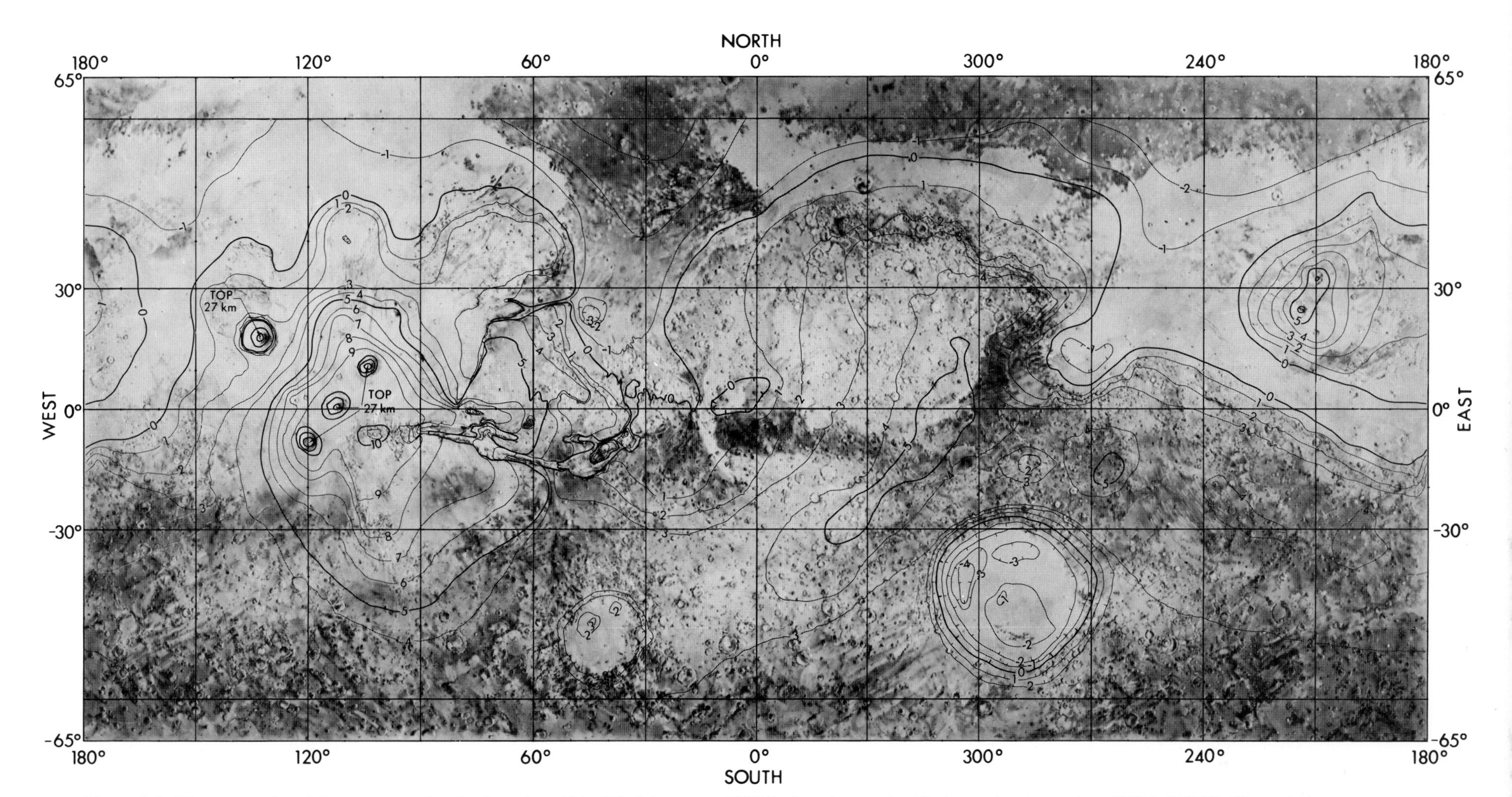

Figure 2.3. The topography of the martian surface for latitudes within 65° of the equator. Contour interval is 1 km. The main characteristics of the topography are high elevations in the old cratered terrain to the south, low elevations in the northern plains, and two bulges, one centered on the volcanic region of Tharsis (0°, 105°W), the other on the Elysium volcanic province (30°N, 210°W). The techniques used in compiling the map are described in the text. (Reproduced with permission from Wu 1978.)

mobilized by the wind (see chapter 11). In bright areas the cover is essentially continuous; in dark areas the underlying bedrock or coarser debris is probably showing through. The most prominent dark markings are in the 0° to 40°S latitude belt, which includes Sinus Sabeus, Sinus Meridiani, Margaritifer Sinus, and the Maria Sirenum, and Tyrrhenum. Nearly all these features are within densely cratered terrain, and for the most part their boundaries cannot be distinguished on the basis of topography. Some markings are related to topography, however. Syrtis Major, for example, coincides with an area of steep regional slopes and Juventae Fons with a box canyon. In the north, Mare Acidalium coincides with low-lying plains, and the dark collar around the poles is largely coincident with a vast dune field. Similarly, some light markings, such as Hellas, have obvious topographic control, whereas others, such as Arabia, do not. It appears that the distribution of superficial debris is controlled by the general circulation pattern of the atmosphere, and only insofar as the circulation is perturbed by local topography is there any obvious relation between albedo and relief. The general pattern probably changes slowly, as wind regimes change in response to climatic variations. The albedos range from a minimum of 0.089, in the middle of Syrtis Major, to a maximum of 0.429, just to the north of Ascraeus Mons (Kieffer et al. 1977), although the latter measurement may be affected by clouds. The albedo is markedly bimodal, with peaks at 0.13 and 0.26. Estimates of the ratio of bright to dark areas range from 1.8 to 3.0 (McCord and Westphal 1971; Binder and Jones 1972).

The deep red color of Mars is one of its most striking characteristics as viewed from Earth and has long been ascribed to iron oxides. Early telescopic observers imagined the martian deserts to be much like those of Earth, with iron being the main cause of the rich coloration. The red color shows up dramatically in the reflection spectra of the planet (Adams and McCord 1969; McCord and Westphal 1971). At wavelengths below 0.5 μm, the reflectivity is very low, ranging from around 4 percent at 0.35 μm

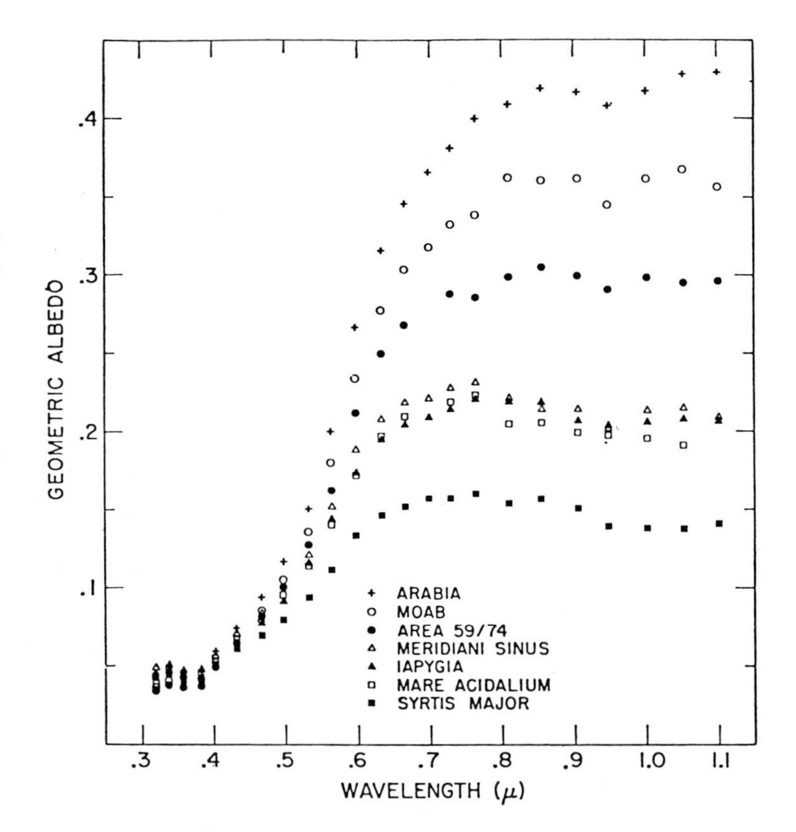

Figure 2.4. The reflectivity of different regions of Mars as a function of wavelength. Most of the contrast between dark (Syrtis Major) and bright (Arabia) regions is in the red. Because of the lack of contrast at shorter wavelengths, the planet looks featureless in the blue, an attribute that was formerly ascribed to a blue haze. (Reprinted with permission from McCord and Westphal 1971 and the *The Astrophysical Journal,* published by the University of Chicago Press; © 1971 The American Astronomical Society.)

to approximately 10 percent at 0.5 μm. At these wavelengths the dark and light areas have similar reflectivities (figure 2.4), which accounts for the poor definition of surface markings in blue light, a phenomenon that early observers ascribed to a blue haze in the atmosphere. Differences between dark and light areas are largely due to differences at the red end of the spectrum. At wavelengths above 0.5 μm, the spectra from different areas diverge so much that the bright areas have reflectivities of around 30 percent at 0.85 μm, compared with values of 15 percent for dark areas. At wavelengths greater than 0.85 μm, the reflectivity is fairly flat out to 2.5 μm, except for a possible shallow absorption feature at 0.95 μm. SImilar results were obtained with the lander cameras for the materials around the two Viking landing sites (Huck et al. 1977).

The characteristics of the spectrum are due mainly to different iron adsorption bands. Strong Fe^{3+} adsorption dominates the spectrum between the ultraviolet and a wavelength of 0.75 μm, giving the planet its red color. A much weaker Fe^{3+} adsorption at 0.87 μm results in a reflectance peak between the two bands at around 0.8 μm (Singer et al. 1979). The two Fe^{3+} bands, which are probably from iron oxides, are most prominent in spectra from bright areas. In spectra from dark areas, other more subtle features show through, such as a Fe^{2+} adsorption close to 1 μm, which may be due to iron in silicates rather than oxides. Direct chemical analyses show that the surficial debris at the Viking landing sites is rich in Fe and poor in Si and Al, as compared with most terrestrial rocks. The favored interpretation is that the debris consists of Fe-rich smectite clays, with minor amounts of sulfates, carbonates, and oxides (see chapter 13).

SURFACE TEMPERATURES AND THERMAL INERTIAS

Temperatures at the martian surface depend on latitude, season, time of day, and the properties of the surface itself, mainly its albedo and thermal inertia. The effect of the atmosphere is negligible because of its low heat capacity. Temperatures are at their lowest just before dawn, rise rapidly in the morning to a maximum just after noon, then fall rapidly in the afternoon and more slowly during the night to their predawn low (figure 2.5). The Northern and Southern hemispheres have different temperature regimes because of the effects of precession that were referred to in the previous chapter. The lowest temperatures that occur on the planet are at the south pole during winter, where they fall as low as 148°K, the frost-point of carbon dioxide. The highest temperatures are at southern midlatitudes in summer, when midday temperatures may reach as high as 295°K.

Thermal inertia is defined as $(K\rho c)^{\frac{1}{2}}$, where K is the thermal conductivity, ρ is the density, and c is the specific heat. It is a measure of the responsiveness of a medium to changes in the thermal regime: if a material has a low thermal inertia, its temperature will respond rapidly to any change in heat input or output; if the thermal inertia is high, it will respond slowly. The dominant cause of variations in thermal inertia is almost certainly the existence of differences in the grain size of the material at the surface, although variations in thermal conductivity, such as might be caused by vesicularity of near-surface rocks, could also have an effect. Grain size has a strong effect because, with decreasing grain size, the number of interfaces between grains increases, and, since the interfaces generally conduct more poorly than the individual grains, the thermal conductivity decreases. Thus, data on thermal inertia have the potential of telling us where the material at the surface is relatively coarse-grained and where it is relatively fine-grained.

Predawn temperatures of the surface are sensitive measures of thermal inertia. If latitude, season, and albedo are the same, then predawn temperatures are highest where the thermal inertia is highest—that is, where the surface materials are coarse-grained. In these areas the ground cools less rapidly during the night because of the ease of conduction of heat from below. Conversely predawn temperatures are lowest where the thermal inertia is lowest. In these areas conduction of heat from below is impeded, probably because of the fine grain size. Using the infrared thermal mappers

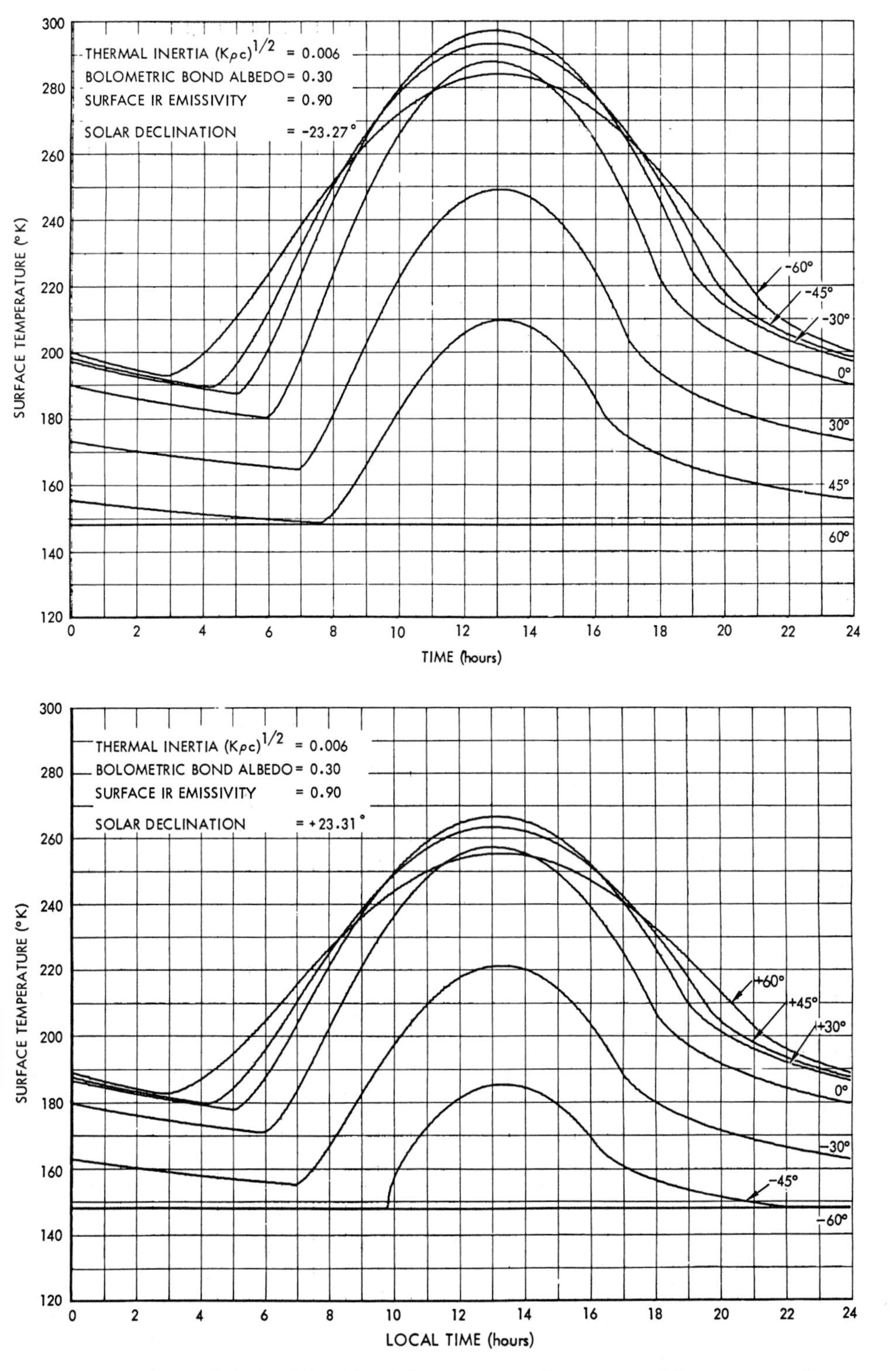

Figure 2.5. Models of the daily temperature fluctuations of the martian surface as a function of latitude. The upper diagram shows the situation at perihelion, when it is summer in the south; the lower diagram is for aphelion, when it is winter in the south. (From Michaux and Newburn 1972, NASA/JPL.)

(IRTM) on the Viking orbiters, Kieffer and colleagues (1977) measured surface temperatures for a wide range of conditions. From these data they constructed a planetwide map of thermal inertias, by subtracting the actual predawn temperatures from those predicted by an idealized model of the planet. In effect the subtraction corrects for latitude and season. The residuals were then corrected for albedo, which was also measured by the IRTM, to give thermal inertias.

The thermal inertia ranges over an order of magnitude and shows only partial correlation with physiographic features (figure 2.6). Much of the Tharsis region, including the large volcanoes Elysium and Amazonis, has a low thermal inertia (1.5 to 4×10^{-3} cal cm^{-2} sec$^{-\frac{1}{2}}$ K^{-1}), indicating that the surface materials in this area are relatively fine-grained. Another broad region of low thermal inertia occurs north of the equator on either side of the 330°W meridian. This relatively bright area, named Arabia on the telescopic maps, is physiographically indistinguishable from old cratered terrain, which elsewhere has a high thermal inertia. Especially high thermal inertias ($> 8 \times 10^{-3}$ cal cm^{-2} sec$^{-\frac{1}{2}}$ K^{-1}) occur in the canyons and the channel-chaos regions just to the east, at 30°W. In these regions more blocks are likely to be present and more bedrock exposed than the average. A strong anticorrelation exists between thermal inertia and albedo. Bright areas have low thermal inertias so are probably fine-grained; dark areas have higher thermal inertias, so are probably more coarse-grained. Kieffer and co-workers (1977) suggested that areas with high albedo (> 0.27) are associated with mean grain sizes less than 100 μm. This conclusion is consistent with the interpretation of bright albedo features as covered with a thin layer of relatively mobile bright debris and dark areas as swept partly free of this surficial debris (Sagan et al. 1973a; Soderblom et al. 1978).

THE VIEW FROM THE SURFACE

We have a close-up view of the surface of Mars only at the two places where the Viking spacecraft landed. One might argue that the sites cannot be taken as representative of Mars because of the selection process. They were deliberately chosen because they were free of hazards as seen from orbital altitudes or as predicted by remote sensing data. As a result both landings were on plains that appeared relatively featureless from orbit. However, the prelanding projections as to what each site would be like were not overly successful, and it may well be that despite our best efforts to the contrary the sites are typical of much of Mars.

On July 20, 1976, a Viking spacecraft sat down on the rock-strewn plains of Chryse Planitia and gave us our first close-up view of the martian surface (Mutch et al. 1976). The lander is located at 22.5°N, 48.0°W (Morris et al. 1978), in an area which, from the orbiter, looks like a lunar mare. The region is almost featureless except for impact craters and wrinkle ridges that resemble in almost every detail those on the Moon. The density of impact craters in the 100 m to 2 km range is close to half the average for

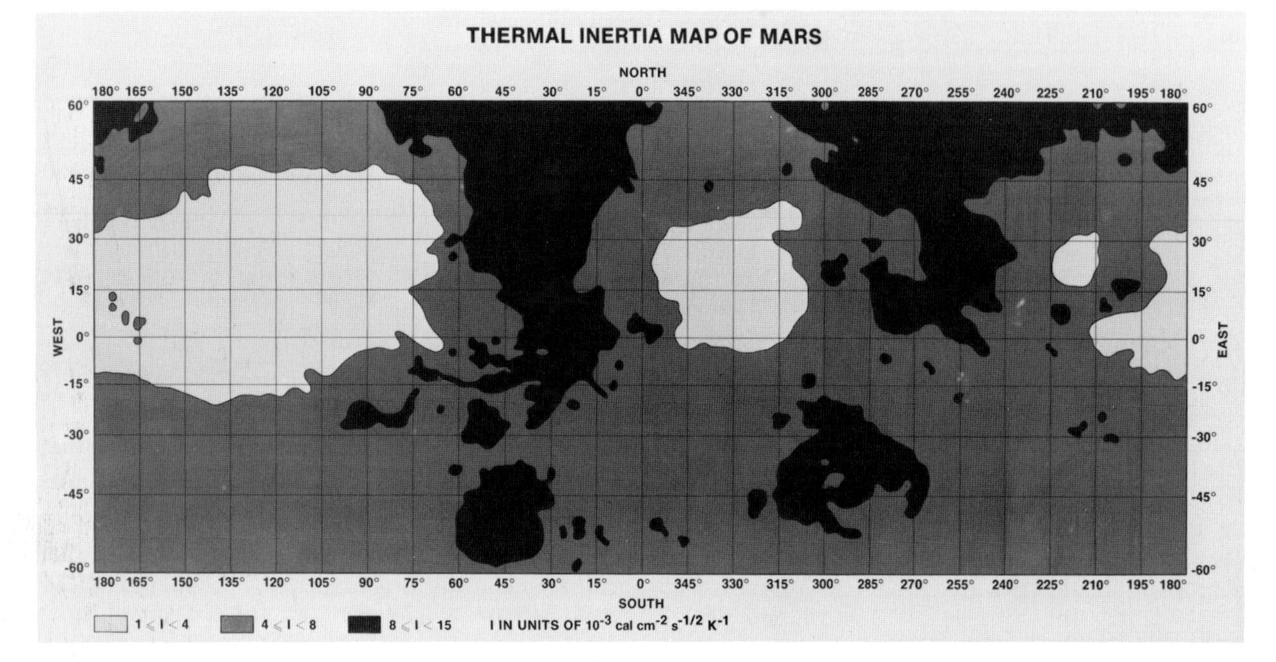

Figure 2.6. Generalized thermal inertia map of Mars. A broad area around Tharsis and the Arabia region has low thermal inertias, suggestive of fine-grained materials at the surface. The canyons and chaos region, many of the northern plains, and the centers of the Hellas and Argyre basins have high thermal inertias, suggestive of relatively coarse-grained surface materials. (From Palluconi 1979.)

lunar maria (Dial 1978). No traces can be seen of the large channels that scour the surface of Chryse Planitia 200 km to the southwest and trend toward the site (Carr et al. 1976). The plains appear volcanic, and the absence of fluvial features suggests either that the channels did not reach as far as the site or that fluvial features are covered by younger deposits.

The view from the lander shows that the site has a generally rolling topography. The surface is strewn with rocks in the centimeter to meter size range. Between the rocks is mostly fine-grained ($< 100\ \mu m$) material, which is also present as drifts ranging in size up to 10 m across (figure 2.7) and as tails behind rocks (Binder et al. 1977). Several small areas, inter-

Figure 2.7. Looking northeast from the Viking 1 lander in Chryse Planitia. The landscape is dominated by drifts of fine-grained materials. Exposure of an internal stratification suggests that the drifts are not actively growing but are being eroded. (Viking lander 1 event no. 11A097.)

Figure 2.8. Looking south-southeast from the Viking 1 lander. Drifts are sparsely distributed across a blocky terrain. In the right foreground is a possible exposure of bedrock. (Viking lander 1 event no. 12A010.)

preted by Binder and colleagues as bedrock, appear to be free of surficial debris (figure 2.8). The rocks have a wide range of color, shape, and texture, probably reflecting variations in lithology and length of exposure at the surface. Most are angular with a coarsely pitted surface. Binder and co-workers (1977) suggested that the grain size of most of the rocks is probably close to the size of the pits (0.3–1 cm), which is quite coarse for surface lavas. They argued, however, that the coarse grain size does not necessarily indicate a deep origin but more probably is the result of a high volatile content and the low viscosity of the lavas. Low viscosities are anticipated because of the high iron content of the surface materials. An array of distinctively different rocks occurs in the midfield, approximately 15–20 m from the spacecraft. These rocks are bluish in color and have flat angular faces, much like terrestrial basalts (Binder et al. 1977). Layering and vesicles are visible in other rocks.

Figure 2.9. Close-up view of the ground beneath the Viking 1 lander. Removal of surficial debris by the terminal descent engines exposed a calichelike hard pan, or "duricrust." This proved to be richer in sulfur, and presumably sulfates, than the loose debris. The etched appearance of the rocks is typical. (Viking lander 1 event no. 12C158.)

The origin of the strewn field is not known. An obvious explanation is that the blocks result from impact and that the various types originated at different depths below the surface and at different locations. The block frequency is comparable to that on the rim of the lunar crater Tycho (Binder et al. 1977) and is considerably greater than that on typical lunar mare. The enhanced abundance of coarse debris may result from the protective effects of Mars's atmosphere. Gault and Baldwin (1970) suggested that craters < 50 m in diameter should be absent from the surface because of ablation and breakup of meteoroids in the martian atmosphere, which may account for the almost total absence of craters in the lander field of view. The atmosphere thus prevents microerosion by small impacting particles, and by preventing production of small craters, may bias production of impact debris toward larger sizes as compared to the Moon. Wind may also have a winnowing effect, culling out some of the fine-grained material and leaving the coarser debris.

The fine-grained debris close to the surface at both sites is cemented to form a crust a few millimeters thick (figure 2.9). The cemented material, termed duricrust (Binder et al. 1977), was exposed beneath the spacecraft, where it was swept free of fine debris by the rocket exhaust. Many of the smaller-sized fragments within the field of view are also formed of duricrust, as indicated by their chemistry and mechanical properties. The crust is identical in composition to the loose fine-grained debris except for a significantly higher sulfur content and possibly slightly more chlorine (see chapter 13). It is probably a calichelike formation, with a cement of soluble, mobile salts, mainly magnesium sulfate and possibly sodium chloride.

Accumulations of fine-grained materials occur as streaks of up to a few meters long in the lee of rocks. Some of the larger drifts appear to be trapped among larger boulders. A distinct layering is visible in some of the drifts, indicating that the dunes are not active but are stable and partly eroded. The general morphology of the drift deposits did not change perceptibly over an entire martian year, despite exposure to the effects of two relatively severe dust storms that were almost global in extent. A slump on one accumulation close to the lander may have been induced by the lander itself. It did indicate, however, that the drift materials have an indurated crust, much like the surface beneath the lander, and it confirmed that the present configuration is quite stable.

The second Viking lander (figure 2.10) is located in Utopia Planitia, at 48.0°N, 225.6°W. The region is part of the vast plains that cover much of the northern hemisphere. Common characteristics of these plains are polygonal fractures, complex albedo patterns, and craters within raised platforms (pedestal craters), all of which are found around the Utopia site, but somewhat muted, as though there were a thin cover masking the more general characteristics of the plains. At the time of site selection, the muting was attributed to an eolian debris blanket, the existence of which had been postulated for high latitudes by Soderblom and co-workers (1973), largely on the basis of crater frequencies. This interpretation was partly supported by the presence at the limiting resolution of the orbiter photographs of a faint linear texture, which was tentatively attributed to dunes (Masursky and Crabill 1976). No dunes were observed at the landing site, however, although they could well be present in the general area. It now appears that the muted topography at the second Viking lander site is due mainly to the 90-km–diameter crater Mie, whose rim is 170 km to the east of the site. The polygonal fracture pattern is especially subdued around Mie; moreover, faint, tonguelike lobes extend radially outward from the crater, some for distances over 300 km. The lander may thus have set down on a flow of ejecta from Mie.

The topography of the Utopia site is remarkably flat, with slopes in the near field of less than 1° and those on the horizon all less than 2°. Several flat-topped bluffs to the east of the spacecraft may correspond to some of the tongues of debris from Mie. Within a few meters of the spacecraft are some interconnected, drift-filled troughs, 1 m across and 10 cm deep (figure 2.11). They appear to form a polygonal pattern that mimics the pattern observed on a much larger scale in the orbiter photographs. Their origin is not known. Mutch and co-workers (1977) suggested that the cracks could form in four possible ways: (1) by contraction of lava on cooling, although the lack of identifiable flows at the surface argues against this hypothesis; (2) by dessication of water-saturated clays exposed at the surface, a hypothesis that gains some support from the apparent predominance of clays in the fine-grained material at the site (Toulmin et al. 1977); (3) by fluvial action, in which case the troughs are unrelated to the polygons seen from orbit—a hypothesis that appears unlikely, however, in view of the lack of supporting evidence of fluvial activity from the orbiters and the absence of fluvial features such as terraces and bars within the troughs; or (4) by ice wedging, in a way analogous to patterned ground in arctic regions of Earth. Some of these possibilities will be discussed further in chapter 6, where polygonal ground in general is discussed.

While the blocks at the Viking 1 site show a wide variety of shapes, colors, and textures, those at the Viking 2 site are very similar to one another and are generally subangular, equidimensional, and profusely pitted. The blocks at the second site are distributed evenly across the surface, with no bedrock outcrops or large drifts. The combination of these characteristics gives the scene a remarkably uniform aspect, with similar blocks being evenly distributed out to the horizon on an otherwise featureless plain. The debris around the site has a bimodal distribution with respect to grain size, one mode being less than 100 μm, the other being in the 10–20 cm size range. The areal density of blocks in the larger size range is about twice that at the first site. Some of the blocks are distinctly fluted as though eroded by the wind; other more massive blocks appear to be conchoidally fractured. The most striking characteristic of the blocks, however, is the presence of numerous pits, which range in size from a few millimeters to a few centimeters across. Some blocks are pitted on one side and not on the other; in some the pits delineate a distinct layering. The two most obvious explanations of the pits are either that they are primary and caused by vesicles in the volcanic rocks, or that they are secondary and caused by preferential etching of softer minerals in coarse-grained rocks.

Mutch and colleagues (1977) discussed several possible explanations of the generally uniform scene at the Utopia site and indicated a preference for two models. The first is that the strewn field is the residue left after eolian winnowing of a layer of ejecta from Mie. Mutch's group postulated that the ejecta was deposited over a former eolian deposit as a surface debris flow, in a manner similar to that suggested by me and my colleagues (Carr

Figure 2.10. View from the Viking 2 lander in Utopia Planitia. The horizon appears at an angle because of the tilt of the spacecraft. The blocks, which have a narrower size range and a more uniform distribution than those of the Viking 1 site, may be part of a lobe of ejecta from the 90-km–diameter crater Mie, whose rim is 170 km to the east of the site. (Viking lander 2 event no. 21A024.)

Figure 2.11. The near field at the Viking 2 landing site. A trough filled with fine-grained sediment runs from the upper left to the lower right. Most of the rocks are deeply pitted, as a result of either a primary vesicularity or etching by the wind. The more massive rock to the upper right is about 1 m across. (Viking lander 2 event no. 21A024.)

et al. 1977b) for martian impact craters in general. They expressed concern, however, about the even distribution of the debris, the uniform lithology, and the lack of brecciation. The alternative explanation is that the strewn field is the remnant of a thin lava flow that was also deposited over a former extensive eolian deposit. Mutch and co-workers suggested that the flow was subsequently eroded by the wind to leave the block field we now observe. They acknowledged that the difficulties with this hypothesis are almost insurmountable, since it requires either removal of most of the flow by wind erosion or dispersal of the flow remnants, which is very difficult to achieve on a level plain. The debris-flow hypothesis appears far more likely and is consistent with what is observed around large fresh martian craters where the impact-related phenomena can be more clearly recognized than around Mie (see chapter 4).

The main change observed at the Viking 2 site in the years after landing was the deposition of a thin veneer of water-ice frost, which lasted for a third of the martian year. The frost first appeared at $L_s = 243°$ toward the end of northern fall and after the occurrence of a major dust storm in the southern hemisphere (Jones et al. 1979). At this time temperatures at the site were too low (150–170°K) for the atmosphere to contain an appreciable amount of water vapor. Jones and co-workers proposed, therefore, that the water originally condensed on dust particles in the warmer southern latitudes and that the dust was transported northward by the storm and finally settled in the latitude belt of the second site, as carbon dioxide–ice nucleated on the particles at night. The carbon dioxide sublimed during the day, leaving behind water-ice and dust. The water-ice disappeared at the end of northern winter.

3 THE ATMOSPHERE

In many respects the martian atmosphere is similar to our own. It has regional and seasonal weather patterns, clouds, global wind regimes analogous to those on Earth, and is commonly close to saturation with water vapor. There are, however, fundamental differences, partly because of its distinctively different composition, and partly because it is so much thinner than the atmosphere of Earth. The composition (table 3.1) is now known to be 95.3 percent CO_2, 2.7 percent N_2, and 1.6 percent Ar, with only small amounts of other constituents (Owen et al. 1977). The average surface pressure is 8 mb, less than one-hundredth that of Earth. The preponderance of CO_2 was not recognized until the mid-1960s. CO_2 had been detected spectroscopically in the 1940s (Kuiper 1952), but at that time the atmosphere was thought to be considerably thicker. Early estimates of pressure were based on scattering of sunlight by the atmosphere and were too high because of the effects of dust and haze. The first good pressure estimates were made from broadening of the CO_2 adsorption bands, as determined from telescopic observations (Owen 1966; Belton et al. 1968).

TABLE 3.1. COMPOSITION OF THE ATMOSPHERE AT THE SURFACE

Gas	*Proportion*
Carbon dioxide (CO_2)	95.32%
Nitrogen (N_2)	2.7%
Argon (Ar)	1.6%
Oxygen (O_2)	0.13%
Carbon monoxide (CO)	0.07%
Water vapor (H_2O)	0.03%
Neon (Ne)	2.5 ppm
Krypton (Kr)	0.3 ppm
Xenon (Xe)	0.08 ppm
Ozone (O_3)	0.03 ppm

Isotope Ratios

Ratio	*Earth*	*Mars*
$^{12}C/^{13}C$	89	90
$^{16}O/^{18}O$	499	500
$^{14}N/^{15}N$	277	165
$^{40}Ar/^{36}Ar$	292	3000
$^{129}Xe/^{132}Xe$	0.97	2.5

SOURCE: Owen et al. 1977.
NOTE: Uncertainties in the Mars values are presently ± 10% except for Ar and Xe.

The low values were later confirmed by Kliore and co-workers (1969), who derived surface pressure from the decay of the radio signals from the Mariner 6 and 7 spacecraft as they passed behind the planet. (The rate of decay depends on the vertical profile of the refractive index in the atmosphere, which in turn depends on the pressure.) When the tenuous nature of the atmosphere was ultimately realized, it was predicted that the surface pressure would vary with the seasons because of the condensation of a significant fraction of the CO_2 on the poles in winter (Leighton and Murray 1966). The prediction was subsequently confirmed by direct measurements on the surface (Hess et al. 1980).

The pressure at the surface also varies according to elevation, and since there is considerable relief on Mars, the effect is substantial. The floor of Hellas, for example, is at an elevation 4 km below the reference datum, and the summit of Olympus Mons is 27 km above, which gives an elevation range of 31 km. The rate at which pressure decreases with height is defined in terms of a "scale height"—the vertical distance over which the pressure decreases by a factor of $1/e$ (0.368).* The scale height of the martian atmosphere is close to 8 km (Seiff and Kirk 1977), which gives a range of surface pressure of almost a factor of ten. It was because of this that the Viking landers were constrained to land in relatively low areas. Only in such areas was the atmosphere judged to be thick enough to provide the drag required to slow the landers' descent to the surface.

The atmosphere is discussed in detail in this chapter because of its influence on the geology of Mars and because of the clues it provides about the amount of water that may be near the surface to participate in geologic processes. Water is of special importance, since fluvial features are observed on the planet, and water-rich weathering products were detected at the Viking landing sites. The atmosphere contains only minute amounts of water (< 100 pr μm).† It is, nevertheless, close to saturation almost everywhere, at least for night temperatures, and large amounts of water may exist close to the surface as water-ice at high latitudes (see chapter 13). Wind is also of geologic interest because of its potential for erosion and deposition. The present wind regime is described here, but the potential

*$P = P_o e^{-h/\tau}$ where P is the pressure at height h above the surface, P_o is the pressure at the surface, and τ is the scale height.

†Water vapor abundances in Mars's atmosphere are measured in precipitable microns (pr μm), or the depth of water on the surface, measured in micrometers, that would result if all the water in the air above the surface precipitated onto it.

effects of the wind on the geology of the surface are discussed in chapter 11. The large dust storms are described in some detail for, as we shall see in chapter 12, they may be largely responsible for layered deposits at the poles. The chapter concludes with a brief discussion of the outgassing history of the planet, as inferred from the atmosphere's chemical and isotopic composition. Again, the main focus of interest is in water, on whether large amounts of water could have outgassed and could be stored undetected close to the surface, where they could participate in different geologic processes. The discussion is quite detailed in places. This is particularly true of the section on meteorological conditions at the Viking landing sites, and readers interested more specifically in the geology may wish to pass over this section.

WINDS

Winds arise from a variety of causes. A planetwide circulation is controlled by seasonal variations in surface temperatures, condensation and evaporation of carbon dioxide at the poles, and Coriolis forces. If the unfrosted parts of the planet had the same albedo and elevation everywhere, then all points at the same latitude would experience the same wind regime. Albedo and topographic differences have a significant effect, however. Topography, for example, may directly divert the regional flow or create temperature and pressure anomalies that perturb the circulation pattern. Additional complications are caused by the daily temperature cycle. At each terminator tidal winds tend to blow from the cold nightside to the warm dayside. Diurnal effects may become extreme at the height of southern summer, when tidal winds can almost entirely disrupt the planetwide circulation, as large amounts of dust are lifted into the atmosphere, and the entire planet is engulfed in a violent dust storm.

GENERAL CIRCULATION. The circulation of the martian atmosphere has been modeled theoretically, using techniques that were developed to model Earth's atmosphere (Leovy and Mintz 1969; Blumsack 1971; Pollack et al. 1976; Webster 1977). While the circulation of Mars is generally simpler than that of Earth, because of the absence of oceans, it is more complicated with respect to topographic effects, which are considerably larger than on Earth. The circulation is driven by seasonal temperature gradients and movement of the atmosphere from pole to pole, as 25–30 percent of the atmosphere condenses on the winter pole. The overall circulation is such that most of the activity other than dust storms takes place in the winter hemisphere. During winter large temperature gradients around the poles create baroclinic instabilities in the mid to high latitudes, with the result that cyclonic and anticyclonic storm systems develop (figure 3.1). The winter regime at these latitudes is accordingly much like that on Earth, with prevailing westerlies, high-altitude jet streams, and traveling storm systems. Summers in the two hemispheres are quite different, both from each other and from the winters. Northern summers are relatively quiet, with no storm systems and only small zonal (east-west) winds. In contrast, during summer in the south dust storms are common, and these may become so extensive as to affect the entire planet and disrupt the circulation established by the north-south temperature gradients. During spring and fall both hemispheres are quiet, with westerly flow at high latitudes and easterly flow at the equator. As a result of this general pattern, local slope winds and diurnal winds dominate at most locations for much of the martian year. It is only during the dust-storm period and during winter that the wind regimes are greatly different.

Figure 3.1. During late summer and fall, cyclonic disturbances are common around the edge of the polar cap. The storm system in evidence here occurred in mid-1978 at the edge of the northern cap. Frost cover can be seen in the foreground. (738A27)

The pattern of circulation just outlined is considerably modified by the large-scale relief of the martian surface. Standing waves develop around topographic highs and lows, and the configuration of the waves changes slowly with season (Webster 1977). Pollack and colleagues (1976) modeled the planetwide wind regime at the time of the arrival of the Viking spacecraft at Mars (northern summer solstice). Prominent features of this model are cyclonic flow around Olympus Mons and anticyclonic flow around Isidis and Hellas.

TIDAL WINDS. Because the atmosphere of Mars is so thin, it has a small heat capacity, so it cools and heats more rapidly than our own atmosphere.

This is particularly important with respect to diurnal and slope effects. Daily temperature fluctuations at the surface are considerably larger than those on Earth. Their magnitude varies with season and location, but daily fluctuations of 50°C are not uncommon. In the absence of dust the atmosphere absorbs little sunlight directly; its temperature profile is controlled mainly by transfer of heat from the ground by conduction and convection. Thus, while large temperature variations may occur close to the surface, they damp out rapidly with increasing elevation (figure 3.2), and, because only the boundary layer is affected, the tidal winds tend to be small. In contrast, a dusty atmosphere absorbs radiation and heats itself directly. Because the emissivity is also enhanced, the atmosphere cools more efficiently at night. The result is a more isothermal temperature profile, with narrower oscillations of temperature in the boundary layer but more extreme diurnal temperature variations at higher elevations, as compared with a clean atmosphere (figure 3.2). Because so much of the atmosphere's profile experiences the diurnal changes, tidal winds can increase substantially.

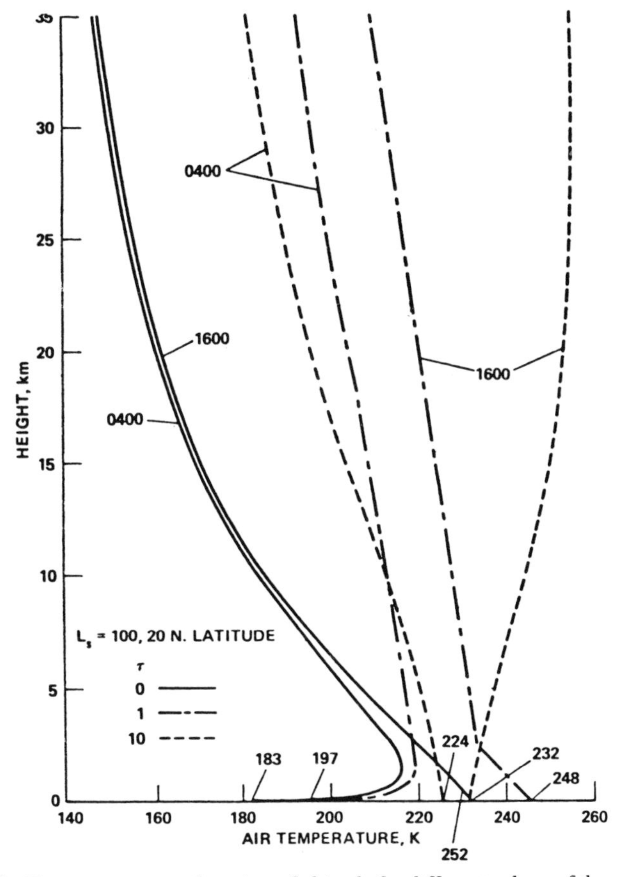

Figure 3.2. Temperature as a function of altitude for different values of the optical depth τ. For each optical depth separate curves are shown for two times of day. With a clear atmosphere ($\tau = 0$) temperature fluctuations are restricted to near the surface. As the atmosphere becomes loaded with dust, the vertical profile becomes more isothermal, daily fluctuations high in the atmosphere are amplified, and the oscillations near the surface are reduced. (From Pollack et al. 1979, copyrighted by the American Geophysical Union.)

SLOPE WINDS. Slope effects are also enhanced by the atmosphere's low heat capacity. On Earth the atmosphere responds only sluggishly to local topography. The temperature on top of a mountain is controlled more by the regional vertical temperature profile within the atmosphere than by daily heating of the ground directly by the Sun and the cooling off at night. Atmospheric temperatures on top of a mountain are therefore lower than at the bottom. On Mars, however, because of the atmosphere's rapid response, near-surface temperatures are controlled more by local ground temperatures. Since these are mainly a function of albedo and the local insolation cycle, they may be the same at the top of a mountain as at the bottom. Consequently, large horizontal temperature (and therefore pressure) gradients may be established, giving rise to large slope winds. The winds are upslope during the day and downslope at night.

METEOROLOGICAL MEASUREMENTS AT THE VIKING LANDING SITES

The Viking spacecraft landed in early northern summer, Viking lander 1 at 22.5°N, 48°W, and Viking lander 2 at 48°N, 226°W. The meteorology boom, containing the temperature, pressure, and wind sensors, was deployed at an elevation of 1.3 m above the ground, well within the boundary layer, which undergoes large diurnal oscillations in temperature (Hess et al. 1977). At the time of writing, observations had been made for more than one full martian year, although not all the results were in print. As expected from modeling of the global circulation pattern, conditions were relatively stable at both sites at the start of the mission. After autumnal equinox, however, conditions changed as strong meridional (north-south) temperature gradients were established in the northern hemisphere. By midfall the polar hood had developed, and storms were passing regularly to the north of the Viking 2 site, affecting pressures and winds at both sites, but especially at the second. At midfall the slowly changing seasonal patterns were abruptly interrupted by the first major dust storm of 1977, which originated in the southern hemisphere. For the next eight Earth-months—a third of the martian year—the meteorology at the landing sites was dominated by the effects of the dust storms, and it was not until late in 1977 that the regular seasonal patterns, free of dust effects, were reestablished.

TEMPERATURE. The temperature of the atmosphere at the surface is governed largely by the temperature of the ground and the amount of dust in the atmosphere. At the start of the Viking mission, in northern summer, with the atmosphere relatively free of dust, the diurnal temperature cycles at both landing sites were repeated from day to day. At the first site the temperature cycled from a minimum at dawn of just under 190°K to a maximum close to noon of 240°K. Temperatures at the second site were 5–10° cooler (Hess et al. 1977; Ryan and Henry 1979). The minimum air temperatures were close to the minimum surface temperatures, but the maxima fell approximately 20° short of the peak surface temperatures. As nothern winter approached, there was a slow seasonal cooling, which was most marked at the second site (figure 3.3). This steady cooling pattern was sharply interrupted by the two major dust storms of 1977. Increased opacity of the atmosphere resulted in a considerable narrowing of the diurnal temperature range, especially at the second site, where the diurnal range was reduced to 10°K, as compared to over 50°K at the start of the mission, when the atmosphere was clear.

PRESSURE. At the Viking 1 landing site the pressure ranged from a 6.7-mb minimum in northern summer to an 8.8-mb maximum at the start of northern winter (Hess et al. 1977; Ryan et al. 1978). The comparable figures for the Viking lander 2 site, 7.4 mb and 10 mb, respectively, are somewhat higher because of the lower elevation (figure 3.4). The cause of the seasonal variations is reasonably well understood and was predicted before acquisition of the Viking data (Leighton and Murray 1966; Briggs 1974; Dzurisin and Ingersoll 1975; Pollack et al. 1976). The pressure varies, because a significant amount of carbon dioxide in the atmosphere freezes out onto the polar caps during winter. Since winters in the south are longer and colder than those in the north, the south polar cap is more extensive and incorporates more of the carbon dioxide from the atmosphere. Moreover, summers in the south are shorter and hotter than those in the north, leading to more complete sublimation of the southern seasonal cap. Because of their magnitude the variations in the south tend to dominate the seasonal pressure cycle. Atmospheric pressure is thus at a minimum when the southern cap is at its maximum, and vice versa. It appears that adsorption and desorption of carbon dioxide in the regolith plays a relatively minor role in the seasonal cycle (Hess et al. 1979). As will be discussed in chapter 13, however, the regolith may still be a major reservoir of carbon dioxide and may interchange with the atmosphere slowly over many years.

Superimposed on the seasonal variations are pressure changes over time spans of hours to days. The magnitude of these changes depends on the season. During summer at the Viking landing sites there were only minor pressure variations from day to day and small diurnal changes, probably due to tides and/or local slopes (Hess et al. 1977). The pattern changed during fall at both landing sites. Tidal effects increased, probably as a consequence of increasing atmospheric opacity, and completely dominated the local slope effects (Ryan et al. 1978). The daily, tidally induced pressure variations at both landing sites increased dramatically with the onset of the major dust storms in 1977. Before the dust storms the diurnal pressure range was less than 0.05 mb; after the dust storms the diurnal variations ranged as high as 0.4 mb. The jump in the diurnal range was

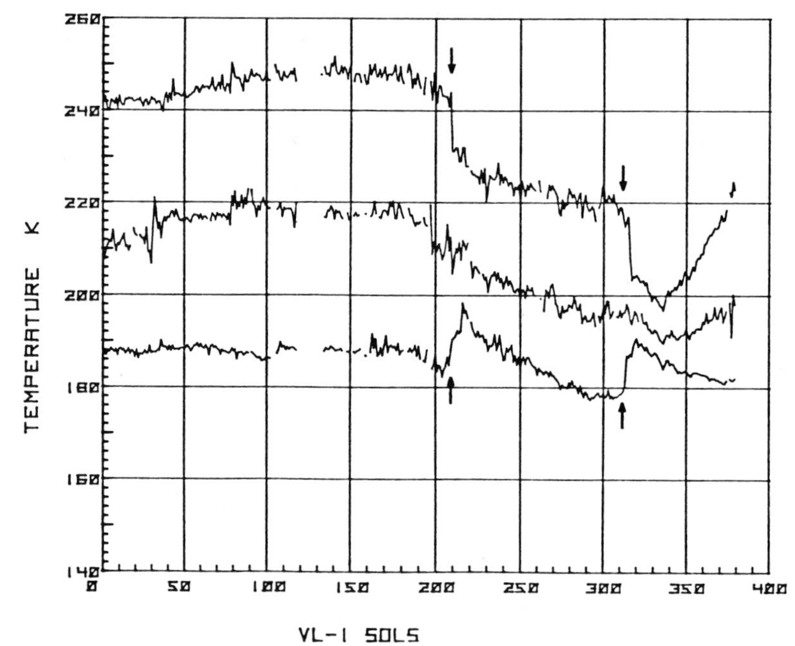

Figure 3.3. Maximum, minimum, and mean temperatures at the Viking 1 landing site, from the time of landing until after the second 1977 global dust storm. Arrows indicate the first arrival of dust from each storm. Time is given in martian days (sols). (See the next figure for the correlation of sols with season.) The dust reduced diurnal temperature fluctuations from in excess of 60°C before the storms to 14°C after the second storm. (From Ryan and Henry 1979, copyrighted by the American Geophysical Union.)

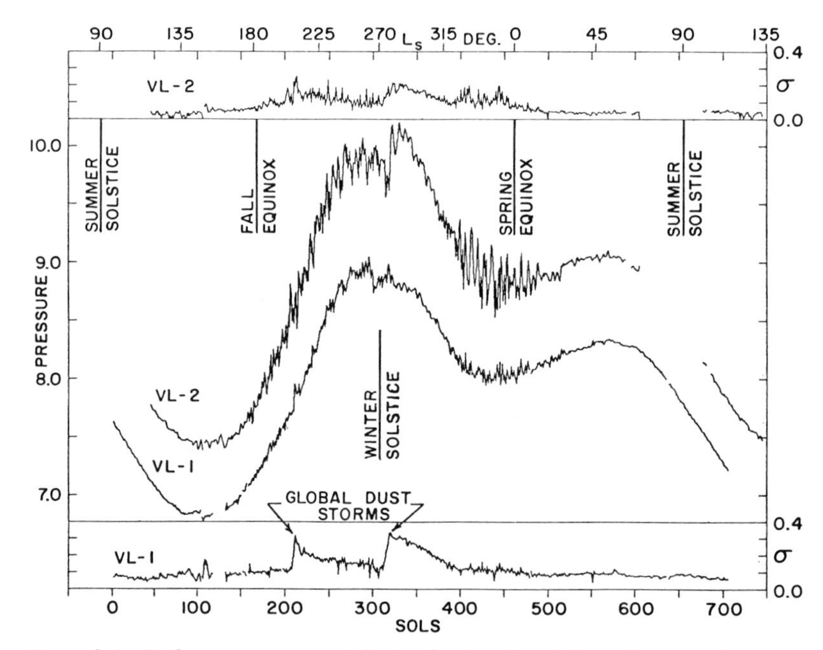

Figure 3.4. Daily mean pressure at the two landers for a full martian year. Seasons are for the northern hemisphere. The upper and lower blocks show the standard deviations of pressure within each sol. (From Hess et al. 1980, copyrighted by the American Geophysical Union.)

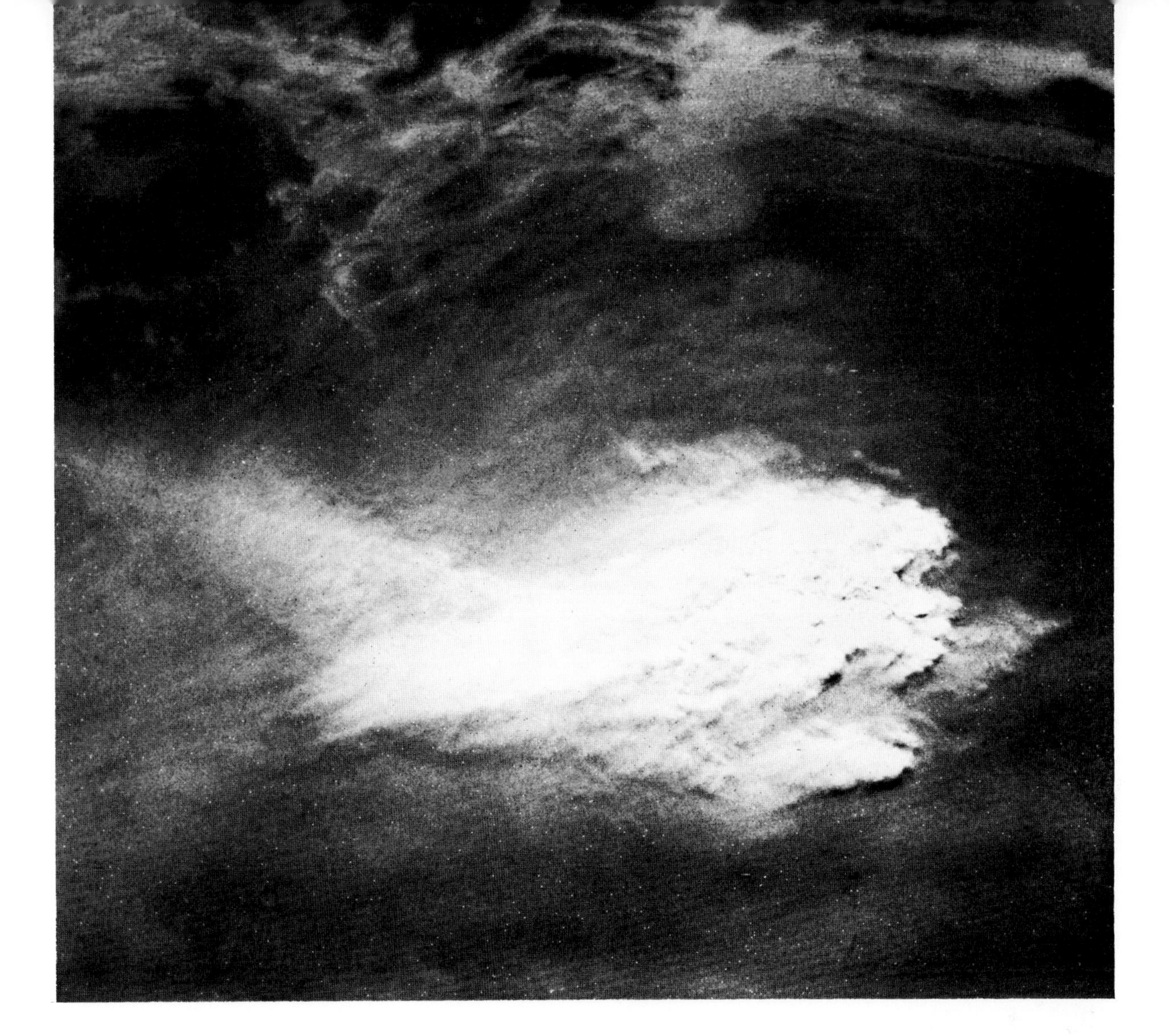

Figure 3.5. Dust storm in the Solis Planum region at L_s = 227°. The season is midsouthern spring, between the two 1977 global storms. (221B24)

probably caused by the large diurnal temperature variations throughout the atmosphere above the boundary layer due to the presence of dust (Ryan and Henry 1979). At approximately the same time as the arrival of the first dust storm—midfall in the north—cyclonic storms were detected at the Viking 2 site at 48°N. These are believed to have been associated with the formation of the polar hood. Such storms passed over the Viking 2 site once every 3.3 days on an average, for a period of approximately 100 days (Tillman et al. 1979), but ceased on arrival of the second dust storm of 1977 at L_s = 284°, just after winter solstice in the north.

WINDS. During the early part of the Viking mission, in northern summer, the winds at the two Viking landing sites were generally slight, with speeds of 2 m/sec at night and up to 7 m/sec during the day (Hess et al. 1977; Ryan et al. 1978). Tidal and local slope winds appeared to dominate. As summer progressed, the directional patterns changed, implying an increasing dominance of tidal over slope winds. During the early fall the pattern changed again, particularly at the second site, where pressure oscillations and changes in wind direction indicated the passage of storm systems in approximately 3-day cycles. Winds were highest and from the north during times of rising pressure (Ryan and Henry 1979). Average wind speeds were around 5 m/sec, with daily maxima close to 10 m/sec. With the arrival of the first global dust storm just after the autumnal equinox, at L_s = 205°, significant increases in wind speed occurred at both landing sites.

DUST STORMS

The occurrence of great dust storms on Mars has been known for many years from telescopic observations, and they were discussed briefly in chapter 1. They tend to recur annually close to perihelion, originating in the southern hemisphere and often expanding to cover much of the planet. During 1977 the Viking spacecraft, both orbiters and landers, made extensive observations of dust-storm activity. Prior to and during the major storms, numerous local dust storms were observed by the orbiter cameras (figure 3.5) and the infrared thermal mappers (Briggs et al. 1979; Peterfreund and Kieffer 1979). Most of the storms moved with velocities of 14–32 m/sec and dissipated within a few days. Storms were most frequent where local temperature gradients are expected to cause high winds, as around the receding edge of the south polar cap, or where there is high

Figure 3.6. Synoptic view of the second 1977 global dust storm. The view is of the southern hemisphere south of the canyons, which are just visible in the upper part of the picture. Almost the entire surface is covered by thick, turbulent, dust clouds, which are spreading northward. (211-5379)

surface relief, as in Argyre, Hellas, and the Claritas Fossae region. In 1977 two dust storms, separated by four months, became almost global in extent, although neither was global in the sense of the 1971 storm. The first started in the Thaumasia region early in southern spring (L_s = 205°). At this time the atmosphere was clear over much of the planet, but in the following days the storm grew to obscure a large fraction of the surface (figure 3.6). After approximately four to five weeks the atmosphere started to clear slowly. At L_s = 275°, however, another major storm arose. It is unclear where it started, but pictures taken at this time show large obscurations west of Argyre (Briggs et al. 1979). This storm also spread rapidly over much of the planet, raising large amounts of dust into the atmosphere, which took three to four months to clear.

Although the Viking landing sites were far removed from the storm initiation centers in the southern hemisphere, the two dust storms had major effects on the meteorology at the two landing sites (Ryan and Henry 1979). The arrival of the first storm at the first site caused winds averaged over periods of one hour to increase to 17 m/sec, with gusts up to 26 m/sec, much larger than any winds previously recorded. The arrival at the second site was less clearly defined, partly because of the effects of cyclonic storms passing through and partly because of the larger distance between this site and the storm center. The maximum wind speed recorded at the second site at this time is 14 m/sec. The arrival of the second storm at the first site was more gradual than that of the first but was likewise marked by wind speeds in the 17–26 m/sec range. The arrival of the second storm at the second site was even less obvious than that of the first.

Storm arrivals had much greater and more lasting effects on surface temperatures and atmospheric opacity. At the first site, before the dust storms, the optical depth of the atmosphere* had a value close to 1 (Pollack et al. 1979), and the difference between the daily maximum and minimum temperatures was in excess of 50°C (Ryan and Henry 1979). Within a few days of arrival of the first storm, the optical depth had increased to about 3, and the difference between the daily maximum and minimum temperatures had been reduced to close to 30°C. This range remained fairly constant until the arrival of the second storm, when the difference was reduced to approximately 15°C, and the optical depth increased to a value of almost 6. The temperature effects at the second site appeared to be smaller and were more difficult to detect because of the larger daily and seasonal changes. As expected from the increased dust loading, the diurnal tides, as indicated by the daily pressure variations, increased greatly with the onset of each storm.

A number of theories have been proposed to explain global dust storms. The most plausible appears to be a feedback mechanism that develops

*The optical depth of the atmosphere is a measure of its opacity. For vertical illumination, $I = I_0 e^{-d}$, where I is the intensity of light at the surface, I_0 is the intensity of light entering the upper atmosphere, and d is the optical depth.

between dust storms and diurnal tides (Leovy and Zuruk 1979). The preferential location of local dust storms in areas of known slopes and around the retreating seasonal polar cap suggests that dust is initially raised into the atmosphere as a result of slope winds or winds that develop as a result of the large temperature gradients adjacent to the polar cap. We have already seen how the presence of dust increases the absorption of insolation in the atmosphere, thereby enhancing the daily excursions of temperature in the atmosphere above the boundary layer. The result is increased tidal winds of the kind that were observed at the Viking landing sites. Under certain circumstances, the augmentation of local winds by tidal winds may be sufficient to raise more dust, further amplifying the tidal winds. This causes a runaway situation, in which tidal winds amplify themselves by raising more and more dust into the atmosphere over a wider and wider area. The mechanism is most efficient close to perihelion, because at that time the amount of daily insolation available to drive the system is at a maximum. The dust storms may turn themselves off by raising so much dust into the atmosphere that the near-surface temperature gradient during the day decreases drastically. As a consequence, convective coupling with the strong tidal winds aloft is diminished, and the near-surface winds drop so much that they can no longer pick up dust.

WATER VAPOR

The water-vapor content of the atmosphere varies considerably with latitude and season (Farmer and Doms 1979). An analysis of the amount of water vapor detected over the Viking lander 1 site at different viewing angles suggests that the vapor is evenly mixed vertically throughout the atmosphere and is not concentrated near the surface (Davies 1979). The distribution of water vapor over the planet has been tracked for one full year by the Mars atmospheric water detectors (MAWD) on each of the Viking orbiters, and the results have been summarized by Farmer and Doms (1979).

At the start of northern summer in 1976 the water-vapor distribution showed a strong dependence on latitude (figure 3.7). The winter hemisphere was essentially devoid of water vapor, but the abundance increased rapidly across the equator, peaking at about 100 pr μm around the edge of the north polar cap. The air over the cap was at 205°K and was close to saturation, indicating that the residual cap was water-ice. As northern summer progressed, the peak at the poles declined, and the water vapor became more evenly distributed by latitude. By the end of northern summer the distribution was more nearly symmetrical about the equator, but most of the water vapor was still in the northern hemisphere. Through northern fall and midnorthern winter the distribution by latitude was fairly flat, averaging around 5 pr μm at all latitudes. This result may have been affected, however, by the increase in atmospheric opacity caused by the global dust storms. In late northern winter the bulk of the water vapor

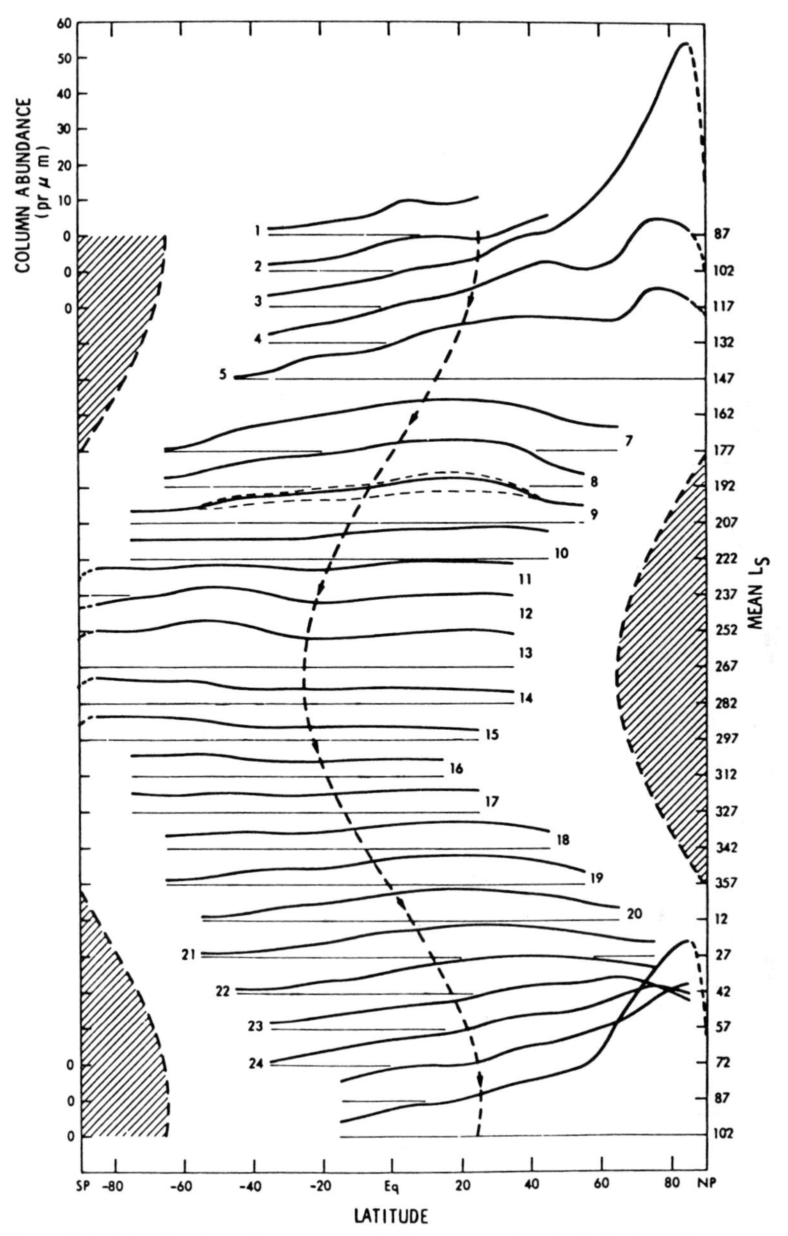

Figure 3.7. Distribution of water vapor in the atmopshere by latitude and season. The central dashed line shows the latitude where the Sun is vertical at noon. The shaded areas are schematic representations of the extent of the annual caps. The main feature is the large amount of water vapor over the residual northern cap in northern summer. (From Farmer and Doms 1979, copyrighted by the American Geophysical Union.)

shifted again to the northern hemisphere, and toward the end of northern spring the sharp buildup in the northern polar regions began again.

The total amount of water vapor in the atmosphere can be derived by integrating the observed values over all latitudes. The integration shows that the total abundance in the atmosphere peaks during northern summer and rapidly declines with the onset of dust storms. The observational effect may be larger than the actual effect because of dust interfering with the detection of water, but even after the storms, when the dust had largely cleared, the atmosphere contained only 60 percent of its pre–dust storm water content. The storm may have swept some of the water vapor into the northern hemisphere, where a fraction was precipitated out as ice in the polar regions. As we saw in the previous chapter, ice observed at the Viking 2 site during winter was interpreted as water-ice that originally nucleated on dust particles in the southern hemisphere. Because of the apparent efficacy of dust in scavenging water from the atmosphere, Farmer and Doms (1979) suggested that there is an annual net transport of water vapor from the southern to the northern hemisphere. The extent to which this is counterbalanced by transport southward, as crystals in clouds, for example, is not known.

CLOUDS

Although the martian atmosphere contains only traces of water at its usual temperatures and pressures, it is rarely far from saturation. Clouds are therefore common and are always present in some area or another. Larger regional cloud masses cause transient brightenings that are visible at the telescope, and these were discussed in chapter 1. With spacecraft imaging, individual clouds of various types can also be seen (Leovy et al. 1973; Briggs et al. 1977).

1. *Polar hood.* In the fall, a general haze of water-ice, and perhaps carbon dioxide–ice, forms over the polar regions. In the north the hood is pervasive and extends down to 50°N. The presence of a fall hood in the south is less certain; the Viking spacecraft saw only discrete clouds and streaky hazes in the southern high latitudes at this season. In early winter most of the water-ice over the poles precipitates out, and the atmosphere clears. In the south an extensive hood forms for a period toward the beginning of spring, when the retreat of the cap begins.

2. *Wave clouds.* These form in the lee of large obstacles such as craters and are particularly common around the edge of the polar cap when the polar hood is present. Here, low-level cooling of the atmosphere creates a stable vertical temperature profile. Air deflected vertically by the strong westerlies undergoes a wavelike oscillation in the lee of the obstacle. The air cools as it rises in the waves and warms as it descends. Water and/or carbon dioxide may condense at the crest of the waves, to leave in the lee of obstacles a regular train of clouds aligned roughly at right angles to the prevailing winds.

3. *Convective clouds.* These generally form in high areas at midday, when strong surface heating creates atmospheric instability. Cloud heights are in the 4–6 km range, and the intercloud spacings are of comparable dimensions. Much of Syria and Sinai Planum were covered by such clouds early in the Viking mission.

4. *Orographic clouds.* These form as a result of the uplift of air over large-scale topography, such as the large Tharsis volcanoes and Olympus Mons. They are most common in spring and summer when the water-vapor content of the local air is at its highest. They develop slowly in the morning and reach their peak in the afternoon, when upslope winds are at their maximum; they may occur on the lee or windward side of the obstacles, according to the strength of the prevailing winds (Briggs et al. 1977).

5. *Ground hazes.* These are seen in low areas at dawn and dusk, the coolest parts of the day, and probably consist of water-ice. Morning hazes are common within the equatorial canyons and are probably responsible for the brightenings in Memnonia observed early in the Viking mission (Briggs et al. 1977).

6. *High-altitude clouds.* Occasionally wispy clouds are seen on the dark side of the terminator as they catch the sun shortly after sunset or before sunrise. High hazes are also commonly seen on limb photographs. They occur at altitudes of a few tens of kilometers and probably include both water-ice and carbon dioxide ice.

OUTGASSING HISTORY

The present atmosphere may represent only a fraction of the total volatiles that have outgassed from the planet. As we shall see in chapter 13, a significant fraction, particularly of carbon dioxide and water, may reside in the regolith or in near-surface rocks. Large amounts may also have escaped from the planet as a result of processes at the top of the atmosphere. There remain clues, however, from which an assessment can be made of the total amounts of volatiles outgassed. Estimates can be made in two main ways: first, by reconstructing the history of isotopic fractionation so as to arrive at the presently observed isotopic compositions of gases within the atmosphere; and second, from geochemical arguments based on condensation and outgassing models. Recent Pioneer Venus results (Pollack and Black 1979) have cast doubt on some of the previous geochemical models, but the main model (Anders and Owen 1977) is presented here as an example of geochemical reasoning, and also because elements of the argument are probably still valid.

McElroy and colleagues (McElroy et al. 1977; McElroy and Yung 1976) used oxygen and nitrogen isotopic ratios to determine the evolution of volatiles in the martian atmosphere. Both oxygen and nitrogen can escape from Mars's gravity field as a result of processes in the exosphere—the very tenuous upper reaches of the atmosphere. Temperatures at 100 km to 200 km altitudes are in the 150–450°K range, depending somewhat on season. At these altitudes the mean free path of molecules between successive collisions is long and molecular speeds are dispersed about a mean, which is governed by temperature. A small fraction of molecules may achieve such high velocities that they are lost from the planet. For the temperatures measured in the upper atmosphere, significant loss by this mechanism occurs only for hydrogen. Velocities may be increased, however, as a result of recoil during photodissociation of molecules such as CO and NO (McElroy and Yung 1977). By this means significant amounts of carbon, nitrogen, and oxygen can also escape. The process is still very slow, requiring billions of years of sustained loss to change the atmosphere's composition significantly.

In the upper parts of the atmosphere mixing takes place by diffusion rather than convection. The exosphere should therefore be enriched, and escape should be biased toward the lighter isotopes, specifically ^{16}O and ^{14}N. The remaining atmosphere should be correspondingly enriched in the heavier isotopes. The present martian atmosphere is enriched in ^{15}N relative to ^{14}N, as compared with Earth's atmosphere. McElroy and co-workers (1977) estimated that a substantial fraction of the planet's nitrogen must have been lost to result in a fractionation of this magnitude, and calculated that the initial amount present must have been equivalent to a surface pressure of nitrogen in excess of 1.3 mb. Estimates of the actual values ranged from 20 mb to 50 mb, depending on the values taken for the escape efficiency, the outgassing history, and the exchange of atmospheric nitrogen with nitrates or nitrites at the surface. From the amount of nitrogen outgassed according to their preferred models, and assuming terrestrial values for the hydrogen/nitrogen ratio, they calculated that 3×10^{-5} gm of water have outgassed per gram of the planet, or the equivalent of 130 m averaged over the whole planet.

Despite the enrichment in ^{15}N and the potential for fractionation of the oxygen isotopes, no significant enrichment of ^{18}O is observed. McElroy and colleagues (1977) attributed this lack of enrichment to the presence of a large reservoir of oxygen near the surface, which is interchangeable with the oxygen in the atmosphere. Such a reservoir must be large enough to dilute the fractionation effects so much that they are not detectable even after 4.5 billion years. Assuming the oxygen isotopes are fractionated by 5 percent (a larger value could have been detected), McElroy and co-workers estimated that the reservoir, which is probably mostly water, must be equivalent in size to 13 m of water averaged over the whole planet. The actual fractionation may be considerably less than 5 percent and the amount of water outgassed correspondingly higher. Such a reservoir is thus consistent with the 130 m of water that is estimated on the basis of the nitrogen data to have outgassed.

Anders and Owen (1977), through analysis of elemental abundances in different components of meteorites and on Earth and the Moon, maintained that certain groups of elements behaved similarly during the condensation of the early solar nebula to form the materials from which the planets and meteorites ultimately accumulated. They argue that if the behavior of certain key elements is understood, then the behavior of the other elements within that group can be predicted with reasonable certainty. Of particular importance for the atmosphere are the elements potassium (K) and thallium (Tl). Potassium provides the key to the group of elements that condense at temperatures between 1,200°K and 600°K, including copper, zinc, tin, and sulfur. Thallium is the key to the group of volatiles that condense at temperatures less than 600°K. In this group are most of the atmosphere-forming elements, including carbon, nitrogen, chlorine, bromine, and the noble gases. From examination of the abundances of K and Tl in meteorites, Earth, and the Moon, Anders and Owen tentatively estimated that Mars contains 100 ppm K and 0.14 ppb Tl. The figure of 100 ppm K comes from the low K content of the material analyzed at the martian surface and a guess that the K content of Mars is somewhere between those of Earth (170 ppm) and the Moon (96 ppm). The Tl value, which is considerably lower than the 4.9-ppb value for Earth, is derived mainly from the measured value of the ratio of the two argon isotopes ^{40}A and ^{36}A. For Mars the value of this ratio is 3,000, much higher than the terrestrial value of 296. Anders and Owen suggested that the high ratio cannot be accounted for by more effective outgassing of ^{40}A, because this would also release ^{36}A. They argued that the abundance of ^{36}A must have been low initially, and with it, all the other elements of the Tl group.

The next step is to estimate what fraction of the total inventory of an element in the planet has been released to the crust or the atmosphere. Anders and Owen called this fraction the "release factor." They applied Earth values to Mars to arrive at a "predicted" value for the crustal abundance of each of the volatile elements. They then compared the "predicted" value with the actual value for those elements on which there is data, mainly those in the atmosphere. In general, the predicted values were too high, indicating that terrestrial "release factors" are higher than those for Mars, presumably because the Earth has outgassed more completely. To estimate the actual outgassing of Mars, Anders and Owen used the ^{36}A abundance in the atmosphere, since almost all the ^{36}A that has outgassed should still be present. The ^{36}A abundance suggests that the outgassing efficiency of Mars is only 0.27 that of Earth. Assuming this value, Anders and Owen estimated that enough carbon dioxide had outgassed that if it were all in the atmosphere simultaneously, it would exert a surface pressure of 140 mb. If outgassing were as efficient on Mars as on Earth, the value would be 520 mb. The best estimate of the amount of water outgassed is that it is equivalent to a 9-m–deep layer over the entire surface. Anders and Owen further concluded that chlorine and sulfur could be abundant near the surface, possibly as sodium chloride and magnesium sulfate. They ascribed the smaller-than-predicted value for nitrogen to exospheric loss.

The Anders and Owen model now seems unlikely, because of the composition of the Venusian atmosphere, as determined by Pioneer Venus (Pollack and Black 1979). The ratios of primordial rare gas species (for example, $^{20}Ne/^{36}A$) are the same for all the terrestrial planets, and the absolute abundances of nitrogen and carbon dioxide are essentially the same on Venus and Earth. By contrast, the absolute abundance of ^{36}A and the ratios $^{36}A/^{14}N$ and $^{36}A/^{12}C$ decrease systematically by several orders of magnitude from Venus to Earth to Mars. Indications are that the non-radiogenic rare gases acted independently of other volatiles during accretion of the planets. The rare gases maintain a constant ratio to each other, but their ratio to other volatiles varies greatly, being largest at Venus and decreasing outward. Pollack and Black suggested that the volatiles were incorporated into grains and planetesimals, which subsequently accreted together to form the planets. They proposed that non–rare gas volatiles, such as carbon, nitrogen, and water, were mostly chemically bound within the grains, whereas the rare gases were adsorbed or were in solid-state dissolution in amounts that were highly sensitive to pressure-temperature conditions within the condensing solar nebula. The particles that accreted to form Venus apparently incorporated three to four orders of magnitude more of the rare gases than those that formed Mars, largely because of the higher nebular pressure at Venus.

Irrespective of whether this model is correct, the large variation in the $^{36}A/^{14}N$ ratio among the terrestrial planets indicates that ^{36}A cannot be used as a measure of the abundance of volatiles other than rare gases. The Anders and Owen reasoning, therefore, breaks down. Pollack and Black, on the other hand, assumed that the rare gas content of the atmospheres of the terrestrial planets is a measure of the nebula gas pressure, and that the abundance of other volatiles is a measure of the efficiency of outgassing. They were best able to match the observed abundances by assuming a nebular gas pressure at Mars that was 1/40 to 1/20 that at Earth and an outgassing efficiency 1/5 to 1/20 that of Earth. From these matches they estimated that 1 bar to 3 bars of carbon dioxide and 80 m to 160 m of water have outgassed on Mars, averaged over the whole planet. Their water estimates were thus close to those that McElroy and colleagues (1977) derived from the nitrogen and oxygen isotopic data.

4 IMPACT CRATERS

Impact craters are the most distinctive landforms on solid planetary bodies other than Earth. Almost every solid surface on every planet and satellite observed so far is cratered to some degree. The only exception is the Jupiter satellite Io, on which impact craters are destroyed rapidly by high rates of volcanism. To most people craters are a dull subject. We all know what a crater looks like, and to the layman one crater looks much like another, whether it be on the Moon, Mercury, or Mars. There is a natural inclination in the case of Mars, therefore, to ignore the craters and focus on other, more distinctive features. Nevertheless, craters are crucial to understanding how planets evolve: they give an indication of the nature and configuration of the surface material, they retain a record of erosional and depositional events that have occurred in the past, and, most important, they provide a means of determining relative ages. Older surfaces are more cratered. This simple relationship is the basis for much of what we know of the geologic history of planets other than Earth. It is by counting craters and studying their stratigraphic relations that we are able to arrange features on the surface of a planet in a time-ordered sequence and thus determine the succession of events that have led to its present configuration.

Craters form as a result of collisions between the planets and asteroidal and cometary debris in orbit around the Sun. The rate of impact today is low and is strongly dependent on the size of the impacting object. For every factor-of-ten increase in size, the number of impacts decreases by a factor of about a hundred. As an indication of how low the cratering rates are on Earth, a crater larger than 1 km forms only every two to three hundred thousand years in an area the size of the United States (10 million km^2). A crater larger than 10 km across is expected to form every ten to twenty million years, and one larger than 100 km once every billion years (Hartmann 1977a). The rates on the other inner planets are probably within a factor of two or three of these (see chapter 5).

The residence time of craters on Earth is relatively short, because they are destroyed rapidly by infilling and erosion. There are therefore relatively few observable impact craters on Earth. There is, however, a sizeable record of impacts because, although a crater may disappear quite rapidly, an impact scar of deformed and recrystallized rocks may remain for a long period of time, or a crater may be buried and subsequently exhumed. On all other planets and satellites except Io, resurfacing rates are lower than on Earth, so more craters are present. In the absence of atmospheres, rates of erosion on the Moon and Mercury are extremely low, and large numbers of craters survive in a relatively pristine form. Mars is somewhere between Earth and the Moon in crater retention. Large numbers of craters survive, but they are greatly modified. The rates of cratering on all planetary bodies are so low that if a surface appears cratered in a spacecraft image, it is almost certainly old compared with most features on the surface of Earth. If several craters larger than 100 km in diameter are present on a planetary surface, then it almost certainly dates back to very early in the history of the planet, when impact rates were much higher than now.

Although martian impact craters look much like those on other planets, the patterns of ejecta around them are quite different. Fresh lunar and mercurian impact craters typically have a hummocky texture close to the rim, which changes to a radial texture further out. The ejecta around most martian craters, however, are disposed in discrete lobes, which extend out from the rim crest for several radii and appear to have been deposited by flow along the ground instead of directly from ballistic trajectories, as on the Moon and Mercury. The cause of the different mode of deposition on Mars is not known, but it may be connected with the presence of water or water-ice close to the surface. Modification of martian craters subsequent to their formation also produces a variety of crater shapes not found on other planets. The type of modification appears to be related to some degree to latitude, possibly indicating climatic control.

In this chapter we shall examine the process of crater formation. The intent is to understand how different types of craters form, and what the configuration of the near-surface materials is afterward. The chapter draws heavily upon comparisons with lunar and terrestrial craters and with those produced experimentally. Many lunar craters are well preserved. They give the best indication of what pristine impact craters look like and how their appearance changes with size. Lunar craters also provide the best examples of what ballistically deposited ejecta looks like and serve as a contrast to the quite different martian patterns. Drilling and erosion have provided information on the configuration of rocks beneath and around some terrestrial craters. The data are discussed here briefly, since the configuration for martian craters is probably similar. The subsurface information is particularly important for understanding large, complex craters, for such craters are more difficult to reproduce experimentally and are thus less well understood than the simple types. Experimental and modeling studies are also reviewed, because they give a direct indication of what happens during impact. It is hoped that these comparisons will provide us with a reasonably accurate picture of how impact craters form on Mars, since the same sequence of events appears to occur no matter where craters form.

CRATER MORPHOLOGY AND FORMATION

Lunar craters

The shapes of lunar craters depend strongly on size. The smallest craters, those less than 10 km in diameter, generally have a simple bowl-shape and a circular outline. At somewhat larger sizes, flat floors, terraces, and central peaks start to appear, so that essentially all fresh craters larger than 40 km in diameter have these features. Another transition occurs in the 200 km to 300 km size range, in which rings begin to occur within craters, and craters larger than this are generally referred to as multiringed basins.

The walls of craters less than 10 km in diameter are generally smooth and close to the angle of repose and may have radial streaks caused by movement of material downslope. The walls may converge on a point at the center of the crater or become shallower at their base and merge smoothly at the center. Occasionally, there is a separate, sharply defined floor. Many mare craters in the size range from twenty to a few hundred meters in diameter develop concentric benches, or breaks in slope on the crater walls. The benches may result from layering in the near-surface materials, particularly at the boundary between the upper, unconsolidated, impact-gardened material and deeper, more massive, basalt layers (Quaide and Oberbeck 1968). The concentrically benched craters are also more blocky inside the crater than the bowl-shaped types. Craters smaller than approximately 20 m in diameter generally lack terraces, presumably because they form wholly within the near-surface, fragmented materials.

At diameters of around 10 km the morphology of lunar craters starts to change significantly, from simple bowl-shapes at smaller diameters to complex types with slump terraces, flat hummocky floors, and central peaks at larger sizes (Howard 1974; Hartmann 1972b; Smith and Sanchez 1973; Pike 1974). The transition in morphology is accompanied by a transition in dimensional relations, notably between rim-crest diameter and crater depth (figure 4.1). For craters smaller than about 20 km depth and diameter are proportional, but above 20 km the depth increases only slightly, so that for every factor-of-ten increase in diameter, the depth increases by only 50 percent. Plots of depth against diameter thus show a sharp break in slope at a diameter of 20 km. Clearly defined central peaks and terraces also start to appear at about 20 km, and with increasing size, through the 20–40 km size range, more and more craters have these features. Craters at the lower end of the range may have only incomplete terraces, and central peaks, if present, are single. At the upper end of the range, multiple terraces completely encircle craters that all have multiple central peaks and hummocky floors. These characteristics typify craters up to diameters as large as 200 km. The crater Copernicus is a familiar example. Features believed to result from viscous downhill flow of material

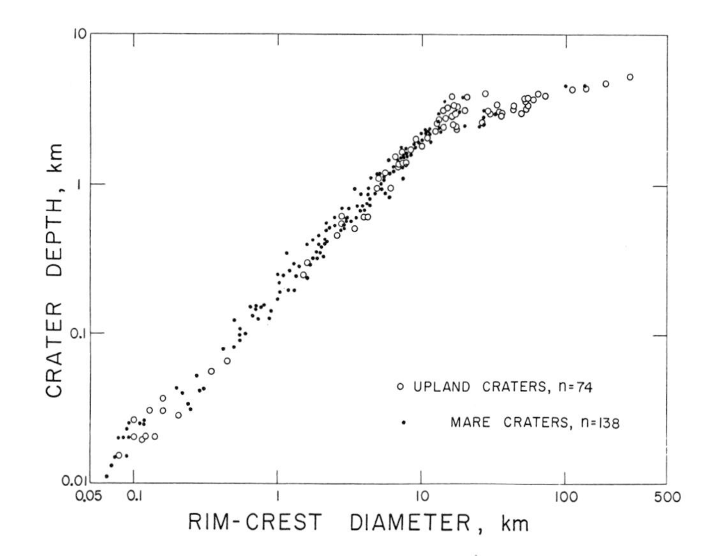

Figure 4.1. Crater depth as a function of diameter for the Moon. Open circles are upland craters; closed circles are mare craters. For diameters less than 15–20 km crater depth is proportional to crater diameter. At larger diameters an increase in crater diameter is accompanied by only a small increase in crater depth. The transition in depth/diameter relation occurs at approximately the same diameter as the transition from simple to complex craters. (Adapted from Pike 1980a.)

start to appear in craters larger than about 30 km. These features include pools of material in hollows on the floor, on the walls between terraces, and occasionally on the rim. Flow lobes and leveed channels may also be present (Howard 1974).

Large lunar craters with inner concentric rings are generally referred to as basins. There appears to be a smooth transition from central peaks to rings. With increasing crater size central peaks tend to become increasingly complex: single peaks are replaced by clusters of peaks, and the peak clusters are ultimately replaced by rings (Hartmann and Wood 1971; Hodges and Wilhelms 1978). The transition takes place over a range of diameters, from about 140 km to 300 km. At the lower end of the range few craters have rings; at the upper end most craters have them. Apart from the rings themselves craters and smaller basins are identical. The best-preserved large lunar basin is Orientale (figure 4.2). Its outer escarpment, the Montes Cordillera, is 930 km across, and two inner rings, together called Montes Rook, are about 620 km and 480 km across. The outer rings are better defined than the inner. Whereas the Montes Cordillera is an almost continuous, inward-facing escarpment, the Rook rings are discontinuous, consisting in part of individual massifs with steep slopes. At the crest of the cordillera is a sharp change in surface texture. On the back slope is a radial and concentric pattern, which grades outward into prominent radial ridges and grooves; on the basin side is a coarse hummocky and somewhat chaotic texture, and the radial and concentric elements are less obvious.

The mode of formation of basin rings is controversial. Different models

Figure 4.2. (*opposite*) The lunar multiringed basin Orientale. The outer scarp forms the Cordillera mountains and is roughly 900 km in diameter. The two inner rings constitute the Rook mountains. Similar rings are observed in martian basins of intermediate size, such as Lyot (200 km) and Lowell (190 km), but are difficult to discern in the larger basins, such as Hellas (2,000 km) and Argyre (1,200 km). (Lunar orbiter 4, M-187.)

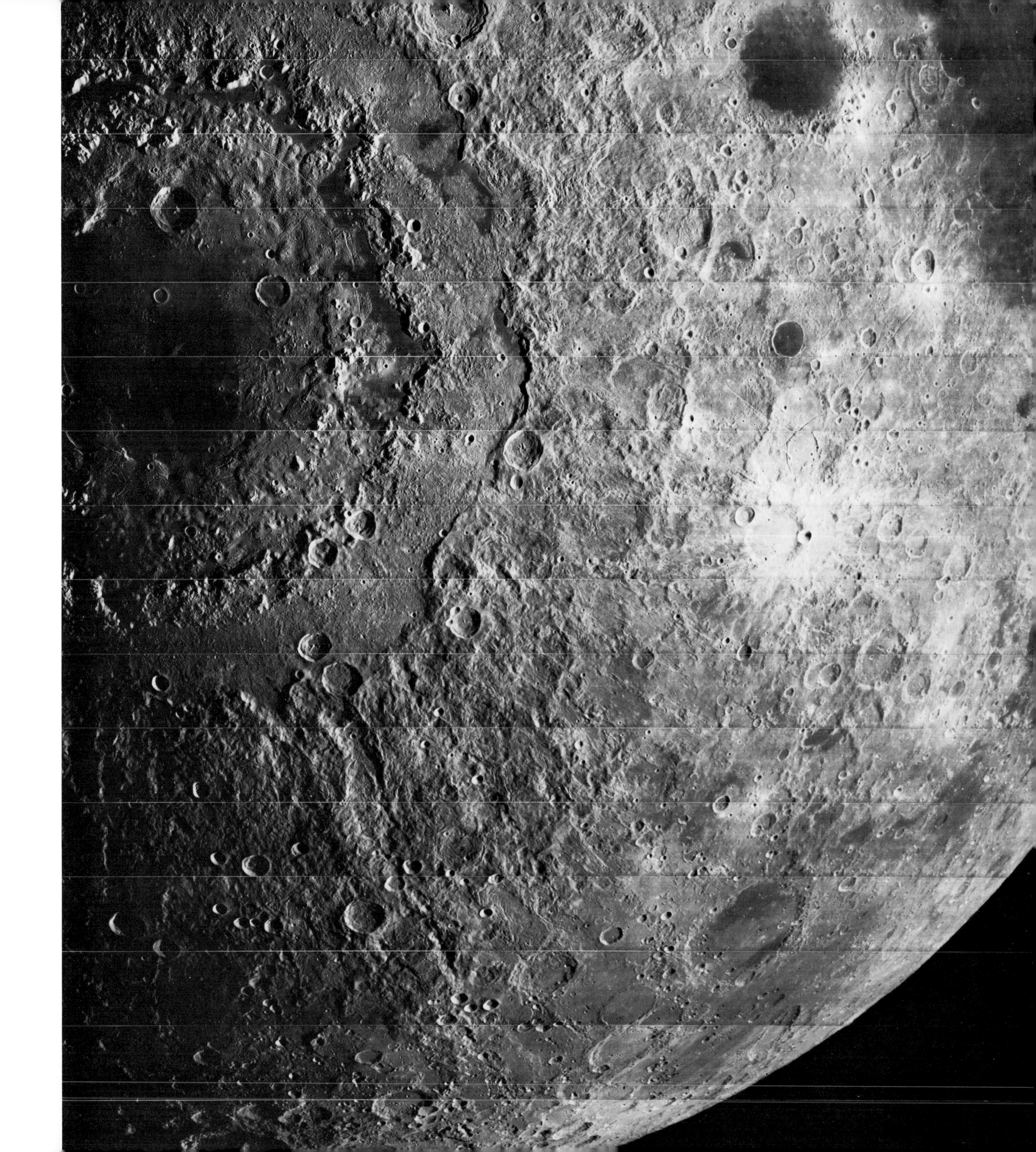

imply different sizes for the cavity that exists immediately after the excavation phase, and hence they differ on the amounts of ejected materials and the size of the impacting bodies. The two main models are the "megaterrace" model and the "nested crater" model. According to the megaterrace model, one of the inner rings represents the rim of the transient cavity that formed at the climax of the impact event. The rings external to this are fault scarps along which the rim has slumped toward the basin. Dence (1973) and Short and Foreman (1972) have argued that the innermost ring of Orientale is closest to the ring of the transient cavity. They suggested that the rim collapsed toward the center on arcuate faults, filling the transient cavity and forming a central uplift. McCauley (1977) and Moore and colleagues (1974) envisaged a similar model but suggested that the outer Rook ring marks the rim. They suggested that the inner Rook ring is part of a central uplift caused by the centripetal collapse of the rim, mainly along the cordilleran scarp. In an alternative "nested crater" model, Hodges and Wilhelms (1978) suggested that the Cordillera best represents the position of the transient cavity rim. They argued, partly on the basis of the continuum between lunar central peaks and rings and partly from comparisons with experimental ringed craters, that all the inner rings are part of a broad central uplift, and that the different rings form as a result of layering in the lunar crust, much in the way that benches form in small craters.

Terrestrial impact features

Although the primary morphology of terrestrial impact craters is preserved in relatively few examples of modest size—for example, Meteor Crater, Arizona—some sixty to seventy circular structures are now widely regarded as being of impact origin. These range from a few meters to over a hundred kilometers in diameter. In many cases the original crater has been eroded away, and all that is left is a central uplift and a surrounding depression or a deformed zone beneath the once-present crater. Initially there was great reluctance in accepting these as impact structures, but the presence of a variety of high-pressure mineral polymorphs, shock-induced petrographic features, and shatter cones appears convincing. Here, the main interest in terrestrial craters is that they give a three-dimensional view, which is not possible for impact features on a remote planet. Terrestrial craters pass through the same morphologic transitions with increasing size as impact craters on the Moon: small craters have a regular bowl-shape and are referred to as simple; larger craters have a central uplift and are referred to as complex (Dence 1965). The transition takes place at diameters of 1.5 km to 2 km in layered sedimentary rocks and at about 4 km in crystalline rocks (Dence 1972), as compared with around 20 km on the Moon. At larger sizes ring structures develop: in crystalline rocks, ring structures are present at crater diameters in excess of 30 km, but in sedimentary rocks, the threshold diameter may be smaller.

Simple craters are exemplified by Meteor Crater, Arizona (1.2 km), and Brent Crater, Ontario (3.5 km). For a simple crater, the depth is about one-eighth of the diameter. Below the crater is a breccia lens that extends to a depth about twice that of the crater. The most strongly shocked materials are in the upper central part of the lens and near the base. The basal shocked material may include shock-melted rocks and traces of the impacting meteorite either in the form of fragments, as at Meteor Crater (Shoemaker 1960), or as a nickel anomaly (Dence 1971), as at Brent. The breccias persist below the highly shocked layer for a few tens of meters before giving way to fractured country rock. The crater rim is slightly upturned and is covered with ejecta in which the fragments are in their inverse stratigraphic sequence, as though they had been peeled back from the crater, somewhat like the petals of a flower (Shoemaker 1963).

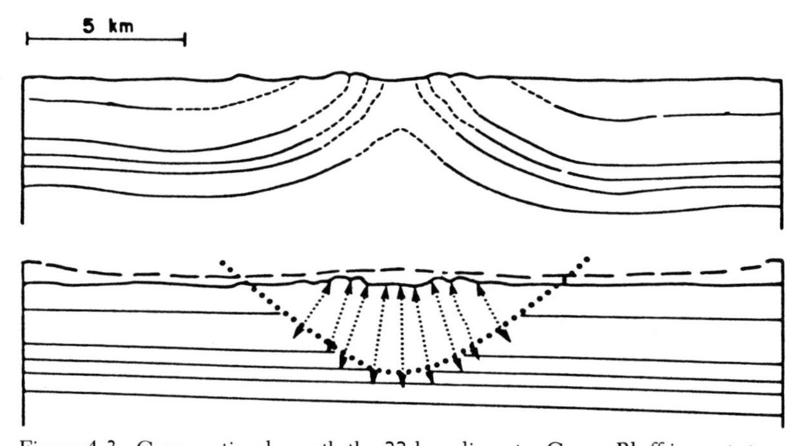

Figure 4.3. Cross section beneath the 22-km–diameter Gosses Bluff impact structure in Australia. The upper diagram shows the present configuration of the sedimentary rocks, with the central uplift and the shallow peripheral depression. The lower diagram shows a restoration of the beds, in accordance with shatter-cone data. (After Milton et al. 1972. Reprinted with permission from Dence et al. 1977, copyright 1977, Pergamon Press.)

Most known terrestrial impact features are of the complex type with a central peak, since complex craters are larger and are more easily preserved than the smaller bowl-shaped craters. In some cases, parts of the original crater can still be seen, but more commonly the structure is deeply eroded, and the most prominent feature is the central uplift. The former site of the rim may be marked by circular normal faults. Where the rocks are well stratified, their deformation provides direct evidence of uplift at the center and downdropping of the rim. Where the rocks are crystalline the evidence of uplift is indirect and comes from the shock metamorphism. The complex craters all appear to be proportionately shallower than the simple types. Two examples of complex craters in sedimentary rocks are the 13-km–diameter feature Sierra Madera (Wilshire et al. 1972) and the 22-km–diameter Gosses Bluff (figure 4.3) (Milton et al. 1972). The central uplift at Sierra Madera is about 4 km in diameter, that at Gosses Bluff about 10 km. Rocks within the uplifts appear to be 1/10 to 1/8 of a crater diameter above their normal stratigraphic position. Movement around the uplifts has been both downward and inward, as with the experimentally produced complex craters discussed below. The rocks within the uplifts are strongly faulted, fractured, and locally crushed. At Sierra Madera, rocks at the uplift center have been shocked to pressures of 10 kb to 20 kb (Wilshire et al. 1972).

The best-studied terrestrial ring structures are the 32-km–diameter West Clearwater Lake and the 70-km–diameter Manicouagan structure. According to Dence and co-workers (1977), the circular shore of West Clearwater Lake is close to the rim of the original crater. A 19-km–diameter inner ring of islands marks a second ring. A third ring is defined by an outer fracture zone about twice the diameter of the crater. The trough between the island ring and the crater rim is a graben, within which are preserved sedimentary rocks that formed a veneer over the crystalline rocks at the time of impact. The veneer is absent on the island ring and the central peak, suggesting that they were uplifted, in contrast to the margin, which was depressed. A thick sequence of breccias and meltrocks covers the island ring, but not the central uplift, and was probably deposited on the original crater floor. Peak shock levels in the central uplift are in the 20 kb to 30 kb range. Similar relationships occur at Manicouagan. Shock metamorphism of the basement rocks becomes more intense toward the center of the structure, in the central uplift. Around the uplift is an annulus of breccias and meltrocks, which lined the floor of the original crater. A peripheral trough occurs just inside the crater rim.

In summary, terrestrial impact craters go through a similar sequence of morphologic changes with increasing crater size as craters on the Moon, but the threshold diameters at which specific features appear are smaller. We will see later that the transition diameter for martian craters is intermediate between those for the Moon and Earth. The three-dimensional view of central peaks of terrestrial craters shows that they are formed by the uplift of country rocks to heights well above their normal stratigraphic position. The central peaks of martian craters may be similarly composed of materials brought up from depths below the surface.

Experimental craters

Experimental techniques provide a means of observing craters as they form. In natural craters we see only the end result of a complex sequence of events and must infer how the final configuration was achieved. With experimental craters, however, we can often observe the events directly. Moreover, the experimental conditions can be changed in order to determine the effects of different parameters. Experimental studies are of two basic types. In the first, craters are produced by impact. Most of the pioneering work in this area has been done at Ames Research Center (Gault et al. 1968). In these experiments a small projectile is fired from a light gas gun toward a target in a vacuum chamber. Projectile velocities are generally in the range of several kilometers per second. Very high-speed framing cameras photograph the craters as they form. The results are then modeled on the basis of the known shock deformation characteristics of the materials involved. One drawback of the method is that only small-scale experiments can be performed, and the resulting craters are generally only a few centimeters across or less. Scaling over several orders of magnitude is therefore required to make the results applicable to natural craters, and we have already seen that crater shape, and presumably the crater-forming process itself, is sensitive to size.

The second way of studying craters experimentally is to use large explosive charges. Craters as large as 1 km in diameter have been produced by nuclear explosions (Cooper 1977), so that the scaling problem is less severe than with craters produced experimentally by impact. However, the phenomenology of explosion craters may be somewhat different from that of impact craters, and the results may be applicable to natural impact features only under certain circumstances. Unfortunately, the differences between large-scale impact and large-scale explosion craters are not fully understood; however, it appears that explosion craters best mimic impact craters when the center of the burst is at, or close to, the surface. Another drawback of large explosions is that the scale of events precludes as detailed a look as is possible with experimental impacts, so that movements of materials have to be inferred somewhat indirectly from markers and from short-lived motion-detection devices.

Experimental studies of high velocity impacts suggest that crater formation can be perceived as a process consisting of three stages—compression, excavation, and modification (Gault et al. 1968). The compression stage starts with contact between the target and the projectile (figure 4.4). Two

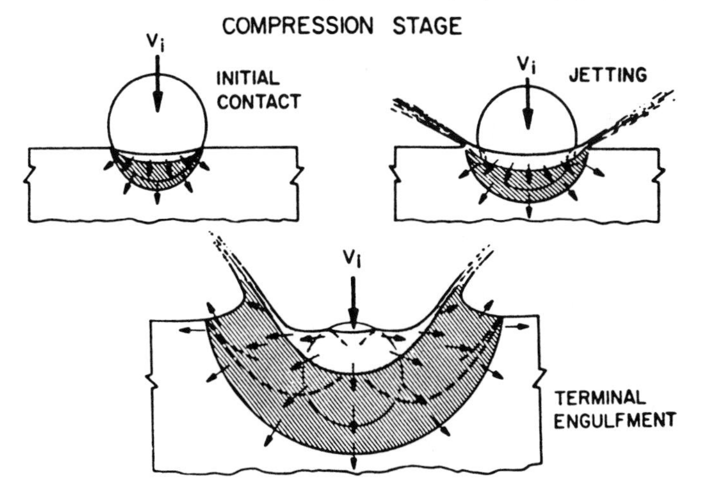

Figure 4.4. Schematic diagram showing three early stages in the production of an impact crater. In the first stage, immediately after impact, shock waves are moving out into the target and back into the projectile, although the net movement of the shock wave moving back into the projectile is still downward, because its velocity is less than the impact velocity. In the second stage a highly shocked fluid mass is caught between the projectile and the target, and material is jetted out laterally around the periphery of the projectile. In the third stage the projectile has been almost completely engulfed, and rarefaction waves generated at the free surfaces cause the initially outward motion of shocked materials to be deflected upward to form a conical spray of ejecta. (Reproduced with permission from Gault et al. 1968.)

shock waves are produced: one that travels upward into the projectile, and another that travels downward into the target. Between the two shock waves material is compressed to pressures that are in the megabar range for impact velocities of several kilometers per second. Within the shocked region the stresses greatly exceed the strengths of the materials involved, so the materials behave like fluids. Jets of material are hurled laterally outward at extremely high velocities from the interface region between the target and the projectile. This material is the most intensely shocked material ejected from the crater during the entire event. As the shock waves penetrate deeper into both the target and the projectile, rarefaction waves are generated at all free surfaces as the shocked material decompresses and expands toward the surface. Initial passage of the shock wave gives the particles within the target material a radially outward motion. Subsequent rarefaction waves cause decompression in the direction of the free surface, imposing a velocity component toward the surface, which increases in magnitude as the material is further decompressed. The net result is that particles follow curved trajectories, starting downward in a radial direction close in and curving upward toward the surface with increasing radial distance. At the end of the compression stage the upward-moving shock wave has completely engulfed the projectile, and a spherical shock wave is penetrating ever deeper into the target, followed by a train of rarefaction waves.

The bulk of the material is ejected in the next phase, the excavation phase, while the shocked material is being decompressed. As the shock wave expands to greater distances, the impact energy is distributed over ever larger volumes of the target, so that the amplitude of the shock wave declines rapidly. The rotational motion induced by the rarefaction waves causes material to be ejected outward at an angle, to form a conical curtain of debris around the growing crater. The curtain leans outward at angles that are generally close to 50° from the horizontal. The debris curtain is essentially an extension of, and is continuous with, the crater in the early stages. As the rim of the crater enlarges, the curtain of debris moves outward with it. This continues out to radial distances at which shock pressures have declined so much that the rarefaction wave is very weak and only just exceeds the tensile strength of the target materials. At this point passage of the shock wave and its trailing rarefaction wave will cause the rim materials to just roll over on themselves. The upward velocity imparted is insufficient to launch the materials into ballistic trajectories, and growth of the crater then ceases. However, the debris curtain continues its outward motion, since its constituents are on ballistic trajectories. Close to the rim the curtain is deposited at fairly low velocities and low angles as continuous ejecta. Further out the debris strikes the target at higher velocities and higher angles and may disturb the surface and cause secondary craters. In the final modification stage the walls generally slump or collapse to produce a stable configuration.

Although the sequence of events in a large explosion crater cannot be followed as closely as that just described for impact craters, explosion craters are of considerable interest in that many of the features of complex craters, such as central peaks and rings, can be simulated. For bursts where the explosive charge is at, or near, the surface, crater development is similar to that during an impact; particle motions are radially downward and outward from the blast center and then curve upward as rarefaction waves unload the shock-compressed material toward the surface (figure 4.5). A debris curtain and overturned rim flap also form, as with impact. The effects of two large surface bursts are discussed here as illustrations of

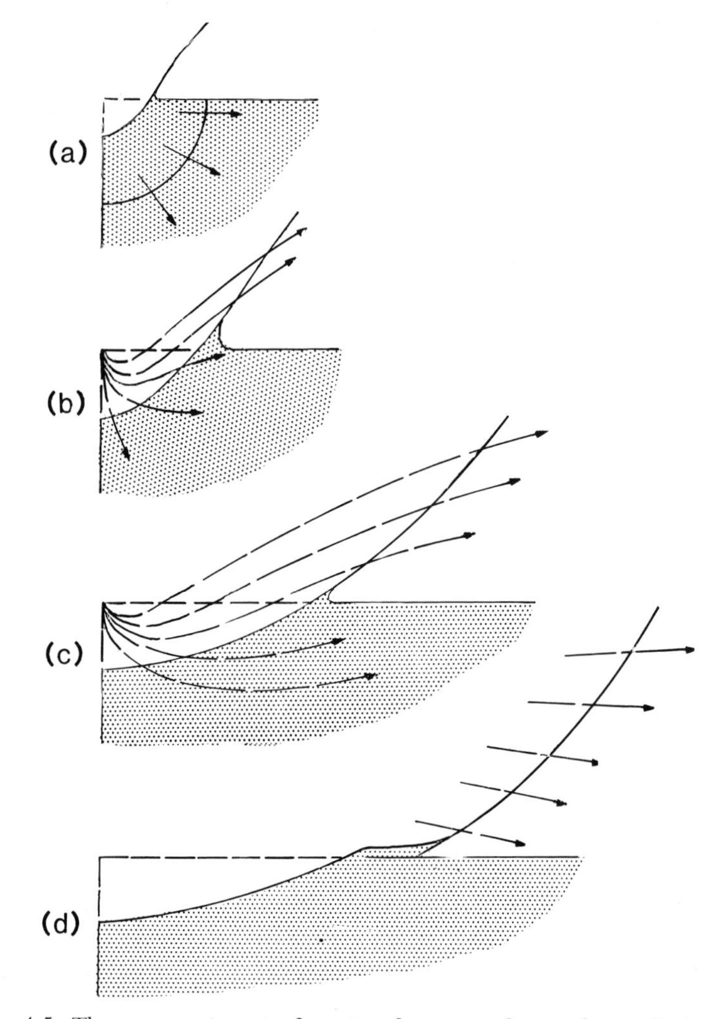

Figure 4.5. The sequence in crater formation for near-surface nuclear explosions: (a) A shock wave is moving into the surface as a crater is starting to form, and ejecta is flowing outward to form a skirt. This step is almost identical to the last stage in the previous figure. (b) The ground shock has moved far out of the picture. Rarefaction waves off the free surfaces have caused the initial radially outward motion to be deflected upward, and material is being ejected outward into ballistic trajectories. (c) The ejecta skirt has traveled further out, and the rarefaction wave at the base of the skirt has become so attenuated that it only just exceeds the tensile strength of the rocks. This is the final stage in crater formation. (d) The ejecta skirt continues to move outward on its ballistic path, laying ejecta on the surface, but no more material is ejected. (After Cooper 1977.)

experimental simulations of features of complex craters, such as central peaks and rings. For more complete discussions of explosion craters, see Cooper (1977) and Roddy (1977) and other papers in the same volume.

In June 1964, a 500-ton hemispherical charge of TNT lying directly on a sequence of unconsolidated sediments was detonated at the Canadian Defence Research Establishment at Suffield, Alberta, to produce a shallow, flat-floored crater (figure 4.6), 108 m in diameter and 7.5 m deep, named Snowball (Roddy 1977). The crater that formed had a broad, relatively flat floor and a prominent central peak 21 m in diameter, composed mainly of folded and faulted clay beds. Many of the clays were from a horizon 10 m below ground zero and appeared to have been uplifted by typically 6 m to 10 m. Marker cans buried before the event indicated that the net movement was downward and slightly inward at the rim, horizontally inward under the flat part of the floor, and diagonally upward under the peak itself. Subsurface motions were almost certainly enhanced by the water-saturated condition of the alluvium. The overturned flap near the rim crest merged outward with a continuous ejecta blanket that extended to about 130 m from the rim, forming an irregular hummocky surface. Further out were traces of rays and secondary craters. Features of special interest, because of their resemblance to features that occur around large martian basins, particularly Hellas and Isidis, are concentric fractures around the crater beyond the rim. The fractures formed mostly during relaxation, just after passage of the rarefaction waves, but they continued to widen slightly for several days after the event.

Another explosion crater produced at Suffield, named Prairie Flat, is of interest because a multiringed structure formed (figure 4.7). The test was in unconsolidated alluvium, with the water table at a depth of 7 m. The resulting crater was 85 m across and 5 m deep. Within the crater a shallow central depression was surrounded by low, concentric ridges 1–2 m high. The rings were caused by concentric anticlines and synclines that were locally broken by faults. In the outermost ridge, extension fracturing occurred at the crest of the anticline, and fragments of the underlying breccia were injected into the core. The entire floor was uplifted about 5 m above its normal stratigraphic position. As in the case of Snowball the surrounding rocks dipped inward toward the crater, then curled over beneath the crater walls to form a large overturned flap. Veloc-

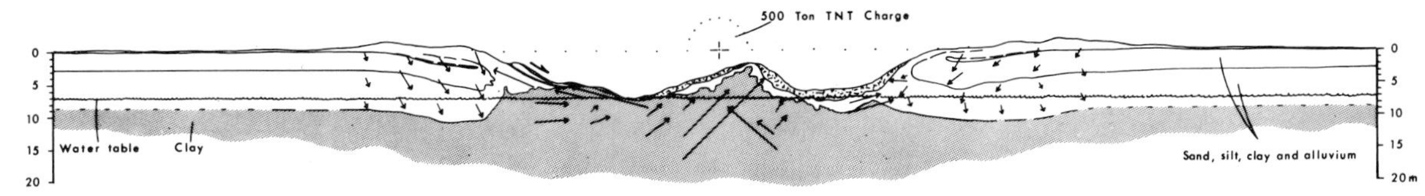

Figure 4.6. The 108-m-diameter Snowball crater produced by a 500-ton surface charge of TNT. The crater has a central uplift, terraced walls, and exterior concentric fractures. The cross section shows the downward and inward movement around the crater edge and the upward movement to form the central peak. (Reproduced with permission from Roddy 1977, copyright 1977, Pergamon Press.)

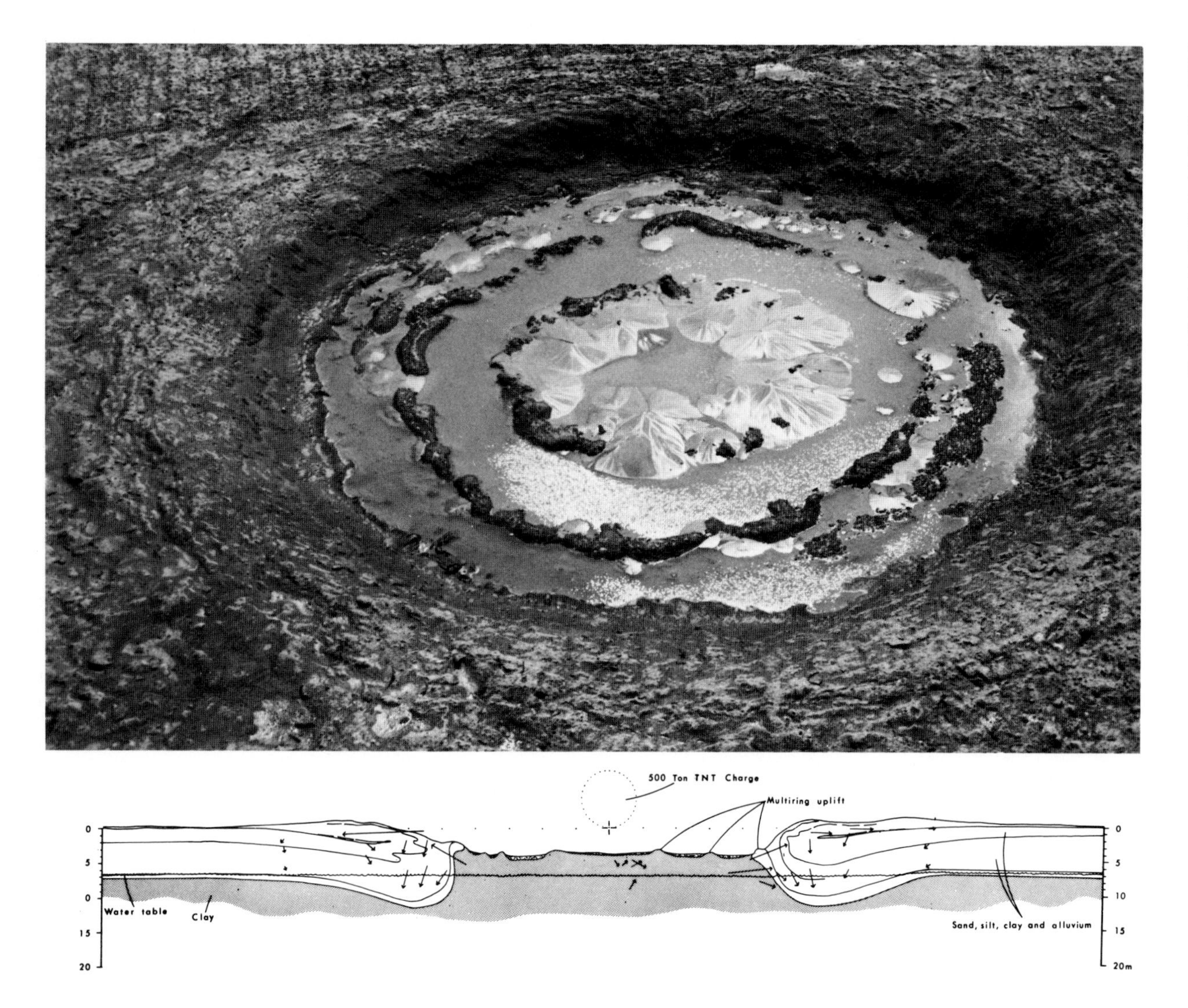

Figure 4.7. The 85-m-diameter Prairie Flat crater was produced by a 500-ton sphere of TNT on the surface. The concentric rims on the crater floor resemble those within large impact basins. The cross section shows that, as with Snowball in the previous figure, the peripheral movement is down, and the materials within the crater are above their normal stratigraphic position. (Reproduced with permission from Roddy 1977, copyright 1977, Pergamon Press.)

ity- and acceleration-gauge data showed that the craterward tilt of the sequence under the overturned flap occurred during the explosion, in compensation for the inward and upward movement of the floor materials.

The significance of these two experiments is that they demonstrate the influence of target strength on crater shape. Numerous craters of comparable size or larger, formed by nuclear explosions in consolidated alluvium and basalts at the Nuclear Test Site, Nevada, are simple in form (Cooper 1977). Lowering the strength appears to lower the threshold diameter for formation of complex craters.

Martian craters

Craters on Mars are much less well understood than those on the Moon, partly because fresh craters are relatively rare, owing to the higher erosion and infilling rates, and partly because comparable imaging data have only recently become available for Mars. Data on crater depths and rim heights are particularly sparse. Martian craters go through the same transitions with size as craters elsewhere. The smallest craters are simple, bowl-shaped types; larger craters are complex, with central peaks and terraces; the largest are multiringed basins. The descriptions of lunar craters, therefore, generally apply to Mars. A possible exception is the larger proportion of martian craters that have central pits or peaks with central pits (Smith 1976). Wood and co-workers (1978) showed that while central pits occur in the majority of fresh martian craters in the 30–45 km size range, they are absent in craters smaller than 5 km. Such pits are rare on the Moon and Mercury. Wood and colleagues (1978), following earlier workers (Hodges 1978; Smith and Hartnell 1977), suggested that the presence of pits may be due to large amounts of ground-ice, which volatilizes from the core of the central uplift.

Other differences that may exist between Mars and other planets concern the onset diameters for complex craters and their depth/diameter ratios. Pike and Arthur (1979) estimated the depths of martian craters by determining the precise positions of shadows within craters from the digital imaging data. In a comparable study Cintala and Mouginis-Mark (1980) estimated depths directly from the images. Both studies showed that complex craters on Mars are relatively shallow compared to those on the Moon but are deeper than those on Earth. This confirms the qualitative impression derived from early pictures of the planet (Leighton et al. 1969a,b). The two studies differ in their estimates of the onset diameter for complex craters. Pike (1979) estimated it to be between 5 km and 8 km, whereas Cintala and Mouginis-Mark (1979) estimated that the transition takes place at 10 km. The difference between the two estimates is probably not significant, since the transition between the two types of craters is gradual over a range of diameters, and 5–10 km is probably a good indication of that range (figure 4.8). Pike (1979) and Wood and co-workers (1978) showed that flat floors and central peaks occur at somewhat smaller diameters than terraces on the walls, suggesting that, initially at least, the central peaks form by rebound and not by slumping of the walls toward the center of the crater during terrace formation.

Gravity appears to be the predominant factor controlling crater depth. Pike (1979) has shown that there is an inverse relation between the depths of complex craters and the acceleration due to gravity. The larger the gravity field at the planet's surface, the shallower the crater for a given diameter, and the smaller the diameter for transition from simple to complex craters. Pike noted that the relation holds well for the Moon, Mars, Mercury, and Earth. On Phobos, where gravity is extremely small, all the craters are simple, and the transition diameter is never reached. Higher gravity probably affects the cratering process by increasing the work that has to be done against gravity during the excavation phase and by causing earlier and more complete collapse of the transient cavity. Other factors that could cause variations in the depth/diameter ratio from planet to planet are differences in impact velocities and in surface properties. However, the depth/diameter ratios for the planets show no systematic variation

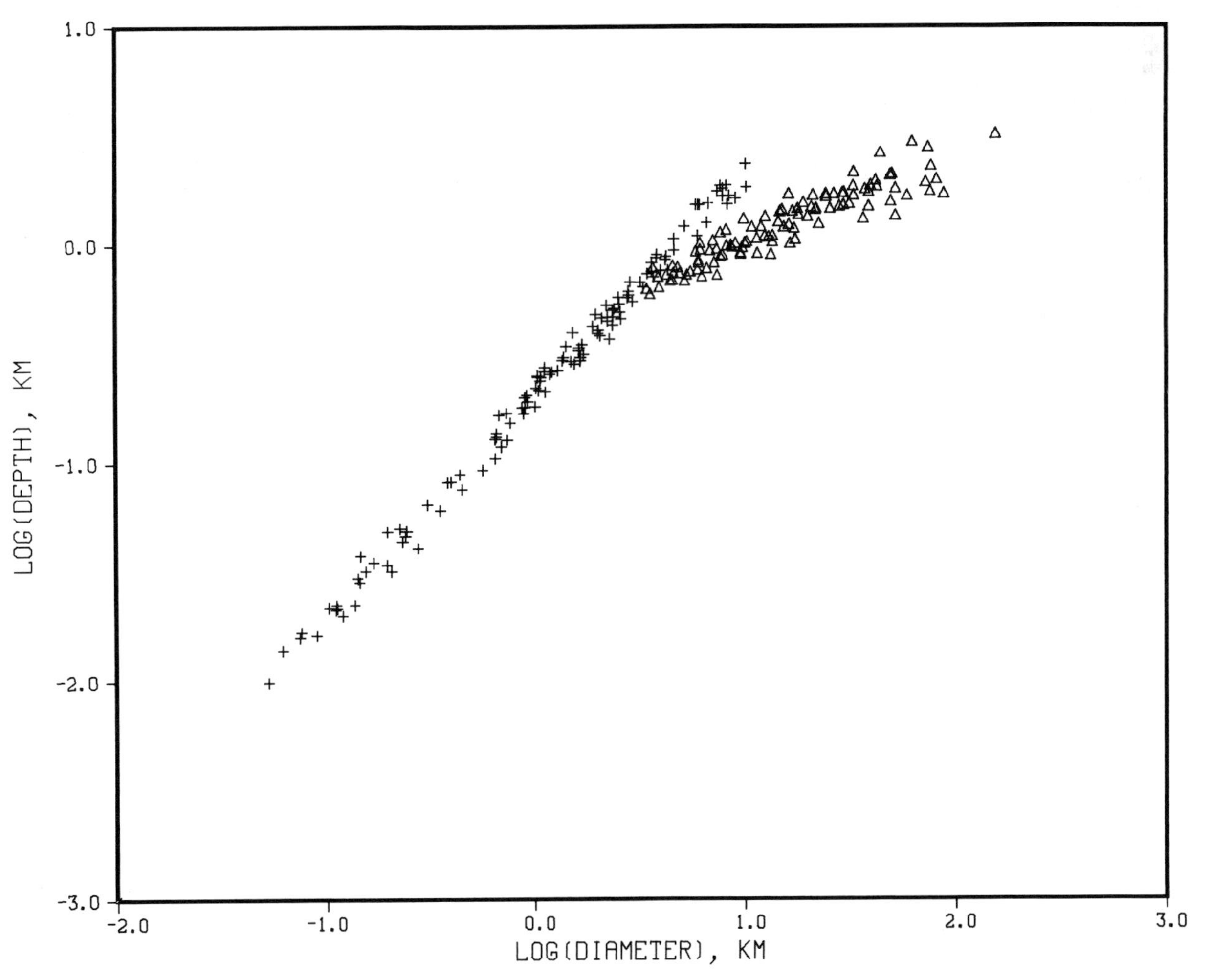

Figure 4.8. Crater depths as functions of diameter for Mars. The shape of the curve is similar to that for the Moon (see Figure 4.1), but the break in slope is at a smaller diameter, probably because of the larger gravity field. Triangles represent complex craters; crosses represent simple craters. (Adapted from Pike, 1980b.)

according to position within the solar system, so that any velocity effect, if present, must be small. The properties of the martian surface may have a larger effect. Boyce and Roddy (1978), on the basis of comparisons with experimental craters such as Prairie Flat, pointed out that if large amounts of water are present in the target, then a larger, shallower crater results. They argued that fluidized ejecta patterns around martian craters indicate that large amounts of volatiles are involved, and that the craters should accordingly be proportionately shallower than those on the Moon and Mercury. Thus, although gravity is the predominant factor controlling depth/diameter ratios, the spread of the data is large, and a secondary factor contributing to the shallow depths of martian craters may be the presence of large amounts of volatiles close to the surface.

Impact basins on the Moon exert strong control over the surface geology. On Mars the effects of basins are much less obvious. Wood and Head (1976) and Wilhelms (1973) have identified 19 ringed basins, ranging from the 135-km–diameter Liu Hsiu to the 2,000-km–diameter Hellas basin (table 4.1). Of these only four have three or more rings. Some large basins may have been destroyed: Wilhelms (1973) suggested, for example, that basins may once have been present in Chryse Planitia and southwest of Hellas. Wood and Head (1976) showed that the density of basins on Mars (1.3 per 10 million km^2) is considerably lower than that on the Moon (7.6 per 10 million km^2) and Mercury (13.7 per 10 million km^2). The low figure for Mars is unlikely to be due solely to erosion, for the density on the relatively well preserved southern hemisphere (2.2 per 10 million km^2) is still low, and enormous amounts of infilling and erosion would be required to change this number significantly. The scarcity of basins on Mars suggests that it was impacted by proportionately fewer very large bodies around 4 billion years ago, as compared with the Moon. Wood and Head also showed that the onset diameter for ringed basins (135–145 km) is similar on the Moon and Mars, although multiringed structures—those with three or more rings—start occurring at much larger diameters on Mars (1,200 km) than on the Moon (350 km).

The largest martian basins differ from typical lunar basins in that inner rings of peaks, such as those that form the Montes Rook in Orientale and the Montes Spitzbergen in Imbrium, are missing. Hellas is the largest basin on Mars. It is about 1,800 km across, as measured from the inner edge of the rim. Its precise size is difficult to determine, because large portions of the rim are missing to the northeast and southwest. The crater rim is relatively low and is best preserved to the west. Outside the rim are several large, arcuate, inward-facing escarpments, which are best developed to the west, where they occur at radial distances of approximately 1,200, 1,600, and 2,000 km. The uncertainty is the result of not knowing the exact center of the basin. The basin thus somewhat resembles the experimental Snowball structure, in having a low rim and concentric fractures around the crater at large radial distances.

The partly preserved Isidis basin is about 1,200 km across, as measured from the inner edge of the rim. Only to the south and northwest is the rim still preserved, although to the west the edge of the basin is marked by a distinct escarpment. Where present the rim is composed largely of numerous, closely spaced massifs, in contrast with the more or less uniformly uplifted rims that form Montes Apenninus (Imbrium) and Montes Cordillera (Orientale) on the Moon. Outside the rugged rim are numerous concentric fractures, which form the Nili Fossae to the northwest and the Amenthes Fossae to the southeast. To the southwest, at a distance of 1,600 km from the basin center, is a prominent inward-facing escarpment.

A third large basin, Argyre, is approximately 600 km across, as measured from the inner edge of the rim, and is the best preserved of the martian basins (figure 4.9). As with Hellas and Isidis, the original floor appears to be buried, and no inner rings are visible. Argyre has a broad rugged rim, extending from 300 km to 800 km from the basin center, and is composed of numerous individual, closely spaced massifs that form vague concentric and radial patterns. Outside the rim are concentric fractures with radii of approximately 1,200 km and 1,600 km. Judging from the preservation of the rim, Argyre is the youngest of the three large basins. However, it is still ancient, as indicated by the superposition of several large craters, including Galle (220 km), Hooke (140 km), and Wirtz (120 km). Smaller basins have inner rings of islands: the 190-km Lowell (52.3°S, 81.3°W) is a particularly well preserved example; others are Schiaparelli (3.2°S, 343.6°W), Huyghens (14°S, 304.2°W), and Antoniadi (21.7°N, 299.1°W).

The reason for the absence of rings within the larger martian basins is

TABLE 4.1. BASINS OF MARS

Name	*Latitude*	*Longitude*	*Diameter*
Liu Hsin	−54.5	171.5	135
Gale	−5.3	222.0	150
Nr. Columbus	−25.2	164.5	145
Ptolemaeus	−46.2	157.5	150
Phillips	−66.5	44.5	175
Molesworth	−27.7	210.5	180
Lowell	−52.3	81.3	190
Lyot	50.0	330.5	200
Kaiser	−46.4	340.5	200
Kepler	−47.0	218.5	210
Galle	−51.0	31.0	220
Herschel	−14.6	230.2	290
Antoniadi	21.7	299.1	400
Huyghens	−14.0	304.2	460
Schiaparelli	−3.2	343.6	470
South Polar?	−82.5	267.0	850
Argyre	−49.5	42.0	1,200
Isidis'	16.0	272.0	1,900
Hellas'	−43.0	291.0	2,000

SOURCE: Wood and Head 1976.

Figure 4.9. An oblique view looking northeast across the martian basin Argyre. As measured from the outer edge of the plain, the basin is 740 km across. The relief in the rim appears to be largely caused by dislocation of the preexisting terrain. The large crater on the far side is the 210-km-diameter Galle. (P-17022)

not known, but a possible explanation is provided by the "nested crater model" of Hodges and Wilhelms (1978), outlined above for lunar basins. In this model the inner rings are caused by layering in the crust. Application of the model to Mars suggests that a discontinuity occurs in the martian crust at a depth of 15–30 km, which causes rings within 150–450-km-diameter craters. No significant discontinuity occurs below this depth. In very large basins the effect of the 15–30-km discontinuity is masked by the size of the craters, and there being no deeper discontinuity, no rings form. This is speculative, admittedly; nevertheless, the large basins may be providing us with clues as to the structure of the crust very early in the planet's history.

EJECTA MORPHOLOGY AND MODE OF DEPOSITION

As previously noted, the pattern of ejecta around martian craters is different from that on other planetary bodies. In this section the patterns around lunar craters are described, as representative of what is normal for planets other than Mars. The martian patterns are then described, differences are noted, and speculations are made concerning why the patterns are different. The characteristics of martian ejecta are still poorly understood, not so much because the data are not available, but rather the reverse. So much information is available in the Viking orbiter pictures that the task of classifying and categorizing the many different ejecta types and mapping their distributions is overwhelming. No satisfactory classification scheme exists. A number of informal terms, such as splosh, rampart, pancake, pedestal, flower, and mound, have been used to describe the ejecta. Unfortunately, they have been used in different senses by different authors. In the summary presented here I have tried to explain some of the terms and to present a simple picture by describing a few broad categories, rather than elaborating upon the variety of patterns. In so doing I may well have presented too simplistic a view.

Lunar crater ejecta

Lunar craters are surrounded by a continuous blanket of ejecta, which merges outward with the surrounding terrain. The ejecta is normally continuous out to a distance of one crater radius from the rim crest, although its precise extent can rarely be established. For craters smaller than 1 km in diameter the rim deposits are relatively featureless except for abundant blocks. Craters in the 1 km to 20 km size range, however, are surrounded by a variety of recognizable features (Howard 1974; Oberbeck 1975): (1) Close to the rim crest, generally within half a crater radius, the ejecta surface is smooth or has a faint hummocky texture. (2) Just outside the smooth or faintly textured rim deposits, and mostly within one crater radius of the rim, some craters have arrays of concentric dunes or low ridges, whereas others have radial ridges and grooves. (3) The zone of radial or concentric ridges merges outward into a zone of secondary craters, which may be arranged in loops or radial chains. (4) A braided or herringbone pattern may develop in association with the secondary craters. The pattern is created by numerous faint ridges that extend from secondary craters outward at a small angle on both sides of the radial direction, to produce a V-shaped configuration that points toward the primary crater. (5) Individual rays may extend outward from the crater for large distances, but they generally merge and become almost continuous within two crater radii of the impact point. The rays are

Figure 4.10. The 30-km–diameter lunar crater Timocharis. Immediately outside the rim are gentle hummocks. Further out is a marked radial pattern, beyond which are strings and loops of secondary craters. (Apollo 15 metric frame 1005.)

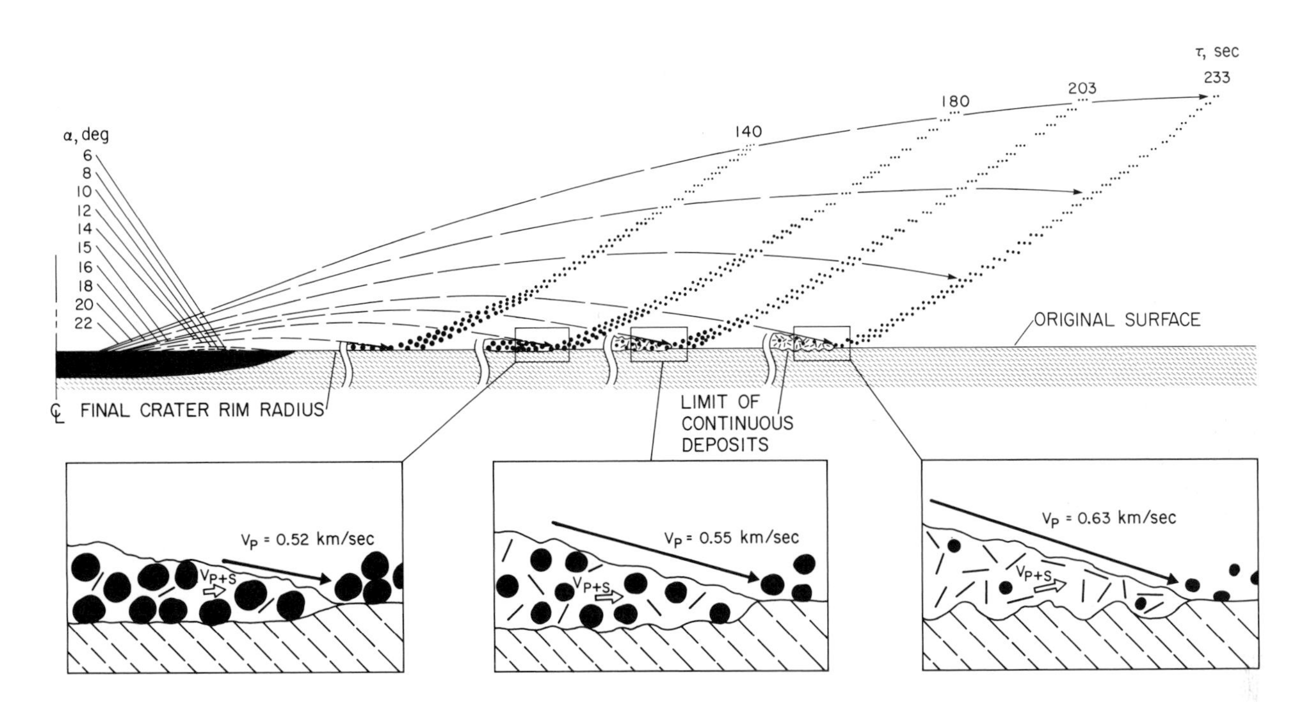

Figure 4.11. Ballistic deposition of ejecta. Formation of a crater results in an outward-moving curtain of debris (see Figure 4.5). With decreasing height within the curtain the materials tend to be coarser-grained and to move at lower velocities. Those ejecta that fall close in do so at low velocities and form a nearly continuous deposit that barely disturbs the former surface. Ejecta that fall further out have higher velocities, crater the surface, and result in extensive mixing of the primary crater ejecta with local surface materials. (From Oberbeck 1975, copyrighted by the American Geophysical Union.)

generally bright, but dark streaks and patches are common. Secondary crater chains often lie within rays or have bright plumes downrange that merge to form the rays.

The change in ejecta morphology with increasing size is less dramatic than the change in crater shape. The zones just outlined increase in diameter with the size of the crater (figure 4.10). The "dunes" found in smaller craters, however, are generally absent around craters larger than about 20 km in diameter. Around some craters are what have been termed deceleration lobes (Howard 1972). These are lobes of debris that are generally on inward-facing slopes toward the edge of the continuous debris blanket. They are of interest here because of their superficial resemblance to the lobate features around martian craters.

The configuration of ejecta around the 930-km–diameter Orientale basin is somewhat different from that around smaller features. On the backslope of the Montes Cordillera is a distinctive wavy texture, caused by the interplay of closely spaced transverse (concentric) ridges with shallow, more widely spaced radial ridges and grooves. The transverse ridges occur preferentially on basin-facing slopes and are mostly restricted to within 250 km of the Cordillera scarps. Further out only the radial fabric is visible. This may range in scale considerably, from valleys tens of kilometers across down to lineations a few tens of meters wide. Similar sculpture is visible around other lunar basins, such as Imbrium and Nectaris; that around the Imbrium basin is visible well over 1,000 km from the basin rim. The sculpture is now generally believed to result from lines of secondary craters (Howard et al. 1974), rather than from fracturing, as was formerly thought.

The mode of deposition of ejecta around lunar craters has been the subject of considerable attention in recent years. The most generally favored model is that of ballistic sedimentation (Oberbeck et al. 1975a,b; Oberbeck 1975; Morrison and Oberbeck 1975), which is, in part, an elaboration of the classic work by Shoemaker (1962) on the ballistics of ejecta from Copernicus. According to the ballistic sedimentation model, the impact causes a conical curtain of ejecta to move radially outward, away from the crater, as is commonly observed during experimental high-velocity impacts (figure 4.11). Ejecta velocities and ejection angles are lowest at the base of the curtain and increase upward. In addition, the mean grain size tends to decrease with height in the curtain. At the start of its outward sweep the lower portion of the curtain is deposited close to the rim to form a continuous ejecta deposit with an inverted stratigraphy. This deposit is relatively coarse, even coherent, and contains relatively little secondary debris. As the curtain moves radially outward, the material intersects the ground at higher and higher impact velocities because of the increasing range. As a result, secondary craters form, and a ground-hugging cloud of primary and secondary debris is created behind the curtain. It seems to be this mixed debris that forms the smooth plains peripheral to many large impact basins. The proportion of secondary debris (local) increases with increasing distance from the primary crater because of the increasing impact velocities. Craters formed by primary ejecta at intermediate distances may be almost completely erased by the debris cloud behind the curtain, but ultimately the distance from the crater becomes so large that the curtain loses its continuity and breaks up into strings of debris, which form lines of secondary craters. A similar sequence of events may occur around martian craters, except that once the curtain debris is deposited, it probably continues its radial motion outward, as a debris flow.

Martian crater ejecta

In contrast to lunar ejecta, in which ballistic features dominate, martian ejecta show numerous indications of having been emplaced by flow across the surface. Craters that exhibit these characteristics have been informally termed "fluidized ejecta craters," "fluidized craters," or "splosh craters." Most workers have concluded that liquid water, water-ice, or other volatiles entrained in the ejecta are responsible for its fluid properties. Although the pattern of deposition appears to depend to some extent on factors such as latitude, elevation, and rock unit, little systematic work has been done to define the relations precisely. Because much of the discussion that follows is based on pilot studies, the conclusions are tentative. An additional factor adding to the uncertainty is the difficulty of knowing which features are primary and which result from modification after impact. This problem is especially severe in high latitudes, where modification rates are high.

The ejecta patterns are discussed here in some detail because of their potential for revealing information on properties of the surface and past climatic conditions. If the flow patterns are due to incorporation of ground-ice or groundwater into the ejecta, then the smallest crater around which the patterns occur should indicate the depth at which water or ice is encountered, and areal differences in the onset diameter or the ejecta style could indicate differences in the distribution of water and ice. Moreover, since the depth of the permafrost varies with latitude, we ought to see latitudinal differences. If the cratering style changes as a function of present-day latitudes, then major shifts of pole position, such as Ward and co-workers (1979) suggested might have happened following formation of the Tharsis bulge, are unlikely over the time span represented by the craters.

The pattern of ejecta around craters smaller than about 5 km in diameter is similar to that on the Moon. The rims are either smooth or slightly hummocky at their crests and merge imperceptibly outward with the surrounding terrain. The pattern around craters larger than 5 km, however, is generally quite different, and only a few have patterns that resemble those on the Moon. Most of these lunar types occur within 20° of the equator; they are rare, if present at all, poleward of the 40° latitude. The overwhelming majority of craters larger than 5 km have distinctively nonlunar, "fluidized ejecta" patterns. The precise pattern depends on size and possibly latitude. Craters in the 5 km to 15 km size range generally have a single ejecta sheet that extends about one crater radius from the rim. The edge of the sheet has either a low ridge or an escarpment. It is usually nearly circular, with a jagged or slightly lobate margin (figure 4.12). The surface of the sheet is generally crinkled, with a roughly concentric pattern, although closely spaced radial striae may be present in places. The ejecta is thick enough to mask features such as low ridges and graben on the underlying terrain. Beyond the inner ejecta sheet, there may be low hummocks, rays, and secondary craters, as around lunar craters. The presence of a low ridge around the ejecta of some martian craters was noted in the Mariner 9 pictures. The craters were referred to as rampart craters, and their peculiar morphology was attributed to the action of wind (McCauley 1973). However, the preservation of very fine features, such as rays and secondary craters, adjacent to the ejecta sheet shows that the rampart is a primary feature and not the result of subsequent modification.

With increasing crater size the ejecta patterns become increasingly complex. My colleagues and I (Carr et al. 1977b) recognized three basic types: those with a lunarlike configuration, those with a very strong radial fabric, and those with various fluidized ejecta patterns. The first two classes

Figure 4.12. A 13-km–diameter crater in Chryse Planitia. The crater is surrounded by a continuous flap of ejecta with a steep outward-facing escarpment. The surface of the flap is wrinkled with a faint concentric pattern. Radial striae are also just visible toward the outer margin. Beyond the escarpment the somewhat rubbly appearance is more lunarlike (compare with Figure 4.10). The flap is believed to have flowed radially outward after ballistic deposition. (10A61,62)

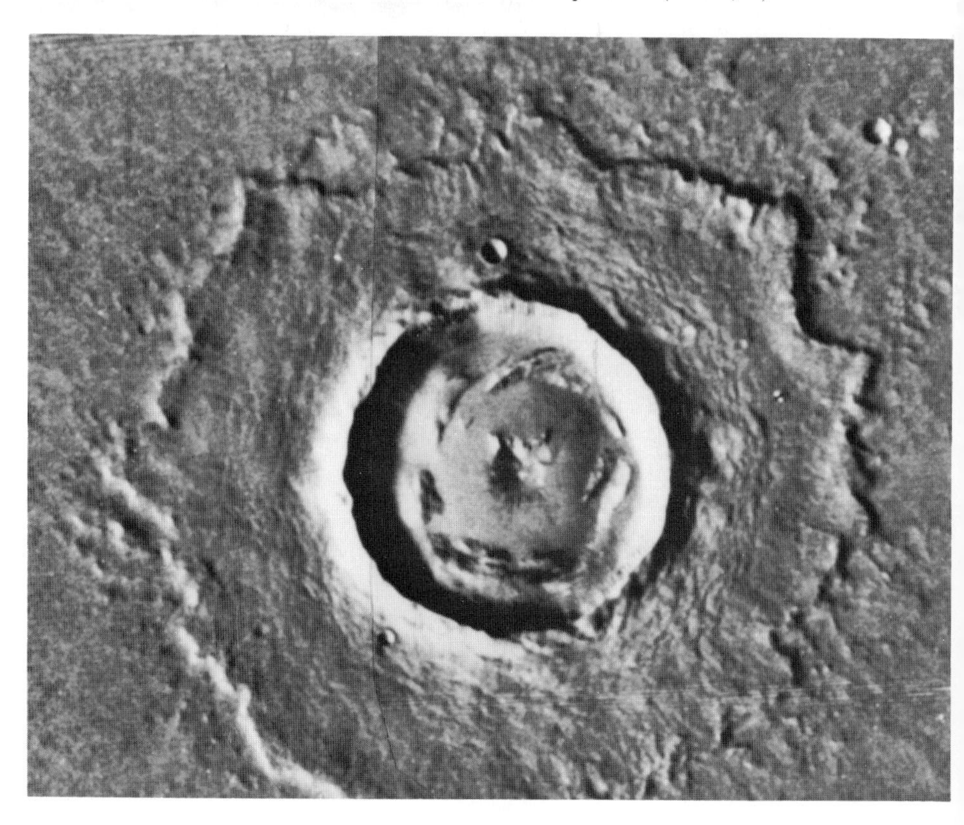

Figure 4.13. The 18-km-diameter crater Yuty (22°N, 34°W), with several sheets of ejecta, each complexly lobed. This pattern is typical of craters larger than approximately 15 km that lie within 30–40° of the equator. Burial of a preexisting crater near the rim suggests that the ejecta layers are thin. (3A07)

represent only a small fraction of ejecta patterns and will not be discussed further. Johansen (1979) and Mouginis-Mark (1979b) have made exploratory attempts to identify different types of "fluidized ejecta" craters on the basis of the presence or absence of ramparts, the number of ejecta sheets, and the presence and steepness of peripheral escarpments around the ejecta. There appear to be two basic types and the possibility of a third. The first type is exemplified by the 18-km-diameter crater Yuty (figure 4.13). It has been referred to as the "flower" type by Johansen (1979) and occurs mostly within 40° of the equator. Several ejecta layers are present, and each has a lobate outer margin with a distal ridge. On large craters many ejecta layers may be present, with their lobate margins and ramparts at different distances from the crater. Within a distance of about one-quarter crater radius from the rim, the ejecta may be hummocky, but the ejecta lobes further out generally have a smooth surface or a faint radial lineation. The ejecta sheets appear to be quite thin (figure 4.14), for quite subtle features of the underlying terrain may show through.

Figure 4.14. Flow ejecta craters in Sinai Planum at 21°S, 76°W, showing extreme development of the peripheral ramparts and the large distance to which the continuous ejecta extends. The largest crater is 15 km across. (608A47)

The second major type is exemplified by the 28-km–diameter crater Arandas. Around the crater is an annulus of ejecta about one crater radius wide (figure 4.15). The outer margin of the sheet is roughly circular and is marked by a convex-upward, outward-facing escarpment. The ejecta within the annulus appears thicker than that in the lobate debris sheets, since the textures of the underlying surface are generally masked. The ejecta surface commonly has a distinct texture, caused by the interplay of both radial and concentric elements, although concentric features usually dominate at the outer margin. Outside the annulus, and seemingly overlain by it (figure 4.16), are, in most cases, lobate flow sheets, much like those of the flower-type craters, except that the lobes tend to be much more elongate radially, giving the outer parts of the ejecta field a strong radial fabric (figure 4.17).

Figure 4.15. Close-up of the 28-km–diameter crater Arandas, at 43°N, 14°W. Craters at latitudes higher than 30–40° tend to have a relatively thick platform of ejecta around the rim, rather than thin sheets with peripheral ramparts, although such sheets do occur further out. At the bottom of the picture the ejecta wraps around a former crater, showing that emplacement is by flow, not deposition from above. (32A30)

Figure 4.16. The outer part of the Arandas ejecta. The ejecta has a strong radial fabric and appears to have flowed over the former fractured surface. (9A42)

This type of crater was referred to as "composite" by Johansen (1979). Although present at all latitudes it occurs most frequently poleward of the 30° latitudes.

The third and more questionable class of large crater is similar to that described above, except that the outer ejecta with lobate patterns is missing. All that is present is a roughly circular ring, or platform, of ejecta around the crater. The configuration is similar to that around craters in the 5–15 km size range except for the lack of crater-related features outside the inner ejecta sheet. These craters have been referred to as "mound" craters by Johansen (1979) and may include some of the "pancake" craters of Mouginis-Mark (1979a). Both these authors implied that the configuration is primary. While this may be true in some cases (figure 4.18), most ejecta blankets of this type appear to form by secondary modification. Slight erosion of the second class of crater referred to above would remove the subtle outer lobate features and leave the inner circular ejecta sheet. The crater would then resemble the descriptions given by Johansen and Mouginis-Mark. This type of crater also resembles the "pedestal" type, which is widely believed to form by the action of wind (see section on crater modification). Furthermore, most of the pedestals cannot be primary, because they constitute volumes too large to have originated from craters. Thus, whether "pancakes" or "pedestals" can be produced by primary impact processes remains questionable.

Ejecta emplacement

Several characteristics of the martian ejecta suggest emplacement as a debris flow rather than directly from ballistic trajectories. The abrupt outer boundary is itself strongly suggestive, but obstacles to the flow provide the most convincing evidence (Carr et al. 1977b). The ejecta is commonly deformed into ridges, which wrap around obstacles, such as craters and low hills, as though the ejecta surface buckled as the ejecta flowed around the obstruction. "Shadow zones" are seen on the lee side of some obstacles, and the ejecta may be absent on top of the obstacle, although it is present elsewhere at the same radial distance. Thus, emplacement is almost certainly by radially outward surface flow and not by deposition from above. Additional evidence for flow is the greater radial extent of the continuous

Figure 4.17. Impact craters at 41°N, 260°W, showing the inner flap and outer flow patterns that characterize craters at these latitudes (compare with figure 4.14). The largest crater is 16 km across. (10B81)

Figure 4.18. Pedestal craters at 46°N, 353°W. The ejecta forms platforms that stand above the surrounding terrain. The conventional view is that such platforms form as a result of preferential retention around the craters of layers of debris that formerly covered the entire surface. If this interpretation is correct, then it would take several such episodes of deposition and removal to explain the superposition of pedestals in the upper right. The largest crater is 6 km across. (43A04)

ejecta, as compared with ejecta on the Moon and Mercury. Continuous ejecta around martian craters extend 1 to 2.5 crater radii from the rims, as compared with 0.7 radii for the Moon and 0.5 radii for Mercury (Carr et al. 1977b). The radial distances which the ejecta travel depend slightly on latitude and elevation. According to Mouginis-Mark (1979b) the average extent of the continuous ejecta ranges from 2.0 radii at elevations 2–3 km below the datum to 1.5 radii 7–8 km above the datum. At the equator the average range is 1.3 radii, as compared with 1.7 and 2.1 radii at 50–60°S and 50–60°N, respectively. If these data are valid, then the ejecta are more fluid at lower elevations and high latitudes.

My co-workers and I (Carr et al. 1977b) concluded that the sheetlike ejecta patterns are caused by the fluid consistency of the ejecta. We suggested that the ejecta continues to move radially outward as a ground-hugging debris flow after being deposited ballistically. We noted that incipient flow features also occur on the Moon but on a small scale. We further proposed that the fluidity could be caused by interstitial water- or ground-ice that was partly melted during impact. Entrainment of atmospheric gases may also be a factor. Gault and Greeley (1978) reproduced some of the lobate patterns by firing pyrex spheres into mud at high velocities. They suggested that multiple flow fronts could form as material squirts outward from each part of an ejecta clot, as its different parts successively impact the surface. On the assumption that water is the cause of the fluid behavior, Boyce (1979) postulated that the threshold diameter for onset of the fluidized ejecta patterns could be used as a measure of the depth to groundwater. The onset diameter for the fluid patterns in the equatorial regions appears to be around 5 km. Assuming a depth/diameter ratio of 1/5 for the transient cavity, this hypothesis implies that there is water at a depth of 1 km. This presumably marks the base of the permafrost.

The conclusion that the ejecta on Mars is more fluid than that on Mercury and the Moon seems inescapable, although why it should be so is uncertain. The inclusion of large amounts of liquid water in the ejecta has already been mentioned as the most probable cause, but other possibilities remain. There is also the question of whether the water must exist in liquid form before impact. If the water exists as ice, only a small fraction is likely to melt and thereby impart fluid properties to the ejecta. The ice near the impact point would be volatilized, and that around the periphery of the event would remain as ice. Only in a relatively narrow transition zone would the water be liquid. Thus, if water is the cause of the fluidity, it probably had to exist below the surface before the impact event. The fluidity of the ejecta could be enhanced by other means, such as entrainment of large amounts of gas derived from the shock heating and decompression of hygroscopic or thermally unstable minerals or by mixing with atmospheric gases, but how effective such mechanisms are is not known. The presence of abundant liquid water at depths in excess of about 1 km seems the most straightforward explanation.

In summary, two main classes of craters are recognized: lunar types and fluidized ejecta types. The fluidized ejecta types greatly predominate. Those in the 5–15 km size range have a continuous ejecta sheet, with a distal ridge or escarpment, outside which are lunarlike features, such as secondary craters and rays. Fluidized ejecta craters larger than 15 km in diameter can be divided into two main classes: most of those in equatorial latitudes have multiple flow lobes, each with a distal ridge, whereas most of the craters in high latitudes are surrounded by a continuous ring of ejecta, outside which are radially elongate lobate features. In a possible third class there are no outer flow features but simply a circular platform of ejecta.

CRATER MODIFICATION

As craters are exposed to conditions on the martian surface, they become modified by various processes, such as the action of wind and water, volcanism, and impact. In much of the old cratered terrain the rims of many old craters are gullied, and craters are superposed on one another. On the plains modification by fluvial action and subsequent impacts is relatively minor, however; the main agents of modification are wind and volcanism. The type of modification depends strongly on latitude. At latitudes higher than about 40°, numerous characteristics of the surface suggest repeated deposition and removal of surficial debris layers by the wind (Soderblom et al. 1973). Mantles of fine-grained debris are believed to have been deposited then partly removed, as a consequence of changes in the wind regime. The process is thought to have been repetitive, connected in some way with long-term, planetwide changes in climate. There is no evidence for comparable wind activity at low latitudes. As a result craters evolve differently in the two latitude belts. At low latitudes modification is relatively slow, and the process is simply one of loss of detail. Fresh-looking craters are common, and the original shape of the modified crater can generally be reconstructed with some confidence. In the high latitudes modification rates are significantly higher, and fresh-looking craters are rare. Moreover, repeated deposition and stripping produces crater shapes that are often difficult to reconcile with typical primary configurations.

The shapes assumed by small craters are less dependent on latitude than those assumed by larger craters. A crater smaller than about 5 km in diameter soon loses its faint, lunarlike ejecta pattern, and all that is left is the bowl-shaped crater itself, with a slightly raised rim. Further modification results simply in subdual of the crater topography. As indicated above, craters in the 5–15 km size range have around the rim a single sheet of ejecta 1–2 crater radii wide, beyond which are hummocks, secondary craters, and rays. The first stage in the modification of such craters is the loss of the lunarlike features beyond the edge of the ejecta blanket. All that is then left is a crater within a low, near-circular blanket of ejecta, which may or may not have an outer rampart. This is one of several types of crater that have been referred to in the literature as "pedestal craters." Subsequent modification results in lowering of the rim, infilling of the crater, and subdual of the outer profile of the ejecta blanket.

The latitude effects are most obvious for craters larger than 10–15 km in diameter. At latitudes below 40° the process of crater modification is what one would expect intuitively. Fine-scale features are destroyed first, then features that are progressively more coarse. In the multilobate, Yuty-type of crater the fine radial texture on the surface of the lobes is lost first, leaving ramparts outlining the individual ejecta sheets but little other evidence for the presence of ejecta beyond the low hummocks close to the rim. With further erosion the ramparts become more subdued, until they ultimately disappear, and all that remains is a flat-floored crater with a narrow hummocky rim only about one-quarter of a crater radius across, beyond which no traces of ejecta remain. Modification of the second type of crater, with its inner, relatively thick ejecta sheet and outer, markedly radial flow lobes, is somewhat similar. The first change is loss of the fine texture on the surface of both the inner sheet and the outer lobes. The crater at this stage of erosion is thus surrounded by a near-circular platform, beyond which low ramparts trace the extent of the former ejecta lobes. Increased erosion and burial result in loss of all traces of the outer radial facies, and the crater is left within a roughly circular pedestal that extends 1–2 crater radii from the rim.

At latitudes poleward of 40° crater modification is dominated by the effects of deposition and deflation of surficial debris. One common outcome of crater modification is the formation of pedestals (figure 4.19), which are much wider than those found in the equatorial regions, extending 2–6 crater radii from the rim crest, as compared with only 1–2 crater radii at low latitudes. McCauley (1973) and Arvidson and co-workers (1976) postulated that in the high latitudes, fine-grained aeolian debris formerly covered the surface over large areas, and that this debris has now been partly removed by deflation. Around craters the surface was armored by ejecta, so that deflation was relatively slow, whereas in the intercrater areas deflation was relatively rapid. Consequently pedestals were left around craters as the surface between them was lowered by sustained wind action. The pedestals at high latitude are not, therefore, remnants of the relief of the original ejecta, like those at low latitude, but emerged as modification of the surface proceeded. The evidence supporting this model is as follows: (1) The outer margins of the pedestals are often radially digitate, like the outer margins of the ejecta. (2) A continuum can be seen between craters with fairly fresh-appearing ejecta and the pedestal types. (3) Some craters are left standing well above the level of the surface in the intercrater areas, indicating that the latter have indeed been lowered. (4) Crater statistics suggest the presence of a mantling of debris in high latitudes (Soderblom et al. 1973). The outer edge of the pedestals commonly have a convex-upward slope. With decreasing size, therefore, the pedestals assume a more domical form. As a result small pedestal craters

Figure 4.19. Pedestal craters at 48°N, 349°W. Traces of the original texture of the impact ejecta are just visible on the surface of many of the pedestals, confirming an impact origin. With decreasing size the pedestals become proportionately smaller, so that the smaller pedestal craters in this picture and the last resemble volcanic domes. The picture is 60 km across. (60A53)

closely resemble volcanic domes, and considerable ambiguity exists concerning the origin of many small cratered domes in these regions.

Pedestal craters thus appear to originate in different ways according to latitude. At low latitudes pedestals are relatively uncommon, and many that are present are remnants of thick sheets of ejecta that were arrayed around the crater at the time of impact. At high latitudes the pedestals are common. Some may be remnants of primary ejecta but most appear to have formed slowly, long after formation of the craters themselves, by preferential retention of eolian debris around the craters while the intercrater areas were being swept free.

5 CRATERING STATISTICS AND CRATER AGES

Probably the most important attribute of craters is that they provide a means of determining age. A planet's surface is continually being bombarded with debris, and the number of craters on it depends on the cratering rate, the length of time that the surface has been exposed to impact, and the rate at which craters are destroyed by such processes as erosion and burial. On Earth, the latter processes so dominate that few craters survive. In most parts of Mars, however, erosion rates have been small over most of the planet's history, so crater densities depend only on the age of the surface and the cratering rate, and relative ages can be determined reliably. Complications do arise, such as on a surface partly flooded by lava, where the large craters might date some pre–lava-flooding event and the small craters the lava surface, or in high latitudes where episodic deposition and removal of layers of debris complicates the cratering history. But such complications are usually evident from the forms of the crater size frequency distributions and can be treated accordingly. Determination of absolute ages, however, is much more difficult, for it requires knowledge of the rate of crater production on the planet. Although we have a fairly accurate picture of the cratering history of the Moon, at least for the last 4 billion years, from radiometric dates on returned lunar samples and crater counts on the sampled units, we do not know how to transpose the lunar rates to Mars with any degree of confidence. The problem has been approached from a theoretical standpoint, by analyzing the distribution of different objects within the solar system and predicting how the cratering rates on Mars and the Moon might be related, taking into account the differences in impact velocities, gravities, and the mix of interplanetary objects hitting the two bodies. The problem has also been approached from a more empirical standpoint, by comparing crater statistics on Mars and the Moon and then making some assumptions as to how the two are related. In recent years the two methods have converged, but the uncertainty in the absolute ages is still large and is likely to remain so until datable samples are returned from the martian surface.

We will start with a discussion of the types of interplanetary debris within the solar system. Later in the chapter we will see that knowledge of the proportions of the different types of objects, particularly comets and asteroids, is helpful in comparing lunar and martian cratering rates and deriving absolute ages. We will then discuss crater statistics—in particular, the evidence of an early period of enhanced erosion—and absolute ages, taking three different models that bracket the possibilities, and will end with a discussion of some of the results of crater dating—how the dates were obtained and what specifically was dated.

SOLID OBJECTS WITHIN THE SOLAR SYSTEM

Interplanetary objects range in size from those that weigh as little as 10^{-13} gm to large asteroids that may weigh as much as 10^{23} gm, but only objects larger than 10^{3} to 10^{7} gm cause craters on Mars at present because of the protective effects of the atmosphere (Gault and Baldwin 1970). Knowledge of the distribution of interplanetary debris comes from a variety of sources. The literature is voluminous, and the reader is referred to a comprehensive article by Dohnanyi (1972) for a summary and appropriate references. Most of the information available pertains to the vicinity of Earth (see figure 5.1). In the discussion that follows, the smaller objects are grouped according to the techniques used in detecting them rather than to their origin, since the latter is generally unknown.

1. *Micrometeoroids.* The smallest solid particles in the solar system, those weighing less than 10^{-6} gm, are generally referred to as micrometeoroids. Their presence was first inferred from ground-based observations of the zodiacal light—a band of light in the ecliptic that is best seen at dusk and dawn. The brightening is believed to be caused by the scattering of sunlight by particles around the Sun. While only crude estimates of the number densities of the particles involved are possible, measurements of polarization and brightness have demonstrated that the particles are concentrated in the plane of the ecliptic, and that the concentration falls off with increasing distance from the Sun. The flux of micrometeoroids has been measured more directly on satellites by a variety of techniques, including penetration of thin films, discharge of capacitors by impact ionization, and detection of impact by microphones. Estimates of the flux of micrometeorites onto Earth have also been made from geochemical measurements, such as the abundance of osmium (Os) and iridium (Ir) in deep-sea sediments. Estimates based on most of these techniques are consistent with the fluxes shown in figure 5.1.

2. *Meteors.* Meteoroids entering Earth's atmosphere create a trail of ionized molecules which can be detected by radar. The mass range detectable is from about 10^{-6} gm to about 1 gm. For masses of milligrams or larger, the trails are also bright enough to be photographed. One problem with both radio and photographic meteors is that it is difficult to determine

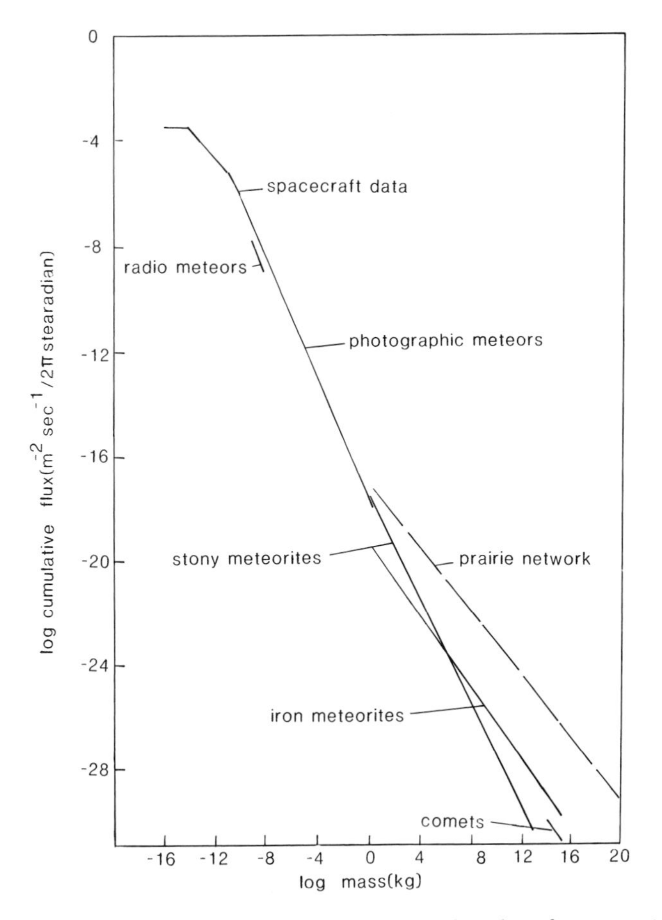

Figure 5.1. Size frequency distribution of objects within the solar system in the vicinity of the Earth. As explained in the text, the fluxes derived from the Prairie network may be too high because of imprecise modeling of the luminous effects of a meteoroid entering the atmosphere. (Adapted from Dohnanyi 1972.)

accurately mass and velocity from the observed ionization and luminescence, and modeling errors may cause significant errors in the estimates of flux. Some meteoroids occur in streams, all the constituents of which have orbits closely associated with known comets and so are presumed to have derived from them. Planets perturb the orbits of particles within the streams, to produce sporadic meteors, and a large proportion of these may also be cometary in origin.

3. *Meteorites and fireballs.* Brown (1960) and Hawkins (1960, 1963) made independent estimates of the influx of meteorites onto Earth (figure 5.1) from the known distribution of falls. Hawkins assumed that 90 percent of the stones and 80 percent of the irons are lost by ablation. Estimates of the flux of objects in the same size range (10^4–10^{15} gm) have also been made from observations of very bright objects by the Prairie Network (McCrosky and Ceplecha 1968). The estimates are considerably higher than those derived from meteorite falls and may be subject to the same errors in calculating masses from photometric data as are encountered in the case of the smaller objects.

4. *Asteroids.* The asteroidal belt between Mars and Jupiter contains several thousand asteroids ranging in size from less than 10^{17} gm up to at least 10^{23} gm. The number density peaks at around 2.6 A.U. from the Sun and decreases both outward and inward. The asteroids are not restricted to orbits totally outside Mars's orbit. On the inner edge of the belt are a number of asteroids with paths that cross that of Mars and a smaller number that cross Earth's path, called Apollo asteroids. Knowledge of the ratios of Earth-crossers to Mars-crossers and of asteroidal impacts to other types of impact is required to relate lunar impact rates to those on Mars. The distribution of orbital periods shows gaps corresponding to resonances with Jupiter, which indicates that these orbits are unstable, and that objects within them are ejected into other orbits. Some groups of asteroids have similar orbital parameters, suggesting that they were derived by fragmentation of the same parent body. Light curves from many asteroids indicate that they are irregular in shape, which is consistent with origin by collision.

5. *Comets.* Comets are believed to be primitive ice/dust bodies that have survived from very early in the history of the solar system when the solar nebula was condensing. They are thought to reside mainly in the Oort cloud (Oort 1950, 1963) that surrounds the solar system, possibly extending as far out as 50,000 A.U. Motions of bodies within the cloud are subject to perturbations by the planets, and at these vast distances even by nearby stars, so that occasionally a body enters the inner solar system and can be observed from Earth. Such bodies normally become visible only within 3 or 4 A.U. of the Sun, although some have been observed as far out as 10 A.U. While in the inner solar system, a comet consists of three parts: a *nucleus,* which is surrounded by a diffuse luminous region, the *coma,* from which extends a long wispy *tail.* Two types of tail are distinguished: type I consists of ionized molecules such as CO^+, CO_2^+, and N_2^+; type II appears to consist mostly of dust grains, which scatter sunlight. Most comets are in orbits with periods in excess of 200 years and aphelia thousands of Astronomical Units from the Sun, so that their paths around the Sun are almost parabolic. Such long-period comets have almost random inclinations, indicating that the Oort cloud is dispersed symmetrically around the Sun and is not confined to the plane of the solar system. Short-period (< 200 years) comets constitute less than one-fifth of the total number observed. They are believed to resemble other comets but were trapped within the solar system as a result of perturbation of their orbits by the planets, particularly Jupiter.

CRATER PRODUCTION

There is a considerable literature on the interpretation of crater statistics on the Moon (Shoemaker 1966; Gault 1970; Marcus 1970), Mars (Chapman and Jones 1977; Jones 1974; Hartmann 1973a; McGill and Wise 1972), and other planets (Murray et al. 1975; Strom and Whitaker 1976; Wetherill 1975). Only an outline of the subject is presented here, partly because the intricacies of crater statistics are beyond the scope of this book, which is intended as an overview, and partly because the present state of knowledge of crater statistics for different parts of Mars does not justify elaborate theoretical treatment. Most of the statistical data available for Mars are still based on Mariner 9, and some of the conclusions drawn, particularly those based on subtle differences in frequencies of craters close to the resolution limit of the Mariner 9 pictures, remain in doubt until confirmed by analysis of the better Viking data.

The cratering history of the Moon has been well established from radiometric ages. The cratering rate was very high around 4 billion years ago, then declined rapidly to a much lower rate around 3.9 billion years ago. Surfaces that formed before the decline (the lunar highlands) are close to being saturated with craters; surfaces that formed after the decline (the lunar maria) are only sparsely cratered. Very few surfaces have intermediate crater densities, because the decline was rapid and the time interval over which a surface with an intermediate crater density could have formed was very short. Soderblom and co-workers (1974) suggested that Mars also has a bimodal distribution of crater densities, which means that it too must have experienced an initially high cratering rate that subsequently declined rapidly to the present low rate. Almost all other workers agree with this sequence of events, but there is disagreement with respect to the timing of the decline and the current cratering rate. Before comparing the different models, it will be useful to examine how the distribution of craters on a surface might evolve when the surface is subject to a continuous rain of interplanetary debris.

The number of craters in any given area on most planetary surfaces increases geometrically with decreasing crater size, following a power law similar to that of the impacting objects (see figure 5.1): $N = k D^b$ where N is the number of craters, k is a constant depending on age, D is the crater diameter, and b is a constant. In practice, as explained below, b is not a constant but is some function of the diameter. Interpretation of crater statistics is based largely on size frequency distribution curves, in which the logarithm of the number of craters is plotted against the logarithm of the crater size. Most of the early cratering data is plotted cumulatively—that is, the cumulative number of craters above a specified diameter is plotted against the diameter. Incremental plots are now being used more, in which the number of craters within a specific size increment is plotted against diameter. The incremental plots have the advantages that they better display differences between counts and that they can be handled better statistically. Incremental plots may be arithmetic or geometric. In an arithmetic plot the counts are normalized to some standard increment, such as 1 km. In geometric plots increments are chosen, so that D_u/D_l is a constant, where D_u and D_l are the upper and lower bounds of the increment. The geometric incremental plots have the advantage that their slopes are the same as on the widely used cumulative plots, so they can be readily cross-referenced. In the following discussion, all values of slopes are for cumulative and geometric incremental plots.

The first step in analyzing crater statistics is to determine the primary production function. What size frequency of crater would be expected if no erosion or destruction of craters took place? The problem can be approached in two ways. The first is to look at the size distribution of interplanetary debris and determine what size distribution of craters would be expected from it. This, however, requires knowledge of the velocities of the impacting objects and of how crater sizes vary with the energy of impact, both of which are poorly understood. Moreover, the curve in figure 5.1 is only a crude representation of the flux of impacting objects, and many variations in slope may be masked. The second approach is to look at areas of the planet that are relatively sparsely cratered, so that formation of one crater does not interfere with formation of another, and where the craters look fresh and appear to be relatively unaffected by erosion. These conditions are satisfied for most lunar maria and for some of the volcanic plains of Mars. This approach has the advantage that it takes into account the production of secondary, as well as primary, craters.

Shoemaker and co-workers (Shoemaker 1966; Shoemaker et al. 1970) derived a production function for lunar craters from the Ranger, Surveyor, and Apollo data. They showed that there is a gradual change in slope close to 3 km on size frequency plots of mare craters. For craters larger than 3 km the slope is close to -2; for craters smaller than 3 km the slope is in the range of -2.8 to -4. There is therefore a disproportionately large number of small craters, which Shoemaker and colleagues attributed to the presence of secondary craters. Neukum and co-workers (1975) also derived a standard production function for craters on the Moon. They examined crater frequency distributions on several sparsely cratered surfaces, including maria, light plains, and basin ejecta. They noted that although the number of craters differed from location to location according to the age of the surface, the shapes of the curves were all the same. They could be superposed simply by moving the curves vertically on the crater frequency diagram. Neukum and colleagues' production curve is similar to that of Shoemaker (1966). For small craters the slope is close to -3, for large craters close to -2, and the transition between the two parts of the curve takes place close to 3 km. In their selection of areas to be counted, Neukum and co-workers avoided regions of obvious secondary craters, so argued that the change in slope around 3 km is not due to secondary craters but is a reflection of the distribution of primary craters. This point is important, for Neukum and Wise (1976) further argued that, since size frequency distributions on the sparsely cratered plains of Mars have a similar shape to those on the Moon, the martian craters were formed by an array of objects similar to that which formed the craters on the Moon. From this conclusion they went on to derive a crater chronology for Mars. According to the Shoemaker model, crater frequencies on Mars and the

Moon look similar not because the size frequency of the impacting bodies is the same, but because large numbers of secondary craters are produced by impacts in both cases.

As a surface ages, the number of superposed craters increases until an equilibrium is reached between the number of new craters that form and the number that are destroyed. If the only way in which new craters are destroyed is by the formation of new ones—that is, if other erosional effects are negligible—then the equilibrium is generally referred to as saturation. On crater size frequency plots the slope of the curve for a saturated surface is always less than the slope of the production curve—that is, a saturated surface has proportionately fewer small craters than originally formed because of their preferential destruction. For the lunar mare saturation is achieved only for relatively small craters (< 1 km in diameter), for which, as we saw above, the production curve has a slope in the −2.8 to −4 range. In this size range the saturation distribution has a slope of close to −2.

As a lunar mare surface ages the crater frequency curve is displaced vertically on the size frequency plot, and the secondary crater curve intersects the saturation curve at what Shoemaker calls the critical diameter (figure 5.2). At this stage the curve has three segments: a primary crater segment above about 3 km diameter with a slope of approximately −2, a secondary crater segment with a slope of approximately −3, and an equilibrium segment below the critical diameter with a slope of −2. The slope of the secondary crater part of the curve, however, may be as high as −4 if dense fields of secondary craters are encountered. Shoemaker and colleagues (1970) showed that the critical diameter—the diameter below which equilibrium is achieved—increases with increasing age. For the Apollo 12 site, for example, the crater population is in equilibrium below about 100 m; for the older Apollo 11 site the figure is 300 m. A similar effect should occur on Mars, except that atmospheric shielding may substantially reduce the number of craters with diameters less than approximately 50 m (Gault and Baldwin 1970).

As the age of the surface increases further, the critical diameter below which saturation is achieved moves to larger and larger values, and the part of the curve above the critical diameter approaches the saturation function. Ultimately the entire surface will become saturated and have a slope of −2 or less on the size frequency plots. The lunar highlands, the mercurian highlands, and the martian highlands for large diameters all appear to be close to saturation, although this has been disputed recently (Woronow 1978; Strom and Whitaker 1976).

The situation on Mars is more complicated and less well understood. Older surfaces, represented by the cratered highlands, have been subject to an early period of intense erosion, so that the primary characteristics of the crater production function are masked. However, many of the plains within 30° of the equator appear to have experienced relatively little erosion, so that the crater frequency curves in these areas may approach a production function. Erosion rates on the equatorial plains have been remarkably low: the crater frequency curve for Chryse Planitia (figure 5.3), for example, has a slope in the 100 m to 1 km size range, which is equal to that for typical lunar maria (Dial 1978). The identical slopes indicate that the cumulative fraction of craters larger than 100 m that have been destroyed on Chryse Planitia is the same as that in the lunar maria—that is, essentially zero. Otherwise, a remarkable coincidence would be required, with the enhanced destruction of craters in Chryse being exactly offset by enhanced production for all size increments. The conclusion that the erosion has been limited is somewhat surprising (see chapter 11). However, low erosion rates are also suggested by the swarms of secondary craters in Chryse that at resolutions of around 100 m appear relatively fresh and uneroded. Many other equatorial plains are similar to Chryse and may also have close to a production distribution. In contrast, plains in the high latitudes commonly have complicated crater frequency curves and appear to have experienced complex erosional and depositional histories.

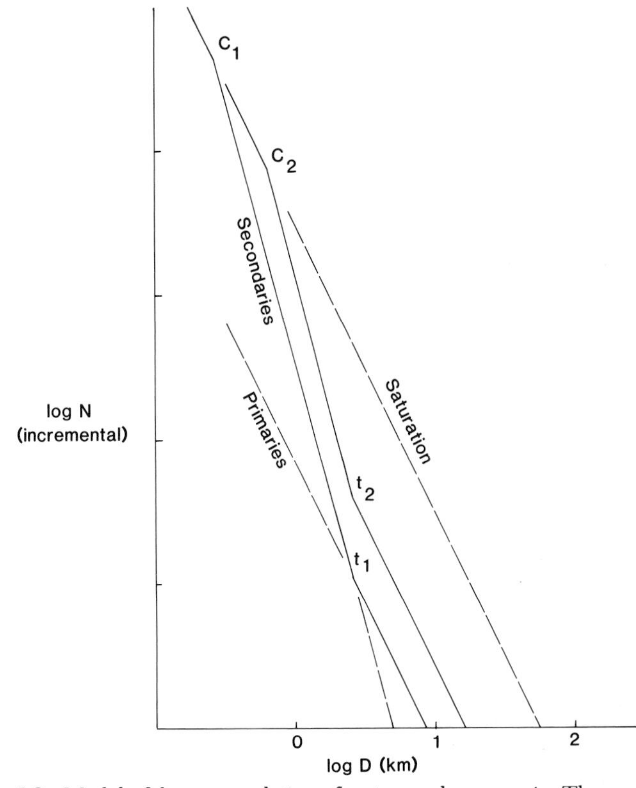

Figure 5.2. Model of the accumulation of craters on lunar maria. The production curves consist of two parts, with craters that are mainly of secondary origin below 2–4 km and those that are mainly of primary origin above. The saturation curve represents a limiting distribution, along which destruction of old craters by new ones just balance one another. As a surface ages from t_1 to t_2, the production curve approaches the saturation curve, and the intersection point, c_1 moves to a larger diameter. (Adapted from Shoemaker 1966.)

Crater curves for the equatorial plains generally have a break in slope close to a diameter of 1 km (Soderblom et al. 1974; Dial 1978). At smaller

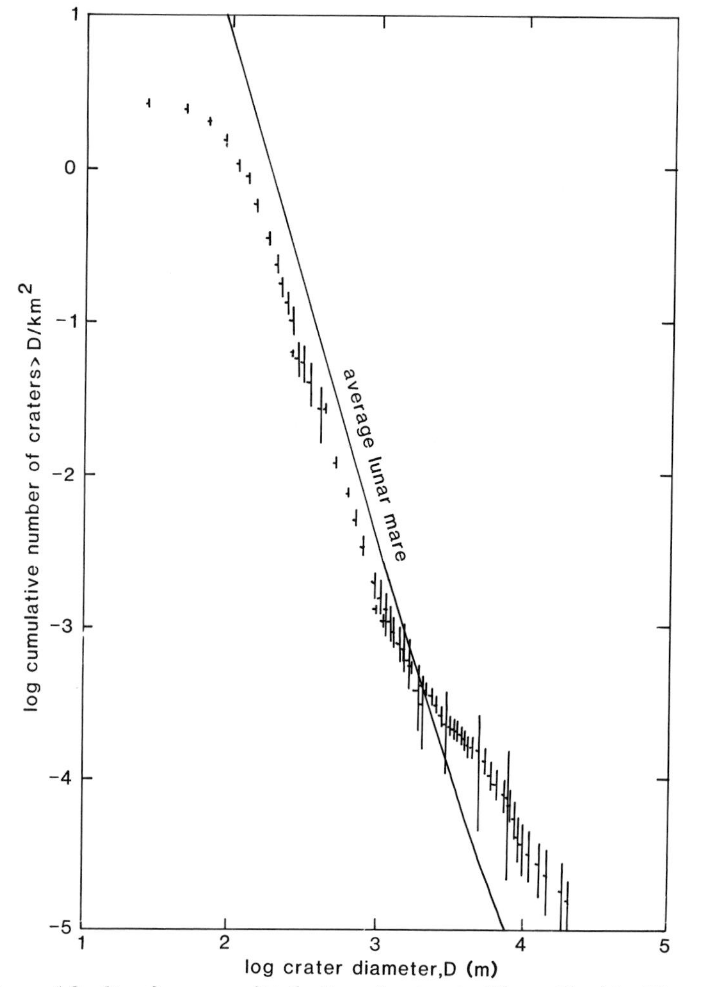

Figure 5.3. Size frequency distribution of craters in Chryse Planitia. The near equality of the slope in the 100 m to 1 km size range to that for typical lunar mare suggests that most of the craters in this size range are secondaries, and that their rate of destruction differs little from that on the lunar maria. The rollover at 100 m may be partly real and partly a resolution effect. Above 1 km the slope is close to −1, whereas the slopes of the lunar curves are close to −2, as shown in the previous figure. The low slopes may indicate survival of an older generation of larger craters or may be caused by an increase in diameter as a result of the transition from craters with lunarlike ejecta to craters with fluidized ejecta. (From Dial 1978.)

diameters the slopes are close to −3; at larger diameters they are in the −1 to −2 range. Soderblom and colleagues suggested that the break in slope is due to the larger proportion of secondaries that are present at the smaller diameters. According to this interpretation, secondaries dominate the production curve for Mars at diameters less than 1 km, as compared with approximately 3 km for the Moon. On the other hand, Neukum and Wise (1976) interpreted the steepening of the crater frequency curves toward the smaller diameters as being due to a change in the size distribution of the impacting objects. Differences between the two interpretations remain unresolved.

Many counts of craters on the equatorial plains show especially low slopes in the 1 km to 10 km size range. The slopes below 1 km are close to −3, as we have just seen. The slopes above 10 km are close to −2, whereas those between 1 km and 10 km range from −1 to −2 (Soderblom et al. 1974; Neukum and Wise 1976; Dial 1978). The effect is not evident in all the counts but, if real, would explain the apparent discrepancy between the smaller number of 1-km–diameter craters observed on plains units, as compared with the number determined by extrapolation from larger diameters (see next section). The low slopes may be caused by the transitions from simple to complex craters and from lunar-type ballistic to fluidized ejecta patterns, both of which occur in the 1–10 km size range. Formation of terraces around a complex crater will result in a larger crater for a specific energy of impact than if a simple bowl-shaped crater forms. Settle and Head (1979), for example, showed that the 34-km–diameter lunar crater Timocharis enlarged its cavity by 40 percent through slumping of its walls. The enlargement may be accentuated by different target properties. As we saw in the previous chapter, the size of a crater may vary by a factor of two, depending on the target properties, with the more easily fluidized targets yielding larger craters. The craters with fluidized ejecta on Mars should be larger than their lunar equivalents. These effects should cause the large-diameter segment of the martian size frequency curves to be displaced to somewhat larger diameters, thereby creating an offset in the crater frequency curves and low slopes in the 1–10 km region. The absence of the low slopes in some counts may be caused by secondary craters toward the lower end of the size range in question, which tend to increase the slope of the size frequency curve and hide the offset.

CRATER DESTRUCTION

Crater statistics provide good evidence for high obliteration rates early in Mars's history. The evidence is entirely from the heavily cratered uplands; the plains all appear to postdate the era of enhanced crater obliteration. The greater intensity of crater destruction in the martian uplands as compared with the Moon was suspected from the first Mariner 4 pictures, which revealed a surface somewhat like the lunar highlands but with more subdued craters (Leighton et al. 1965). Opik (1965) noted a break in slope around 20 km and attributed it to an obliterative process, which he suspected removed many of the smaller craters but affected the larger craters also. Murray and co-workers (1971) pointed out a dichotomy in martian craters as shown by the Mariner 6 and 7 pictures of the cratered terrain. Craters smaller than about 15 km in diameter seemed to be mostly fresh and bowl-shaped, whereas larger craters were mostly shallow with flat floors. The two populations appeared to be distinct, with no transitional types between, and were separated by a discontinuity on the crater frequency curves. Murray and colleagues also noted the lack of large fresh craters analogous to Copernicus and Tycho on the Moon. They postulated that one or more erosional episodes early in the planet's history caused infilling of the large craters and elimination of the smaller ones. The bowl-shaped craters now present, they believed, were formed after the erosional event. We saw in the previous chapter that complex craters on Mars are initially shallower than those on the Moon, so that a shallower profile does not necessarily indicate more erosion. Moreover, if crater depths are plotted against crater diameters for all types of martian craters, both fresh and degraded, they show a scatter that substantially overlaps that of the Imbrian craters on the Moon (Mutch et al. 1976). The evidence of an obliterative event early in Mars's history from crater shapes alone is, therefore, not compelling. Better evidence is provided by the crater statistics.

We saw above that crater frequency curves for the low-latitude plains have slopes around −3 for diameters less than 1 km and slopes between −1 and −2 in the 1–10 km size range. Statistics for craters larger than 10 km are poor, and the slopes are not well known. Crater distributions in the highlands, however, are somewhat different. Hartmann (1973a) showed that the size distribution curves for the cratered highlands could be divided into three segments. Below about 5 km the curves are relatively steep and closely parallel those of the sparsely cratered plains. Above 30 km the slopes are also steep, ranging from −2 to −2.5, but between 5 km and 30 km the slopes are generally low, close to −1 (figure 5.4). Hartmann postulated that the cratering rate on Mars was very high between 3.5 and 4 billion years ago, then declined rapidly to a considerably lower rate, which has been sustained ever since. He further postulated that high obliteration rates were coincident with the early high cratering rates, although most craters larger than 30 km escaped destruction, so that their original near-saturation distribution has been preserved (figure 5.5). Craters in the 5–30 km range, Hartmann suggested, represent an equilibrium distribution between crater production and crater destruction during the intense cratering period. He showed that steady infilling of craters such that a 100-m-deep crater survives only half as long as a 200-m-deep crater will produce an equilibrium distribution with a slope close to −1, as observed. If the rate of infilling is constant, the slope is always −1; differences in the infilling rate merely change the intercept between the equilibrium distribution at small diameters and the saturation distribution at larger diameters. Hartmann further postulated that the period of high erosion terminated early in the planet's history, after which time relatively undegraded craters accumulated. The latter population, with a slope of −3 below 1 km and close to −2 above, is the only one represented on the plains.

In the Hartmann scheme, obliteration and cratering occur simultaneously, creating an equilibrium population of craters. The craters were not destroyed solely by formation of new craters, for this would probably have produced a saturation population with a slope of close to −2 (depending on the crater production function). Other destructive processes, such as infilling by lava and erosion by wind and water, appear to have been at work. If

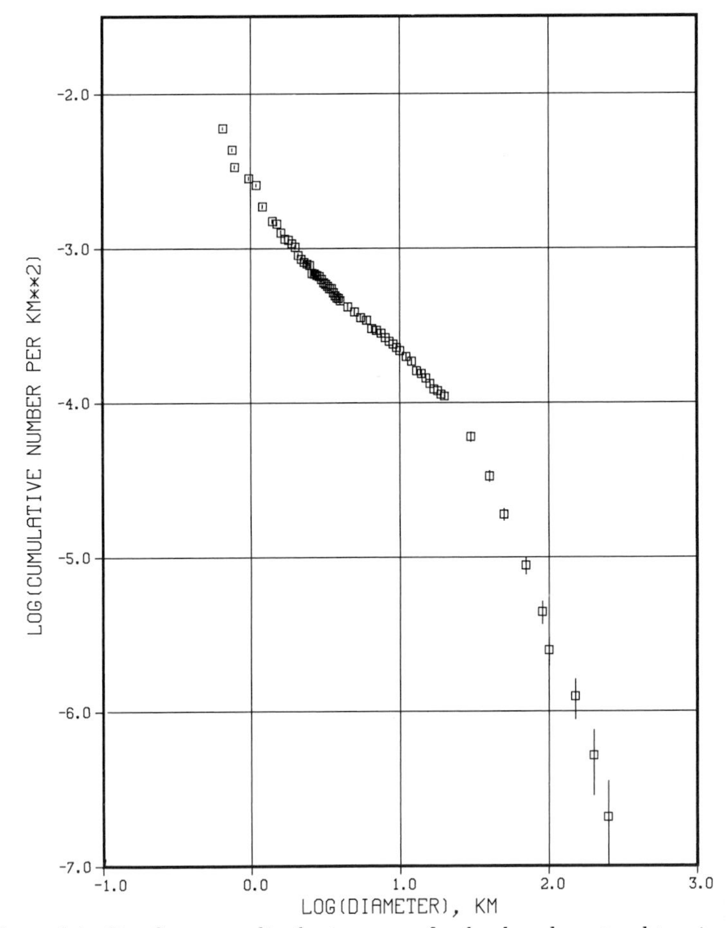

Figure 5.4. Size-frequency distribution curve for the densely cratered terrain. Curves from these regions typically have steep slopes (−2) above 30 km in diameter, shallow slopes (−1) between 3 and 30 km, and steeper slopes again at smaller diameters. Craters less than 10 km in diameter are mostly fresh-appearing; those above about 30 km in diameter are mostly highly degraded. The curve may steepen to −3 for diameters below 1 km, as in the previous figure.

the intense cratering tailed off while erosion was still high, then the slope of the equilibrium distribution would decrease. The effect would show up first at small diameters, with the result that the cumulative distribution curve would roll over at small diameters and could have a zero slope if the cratering rate fell so far that craters were being destroyed as rapidly as they were being formed. This situation approximates the Chapman/Jones model described below. It is doubtful whether such a rollover could be detected in the crater statistics, however, for the postobliteration craters would tend to fill the gap. If a high cratering rate continued after obliteration ceased, then the effect would be merely to supplement the number of relatively fresh-appearing craters in the postobliteration population. Postobliteration craters are, in fact, sparsely distributed, so that the cratering

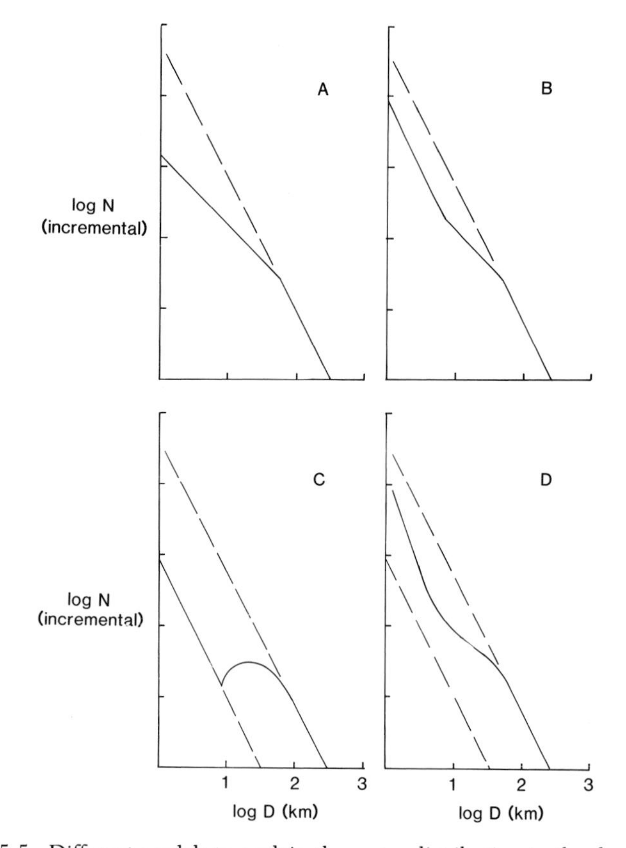

Figure 5.5. Different models to explain the crater distribution in the densely cratered terrain. According to the Hartmann (1973a) model (a, b), an early era of high crater obliteration resulted in an equilibrium distribution for crater diameters below about 50 km with a slope of −1. Above 50 km the distribution curve is unaffected by the high obliteration rates and is close to saturation (partly dashed line), with a slope of −2. After the decline of the high obliteration rates a new population of craters with a slope of −2 was superimposed on the old. According to Chapman/Jones (1977) model, an early era of high cratering rates resulted in a near-saturation distribution (*upper dashed line*). A later discrete erosional episode removed most of the craters below 50 km (c). After the erosional event was over, a new population of craters was superimposed, to form the presently observed distribution (d).

rate appears to have declined before, or contemporaneously with, the decline in obliteration, and not after.

Jones and Chapman (Jones 1974; Chapman and Jones 1977) presented a somewhat different model, in which a major obliteration event occurred after the decline in the cratering rate (see figure 5.5). They suggested that early in the planet's history the surface was essentially saturated at all diameters, so that the distribution had a slope of close to −2 for the full diameter range; then, at some indeterminate time, which could be quite late in the planet's history, a massive event destroyed all fresh craters with diameters less than about 10 km and many of those in the 10–50 km size range. The obliteration event also produced large numbers of modified craters. After obliteration ceased, a new population of fresh-appearing bowl-shaped types was superimposed on the modified surface, which was thus left with a large number of modified craters as compared to fresh ones. The main evidence in support of the Chapman/Jones thesis is the ratio of heavily degraded to slightly degraded craters. In the Hartmann model, while enhanced obliteration is occurring, the proportions of slightly degraded, moderately degraded, and highly degraded craters should be similar in all size ranges; smaller craters merely pass through the different stages faster than larger ones. With a discrete erosional spike, however, all craters below a certain threshold will become highly degraded, and all those below another, higher threshold, will become at least moderately degraded. The smaller sizes will therefore include a disproportionately large number of highly degraded types. It was detection of this anomaly in the 8–20 km size range that led Jones (1974) to suggest the erosional spike model.

The two models just outlined do not differ drastically from one another. Both postulate early high cratering rates, which decline rapidly to the present low levels over a short period of time. Hartmann suggested that erosion and infilling of craters also occurred at a high rate early on and then declined to much lower levels, more or less contemporaneously with the decline in the cratering rate. Chapman and Jones, on the other hand, argued for an erosional spike subsequent to the decline in cratering rates. While they pointed out that there is nothing in the statistics that demands that the erosional spike be early, they acknowledged that an early date is more reasonable on geologic grounds.

I find the erosional spike model unconvincing. The argument is based largely on a subjective discrimination between different levels of degradation in craters close to the limit of resolution of many of the Mariner 9 photographs that were used to collect the statistical data. Moreover, a high rate of erosion and infilling early on, which then tails off, is more plausible geologically than a discrete spike. The early high obliteration rates could readily result from early high rates of volcanism or from a thicker atmosphere. In contrast, a spike requires a more complicated sequence of events, in which the planet evolves in one direction to create the spike and in another to turn it off.

CRATER DATING MODELS

Geology is concerned largely with sequence, with placing events in the history of the planet in their proper succession. Relative ages can be established quite simply by such techniques as determining intersection and superposition relations and from ratios of crater counts. Indeed, it was not until fairly recently that terrestrial geologists were able to determine anything but relative ages, since they were almost totally dependent on fossils for their age determinations. Absolute ages, however, are required to determine many of the processes that have operated on a planet. The planet's internal dynamics and thermal history, for example, are drastically changed depending on whether all volcanic events ceased 3 billion years ago or whether the planet is still volcanically active. Similarly, if channels formed on the surface of Mars 100 million years ago, this places quite different constraints on the evolution of the atmosphere than if the channels ceased to form 3–4 billion years ago. There is, in addition, a psychological satisfaction in being able to pinpoint precisely when an event took place, rather than merely being able to say that it happened before A but after B.

Determination of absolute ages on Mars from crater counts has received considerable attention in recent years. Three different arguments are presented here. Neukum and Wise (1976) predicted that impact rates on Mars are about a factor of 4.5 less than those on the Moon, which led them to conclude that most martian features are old. Soderblom and co-workers (1974) suggested that the impact rate on Mars is about 1.5 times greater than that on the Moon, which results in the events of martian history being spread out fairly evenly in time and indicates that some of the large volcanoes are still active. Hartmann (1977b) described a number of models, the most plausible of which predict impact rates on Mars that are about twice those on the Moon and so give ages close to the Soderblom model. All the models assume that the impact rate over the last 3 billion years approximates the present rate. If this assumption ultimately proves to be invalid, then the conclusions would change accordingly.

The chronology of Neukum and Wise (1976) is based on comparisons of crater production curves for Mars and the Moon. Neukum and Wise noted that for the Moon, the shape of the crater size frequency curve is everywhere the same for sparsely cratered surfaces. The lunar curve has a slope of approximately -1.8 at large diameters (> 4 km) and steepens to -3 or more at smaller diameters. They maintained that the increase in slope is not due to the presence of secondary craters but is for primary impact craters only and is hence a reflection of the size frequency distribution of the impacting objects. Since the shapes of the curves for the lunar maria are all the same, irrespective of age, the size frequency of objects impacting the Moon must have remained unchanged since the earliest maria formed. The same conclusion follows even if one interprets the steeply sloping part of the distribution curve as due to secondaries, for the slope of the primary part of the curve is always close to -1.8. For Mars, Neukum and Wise followed a similar line of reasoning. They counted craters on sparsely cratered surfaces remote from large craters, to compile a standard crater production curve for the planet. They noted that the lunar and martian curves have a similar shape, and from this they drew a major conclusion, that the size frequency distribution of impacting objects has been the same for both bodies.

Neukum and Wise further reasoned that the same size object should produce a smaller crater on Mars than on the Moon because of the smaller impact velocity on Mars (table 5.1). Taking values of 12 km/sec for the average impact velocity on the Moon and 8 km/sec for Mars, they calculated that the same size object should form craters on Mars that are 1.5 times smaller than those on the Moon. They claimed that the good fit of the lunar curve to the martian counts, when the curve is displaced to smaller diameters by this amount, supports the model. Offset of the lunar production curve to smaller diameters results in a lower rate of crater production at a given diameter. Because of the steepness of the curve, the 1.5 offset in diameter translates into a 4.5 decrease in the production of 1-km-diameter craters as compared to the Moon. Neukum and Wise argued, therefore, that the cratering rate on Mars is lower than that on the Moon by a factor of 4.5. The effect is to give old ages (> 3 billion years) to almost every feature on the planet. The most questionable assumption in the model is that Mars and the Moon have been impacted by objects of identical size and at identical frequencies, for the similar shapes of the size frequency curves of craters on the two bodies are equally well explained by a similar mix of primaries and secondaries on the two surfaces.

The chronology of Soderblom and co-workers (1974) is based on a totally different set of assumptions. Soderblom and colleagues pointed out

TABLE 5.1. EFFECT OF IMPACT VELOCITY AND PLANETARY FOCUSING ON THE DIAMETER OF IMPACT CRATERS ON DIFFERENT PLANETARY BODIES

	Local Planetesimals in Heliocentric Circular Orbits	*Asteroids and Short Period Comets*	*Long Period Comets*
Approach velocity (km/sec)			
Mercury	2.1	19	62
Venus	5.1	15	46
Earth	5.6	14	38
Moon	5.6	14	38
Mars	2.5	8.6	31
Impact velocity (km/sec)			
Mercury	4.7	20	62
Venus	11.5	18	47
Earth	12.5	18	40
Moon	6.1	14	38
Mars	5.6	10	31
Impact velocity correction factors (ratio to Moon)			
Mercury	0.73	1.54	1.87
Venus	2.15	1.36	1.29
Earth	2.38	1.36	1.06
Moon	1.00	1.00	1.00
Mars	0.90	0.66	0.78
Effective radius factor			
Mercury	1.28	1.26	0.25
Venus	1.27	0.38	0.26
Earth	1.26	0.42	0.27
Moon	0.29	0.26	0.25
Mars	1.25	0.33	0.26

NOTE: Approach velocities and impact velocities are listed for various impacting objects and targets. The effect of the different impact velocities on crater sizes, as compared to the Moon, are also listed. For example, a typical asteroid hits Mars at 10 km/sec and the Moon at 14 km/sec, and results in a crater that on Mars is smaller by a factor of 0.66. The effective radius factor is $S^2/4R^2$, where S is the effective radius of the planet for impact, and R is the geometric radius. For further details, see Hartmann 1977.

that Mars, like the Moon, has a bimodal distribution of craters, which indicates a high early cratering rate, which later declined rapidly to a much lower rate. They assumed that the decline in cratering rate took place on Mars at the same time as it did on the Moon, 3.9 billion years ago. It is implicit in this assumption that the decline marks a transition from the period of accretion of the planets to a period of nearly steady-state depletion of the asteroid and comet population. The sweeping-up of objects with short half-lives at the tail end of accretion is believed to have occurred more or less simultaneously throughout the inner solar system, so that the drop-off in the cratering rate on Mars occurred at the same time as it did on the Moon. According to this model, therefore, all sparsely cratered surfaces are younger than 3.9 billion years, and all densely cratered surfaces are older. To determine the postdecline cratering rate for Mars, upon which the planet's chronology depends, Soderblom and co-workers reasoned as follows. The histogram of crater densities for different surfaces on Mars is bimodal. One population, of high crater densities, is for surfaces older than 3.9 billion years; the other population, of low crater densities, is for surfaces that are younger. The low point between the two populations represents surfaces that formed at, or just after, the decline 3.9 billion years ago. These rare, intermediate cratered surfaces have therefore been accumulating craters for the last 3.9 billion years. If we assume that the cratering rate has been constant over this period, then the cratering rate on Mars is derived simply by dividing the number of craters on the 3.9-billion-year-old surface by 3.9 billion years. The result is a cratering rate about 1.5 times that on the Moon. The main uncertainty in the method lies in not knowing whether the surface with the 3.9-billion-year accumulation formed entirely after the decline in cratering rate, for, if it did not, the derived production rate would be anomalously high.

Hartmann (1977b) approached the problem of deriving ages for Mars from a third, and probably more fundamental, standpoint. He noted that interplanetary objects can be divided into different families according to their orbits. For each family he computed the relative impact rate with respect to the Moon and then the relative cratering rate, by correcting for such factors as the velocity of approach of the impacting body, the focusing effect of the planet, the different surface gravities, and so forth. The main uncertainty stems from not knowing the proportions of the different types of planetesimals within the inner solar system. Hartmann divided the interplanetary debris into three main families of objects: planetesimals in circular heliocentric orbits near the planet's orbit—a low-velocity family; asteroidal and short-period comets—a medium-velocity family; and intermediate- and long-period comets—a high-velocity family. Objects in the first class would have short half-lives and would probably have produced significant numbers of craters only very early in the histories of the planets, so are of secondary interest for crater dating. The second and third types have been responsible for most of the cratering during the last 3.9 billion years. In the inner solar system their numbers are probably being continually replenished by the addition of asteroids scattered from the asteroid belt by gravity perturbations of the planets and by the addition of comets introduced from the Oort cloud.

TABLE 5.2. SURFACE GRAVITY CORRECTION FACTOR

Planet	*Correction Factor*
Mercury	0.72
Venus	0.51
Earth	0.49
Moon	1.00
Mars	0.72

SOURCE: Hartmann 1977.
NOTE: Objects with the same mass and velocity make craters of different sizes on different bodies because of the effects of gravity. The table shows the correction factor with respect to the Moon.

For a given size of impacting object the size of the crater formed depends on the impact velocity, which in turn depends on the approach velocity of the object to the planet and the acceleration caused by the planet itself. Approach velocities are different for different types of objects: those in circular heliocentric orbits have low approach velocities; comets have much higher velocities (see table 5.1). The size of the impact crater also depends on surface gravity: the lower the surface gravity, the larger the crater that is produced, because less work is done against gravity during excavation (table 5.2). A final factor that must be taken into account is the focusing effect of the planet. A planet's effective cross section is larger than its actual cross section, because objects are deflected toward it by its gravity field. The larger the mass of the planet, therefore, the greater the focusing effect.

Using the data in tables 5.1 and 5.2, Hartmann compared the relative cratering rates on the different planets for different classes of objects (table 5.3). The calculations for objects with known orbits do not involve large uncertainties. The main uncertainty in comparing cratering rates on Mars and the Moon lies in not knowing the relative proportion of the different classes of objects. For example, Earth-crossing asteroids have a higher

TABLE 5.3. ESTIMATED CRATER PRODUCTION RATES, NORMALIZED TO THE MOON

Type of Objects	*Mercury*	*Venus*	*Earth*	*Moon*	*Mars*
Asteroids	0.8	1	0.9	1	2
Comets	5	2	0.7	1	0.3
40% comets/60% asteroids	2	1	1	1	2
Mars-crossers favored (80%)	2	1	2	1	4
Maximum likely	5	2	2.1	1	4
Minimum likely	0.8	0.8	0.9	1	1
Most likely	2	1	1.5	1	2

SOURCE: Hartmann et al. 1981.

cratering rate on the Moon than on Mars, while the reverse is true for Mars-crossing asteroids. To know the relative cratering rates of asteroids on Mars and the Moon, we must know the relative proportions of Earth- and Mars-crossers. Similarly, the proportion of comets to asteroids must be known, for they also have different impact probabilities on the Moon and Mars. If meteorites hitting Earth, and presumably the Moon, are mainly asteroids, as Chapman (1976) believed, then the ratio of the cratering rates on Mars to those on the Moon will be higher than if most of the impacts are by comets. The most probable models have a mix of comets and asteroids, with Mars-crossers being more abundant than Earth-crossers. Most such mixes give cratering rates on Mars about twice those on the Moon, so Hartmann's reasoning results in a model similar to that of Soderblom and co-workers (1974) for which the enhancement was 1.5.

CRATER AGES

Throughout this book, the number of superposed impact craters larger than 1 km in diameter per million km², referred to as the "crater number," is used as an index of age. This is a convenient reference, because the crater size is small enough that even a relatively sparsely cratered surface can provide adequate statistics. Considerable caution must be exercised in using crater numbers. Those used in this book are actual numbers, but many crater numbers in the literature are based on extrapolations from larger diameters, and the extrapolated and actual numbers may differ greatly. In addition the numbers have validity for dating a surface only if a negligible number of 1-km–diameter craters have been destroyed since the surface formed. In many regions, particularly in high latitudes, erosion and infilling have affected the craters in this size range, so their numbers cannot be used for dating.

Wise and co-workers (Wise et al. 1979) initiated the use of crater numbers. Much of their early work was based on the Mariner 9 A frames, which had resolution limits of 2–4 km. To derive a crater number, they extrapolated down to 1 km, using the Neukum and Wise (1976) crater production function. This standard curve has steeper slopes in the 1–5 km size range than many actual counts. Many of the curves given by Soderblom and colleagues (1974) and Dial (1978), for example, have slopes close to -1 in the 1–3 km size range, in contrast to the slope of about -3 for the Neukum and Wise curve. As a result crater numbers derived by extrapolation are considerably larger than those derived by actually counting craters on the surface. For example, Wise and co-workers (1979) gave crater numbers of 14,000 to 20,000 for Lunae Planum, whereas the actual count is 2,500 (Gregory 1979). Neukum and Wise ascribed the discrepancy to preferential destruction of small craters, either during deposition of Lunae Planum or during subsequent erosion, and suggested that the extrapolated numbers might be a better measure of age than the real numbers. However, errors due to the extrapolation process probably also contribute to the discrepancy, so extrapolations are not used here.

Hartmann and colleagues (1981) made a detailed examination of ages on the surfaces of all the terrestrial planets. In the case of Mars, most of the data available to them were for craters 2–4 km in diameter and larger. They assigned ages to different surfaces on the basis of carefully selected crater counts and the cratering model of Hartmann (1977) just described, according to which the cratering rate for Mars is twice that for the Moon (table 5.4). While their data are internally consistent, there is a problem in applying their ages directly to counts at 1 km in diameter. For example, Chryse Planitia is estimated to have 1.1 times as many craters as an average lunar mare and is assigned an age of 3.2 billion years. According to the values of Hartmann and co-workers for an average lunar mare, Chryse Planitia should have 3,200 craters greater than 1 km in diameter per million km², whereas the actual value is 2,100 (Dial 1978). Similarly, the extrapolated value for Lunae Planum is 3,500, whereas the actual value is close to 2,500. In general, the sequence of ages given by the crater numbers at 1 km agrees well with the sequence of ages derived by counts at larger diameters, but the absolute values differ. Again there are two possible explanations. The first is that the smaller craters are being preferentially

TABLE 5.4. CRATER AGES FOR DIFFERENT SURFACE FEATURES

Geologic Province	*Crater Density Relative to Average Lunar Mare*	*Estimated Crater Retention Age (billions of years)* Minimum Likely	Best Estimate	Maximum Likely
Central Tharsis volcanic plains	0.1	0.06	0.3	1.0
Olympus Mons volcano	0.15	0.1	0.4	1.1
Extended Tharsis volcanic plains	0.49	0.5	1.6	3.3
Elysium volcanics	0.68	0.7	2.6	3.5
Isidis Planitia	0.76	0.8	2.8	3.6
Solis Planum volcanic	0.90	0.9	3.0	3.7
Chryse Planitia volcanic plain	1.1	1.2	3.2	3.8
Lunae Planum	1.2	1.3	3.2	3.8
Noachis ridged plains	1.3	1.7	3.3	3.8
Tyrrhenum Patera volcano	1.4	1.8	3.4	3.8
Tempe Fossae faulted plains	1.6	2.3	3.4	3.8
Volcanic plains on south rim of Hellas	1.7	2.6	3.5	3.8
Alba shield volcano	1.8	2.6	3.5	3.8
Hellas floor	1.8	2.6	3.5	3.8
Syrtis Major Planitia volcanic plains	2.0	2.0	3.6	3.9
Heavily cratered plains				
—small D (< 4 km)	1.4	1.8	3.4	3.8
—large D (> 64 km)	13	3.8	4.0	4.2

SOURCE: Hartmann et al. 1981.

destroyed, so that the count at 1 km is not truly representative of the age of the surface. As discussed above, however, the close parallelism between the lunar and Chryse curves down to 100 m argues against preferential destruction. The second possibility is that the ratio of the cratering rates for Mars and the Moon is a function of diameter. At larger diameters (> 4 km) the ratio may be close to 2, as Hartmann (1977) suggested, but at 1 km the ratio may be smaller, closer to 1.5. Such must be the case if the martian curves have lower slopes in the 1–10 km size range due to the onset of fluidized ejecta patterns and the accompanying proportionate increase in crater diameters. If this is the correct explanation of the discrepancy, then the 1-km–diameter craters are still good indices of age.

The crater numbers quoted in the rest of the book are intended only as a guide, but they can be transposed into a rough measure of age by using the data of Hartmann and co-workers (1981). If the age of Chryse Planitia is 3.2 billion years (see table 5.4), then the 2,100 crater number implies that 660 craters larger than 1 km formed every billion years. Other surfaces give similar numbers, ranging from 550 to 800. This cratering rate is clearly only a crude approximation. Hartmann and colleagues (1981) rightly assigned large errors to their ages. The probable ages for Chryse range from 1.2 to 3.8 billion years, with 3.2 billion years as the best estimate. Assuming their best guesses are about right, then the 800 crater number for the flanks of Arsia Mons (see chapter 7) implies an average age of a little over a billion years, and the 2,400 crater number for Tyrrhena Patera implies an age in excess of 3.5 billion years. Such ages are plausible on geologic grounds and give some confidence that the method is approximately valid. One point should be reemphasized. The crater numbers can be used as an index of age only if a negligible number of 1-km–diameter craters have been destroyed since the dated surface formed. They cannot be used to date surfaces at high latitudes, for example, for most such surfaces have experienced complex erosional histories, as we shall see in the next chapter.

6 DENSELY CRATERED TERRAIN AND PLAINS

More than half of Mars's surface is covered with old, densely cratered terrain that superficially resembles the lunar highlands. Its main characteristic is the presence of large numbers of highly degraded craters with diameters in excess of 20 km. The other half of the planet is covered mostly with plains, on which large craters are ten to over a hundred times fewer, and in which crater densities range widely, from close to zero on some of the youngest volcanics to as much as two to three times the density on the lunar maria on some of the oldest surfaces. The distribution of the two kinds of landscape is quite asymmetric. Division of the planet into two hemispheres by a plane dipping 50° to the equator and oriented so that the 50° latitude is intersected at 330°W results in nearly all the densely cratered terrain lying on the more southerly hemisphere and most of the plains on the more northerly hemisphere. Most of the cratered terrain is at a higher elevation than the plains, and the boundary between the two is commonly an escarpment 1–2 km high. The elevation difference is fully compensated, for there is no gravity anomaly along the boundary (Phillips and Saunders 1975). Exceptions to the generally low elevations for plains are some plains on the Tharsis and Elysium bulges that occur at higher elevations than much of the cratered terrain. The cause of the asymmetry in the distribution of the cratered terrain and of the differences in elevation between the heavily cratered and sparsely cratered hemispheres is not known and is one of the major unsolved problems of martian geology.

As we saw in the previous chapter, much of the heavily cratered terrain is probably at least 3.9 billion years old, since it retains a record of the early high cratering rate, whereas most of the more sparsely cratered plains are younger. We saw further that the densely cratered terrain experienced enhanced crater destruction rates at some time in the past. Since the plains show little evidence of obliteration, it appears that the decline in crater degradation and the decline in crater production were roughly coincidental. In this chapter we will examine the morphologic characteristics of the old cratered upland to see if it retains any further evidence of these early events. We will be particularly interested in evidence of a thicker atmosphere.

In contrast to the densely cratered terrain, which all formed early and within a relatively short period of time, the sparsely cratered plains appear to have formed over an extended period that spans most of Mars's history. The youngest lava plains are so sparsely cratered that they may still be forming, although at a slow rate; the oldest date back to around 3.9 billion years ago. The plains differ from each other greatly in appearance and probably in origin and erosional history also. In the Mariner 9 pictures, most of the plains appear almost featureless; all that is visible is an occasional crater and some ridges that resemble lunar wrinkle ridges. The closer look provided by the Viking orbiters shows, however, that the plains are quite diverse and in some areas quite complex. In the equatorial region most have features, such as flow fronts and wrinkle ridges, that suggest a volcanic origin, but in the high latitudes the character of the plains is distinctly different. Here primary volcanic features are rare, and the plains appear to have undergone a complicated history of erosion and deposition. In this chapter the characteristics of the densely cratered terrain and the plains are described in some detail, so as to give an indication of what variation exists within these two major components of the planet's surface. The chapter is sparsely referenced in places, because so little systematic work has been done in this area; of necessity, much of the discussion is based only on impressions gained through examination of large numbers of Viking orbiter pictures. In subsequent chapters, processes such as volcanism, fluvial activity, and wind erosion, which can affect both uplands and plains, will be discussed individually.

DENSELY CRATERED TERRAIN

The main characteristic of the old cratered terrain, the one that distinguishes it from all other surfaces on the planet, is the large number of superimposed craters (figure 6.1). The cratered terrain resembles the lunar and mercurian highlands in this respect and is presumed to be old. However, there are in the Mars uplands two distinct crater populations: an array of relatively large (> 20 km) craters that are mostly greatly degraded, and an array of smaller, relatively fresh-appearing craters. In the previous chapter we concluded that the large degraded craters formed during an early period of intense bombardment, and that early, high crater-obliteration rates resulted in severe degradation of the larger of the old craters and destroyed all the smaller ones. The population of fresh craters was superimposed after crater obliteration and crater production had both declined significantly.

Craters larger than 20 km have a range in morphology. The most fresh-appearing, while not as pristine as the best-preserved lunar craters, have coarse hummocky rims, terraced walls, central peaks, flat floors, and large secondary crater fields. The most degraded are barely visible, shallow,

Figure 6.1. Synoptic view of cratered terrain in the vicinity of 30°S, 15°W. The frame is 500 km across. Most of the craters are rimless depressions with flat floors. Channels are common, although most of the terrain is undissected. Mare ridges occur on the intercrater plains, but a clear distinction cannot be made between the ridged plains and the cratered plateau. (P-18116)

circular depressions. Most craters are intermediate in their state of preservation. They have flat floors at lower elevations than the surrounding terrain and rims that have little, if any, relief and little surface texture (figure 6.2). The crater walls may be just simple inward-facing escarpments, but more commonly they have a texture with strong radial elements, so that the junction of the wall with the flat crater floor is somewhat ragged. Superimposed on the larger craters is the array of smaller (< 20 km) fresh-appearing craters. These generally have distinct rims and are surrounded by fluidized ejecta patterns like those described in chapter 4. Such patterns are less common around the large craters of the older population, but this may simply be a consequence of their usual poor state of preservation. The smaller craters also tend to have more internal relief. For example, a high

Figure 6.2. Detail from the previous figure, showing the scarcity of small (<20 km) craters and the indistinct nature of some of the channels. Most of the craters also lack central peaks, although they are well within the size range where peaks are expected. Ejecta from a 22-km–diameter crater in the lower center has flowed into an adjacent crater. Central pits are present in some of the more fresh-appearing craters. The picture is 250 km across. (84A12)

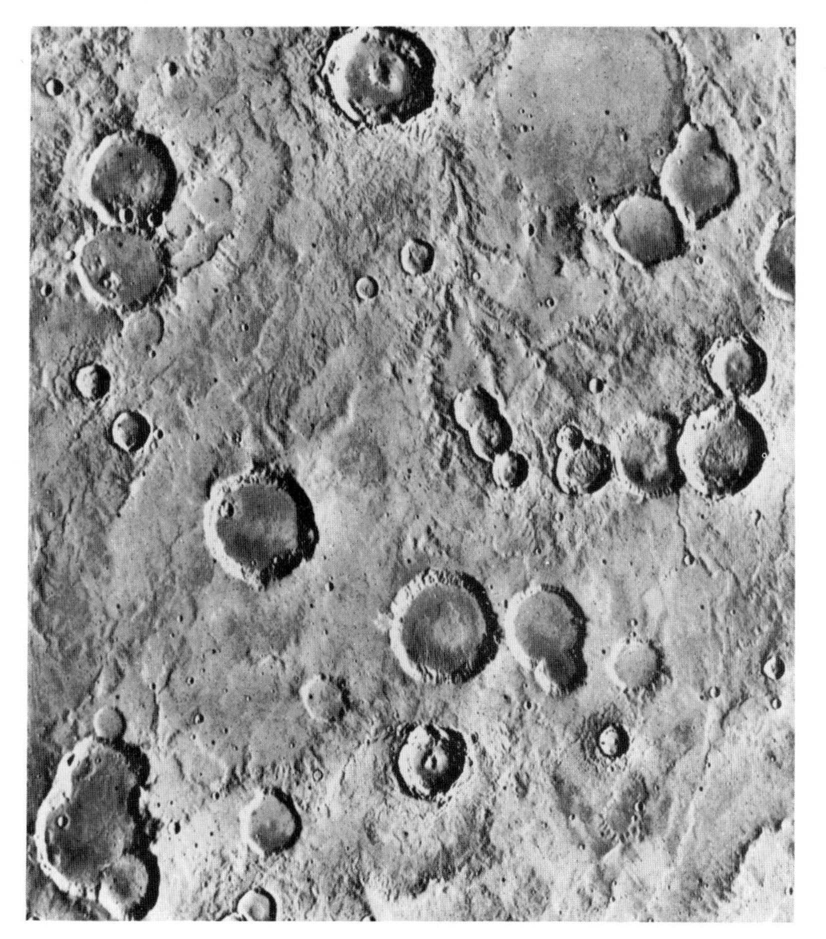

proportion of the craters between 5 km and 20 km in diameter have central peaks, but seemingly a smaller proportion of those between 20 km and 50 km. This again may be a consequence of erosion and infilling. The smaller craters postdate the early era of intense crater obliteration and so retain most of their primary features, while many of the larger craters were formed when destruction rates were still high, so are considerably modified.

Although the density of craters in the uplands is high, the craters are not "shoulder to shoulder," and extensive areas, termed intercrater plains, exist between them. The intercrater plains are complex and poorly understood and have barely been studied. The problem is partly that the information necessary for their study has not been available until recently. Many of the Mariner 9 photographs of the southern hemisphere, in which most of the cratered terrain lies, are poor, because they were taken before the 1971 dust storm had fully subsided. But the intercrater plains are also poorly understood because they are intrinsically difficult to interpret. Just as the early record of Earth's geologic history has been blurred by time, so has that of Mars. The primary evidence of processes such as volcanism and eolian action has been largely destroyed, not only by the pervasive effects of impact, but also as a result of the high burial and erosion rates. Much of what is postulated about the origin of the intercrater plains is therefore based more on indirect reasoning than direct observation. Volcanic deposits appear to be common (Greeley and Spudis 1978), as evidenced by the presence of flow fronts, wrinkle ridges (see figure 6.1), and low cones and domes. These are mostly the youngest deposits of the old cratered terrain, although they still appear to be quite ancient. Evidence of older volcanic deposits has no doubt been destroyed by cratering and erosion. Indications of fluvial activity are also abundant (see figure 6.2). The evidence is mostly of erosion; presumably, the material that was eroded away to form the numerous small channels was deposited in low areas as fans or as lacustrine or playa-type deposits; however, these generally cannot be seen. When the old cratered terrain formed, eolian erosion and deposition were also probably occurring at relatively high rates compared with the present. The old cratered terrain is thus likely to be a complex mixture of impact debris, volcanic rocks, eolian and fluvial sediments, and possibly evaporite deposits.

Although little work has been done on their characterization, the cratered uplands can be crudely divided into four separate components. These tend to merge with one another and are not obvious everywhere; nevertheless, the following subdivision generally holds.

1. *Rugged massifs*. These are rugged, roughly equidimensional hills, angular in plan, and generally less than 20 km across. They occur mostly around the large impact basins of Isidis, Hellas, and Argyre (see figure 4.9), although some massiflike features occur away from large basins, as in Thaumasia and Memnonia. The massifs are closely spaced near basin margins and become widely spaced and have lower relief further out. They are most obvious in the Argyre region, where they form a 600-km-wide annulus around the basin. Those around Hellas are unevenly distributed, extending out from the basin rim for 1,000 km to the east, but only a short distance elsewhere. Most of the massifs clearly formed as a result of the impacts that created the large basins. To what extent they represent deposits of ejecta or merely the preexisting terrain that was fractured and jostled by the impact is unclear. Certainly ejecta patterns comparable to those around the large lunar basins Orientale and Imbrium are absent. All of the basins are very ancient, having numerous large impact craters on their rims.

2. *Rough plateau unit*. This unit generally merges with the rims of large craters and appears to be the old surface in which most of the large craters formed. The surface has a coarse disordered texture at a resolution of 100 m. In most areas the plateau is dissected by numerous small channels, which range in preservation from those that are barely visible to those that are deeply incised and appear fresh at 100-m resolution (see chapter 10). Locally the dissection may be extreme, with the interfluves being sharp-crested ridges. At high resolution (10 m) the surface is sparsely cratered and has numerous closely spaced ridges and valleys (figure 6.3).

3. *Ridged plateau plains*. The ridged plains appear smooth at 100-m resolution, except for widely spaced wrinkle ridges that resemble those on the lunar maria (figure 6.4). They thus contrast sharply with the coarsely textured, rough plateau. Channels are present but are less common than in the plateau, and in many areas the ridged plains transect channels that cut the plateau. The ridged plains are cratered only by the younger population of small, relatively fresh craters, not by the large eroded ones. Greeley and Spudis (1978) estimated that the unit covers 36 percent of the cratered hemisphere.

4. *Crater floors*. The floors of most of the large craters are smooth and resemble the ridged plains, except that wrinkle ridge–like features are less common. At high resolution, most of the crater floors are saturated with small craters (figures 6.5 and 6.6), although some are only sparsely cratered. Many of the heavily cratered floors appear even more densely cratered than similar, flat crater floors in the lunar highlands. Most of the craters are probably secondaries, since they are arranged in loops and strings and commonly have irregular shapes like lunar secondaries.

The massifs are probably the oldest of the four units, since they appear to be surrounded and partly buried by the others. Parts of the rough plateau, however, may be comparable in age. The age relations between the rough plateau and the ridged plateau plains are relatively unambiguous. The former is heavily dissected and heavily cratered by the old degraded crater population. The latter are only sparsely dissected and are cratered only by the younger, more fresh-appearing, crater population. The ridged plains thus appear largely to postdate the decline in crater formation and intense crater obliteration. The materials in the floors of large craters must be younger than the plateau in which the craters lie, but

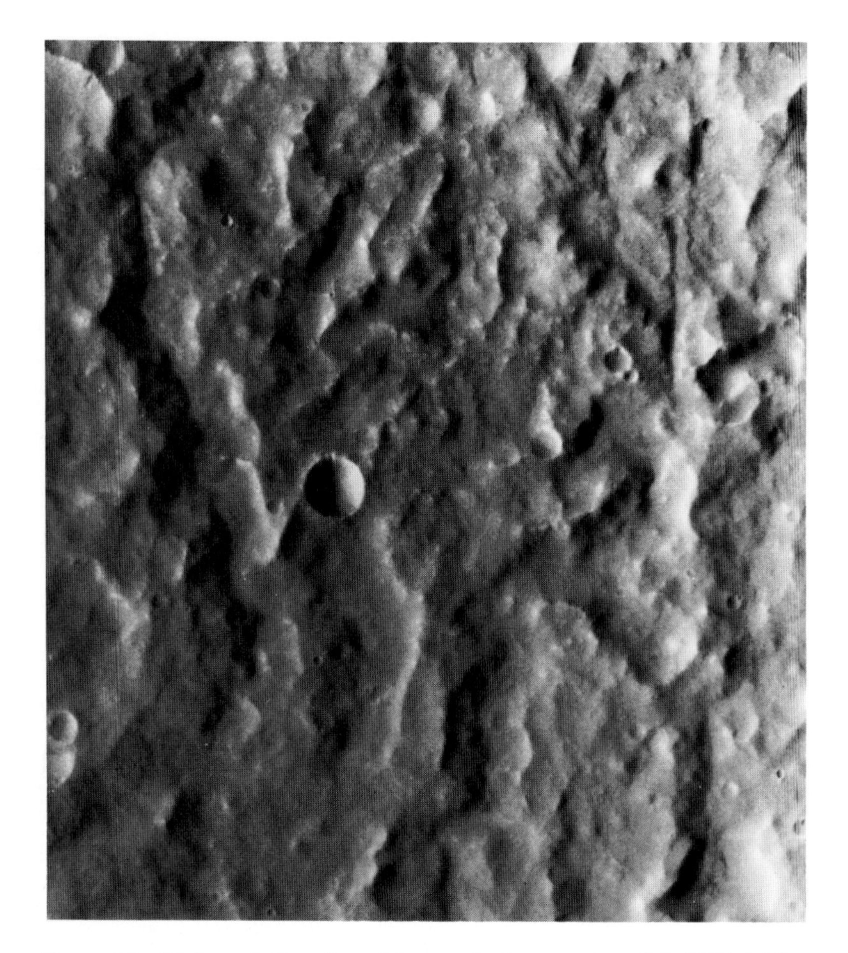

Figure 6.3. High-resolution view of the rough plateau unit at 2°S, 246°W. The unit is probably one of the oldest components of the cratered terrain. It is dissected by vague linear depressions, and small (<1 km) craters are relatively few, almost certainly because of destruction by creep of the near-surface materials. The scene is 15 km across. (124S20)

their age with respect to the ridged plains is unclear. The high density of small craters in the crater floors, as compared with the plateau around the craters, appears to conflict with these age relations. This situation is similar, however, to that which occurs on the Moon, where small craters in rough terrain are destroyed more rapidly than craters in level terrain, by creep of the near-surface materials (Shoemaker et al. 1966). As a result level crater floors in the lunar highlands have higher crater densities than the older, but rough, highland terrain in which the craters lie. A similar explanation appears likely for the martian highlands.

A major distinction between the plateau and the ridged plains lies in the level of dissection. Channels in the rough plains appear to be at all levels of degradation, from those that are barely discernible to those that appear crisp and fresh, at least in the 100-m–resolution photography. The small channels are generally narrow (< 1 km), only a few tens of kilometers long, and commonly have integrated tributary networks. Sharp and Malin (1975) classified these networks as runoff channels and suggested that they formed by surface drainage or seepage. They are distinctively different in origin and age from the large flood features, or outflow channels (see chapter 10), which rarely have tributaries but instead start full-scale, usually in areas of chaotic terrain. While there is debate about the origin of the channels, even the most ardent antifluvialists accept the runoff channels as being formed by water (see chapter 10). The numerous channels in the cratered uplands suggest that the uplands have undergone a long period of fluvial erosion, such that the oldest channels have barely survived, while the younger ones are very obvious. The ridged plains are dissected much less intensely, and the few channels that are present are fresh-appearing. A possible explanation of these relations is that the rate of fluvial erosion had tapered off by the time the younger ridged plains were deposited. Alternatively the ridged plains are more resistant to erosion than the plateau units, so are less dissected.

Figure 6.4. Ridged plains (*right*) resting on older cratered plateau (*left*). The plateau is dissected by numerous ill-defined channels, and many of the craters are highly degraded. In contrast the ridged plains are undissected, and the craters are fresh-appearing. Despite being younger than the plateau, the plains are still old, as evidenced by the large number of superimposed craters. The scene is 245 km across and is centered at 27°S, 200°W. (595A55)

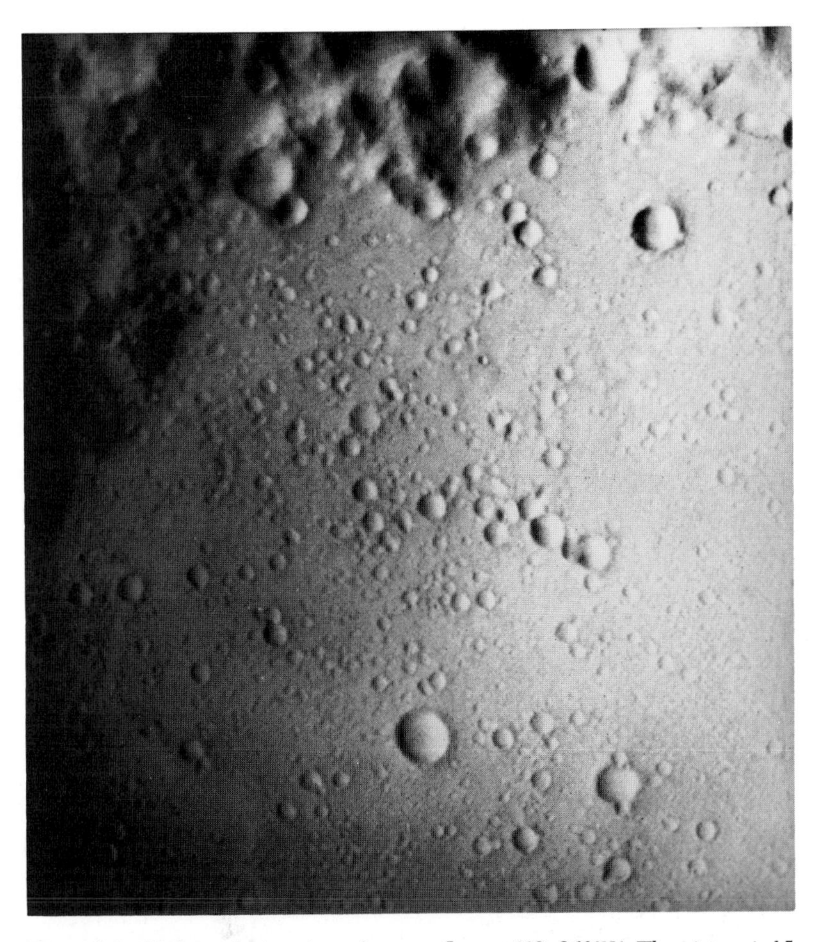

Figure 6.5. High-resolution view of a crater floor at 1°S, 260°W. The picture is 15 km across. The surface is essentially saturated with small (<1 km) craters, of which most are probably secondaries. The surface closely resembles that of cratered floors on the Moon, indicating remarkably little infilling of craters, despite an age that is probably on the order of billions of years. (130S09)

Soderblom and colleagues (1978) examined variations in the color of the old cratered surface. They distinguished rough plateau areas with abundant craters, crater ejecta, and small dendritic channels from volcanic plains with numerous flow ridges. Their two units appear to be identical to the plateau and ridged plains just described. They found that the rough plateaus are redder than the ridged plains, but that the color differences are detectable only in dark areas, where the surface appears to have been largely swept free of surficial debris. Soderblom and Wenner (1978) pointed out the similarity between the reflection spectrum of the upland terrain and that of palagonites—highly hydrated basalts that form by the interaction of basalt and water. While acknowledging that identification is not unique, they showed that a simple process, eruption of basalts into water-rich materials, could produce the spectral features observed.

Malin (1976) pointed out that the intercrater plains are layered horizontally. The layering is evident mainly where the succession is seen in section, such as in channel and canyon walls. Malin also suggested that some of the layers are unconsolidated; the main evidence is exhumation of features such as craters. Malin gave an example where part of a crater protrudes from the bottom of a cliff. He suggested that the crater was unearthed during retreat of the escarpment and could have survived only if the material in the escarpment were much more easily eroded than the crater materials themselves. Layering is particularly evident in numerous high-resolution (30 m) pictures of the cratered terrain north of Syrtis Major that were taken by the Viking orbiters late in the extended mission (Snyder 1979). Highly eroded near-surface layers are present, both between and

Figure 6.6. High-resolution view of a crater floor at 2°S, 4°W. An array of fresh-appearing craters is superimposed on a surface on which vague subdued craters are just visible. It appears that a former cratered surface was almost completely buried, and that a string of secondaries was then superimposed on the fresh surface. The picture is 21 km across. (746A18)

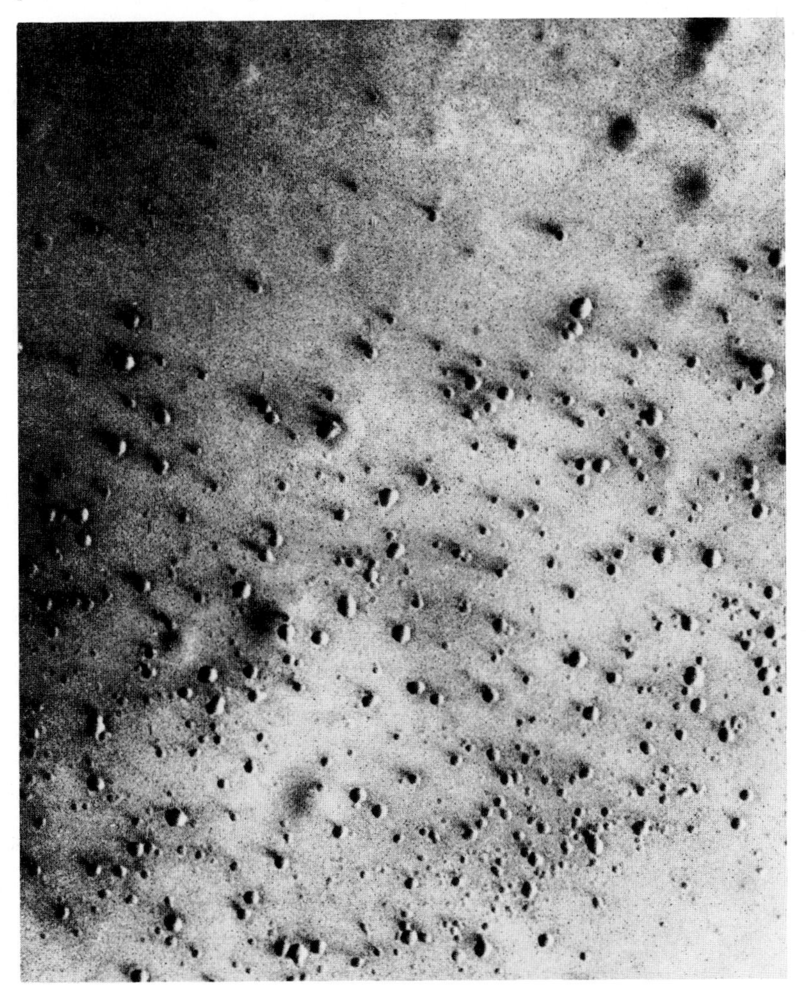

inside craters, as though material had been deposited from above then partly removed. The origin of the layers is not known, but eolian deposition appears likely (see figure 11.10).

The old cratered terrain probably evolved roughly as follows: Before 3.9 billion years ago both cratering and obliteration rates were high. Contributing to the high rates of crater destruction were the cratering process itself and the enhanced levels of fluvial and eolian activity that resulted from an atmosphere that was thicker than at present. The combination of high cratering rates and a significant atmosphere may have resulted in a permanently dust-laden atmosphere and constant planetwide sedimentation. The rates of volcanic activity may also have been high, partly as a secondary consequence of the high impact rates, and may have contributed to the obliteration of craters, although the evidence of early volcanism is largely lost. The old surface was presumably brecciated to depths of several kilometers by the intense bombardment of impacting bodies, much like the lunar highlands. One difference between Mars and the Moon is the smaller number of large impact basins on Mars (see chapter 4), which implies that the highland stratigraphy is less likely to be dominated by the effects of large impact basins than in the case of the Moon. Around the few large basins that are present a veneer of ejecta was presumably deposited, and the surface was severely dislocated. The presence of an atmosphere, however, may have greatly affected the distribution of ejecta, with large amounts being temporarily suspended in the atmosphere, then settling out over much of the planet, rather than immediately around the basins. Most evidence for ejecta has now disappeared, but evidence of the dislocation is well preserved in the numerous massifs.

About 3.9 billion years ago the cratering rate declined rapidly. The surfaces that formed just before the decline (perhaps within 100 to 200 million years) survive as the rough plateau unit. This is widely dissected, indicating that fluvial activity, and presumably a thicker atmosphere, were sustained at least until this time. Such a conclusion is consistent with the crater statistics, which indicate that high rates of crater obliteration must have continued up to, and possibly beyond, termination of the high cratering rate. Thus nearly all the large craters are severely degraded, and almost all the small craters that formed contemporaneously with the large craters are missing. Again, there is no clear evidence for volcanic activity at this time, although the high obliteration rates may be partly due to burial by volcanic deposits.

After the cratering rate had declined to close to its present level, the ridged plains formed. The precise time at which they formed is not known, but the number of superposed craters is high, compared with most of the plains on the opposing hemisphere, and most chronologies indicate that they formed within a few hundred million years of the decline. The ridged plains are probably volcanic, as judged by their resemblance to lunar maria. They are only sparsely dissected by channels and have mainly fresh-appearing craters, so two lines of morphologic evidence—the channels and the preservation of craters—suggest that erosion rates had decreased by the time the ridged plains formed.

If the above reasoning is valid, then the materials close to the surface of the old terrain are of diverse origin but are narrow in age range. Only those features that formed relatively late in the high cratering era, just before the decline in the cratering rate, are likely to have survived. Those that formed earlier would have been destroyed by impact, and their materials would have been redistributed around the planet as impact ejecta. The intercrater plains are unlikely to have formed long after the decline in cratering, because they still have a relatively large number of superposed craters, as compared with most of the plains on the sparsely cratered hemisphere. Since fluvial, eolian, and volcanic processes appear to have been active during the era of intense bombardment, the plateau unit is likely to be quite heterogeneous. However, the ridged plains are probably predominantly volcanic.

PLAINS/CRATERED TERRAIN BOUNDARY AND FRETTED TERRAIN

The boundary between the cratered terrain and the sparsely cratered plains may provide clues concerning the properties of the materials that constitute the old cratered terrain and may give some indication of the processes that caused the planetwide dichotomy between plains and cratered terrain. The boundary varies according to location. In many places the relationship is simply one of overlap of the younger plains onto the older surface. This may result in a sharp transition between the plains and the cratered terrain, as at the eastern boundary of Lunae Planum, along the southwestern and eastern boundaries of Chryse Planitia, and around 15°S, 145°W. In these areas islands of old terrain protrude through the plains close to the boundary, but in general the plains do not deeply invade the cratered terrain, probably because of the higher elevation of the cratered unit. In other areas the onlap is gradual, and the boundary between the two components is diffuse. This is particularly true to the east and southeast of Sinai Planum. Here the proportion of plains to cratered terrain increases gradually to the east and southeast, and a clear line cannot be drawn between the two. The distinction is made more difficult because the plains of Sinai Planum have numerous ridges, like the ridged plains within the old cratered terrain; indeed, they may be similar in origin and age. Within Sinai Planum several large inliers of cratered terrain are completely surrounded by plains hundreds of kilometers to the west of where the cratered terrain is essentially continuous. Most of the inliers are intensely fractured, in contrast to the superimposed plains. In the same region the number of large craters within the plains increases to the east, and many of these are probably in the partly buried cratered terrain. Clearly the cratered surface is at a relatively shallow depth below the plains and is essentially intact, although highly fractured. In other places, such as between 150°W and 160°W, south of Amazonis, and between 220°W and 250°W, south of Elysium Planitia, the boundary between the plains and the cratered terrain is an escarpment, and channels commonly extend from the escarpment deep into the uplands. Between approximately 160°W and 190°W the plains/cratered boundary is masked by a thick sequence of very sparsely cratered, layered deposits of unknown origin (figure 6.7).

One of the most intriguing sections of the boundary is that which Sharp (1973a) termed "fretted terrain." The most extensive occurrence is between latitudes 30°N and 45°N and longitudes 280°W and 350°W. Fretted terrain typically consists of two components: high-standing remnants of the densely cratered terrain, and at a lower elevation, the more sparsely cratered plains (figure 6.8). Separating the two is generally a steep escarpment 1–2 km high. The two components interfinger in complex patterns, with broad, flat-floored, sinuous channels, termed "fretted channels," reaching deep into the uplands. The sinuous channels, the large upland craters, and the embayments of plains in the upland commonly intersect one another to form an accordant surface at a lower level than the surrounding plateau. Their mutual intersection also isolates remnants of old cratered terrain,

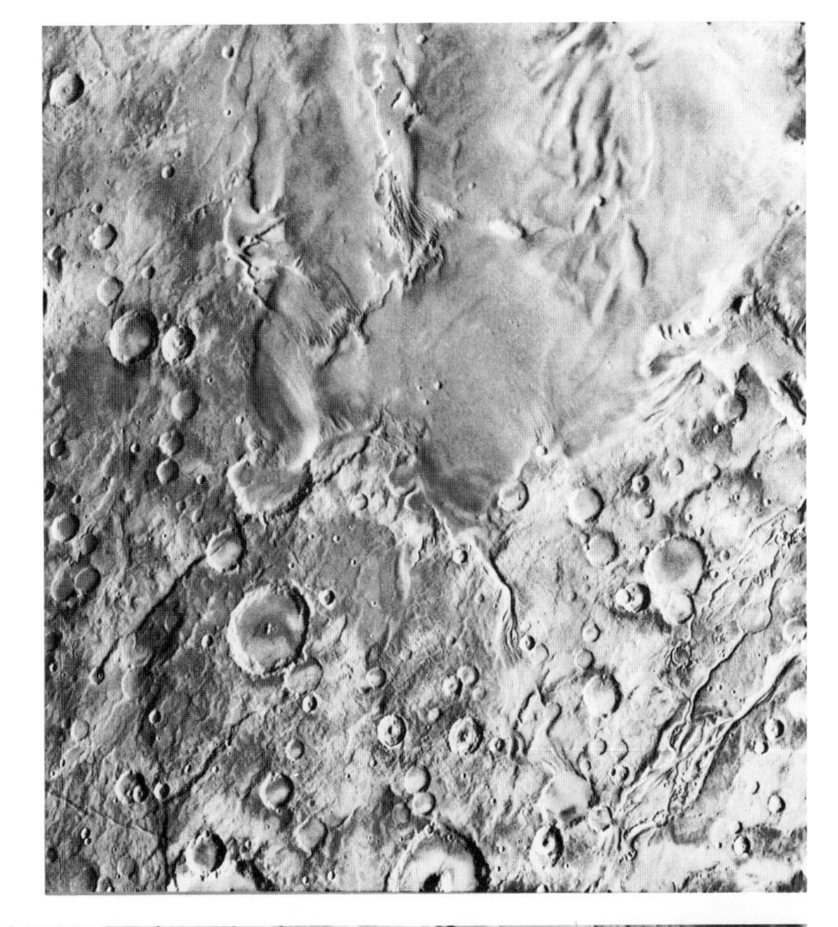

Figure 6.7. (*right*) Synoptic view of the plains/upland boundary south of Amazonis, at 6°S, 157°W. A sparsely cratered deposit (*top*) clearly overlies the older terrain to the south. As we shall see in chapter 11, the sparsely cratered unit is deeply etched by the wind and appears to be friable and easily eroded. The scene is 710 km across. (606A62)

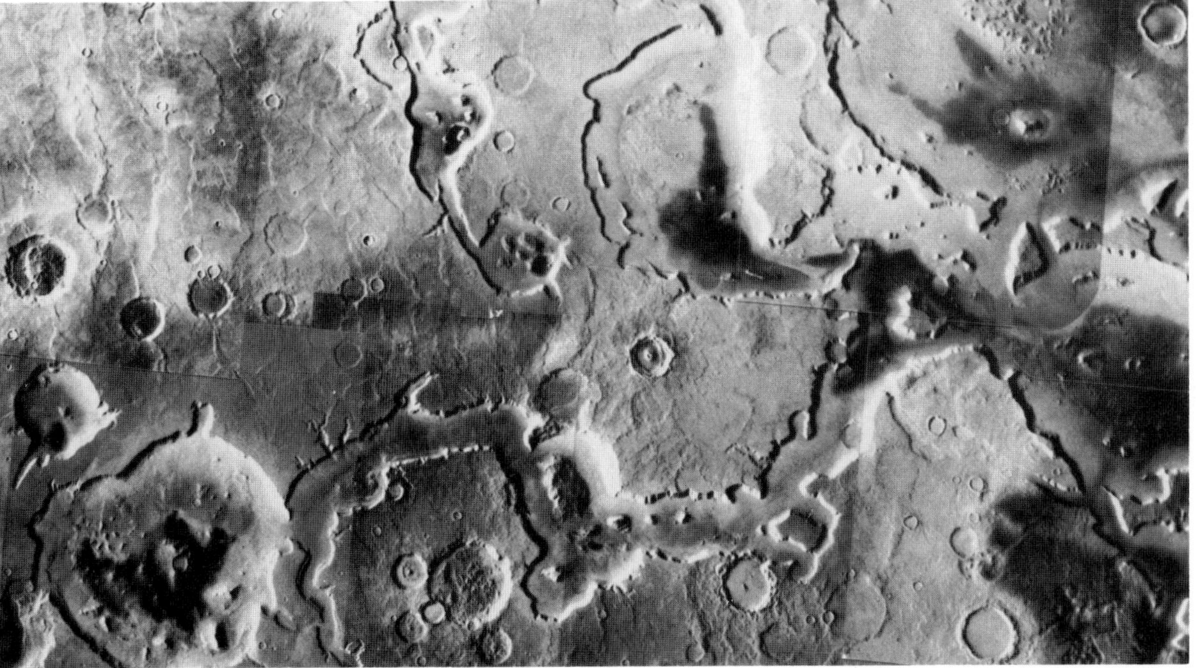

Figure 6.8. Fretted terrain at 38°N, 345°W. The large crater with a dark floor at the lower left is 120 km across. Continuous plains are off the picture to the upper right. A large flat-floored channel extends from the plains deep into the cratered upland and intersects several crater floors. Debris flows are faintly visible as smooth, light bands at the bases of many escarpments. (211-5597)

thereby forming numerous flat-topped outliers surrounded by plains. There is, therefore, a transition northward from undisturbed cratered terrain to cratered terrain that is dissected by broad flat-floored channels to a region where the cratered terrain forms closely spaced, flat-topped mesas with intervening plains and finally to an area of mostly plains with widely spaced outliers of cratered terrain.

Almost everywhere within the fretted terrain, debris aprons occur at the bases of the escarpments that separate the cratered plateau from the plains (Carr and Schaber 1977; Squyres 1978). Where unconfined, such as around plateau outliers, the debris aprons extend 10 km to 20 km across the surrounding plains (figure 6.9). The surfaces of the aprons are almost devoid of craters and are smooth except for a fine surface texture of parallel striae oriented approximately at right angles to the escarpment. The surface striae appear to be flow lines, since they diverge and converge around obstacles, which suggests that the aprons are debris flows. Where the flows are confined, such as within the flat-floored valleys in the uplands (figure 6.10), the striae tend to run parallel to the length of the valleys, indicating flow away from the walls and down the valley. Where valleys meet, the striae merge, much in the manner of median moraines on a glacier. The relation suggests that the old plateau has been eroded by a process of scarp retreat, in which lateral erosion at the face of an escarpment is rapid compared with erosion rates on the upper surface of the plateau. Erosion at the scarps provides materials which feed the debris flows. Such a process is consistent with the general geometry of the fretted terrain, which indicates that escarpments, such as channel walls and cliffs around plains embayment in the upland, have encroached upon the old terrain, intersecting one another and nearby craters.

Figure 6.9. Debris flows around remnants of the cratered uplands at 45°N, 322°W. The flows are sparsely cratered and have surface striae that diverge around obstacles. To the north of the scene are mostly plains; to the south, mostly cratered uplands. (Compare with figure 10.17.) Picture width is 170 km. (211-5266)

Sharp (1973a) and Soderblom and Wenner (1978) have suggested that ice may play a prominent role in the formation of escarpments on Mars, both on the old cratered terrain and in other, younger units. Soderblom and Wenner pointed out that erosional processes appear to have operated to depths of 1–2 km below the preexisting surface, irrespective of the age of the surface or its absolute elevation, and suggested that 1–2 km represents the depth to the base of the permafrost. Above this depth, water, if present, exists as ice, so that induration by solution and cementation is slow. At depths greater than 1–2 km, however, liquid water can exist, so that the rocks become cemented and much more resistant to erosion. Surfaces, therefore, are commonly eroded to depths of 1–2 km but rarely any deeper. Sharp (1973a) and Malin (1976) argued, partly on the basis of terrestrial analogy, that the steeply sloping escarpments and flat-topped plateaus were formed by the undermining and sapping of a poorly consolidated sequence of rocks overlain by a more resistant cap rock. Sharp pointed out that ice could play a major role in the process. If the upper 1–2 km consists of particulate deposits cemented with interstitial ice, then disaggregation could occur at an escarpment face as a result of sublimation and melting of the ground-ice. The disaggregated materials would then be easily erodible and could be removed by wind or mass wasting.

Sharp (1973a) pointed out also that the rate of erosion would depend largely on how efficiently the disaggregated debris is removed from the cliff face. Fretted terrain may form in specific locations, because removal of debris by mass wasting is more efficient in these areas than elsewhere. Squyres (1979b) showed that prominent debris flows, such as are seen in fretted terrain, occur in two latitude bands 25° wide, which are centered at 40°N and 45°S. Squyres (1978) pointed out that in these regions ice is precipitated from the atmosphere onto the ground in midwinter. He postulated that debris flows form because material eroded from the cliff faces becomes mixed with the annual ice deposits to form rock-ice mixtures that resemble terrestrial rock glaciers. Movement of the glaciers is by creep, which is largely accomplished by the flow of the interstitial ice. Although

Figure 6.10. Mosaic of the Nilosyrtis area, at 34°N, 225°W. The mosaic is 180 km across. High-standing areas are remnants of the cratered plateau, which is more extensive to the south. Ridges and grooves within the valleys suggest flow of material away from the valley walls, then down the gentle regional slope to the north. (P-18086)

not explicitly stated by Squyres, such a mechanism may explain why fretted terrain occurs only where debris flows are found. The debris flows provide an efficient way of removing debris that has eroded from an escarpment, thereby allowing erosion to continue. Otherwise a protective talus wedge would form, which would retard further erosion.

In many areas along the plains/cratered terrain boundary are numerous outliers of cratered terrain surrounded by plains. Near the boundary the outliers may be flat-topped mesas, but further away they are mostly small, rounded, equidimensional hills. Where the hills are spaced far apart their origin is unclear, but where close together the knobs outline large craters, indicating that they are remnants of the old terrain (figure 6.11). Areas characterized by numerous closely spaced hills and intervening plains have been termed "knobby terrain" (McCauley et al. 1972; Carr et al. 1973). Such terrain is particularly extensive east of Elysium, between latitudes 10°N and 40°N and longitudes 170°W and 200°W, and north of the Protonilus and Nilosyrtis Mensae, between 280°W and 350°W. Many of the hills in the 40–50°N latitude band are surrounded by smooth, gently sloping debris aprons like those around the mesas of the fretted terrain. Where the knobs are close together, the debris aprons merge to form smooth areas between the hills. At other latitudes, debris aprons are lacking.

The low hills of the knobby terrain have generally been interpreted as remnants of the old, densely cratered terrain, because of the preservation of the rims of large craters and the transition from the old cratered plateau through fretted terrain to knobby terrain (Mutch et al. 1976; McCauley et al. 1972; Carr et al. 1973; Scott and Carr 1978). West (1974), however, suggested that some of the hills may be volcanic, because of the presence of summit craters. This is discussed further in chapter 7. Although the knobs are widely accepted as remnants of plateaus, the precise mechanism by which the former ancient crust has been so extensively destroyed in the northern hemisphere is poorly understood. Evidence of erosion by mass wasting along the present plains/uplands boundary around plateau remnants is persuasive. Erosion alone, however, cannot explain the disappearance of the primitive crust in the north, for there is no sink of sufficient size to accommodate the debris. Mutch and co-workers (Mutch et al. 1976; Mutch and Saunders 1976) suggested that a phase change in the mantle could cause a volume increase and a massive disruption of the crust on one hemisphere. Wise and colleagues (1979) invoked subcrustal erosion by mantle convection, combined with foundering of the partially eroded crust, to explain removal of the primitive crust in the north. Both suggestions are tentative, and the cause of the contrast between the northern and southern hemispheres remains unknown.

PLAINS

Plains occur mostly in the northern hemisphere, although a large area of plains protrudes into the south between longitudes 80°W and 140°W, at the south end of the Tharsis bulge. The plains are probably diverse, both in origin and age. The oldest almost certainly date from just after the early

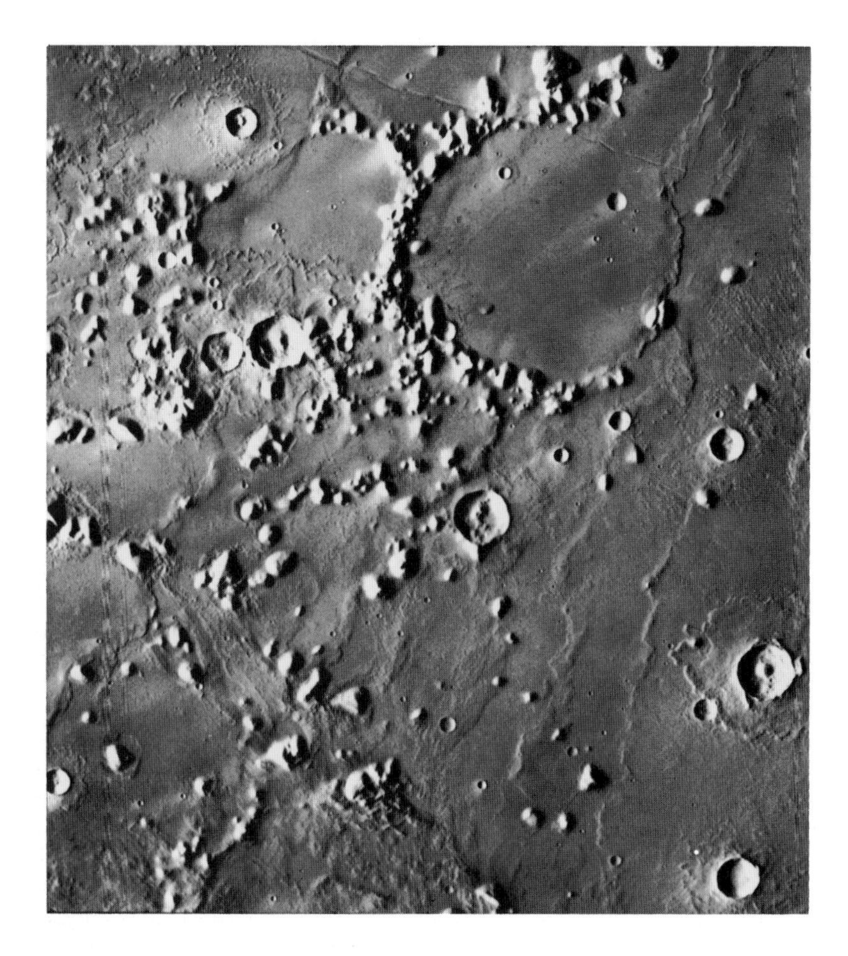

Figure 6.11. Knobby terrain at 19°N, 181°W. The knobs outline large craters and appear to be remnants of an old cratered surface that has been disrupted then flooded with lava to form the intervening smooth plains. The large crater at top center is 47 km across. (545A50)

decline in the cratering rate; the youngest, those close to Olympus Mons and the large Tharsis volcanoes, are probably of relatively recent origin, being less than a few hundred million years old if the cratering rates on Mars are within a factor of two of those on the Moon. The character of the plains changes dramatically between latitudes 30° and 40° in both hemispheres. At lower latitudes their topography is crisp, and surface forms are clearly delineated. Most of the features such as flow fronts, wrinkle ridges, craters, graben, and channels are relatively familiar from lunar and terrestrial experience, even if their precise origin is not known. The albedo patterns also tend to be simple and comprehensible and are mostly confined to crater-related streaks. The plains north of 30°N, however, are quite different. Their topography is complex, and many of their characteristics are difficult to relate either to those elsewhere on the planet or to those on any other planetary surface. The albedo patterns are irregular at all scales, from tens of meters to tens of kilometers, and in places there is a fine overlay of streaks that appears to be caused by circumpolar winds. The net result is some of the most confusing terrain on the planet. In equatorial latitudes most of the plains appear to be volcanic, although some have undergone extensive fluvial and eolian modification. The origin of the plains at higher latitudes is unclear. They may also be volcanic, but if so, the primary volcanic features have been largely destroyed. The northern plains appear to have undergone extensive modification since their formation, and ice may have played a prominent role in the process. The plains of the high southern latitudes, such as those of Hellas and Argyre, are similarly complex and contrast with the simpler plains further north.

The global geologic map of Scott and myself (Scott and Carr 1978) identifies several types of plains and portrays their distribution, but the map was based on Mariner 9 data, in which very little topographic detail is visible on the plains. What follow are Viking-based descriptions of several different types of plains, together with an indication of where they occur. The precise extent of the different types has yet to be established; the locations given are only examples to be used for comparative purposes. Plains merge with one another, and the plains in any one area are not necessarily classifiable into one of the types given, although most are. The intent is to give an impression of the diversity, rather than to present a comprehensive, all-encompassing classification.

Low-latitude plains

Plains within 30° of the equator can be divided into two broad categories on the basis of surface morphology: flow plains, characterized by numerous superposed, clearly delineated flows, and ridged plains, whose surfaces are generally smooth except for wrinkle ridges. The two types are cratered to varying degrees and are variously modified by fluvial and eolian processes. Almost all the low-latitude plains fit into these two major categories. A major exception is the southern part of Amazonis Planitia, which has neither wrinkle ridges nor flows but has a variety of other features that appear to be unique to that area.

FLOW PLAINS. Lava flows are visible on almost all the plains on the Tharsis and Elysium domes. To the east and southeast of Tharsis the flows are visible as far east as 80°W, both north and south of the canyon. To the west of Tharsis flows are mostly restricted to east of the 150° meridian. To the south the flows lap against the old cratered terrain, while to the north the flow field merges with the high-latitude plains or terminates against Alba Patera, which has its own system of radial flows. Around Elysium the flow field (figure 6.12) is restricted to above the zero elevation contour, except to the east, where the flows intermingle with the remnants of the old cratered terrain and their extent is not clear.

Tharsis has the most extensive and best-studied flow plains (Carr et al. 1977a; Schaber et al. 1978). The flows are mostly irregular, finger-shaped

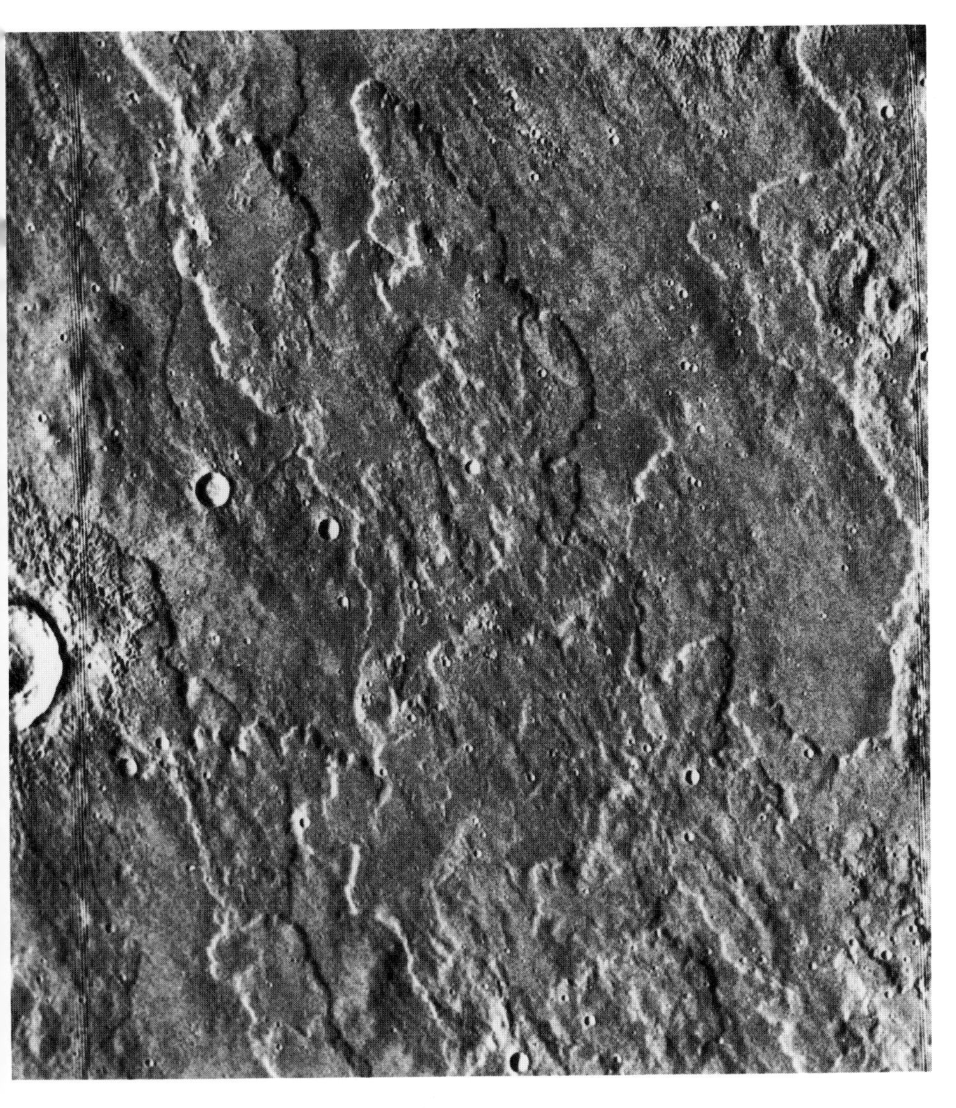

Figure 6.12. Lava plains in Elysium, at 32°N, 213°W. Flow of lava is to the north, away from Elysium Mons, which is to the southeast. The frame is 60 km across. (651A10)

features, with multilobate, outward-facing flow fronts. Flow thicknesses range from 10 m to 80 m (figure 6.13), but most are in the 20–30 m range (Schaber et al. 1978). In some cases a flow may be traced upslope for several hundred kilometers, but the upper parts of most flows are buried by later ones, so that their source and true length can rarely be established. Flow morphology depends somewhat on location. In the plains south of Arsia Mons close to the volcano, flows are long and narrow, generally less than 3 km wide and many tens—even hundreds—of kilometers long. In contrast, a thousand kilometers to the southwest the flows form broad, flat sheets, commonly as much as 40 km across, with irregular outlines, as though the flows had spread out in different directions (Carr et al. 1977a). These differences are almost certainly caused by variations in slope, since lava flows tend to narrow where slopes are steep and to spread out laterally where slopes are low. Flow morphology can also be affected by other factors, such as loss of volatiles and cooling, but at the scale of observation referred to here, these are probably secondary. The long narrow flows commonly include a lava channel with distinct levees, whereas channels are rarely, if ever, seen in the broad tabular flows. On the lunar lava plains

Figure 6.13. Detail of flows in Tharsis, at 2°S, 139°W. Picture width is 17 km. The flows are several tens of meters thick and have surface corrugations presumably caused by crumpling of the solidified near-surface layers as the flow was emplaced. The topography of the older flows is muted adjacent to the later flow front, possibly as a result of the accumulation of windblown debris on the lee side. (731A41)

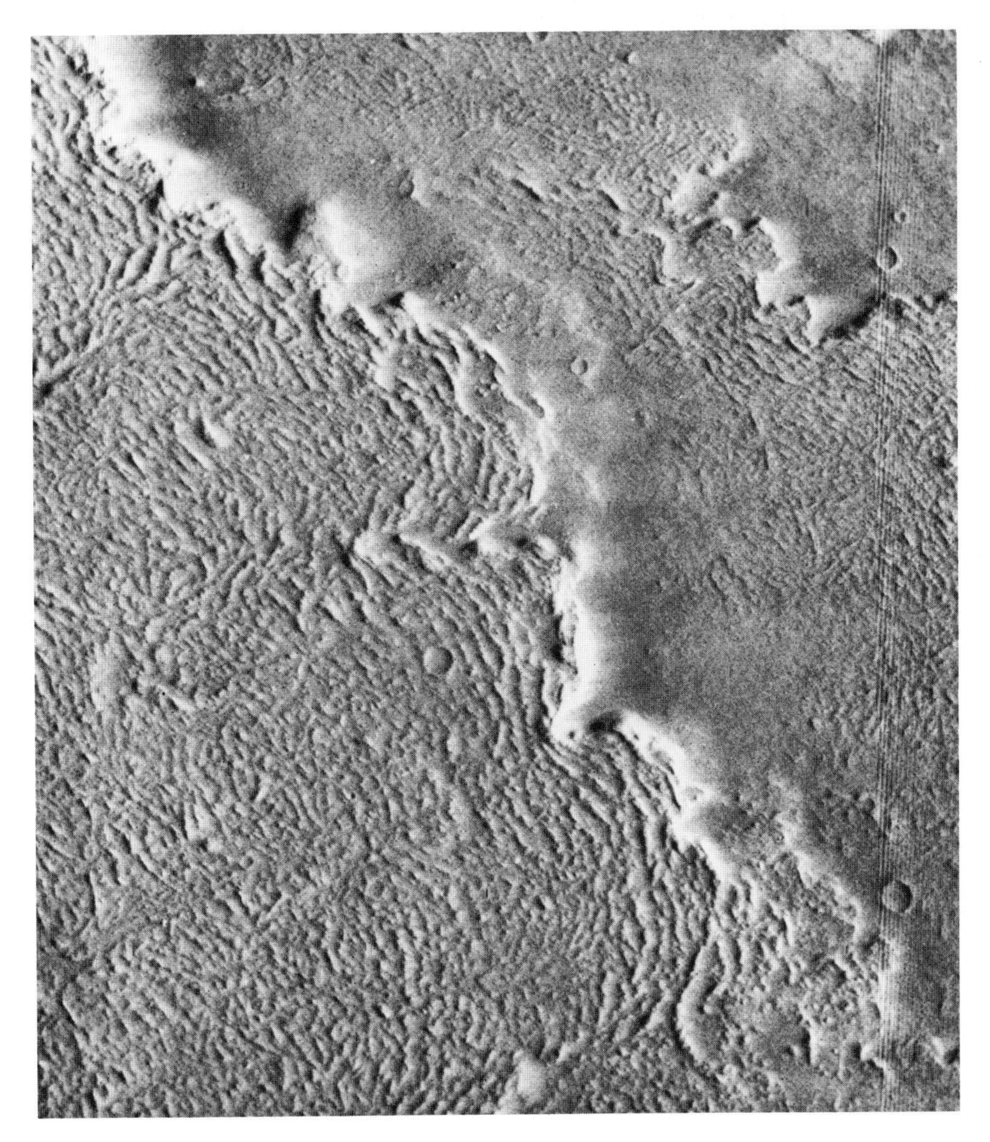

sinuous rilles are common. These are long, winding, steep-walled channels that rarely branch but that maintain their width over distances as long as hundreds of kilometers. They are generally interpreted as the channels along which the lavas that formed the lunar maria were transported. Rilles are rare in the martian lava plains, but there are small ones near the large Tharsis shields and some larger ones to the northeast of Olympus Mons.

The flows that form the lava plains of Tharsis come mostly from unknown sources, but the majority probably erupt through buried fissures that are part of the vast radial fracture system around Tharsis. Exceptions are the flows immediately adjacent to the large Tharsis shield volcanoes, many of which originate in satellitic calderas in the southwest and northeast flanks of the shields and form fan-shaped arrays of flows that diverge away from the volcanoes. Elsewhere, however, particularly around Olympus Mons, flow directions appear to be unrelated to identifiable shield volcanoes (Schaber et al. 1978). Even around the Tharsis shields the more distant flows are unlikely to have traveled the hundreds of kilometers from the shield vents. More probable sources are fractures that are visible occasionally both among the flows, particularly southwest of Arsia Mons, and in older terrain that pokes up through the flows. While most of the fractures are older than the present flow-covered surface, some are penecontemporaneous, in that they cut flows and are in turn partly buried. In Syria Planum and Sinai Planum, south of the canyon, flow directions are mostly to the east and southeast, down the regional slope caused by the Tharsis bulge. Some appear to originate in low, indistinct shields, others in fissures radial to the center of the bulge.

The flows are probably iron-rich and basic to ultrabasic in composition. The material analyzed at both Viking landing sites is iron-rich (see chapter 13), and while it is almost certainly a weathering product, an iron-rich parent rock is likely. Iron-rich rocks are also indicated by the reflection spectra of the planet (McCord et al. 1977; Singer et al. 1979), which have features that have been interpreted as indicating the presence of pyroxenes and olivine or basaltic glass (Huguenin et al. 1978). The younger flows around Tharsis are relatively dark and red as compared with other parts of the equatorial region, perhaps indicative of especially high iron contents (Soderblom et al. 1978). Iron-rich, silica-poor lavas are also suggested by the high density of Mars's mantle (McGetchin and Smyth 1978). The scale of the flows and their morphologies suggest high fluidity, which is consistent with a mafic composition. Although fluidity can be affected by such features as eruption temperatures and volatile contents, as well as by composition, the only flows on Earth and the Moon that are comparable in scale to those of the flow plains on Mars are basaltic. There is thus a convergence of evidence from several sources to suggest iron-rich basalts.

The flow plains show a wide range in ages (table 6.1). Schaber and co-workers (1978) divided the flows of the Tharsis region into 17 units, mapped their distribution, and counted the number of superposed craters. On the oldest units the number of craters larger than 1 km ranges from

TABLE 6.1. AGES OF PLAINS

		Crater Age (billions of years)		
	No. of Craters >1 km/10⁶ km²	*Minimum Likely*	*Best Estimate*	*Maximum Likely*
Lunae Planum	2,400	1.7	3.5	3.8
Chryse Planitia	2,100	1.2	3.0	3.8
Sinai Planum	970	0.4	1.4	3.0
Hellas	2,640	2.9	3.8	3.9
Mare Acidalium	830	0.2	1.2	1.7
Amazonis Planitia	1,940	1.0	2.8	3.7
Noachis	1,740	0.9	2.5	3.6
Hesperia Planum	2,710	3.0	3.9	3.9
Utopia Planitia	1,270	0.6	1.8	2.3
Syrtis Planitia	2,053	1.2	2.9	3.7

NOTE: Crater counts are from Gregory (1979) and Dial (1978). The Mare Acidalium count is new. The Gregory counts were corrected for rollover at the limiting resolution of the photography. Ages are based on the model of Hartmann et al. (1981), adapted as described in chapter 5. Some numbers differ slightly from those in table 5.4 because of the different method of derivation.

2,400 to 3,200 per million km² (figure 6.14). According to the chronology given in the previous chapter, these numbers are equivalent to ages of 3.2–3.8 billion years. The youngest flows are those immediately adjacent to Olympus Mons, which have 90–135 craters larger than 1 km per million km². By the same chronology these range from 130 to 200 million years old. These ages differ from those given by Schaber and colleagues for the reasons given in the previous chapter. Such ages, if valid, imply that eruptions are still occurring, as a result of which lava plains are still being formed; to suppose otherwise would invoke a remarkable coincidence, that after four billion years of accumulation, we happen to be viewing the flow field so shortly after activity ceased. Moreover, the 130-million-year age is an average for a large area and must surely include younger surfaces. Most of the plains in Tharsis are of intermediate age, with crater numbers in the 660 to 1,150 range. The plains of Syria Planum, Sinai Planum, and Solis Planum, to the southeast of Tharsis, are older than the plains that surround the Tharsis shields (Jones 1974; Scott and Carr 1978). This whole region was included under unit 10 of Jones (1974), for which Hartmann and co-workers (1981) gave an age of 3.0 billion years. The plains become progressively older to the east and merge eastward with the older ridged plains. Plains to the southeast and southwest of Alba Patera are of a similar age (Jones 1974). The ages given above are all consistent with the data of Condit (1978), who counted the number of 4–10-km–diameter craters on many of these units.

RIDGED PLAINS. In many areas individual flows cannot be seen, but numerous wrinkle ridges, like those on the lunar maria, are visible (figure 6.15). The ridges are identical in every respect with those on the Moon (Lucchitta 1977). They consist of two components: a broad arch up to 10 km across, which is relatively simple in outline, and superimposed, a narrower ridge,

Figure 6.14. Young plains on old plains, northeast of Olympus Mons. Part of the aureole (*lower left*) and some relatively old, cratered plains (*right*) are flooded by sparsely cratered flows. The cratered plains to the right have more than 2,400 craters greater than 1 km per million km^2, suggesting an age in excess of 3 billion years. The young flows have fewer than 200 craters greater than 1 km per million km^2, suggesting an age of a few hundred million years. Volcanism apparently occurred in the region throughout much of the planet's history. The plains to the right are also clearly older than the aureole. The large crater at upper center is 18 km across. (211-5537)

which has steeper slopes and in general a complex outline, with an echelon of parallel segments. Apart from the ridges and superposed craters the plains have little surface detail at resolutions down to around 50 m. The ridged plains vary significantly in the degree to which they are cratered. Heavily cratered ridged plains occur in a north-south belt, which is about 1,000 km wide around the eastern flank of the Tharsis bulge. The belt extends from about 30°N, 70°W, southward to include all of Lunae Planum, then across Valles Marineris to include the eastern margin of Solis Planum and Sinai Planum. Similarly heavily cratered ridged plains also occur in Syrtis Major Planitia, centered at 10°N, 290°W, and in Hesperia Planum, centered at 20°S, 250°W. As indicated in the previous section, small patches of old ridged plains also occur throughout the old cratered terrain. Less heavily cratered ridged plains occur in Chryse Planitia and south of the Elysium bulge. On most of these plains flow fronts cannot be seen, and wrinkle ridges are almost always present. Isidis Planitia differs from most other plains in having numerous low cratered domes. The domes appear to have formed along the fissures from which the lavas that formed the plains erupted (chapter 7).

Precisely how lunar mare ridges form is not known. They commonly

Figure 6.15. Ridged plains in Sinai Planum, at 23°S, 79°W. Although these plains are probably volcanic, flows are rarely visible. The wrinkle ridges are similar to those on the lunar maria. A sharply defined, crenulated ridge is commonly situated on a broader upwarp. The frame is 250 km across. (608A45)

have preferred orientations both radial and concentric to the large basins and clearly are related in some way to regional deformation. The main controversy is whether the ridges formed simply by postdepositional deformation or whether they formed while the lava was being deposited, as a result of upwelling of lava onto the surface or thin-skin tectonics on a cooling lava lake. Strom (1972) suggested that structures in the lunar highlands controlled the location of fissures from which the mare basalts erupted, and that the ridges form where lava breaks through to the surface over the source regions. Schaber (1973) showed that ridges formed while the mare basalts were accumulating, since some flows are diverted by ridges. Bryan (1973) suggested that the ridges formed by compression of a thin crust during settling of the mare surface in response to the load represented by the lava fill. Howard and Muehlberger (1973) attributed mare ridges in eastern Mare Serenitatis to thrust faulting and pointed out that several ridges can be traced into the adjacent highlands. Lucchitta (1977) proposed a similar origin for ridges in Mare Imbrium. Before the modern era of space exploration Baldwin (1963) suggested that mare ridges formed by deformation of the crust in response to eruption and deposition of the mare lavas, and other workers (Solomon and Head 1979, for example) have recently concurred with this view.

Formation of the ridges on the plains to the east of Tharsis is clearly related to the Tharsis bulge. Phillips and colleagues (Phillips and Ivins 1979; Phillips et al. 1981) modeled the stress distribution within Mars that results from the Tharsis load. They treated the planet as a Maxwellian fluid that deforms in response to the Tharsis load and calculated the surface pattern of stresses on the assumption that the load is supported by an elastic lithosphere several hundred kilometers thick. Their plots show a remarkable orthogonality between the direction of principal stress and the alignment of the wrinkle ridges (see chapter 8). Similar but more qualitative results were obtained by Wise and colleagues (1979) and by Plescia and Saunders (1980). The implication is that the wrinkle ridges are tectonic in origin and were formed not during the updoming of Tharsis but as a consequence of the presence of the Tharsis dome. The question is still open as to whether the deformation postdated the formation of the present surface or was penecontemporaneous. The ridges on the plains of Hesperia Planum, Syrtis Major Planitia, and the patches in the old cratered terrain are even less well understood, but they may be deformational features that formed in response to local stresses that were present when the lavas were being deposited but which have long since disappeared.

The ridged plains of Lunae Planum, Hesperia Planum, and Syrtis Major Planitia are among the oldest plains on the planet (see table 6.1). Hartmann and co-workers (1981) gave ages of 3.2 billion years for Lunae Planum, 3.4 billion years for Hesperia Planum, and 3.6 billion years for Syrtis Major Planitia. The differences between the counts for each of these plains are small, and the plains must be considered to be about the same age. Chryse Planitia and the plains west of Elysium appear to be slightly younger than those just described. Chryse has 2,100 craters larger than 1 km per million km^2 (Dial 1978), and Elysium Planitia has 2,000 (Neukum and Hiller 1981), as compared with 2,500 for Lunae Planum.

AMAZONIS PLAINS. The plains of southern Amazonis Planitia are distinctively different from those in the rest of the equatorial zone. The northern part of the plain immediately to the west of Olympus Mons is relatively featureless except for knobby remnants of the cratered upland. South of around 12°N, however, the character of the surface changes greatly. Much of the surface has a strong north-south to northwest-southeast fabric, which appears to be comprised of broad arches and gentle breaks in slopes and a higher-frequency component of closely packed, curvilinear parallel ridges and grooves. In addition, low escarpments, angular and irregular in plan, outline shallow pits and low mesas, to give the surface an "etched" appearance, which is particularly pronounced around the crater Nicholson, at 0°N, 165°W. Ward (1979) suggested that the parallel ridges are yardangs, formed by the wind. The etched pattern could also be eolian in origin, although why wind scour should be concentrated in this region is unclear.

Also present in this area is a thick deposit of debris that laps onto the old cratered terrain between 5°S and 10°S. The surface of the deposit is mostly smooth (see figure 6.7), giving it a similar appearance to the layered deposits at the poles. Locally, however, the surface is scoured like the adjacent plains, to form arrays of closely spaced grooves approximately 1 km across and up to 50 km long. These again follow curvilinear trends, as though formed by the wind. While most of the deposit's surface is sparsely cratered, in places there are flat-topped mesas, surrounded by low escarpments that are more heavily cratered. The relations suggest that the deposit was at one time more extensive than at present, and that the flat-topped areas are remnants of its former surface. The origin of the smooth deposits is puzzling. Their deep scour, presumably by the wind, suggests that they are unusually friable compared with other deposits in the equatorial zone. Volcanic ash, eolian deposits, or ice-rich materials are possibilities.

High-latitude plains

Between 30°N and 40°N, the appearance of the plains changes greatly, and the twofold classification of the equatorial plains into flow plains and ridged plains no longer applies. In the south the transition is not observed, but the plains of Argyre and Hellas have some of the characteristics of the high northern plains. The high-latitude plains are poorly understood, and there is almost no literature on them. They have a greater areal variation than the equatorial plains and in general a greater complexity. Primary depositional volcanic features, which abound in the equatorial regions, are relatively rare; erosional rather than depositional features usually dominate.

Soderblom and co-workers (1973) postulated that the high northern and southern latitudes were covered with a blanket of debris several tens of meters to kilometers thick. They suggested that the equatorial regions, particularly the dark areas, are largely swept free of debris, and that the "swept" areas move north and south on a 50,000-year cycle, as precessional effects cause alternation of the climatic regimes of the two hemispheres. The boundary between the swept equatorial and unswept high-latitude regions should therefore have experienced repeated stripping and accumulation of debris. The evidence for the debris blankets is largely the distribution of 4–10-km–diameter craters, which appear to fall off in numbers at high latitudes. Soderblom and colleagues also cited a general loss of topographic detail as additional evidence for the blanketing, but this now appears to be an observational artifact. Most of the plains between 30° and 60° have abundant surface detail, in fact, far more than the equatorial plains, at least down to resolutions of 100 m to 200 m. Despite the lack of overall blanketing, the repetitive deposition and stripping model may still be valid, with patterned ground, pedestal craters, and complex disordered textures being among the results.

Ice may have played a more important role in sculpting these surfaces than those closer to the equator. Several features, such as crater ejecta set into the surface, moats around low hills, and arrays of closed depressions, suggest ground-ice. Ice may have been a significant component of the postulated debris blankets, and their deposition and removal may have been modulated by differential sublimation and cementation. For example, the ice may have condensed around dust grains in a manner similar to that proposed by Jones and colleagues (1979) for the ice observed at the Viking 2 landing site during winter. More speculatively, the large outflow channels, most of which drain northward and disappear in the high northern plains, could have been a source of large amounts of ice, which might have been retained near the surface. If ice is a significant component, then deposition and removal of the mantles may be sensitive to climate changes, not only insofar as they affect wind regimes, but also in that such changes could affect the stability of ice near the surface. The occurrence of patterned terrains close to 40°N might then be related to the fact that ground-ice is stable with respect to the present atmosphere north of this latitude but is unstable to the south (Farmer and Doms 1979). Three types of northern plains and a polygonal fracture pattern that is quite common in the 30–50°N latitude belt are described below. Brief descriptions are also given of the plains on the floors of Hellas and Argyre, which in some respects resemble those at equivalent latitudes in the north.

MOTTLED PLAINS. Extensive areas of the far northern plains, especially north of 50°N but extending as far south as 30°N in Mare Acidalium, have a mottled appearance (figure 6.16). The mottling is caused by a sharp contrast in albedo between the ejecta around craters and the much darker surrounding plains. The mottling is best seen in Mare Acidalium, between 30°N and 50°N. Most craters in the mottled plains are of the composite type, in which an inner, relatively thick circular flap of ejecta overlies

Figure 6.16. Mottled plains in Mare Acidalium, at 48°N, 10°W. The light-colored ejecta contrasts sharply with the dark uncratered plains. Most of the craters have an inner circular flap and an outer, radially textured zone, as is typical at these high latitudes. See also figure 4.17. The scene is 245 km across (673B30)

a much thinner sheet of ejecta with a strong radial pattern and jagged outline. The high albedo extends to the farther extremity of the ejecta, commonly as far as six crater radii from the crater rim. Remnants of old cratered terrain, either as equidimensional hills or as flat-topped fractured plateaus, also have a high albedo. The cause of the contrast in albedo between the ejecta blankets and the intervening plains is uncertain. It may be simply a result of the ejecta excavated from below the surface having an intrinsically high albedo, but another possibility is that the ejecta around craters act as traps for fine-grained, wind-blown debris and so acquire a high albedo. The mottled plains between 40°N and 50°N and 10°W and 30°W are broken into polygonal patterns by intersecting fractures. The polygonal fractures also affect other units and are discussed separately below.

PATTERNED PLAINS. Within 500 km of the northern boundary of the old cratered terrain between 280°W and 0°W are some distinctively patterned plains. To the north they merge into the mottled plains and to the south into the fretted terrain along the plains/upland boundary. In most areas these plains can be divided into two components, a relatively smooth unit with gently rolling topography and a stippled or dimpled unit with numerous closely spaced mounds and depressions, which are generally around a kilometer across or less (figure 6.17). The mounds and pits may be arranged in closely spaced lines, to give the ground a striped appearance or thumbprint texture (Guest et al. 1977) that resembles patterns caused by contour ploughing (figure 6.18). More rarely stripes are caused by linear ridges several kilometers long and approximately a kilometer across (figure 6.19). A low escarpment may occur at the junction between the stippled and smooth units, such that the smooth unit forms the higher ground. In other areas the junction is marked by a rounded ridge. Remnants of old cratered terrain occur throughout the patterned plains as steep, knobby hills or flat-topped mesas. In the smooth unit vague circular outlines of craters are common, suggesting that the smooth unit is either a highly eroded remnant of the old terrain, or that the old terrain is in places buried at a relatively shallow depth.

Most craters in these plains are of the pedestal type, in which the extreme outer edge of the ejecta is marked by an outward-facing escarpment. Numerous cratered domes are also present. As we saw in chapter 4 and will see again in chapter 7, such domes can form either by modification of impact craters by the same processes that produce pedestal craters or

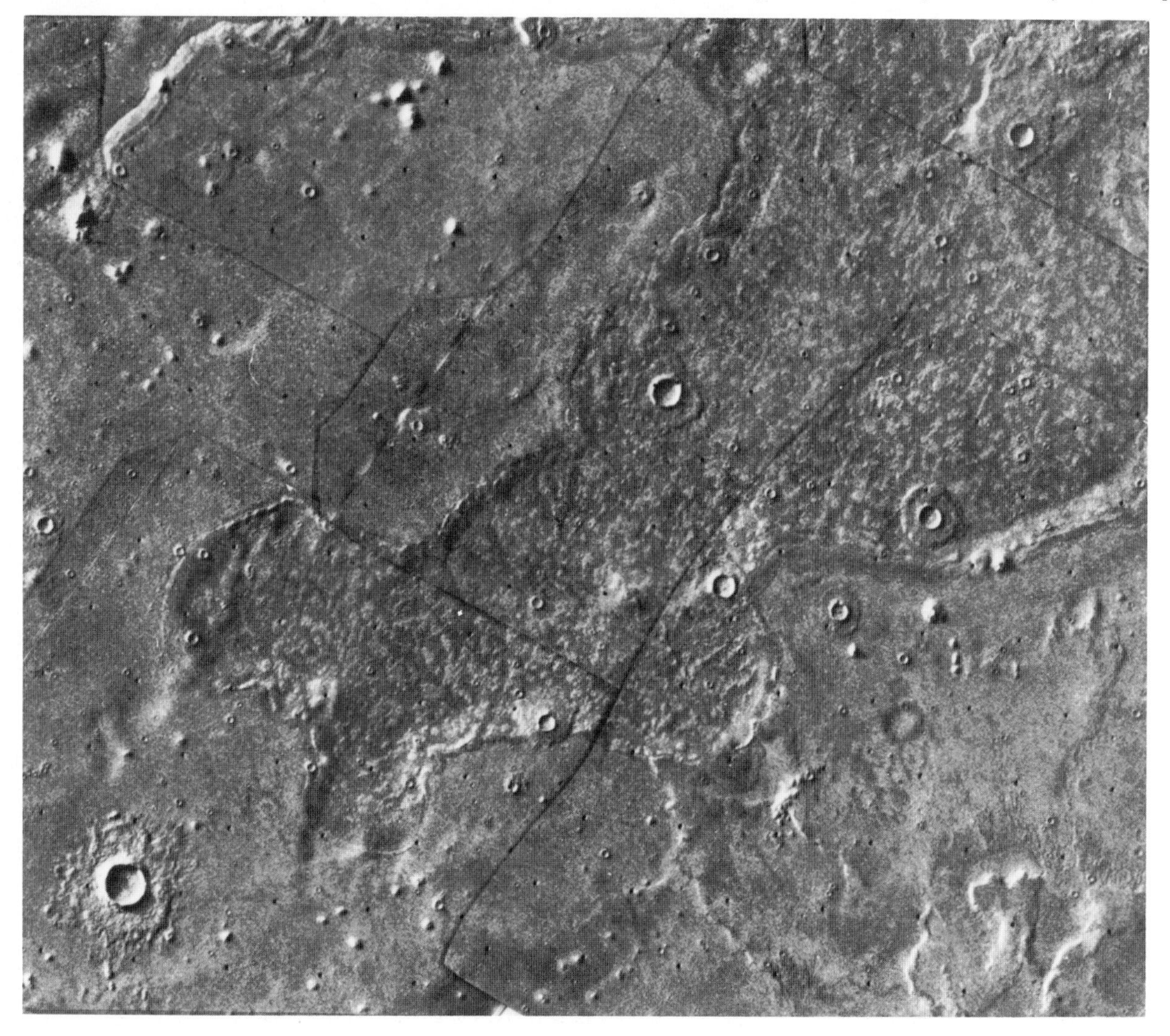

Figure 6.17. Patterned plains at 46°N, 351°W. The terrain consists of two components, a relatively smooth unit and, at a slightly lower elevation, a stippled unit. Also present are hilly remnants of the old cratered plateau, which occurs extensively to the south. The origin of the terrain is not known, but it may result from repeated deposition and removal of debris blankets. The scene is 130 km across. (211-5066)

Figure 6.18. Patterned ground at 45°N, 354°W, close to the patterned plains in the previous figure. Here the stipples are arranged in a regular pattern of parallel lines. The picture is 56 km across. (52A35)

by volcanism. Most of the small domes in this area are believed to be produced by impact because of the continuity between the small ambiguous domical features and the larger, unambiguous impact features and because of the general absence of small craters other than those within the domes. However, some volcanic domes may also be present.

The patterned plains occur close to one of the predesignated landing sites for the second Viking lander and were the subject of considerable debate during the Viking mission, as attempts were made to understand the plains and to assess their potential threat to the safety of landing (Masursky and Crabill 1976). Most of the features were interpreted on the basis of the Soderblom and colleagues (1973) model of repeated episodes of deposition and removal of debris blankets or mantles (Guest et al. 1977). The smooth areas were interpreted as remnants of a former, more extensive cover of debris. The stippled regions were interpreted as areas where the surficial debris layer had been removed. Carr and Schaber (1977) suggested that the striped patterns were somewhat analogous to recessional moraines left by a retreating glacier, and that they marked successive positions of the edge of the debris blankets as they were removed. Indeed, if volatiles are a significant component of the debris blankets, then the analogy is a close one. Some support for this interpretation is the rough parallelism between the

Figure 6.19. Linear features at 50°N, 287°W. The pattern consists mostly of low ridges, but some troughs are also present. The troughs and ridges appear to be relatively old, since most craters are superimposed. Again the origin of the pattern is not known, although the ridges may be analogous to recessional moraines and mark the positions of the edges of debris blankets that formerly covered the region. The frame is 100 km across. (11B03)

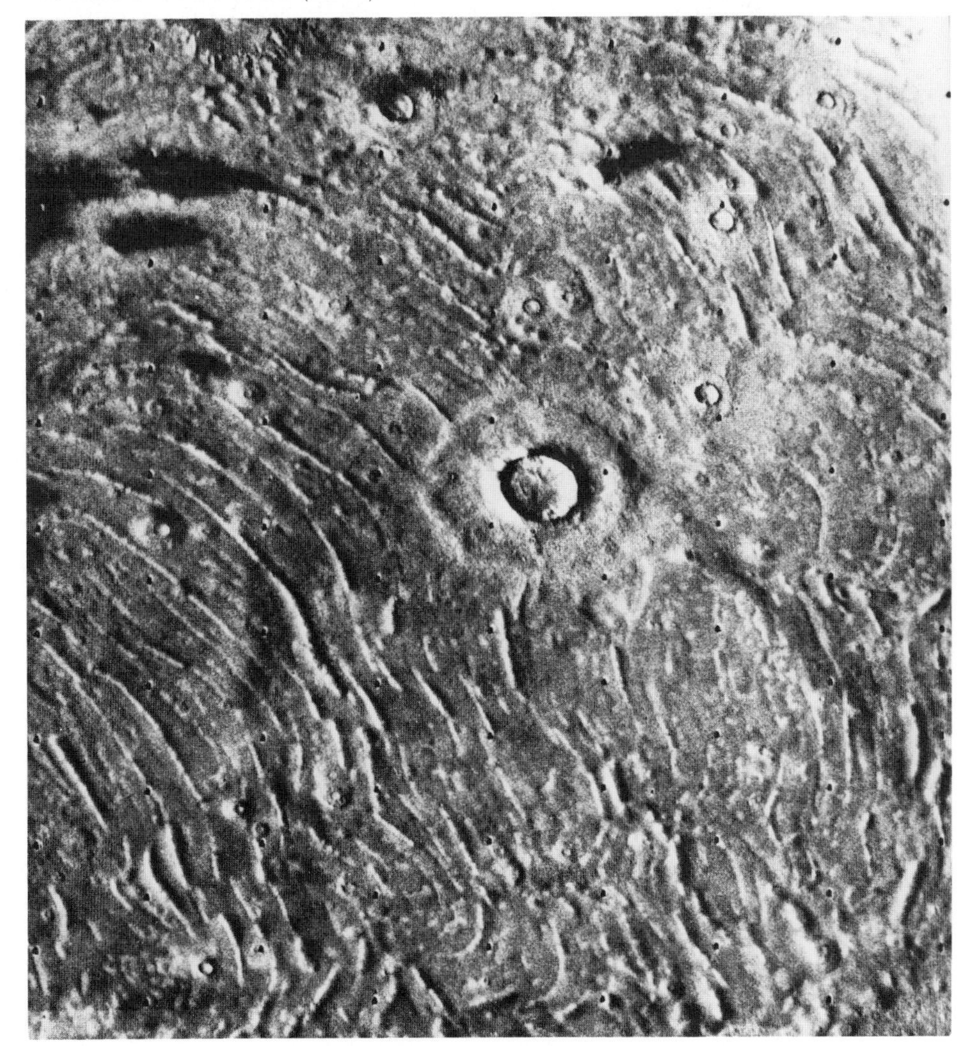

stripes and the present boundary between smooth and striped ground that is seen in some areas.

DISORDERED PLAINS. To the west of the Viking 2 landing site, between latitudes 40°N and 50°N and longitudes 240°W and 290°W, the plains are complex and confusing. At a scale of 1 km or larger they have little obvious relief other than the craters, but at scales below 1 km the surface is highly textured (figure 6.20). Irregular arrays of numerous closely spaced, roughly equidimensional hills and hollows a few hundred meters across are separated by smooth areas, and locally there are swarms of closely spaced, low linear ridges less than 1 km across (figure 6.21). The ejecta around craters is commonly smooth and contrasts sharply with the highly textured intercrater areas. A particularly puzzling phenomenon is that in some places the ejecta appears to be surrounded by an inward-facing escarpment, as though it were set into the surface.

The disordered plains are at the termination of some large channels that originate on the flanks of the Elysium dome, over 1,000 km to the southeast. South of the Viking 2 landing site, at 230°W, 42°N, Hrad Vallis (see

Figure 6.20. Complex terrain at 41°N, 229°W, just south of the Viking 2 landing site. The channels at the bottom of the frame originate close to Elysium Mons, several hundred kilometers to the southeast. The region has clearly undergone a complicated history of erosion and deposition, which may be connected in some way to the area being at the termination of several large channels. The frame is 120 km across. (9B50)

figure 6.20) forms a clearly defined channel, which can be traced southeast to Elysium. Most of the channels in the disordered area, however, are indistinct and are recognizable only as narrow meandering linear features with low albedo and would barely be recognizable as channels if they did not merge to the southeast with channels that are well-defined. Some of the arrays of narrow ridges simulate distributary patterns (see figure 6.21), and similar ridges in northern Amazonis also appear to be at the termination of a large channel. Where the channels die out to the west, the character of the plains changes, so that the peculiar characteristics of the

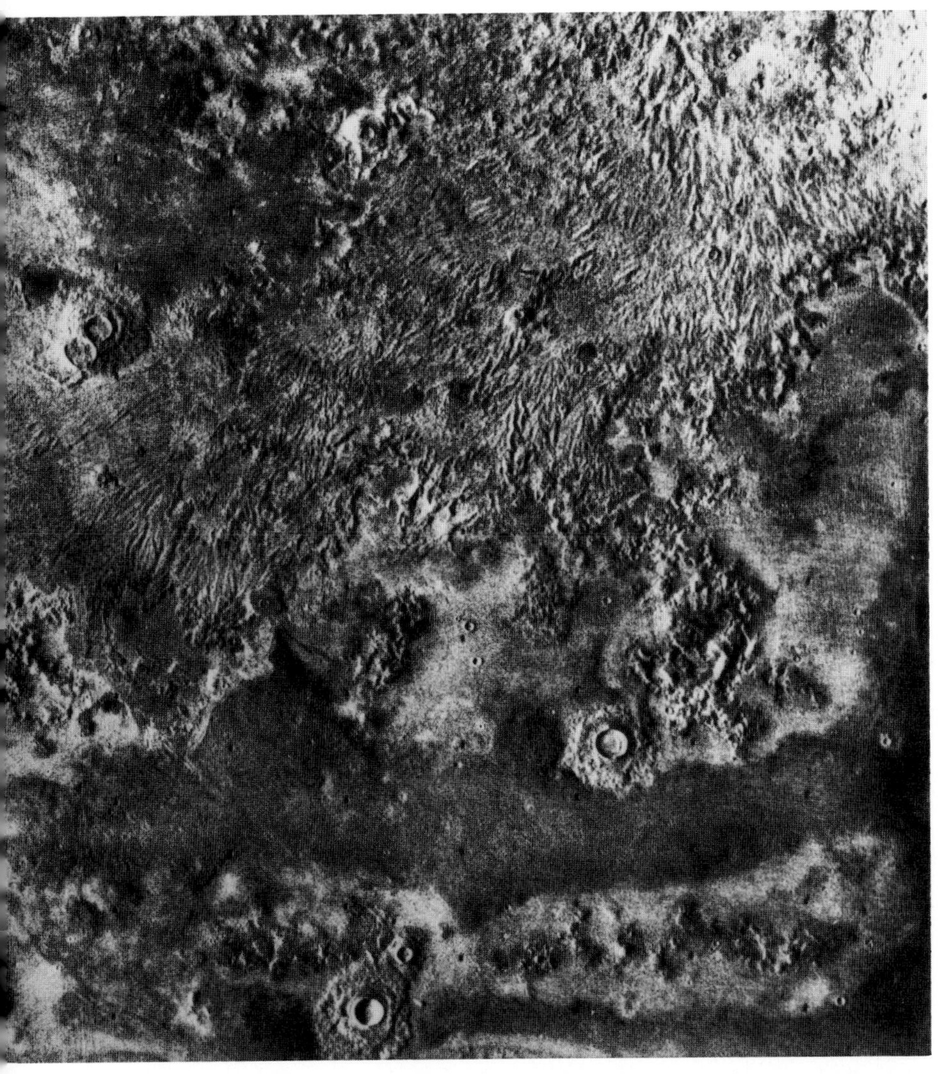

Figure 6.21. Possible distributaries at 40°N, 247°W, 110 km to the west of the complex terrain in the previous figure. Low branching ridges and valleys suggest distributaries, although whether caused by water or lava is unclear. Many of the smooth dark areas, such as that in the lower half of the frame, are linear and continuous, with channels like those in the previous figure. The frame is 110 km across. (10B03)

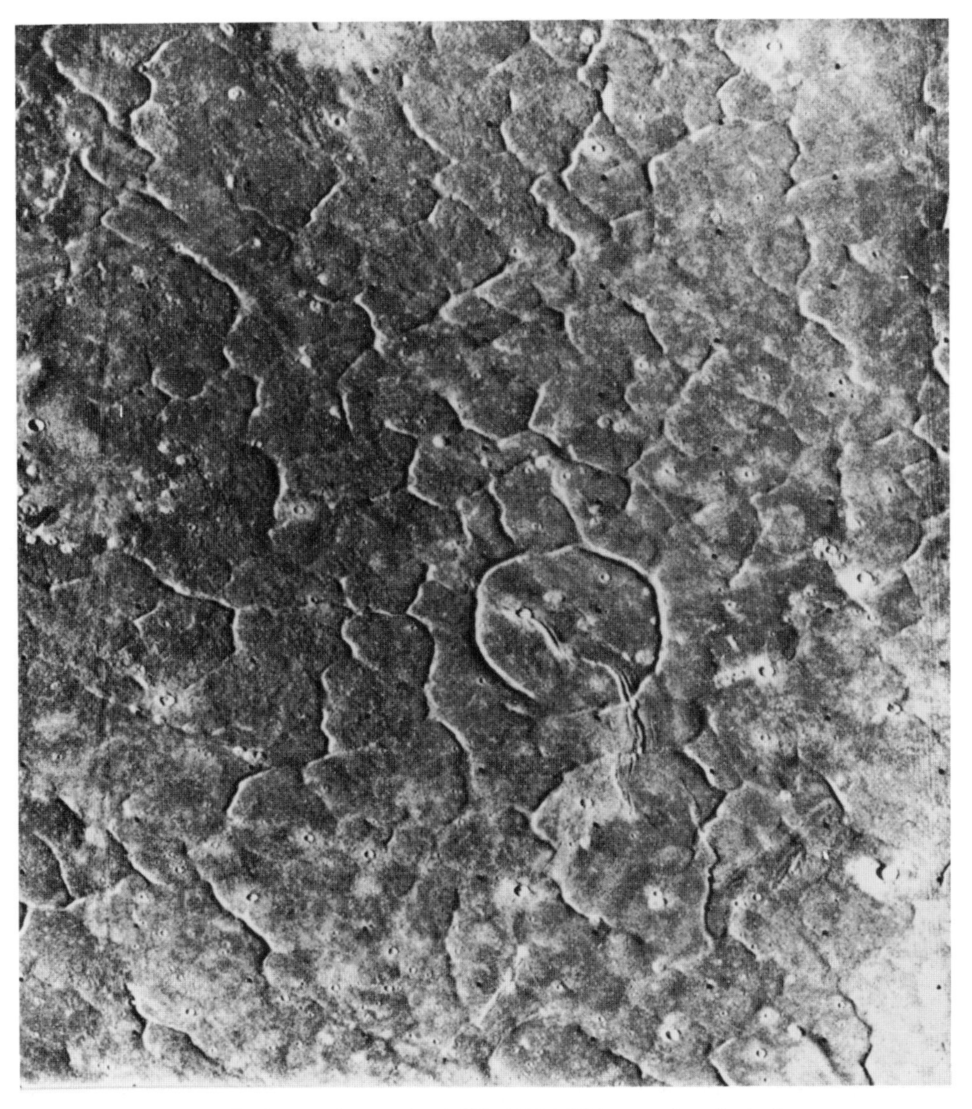

Figure 6.22. Polygonally fractured ground at 44°N, 18°W. This scene is typical of large areas of the northern plains. The pattern resembles that caused on Earth by ice-wedging or by cooling of lava, but is about one hundred times larger in scale. The frame is 53 km across. (32A18)

area appear to be connected in some way with the channels. Whether the Elysium channels were formed by water erosion or through the action of lava is unclear (chapter 10), but if by water, then modification of ice-rich fluvial deposits could be the cause of the strange surface morphology.

POLYGONAL FRACTURES. In extensive areas of the northern plains, particularly in Mare Acidalium and Utopia Planitia, between longitudes 0°W and 30°W and latitudes 40°N and 50°N, the surface is cut by cracks, which partly intersect to form a roughly polygonal pattern (figure 6.22). Such fracturing is rare at low latitudes, although Isidis Planitia is extensively

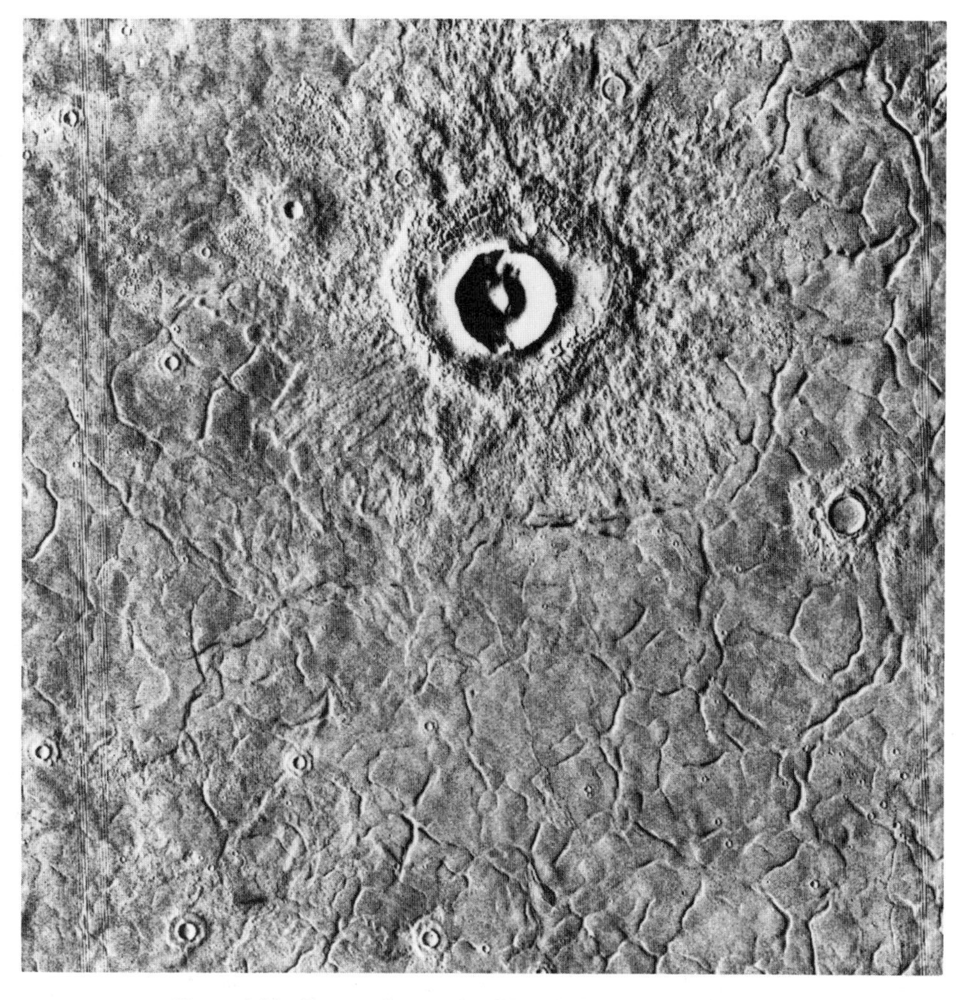

Figure 6.23. Fractured ground at 38°N, 258°W, north of Isidis Planitia. Here the pattern of fractures is less polygonal than in the previous figure, and many of the cracks are clearly graben with flat floors. The picture width is 176 km. (573A08)

fractured between 20°N and 40°N and 240°W and 260°W (figure 6.23). The cracks commonly divide the surface into a mosaic of blocks, ranging in size from a few to about ten kilometers across. The blocks are generally angular, but occasionally the cracks outline a circle, which is probably a reflection of a crater in the surface beneath the plains. Intersection angles of the cracks are generally in the 90° to 120° range. Most of the cracks have steep walls, which are about 300–400 m apart at the surface. The walls close downward to a V-shaped termination at depth, except at intersections or where the walls are further apart at the surface, in which case the troughs are flat-floored. In rare cases a linear feature on the surface is visible within the troughs, as though downfaulted. Toward the boundary of the plains with the old cratered terrain the pattern of cracks becomes coarser. Here the walls are generally over 500 m apart, and almost all the troughs have flat floors.

The origin of the fractured plains is not known. The observation that linear fractures can occasionally be traced from the surrounding plains across the flat floors of a trough strongly suggests that the troughs are graben and were formed by faulting (Pechmann 1980). The pattern of cracks indicates uniform tension in all directions within the plane of the surface. Similar patterns commonly form on Earth where horizontal tensional stresses develop because of cooling or dessication. Examples are the cooling of horizontal sheets of lava and the cooling of permanently frozen ground in midwinter to form patterned ground in arctic regions (Lachenbruch 1962). In the latter case the polygonal cracks generally fill with water in spring, and the process of freezing followed by contraction repeats itself the next winter. Polygonal cracks also form in water-rich sediments as a consequence of contraction following dessication. In all these cases, however, the individual polygons are at least two orders of magnitude smaller

Figure 6.24. Channeled plains at 50°S, 307°W, on the south rim of Hellas. The channels appear to be volcanic and to originate at the low-profile volcano Amphitrites Patera (see figure 7.26), to the south of the region shown. The frame is 70 km across. (578B30)

Figure 6.25. Etched terrain within Argyre, at 52°S, 46°W. Horizontal layers appear to have been partly removed, to leave irregular escarpments around the remnants. Such patterns are common at high latitudes (see figures 7.26 and 11.10). The elongate pit at the center is 6 km across. (568B49)

than the polygons observed on Mars. The scaling problem is so severe that formation of the martian patterns simply by contraction is unlikely. A possible alternative is that the cracks are the result of some form of tensional tectonics caused by large-scale warping of the surface (Pechmann 1980).

HIGH-LATITUDE SOUTHERN PLAINS. The only plains of any significant extent in the high southern latitudes are in, and to the south of, Hellas and within Argyre. The plains of Hellas are rarely seen without frost, clouds, or haze and so are poorly photographed. At the center of the basin is an ill-defined ring about 850 km in diameter. Within the ring the surface has many of the characteristics of the etch-pitted terrain near the south pole (see figure 12.8), which suggests that partly eroded layered deposits are at the surface. Outside the ring the terrain is relatively smooth, although wrinkle ridges are present in places. Along the southern margin of the basin numerous ridges and valleys (figure 6.24) and some channels extend down into the basin from several volcanic centers, such as Amphitrites Patera, that occur just to the south of the southern rim. Several wide, flat-floored channels that extend from the northeast rim down into the basin also appear to start near volcanoes.

Parts of the floor of Argyre also have an etch-pitted appearance (figure 6.25), but probably the most puzzling features of the Argyre plains are some long, interconnected ridges. These occur close to, and run almost parallel to, the southern margin of the basin and maintain a uniform width of 1–2 km for hundreds of kilometers (figure 6.26). Such ridges are rare

Figure 6.26. Linear ridges on the floor of Argyre, at 55°S, 41°W. The ridges form an open network around the southern edge of the basin. The origin of the ridges is not known, but they may be dikes left standing after the removal of intervening material. The frame is 67 km across. (567B33)

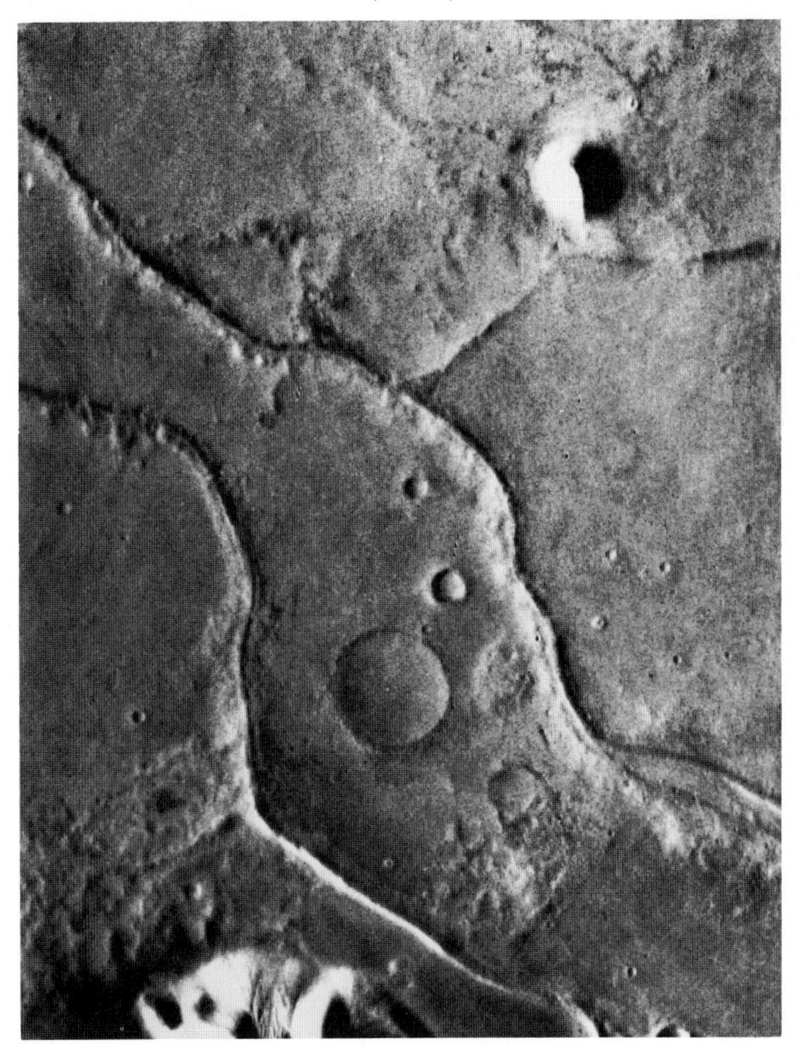

elsewhere on the planet; the only other known occurrence is at 3°S, 207°W. They may be simply volcanic intrusions or dikes that stand out as ridges after erosion of the intervening plains. A less likely alternative is that they are eskers—ridges of debris from a subglacial river, which are left standing after a former ice cover has been removed.

In summary the surface of Mars can be divided into two major components, densely cratered uplands and sparsely cratered plains. The densely cratered terrain, which probably dates from the decline in impact rates 3.9 billion years ago, is close to saturation for craters larger than about 30 km, and most such craters are highly degraded. For craters with smaller diameters the surface is very undersaturated. The almost ubiquitous small channels and evidence of enhanced crater obliteration at the time the larger craters formed suggest an early atmosphere that was thicker than the present atmosphere. The thicker atmosphere may have contributed to dispersion of ejecta from large impact craters and basins and to deposition of the debris over much of the planet's surface. Around 3.9 billion years ago the landscape stabilized, and surfaces which formed after that time are only lightly cratered. The ages of the sparsely cratered plains span the entire history of the planet, from soon after the decline until the relatively recent geologic past. Plains within 30–40° of the equator appear to be mostly lava plains that have undergone little modification since their formation; most have clearly defined flow fronts or wrinkle ridges that resemble those on the lunar maria. Major exceptions are the plains in southern Amazonis, which appear to be extensively modified. Plains poleward of the 30–40° latitude band are complex, and primary depositional features are rare. Their origins are poorly understood, but they are probably formed by an interplay of a variety of processes—volcanic, fluvial, eolian, and glacial.

7 VOLCANOES

Volcanoes are distributed unevenly across the surface of Mars. The largest are in three broad provinces, Tharsis, Elysium, and Hellas, although several lesser provinces in which small volcanic features may be concentrated have been tentatively identified. The largest and youngest volcanoes are in the Tharsis region, toward the summit and around the northwest edge of the Tharsis bulge. Curiously no large volcanoes occur on the southeastern flank of the bulge, although there are extensive lava plains. The Elysium province resembles Tharsis in that the volcanoes are associated with a bulge in the crust. The Elysium bulge, however, is smaller in height and areal extent and has correspondingly fewer volcanoes. The oldest recognizable volcanoes are around Hellas. They include several on the southern rim, Hadriaca Patera on the northern rim, and Tyrrhena Patera, 800 km to the northeast of Hadriaca. The only prominent large volcano outside these three regions is Apollinaris Patera, at 10°S, 185°W.

Most of the large volcanoes resemble terrestrial shield volcanoes, which form largely by eruptions of relatively fluid, basaltic lava. The most familiar examples on Earth are those in Hawaii. The resemblance between the Hawaiian and martian shields is striking. Not only do the volcanoes as a whole resemble one another, but so do the subsidiary features, such as calderas, lava channels, flow fronts, and collapse pits, although, somewhat puzzlingly, the martian features are mostly on a far larger scale. Other martian volcanoes, termed *paterae* (meaning a shallow dish or saucer), are of broad areal extent but have little vertical relief. They have no known terrestrial counterparts. The most common type of volcano on Earth, the stratocone, which typically forms when activity is more explosive than that in Hawaii and when large amounts of ash as well as lava are produced, is much less obvious on Mars. The apparent scarcity may be simply a resolution effect; stratocones are much smaller than shield volcanoes and are correspondingly more difficult to recognize. More likely, however, differences in the tectonic frameworks of the two planets result in differences in the styles of volcanic activity: on Earth, most volcanism occurs along plate boundaries; on Mars, there appears to be no plate motion, so that the distribution and type of volcanic activity are quite different.

THARSIS PROVINCE

The designation Tharsis province is used here in a broad sense to include Olympus Mons and Alba Patera, which lie outside Tharsis proper. The province includes at least twelve volcanoes and probably more. Three of the largest, Arsia Mons, Pavonis Mons, and Ascraeus Mons, are spaced approximately 700 km apart along a northeast-southwest–trending line (figure 7.1). Large fractures occur along the continuation of the line to the northeast and southwest, outside Tharsis. The volcanoes thus appear to be astride a major fracture zone, now buried by volcanic deposits. The tallest volcano on the planet, the 27-km–high Olympus Mons, is 1,600 km northwest of the Tharsis line; the largest volcano in terms of areal extent, the 1,600-km–diameter Alba Patera, is at 40°N, 110°W, 1,800 km north of Ascraeus Mons. Three small volcanoes, Biblis Patera, Ulysses Patera, and Jovis Tholus, lie between the Tharsis line and Olympus Mons, and three others, Uranius Tholus, Ceraunius Tholus, and Uranius Patera, in a cluster 100 km northeast of Ascraeus Mons, are close to the line. Only one volcano, Tharsis Tholus, is to the southeast of the line. The crest of the Tharsis bulge is close to the west end of Labyrinthus Noctis, although there is some uncertainty as to its precise location. The line of the Tharsis volcanoes is thus slightly northwest of the center of the bulge; Olympus Mons and Alba Patera are at its northwest edge and can barely be considered to be on it. Although the volcanoes differ from one another greatly, most have characteristics indicating that they formed from fluid lava, more fluid even than that which built the volcanoes in Hawaii.

The three aligned Tharsis shields

Each Tharsis shield is 350–400 km in diameter, with a summit caldera 24–27 km above the Mars datum and 17 km above the surrounding plains. The flanks slope away from the summits at angles generally less than 5° (Blasius 1979) and have a fine radial surface texture formed by lava channels and long narrow flows. The flows are generally less than 3 km across and may be traceable for several tens of kilometers. Embayments or alcoves occur in each of the shields on their northeast and southwest flanks. The embayments are irregular in shape, being made up of numerous coalescing pits, fractures, spatulate depressions, and rillelike features. In the case of Arsia Mons the embayments extend almost to the summit (figure 7.2), but they are considerably smaller on the other two shields. Each embayment appears to have been the source of vast quantities of lava, which flowed over the adjacent plains, burying the lower flanks of the shields and forming fan-shaped arrays of flows. Radar profiles across the flows that originate

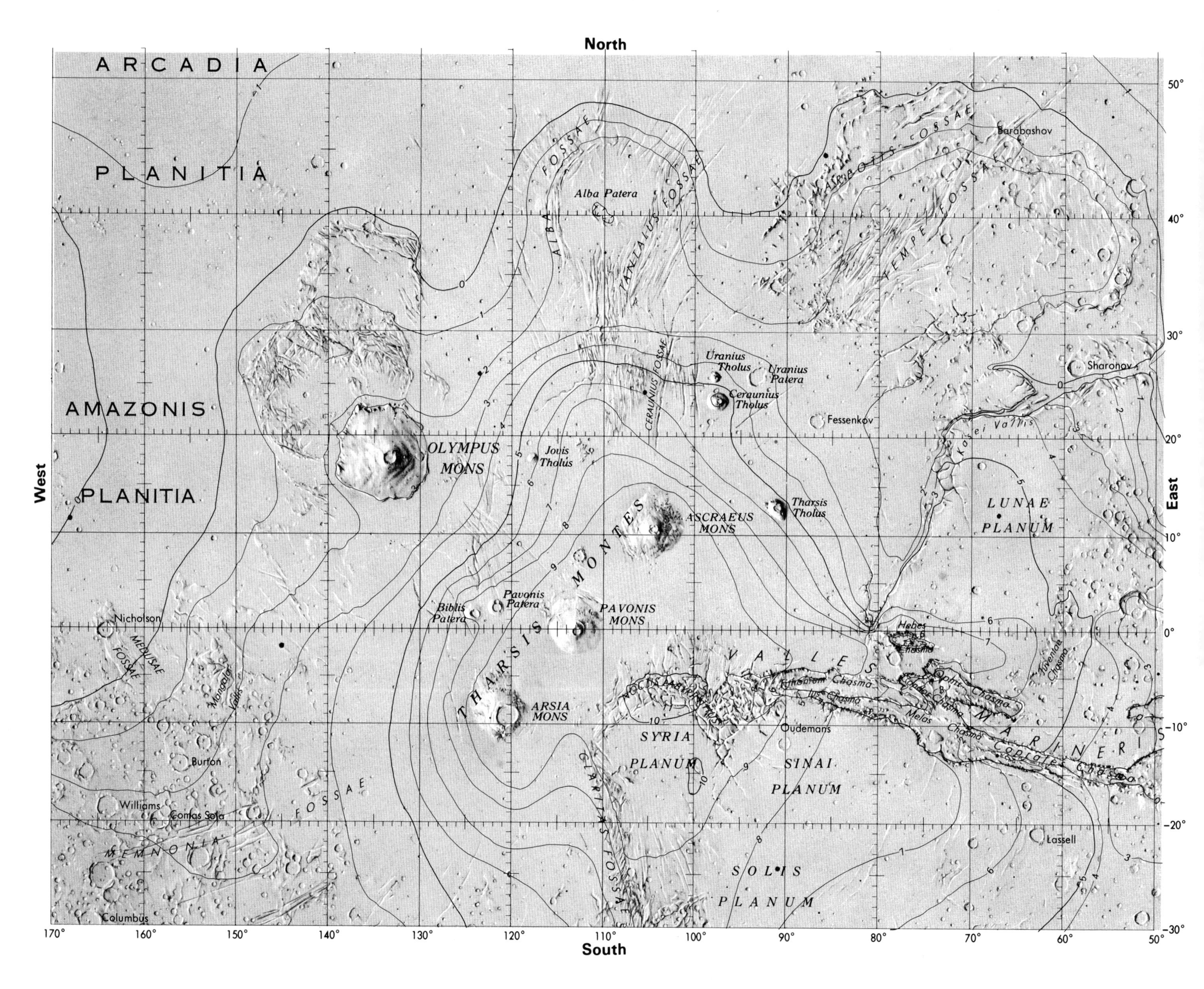

Figure 7.1. Index map of the Tharsis region. Contours are elevations in kilometers above the Mars datum. For scale, 10° longitude at the equator is 590 km. (Courtesy U.S. Geological Survey.)

Figure 7.2. Summit of Arsia Mons with its 120-km–diameter caldera. A line of low mounds on the caldera floor connects the embayments in the northeast and southwest flanks of the volcano. The mounds are probably low satellitic shields along a rift zone. The lower flanks of the volcano appear bright, partly because of increased scattering of light by dust in the atmosphere at the lower elevations.

in the southwest embayment of Arsia Mons show that they form a large spur that extends southward from the volcano (Roth et al. 1980). The spur is several kilometers high close in and decreases in height rapidly to the south. The other volcanoes may have similar extensions, but probably of lower elevations, since the embayments do not extend as far up onto the main edifices as in the case of Arsia Mons.

Arcuate, concentric fractures cut the flanks of all three Tharsis shields (figure 7.3), although they are less obvious on Ascraeus Mons. Commonly the fractures merge into arcuate lines of pits, or pits may occur within graben formed by the fractures. Dislocation appears to have continued as the volcano grew. Although most visible faults are younger than the present surface, some are partly buried, and a faint concentric fabric on the shield suggests older, completely buried fractures. Where the fractures intersect the embayments, they become enlarged. On Pavonis and Ascraeus Mons they are the source of several rilles that extend down the flanks of the volcano, indicating that the fractures have been sites of eruptions. Arcuate fractures that cut the plains north of Ascraeus Mons may result from flexing of the surface under the load created by the volcano (see chapter 8). In addition to the obvious fractures the shields have faint arcuate terraces with convex-upward surfaces. These are most evident on Ascraeus Mons, where they are 20–30 km across. The terraces may have formed by preferential buildup of lava immediately downslope from an arcuate rift zone. Alternatively the terraces may be analogous to the Hawaiian "Pali"—steep outward-facing escarpments that result from outward slumping of the volcano flanks.

The caldera at the summit is somewhat different for each volcano. Arsia Mons has a 110-km–diameter, relatively shallow depression bounded by arcuate faults (figure 7.4). Several flows start at faults in the caldera walls and flow down onto the caldera floor. The floor is sparsely cratered, with only 150 craters larger than 1 km in diameter per million km^2, as compared with 800 on the flanks (Crumpler and Aubele 1978; Plescia and Saunders 1979). In addition to the impact craters there are some elongate rimmed depressions a few hundred meters across, which resemble vents with spatter ramparts—accumulations of clots of lava adjacent to vents—and a northeast-southwest–trending line of low mounds, which connects the embayments on the opposite sides of the shield (see figure 7.2). Pavonis Mons is capped by a simple circular depression 4–5 km deep. The walls have numerous narrow arcuate terraces, as though the caldera had collapsed episodically along closely spaced circular faults. Offset from the caldera to the northwest is a shallow depression outlined by a circular wrinkle ridge. Ascraeus Mons has a multiple caldera, which clearly formed by episodic collapse at different points (figure 7.5), and the caldera walls truncate flows on the volcano flanks, as at the summit of Mauna Loa, Hawaii (figure 7.6). Shadow measurements show that the floor of the main caldera is 3.6 km below the floor of adjacent caldera, so that the main floor is probably close to 4 km below the volcano summit. According to Crumpler and Aubele (1978) the caldera floors of Pavonis and Ascraeus Mons have similar crater counts on the flanks, 300–400 craters larger than 1 km per million km^2. Plescia and Saunders (1979) gave the lower number of 146 for the Ascraeus shield, but the differences may not be significant. Crater counts on volcanoes are subject to sizeable errors because of the difficulty of distinguishing between volcanic and impact craters.

One of the more puzzling features of Arsia Mons is to the west-northwest of the main edifice, where a lobe-shaped feature extends down the regional slope to approximately 350 km from the base of the volcano

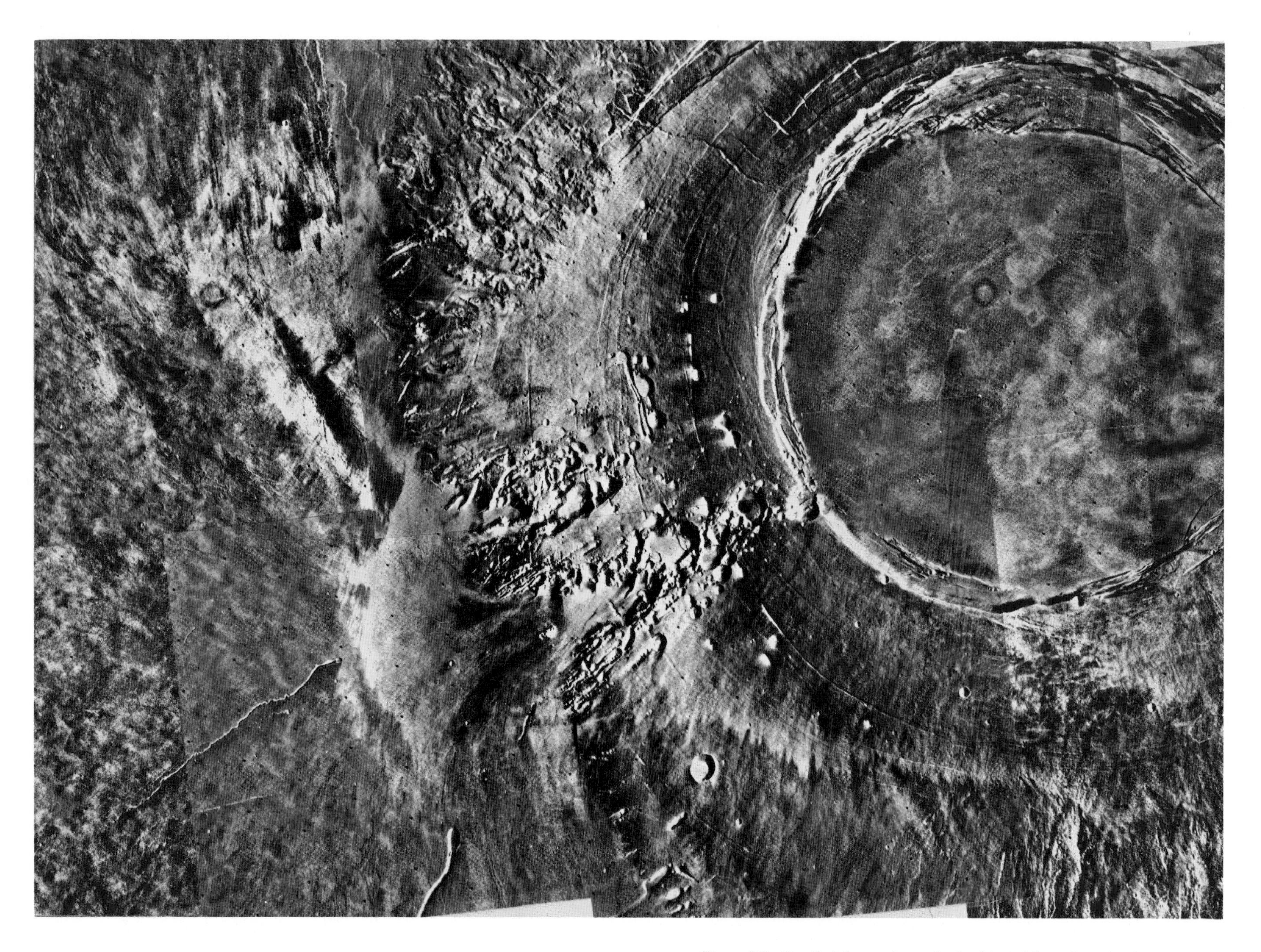

Figure 7.3. Detail of the southwest flank of Arsia Mons. Irregular depressions on the flanks have coalesced to form an embayment in the shield. The embayments have been the source of numerous flows, which fan out over the adjacent plains and cover the base of the shield. Two well-defined channels on the plains resemble lunar sinuous rilles. The shield flanks are cut by concentric fractures, and a faint radial texture is caused by flows. (211-5170)

Figure 7.4. (*above*) Detail on the caldera walls of Arsia Mons. In the upper right, flows from an eruption site on the caldera wall run down into the caldera floor, in the lower right. In the upper left, flows extend down the shield flanks away from the caldera. The frame is 46 km across. (422A31)

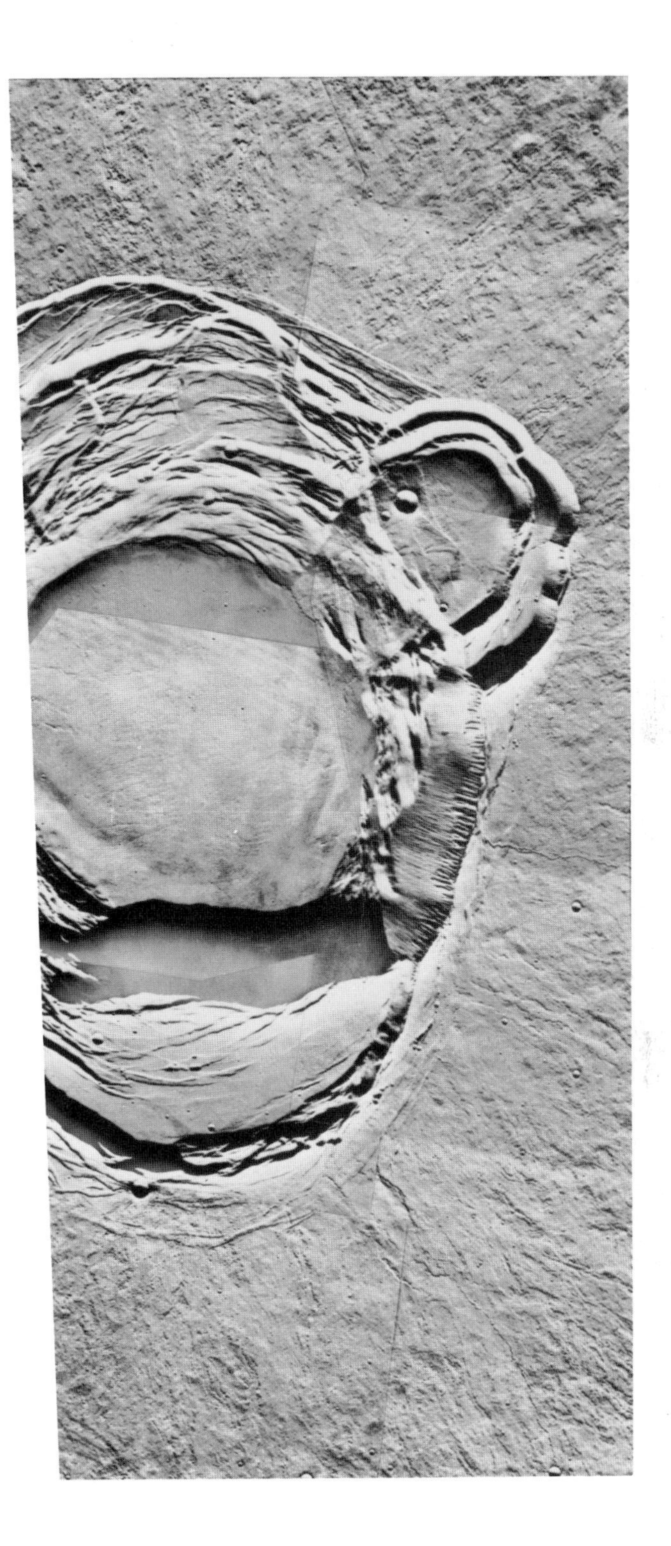

Figure 7.5. (*right*) Summit caldera of Ascraeus Mons. Numerous fractures on the walls suggest repeated episodes of collapse. Faults around the smaller caldera to the right can be seen in section on the walls of the main caldera. Flank flows are transected at the caldera edge, as at the Mauna Loa summit seen in the next figure. The main caldera is 40 km across and approximately 4 km deep. (211-5065)

Figure 7.6. The summit caldera of Mauna Loa, Hawaii, showing truncation of lava flows by the caldera walls, as in the previous picture, although the frame here is only 2.5 km across. The Mauna Loa caldera is 2.7 km across, much smaller than that of the Arsia Mons caldera, and the flows are over ten times narrower than those on the Tharsis volcanoes. (Towill Corp., Honolulu, frame 6679-5.)

Figure 7.7. (*below*) The lobate feature to the northwest of Arsia Mons, whose 120-km–diameter summit caldera is seen at the lower right. The lobe has an inner granular texture and an outer striated margin. The adjacent shield flanks (bottom right) have several jagged, outward-facing escarpments. The lobe is tentatively interpreted as a giant gravity slide, caused by collapse of the volcano's northwest flank. (211-5317)

(figure 7.7). The terrain within the lobe has a coarse "sandpaper" texture, consisting of equidimensional hills 100–500 m across. Along the edge of the lobe are closely spaced ridges, which maintain a remarkable parallelism with each other and with the edge of the lobe for several hundred kilometers, completely ignoring topographic features such as craters and lava flows (figure 7.8). The Arsia shield, adjacent to the lobe, has a ragged appearance quite unlike the smooth, finely striated flanks of the rest of the edifice. My colleagues and I (Carr et al. 1977a) suggested that the lobate feature is an enormous landslide or debris flow that swept over the plains following failure of the volcano flank, and that the adjacent parts of the shield with the ragged texture represent a detachment zone. This suggestion, however, fails to explain satisfactorily the finely ridged outer margin of the lobe. This cannot have been formed directly by flow, since the striae are undeflected by surface obstacles. One possibility is that the landslide formed when the adjacent plains were covered with ice (Williams 1978). The ice would have facilitated flow of the volcanic debris over large distances, but more important, its subsequent removal would have superposed the flow pattern on the underlying topography. A similar, but less well developed, lobe that occurs adjacent to the northwest flank of the Pavonis lobe is also on the downslope side of the volcano, reinforcing the impression that gravity tectonics are involved.

The three volcanoes appear to have gone through similar growth cycles. The first stage was the building of the main edifice by the slow accumulation of fluid lava that erupted both at the summit and along concentric fissures on the volcano flanks. During this stage, eruption must have been fairly symmetrical about the volcano center, with no marked radial rift zones as in Hawaii, since all the main shields are roughly circular and lack radial spurs or ridges. Each volcano grew to approximately the same height, 24–27 km above the Mars datum. This is unlikely to be a coincidence. It may represent the greatest height to which lava can be lifted by virtue of the contrast in density between the lava and the rocks above the source region (see below). This interpretation is supported by the observation that flows tend to be shorter near the volcano summit, as compared with lower down the flanks (Carr et al. 1977a), indicating lower eruption rates. After the formation of the shields, eruption was concentrated along the line of the major northeast-southwest fracture zone on which the three volcanoes lie. Lava may have been transported laterally along the fracture zone from the main conduit under the center of each volcano, to erupt on the northeast and southwest flanks. Here eruptions caused repeated collapse of the surface of the shield, to form the large embayments or satellitic calderas. Eruptions from the satellite vents built rounded ridges of lava to the northeast and southwest of each shield and were the source of numerous flows that spread over the adjacent plains (figures 7.9 and 7.10). Judging from the crater counts (table 7.1), building of the central edifice terminated earlier at Arsia Mons than at the other two shields, although eruptions from the embayments and within the central calderas

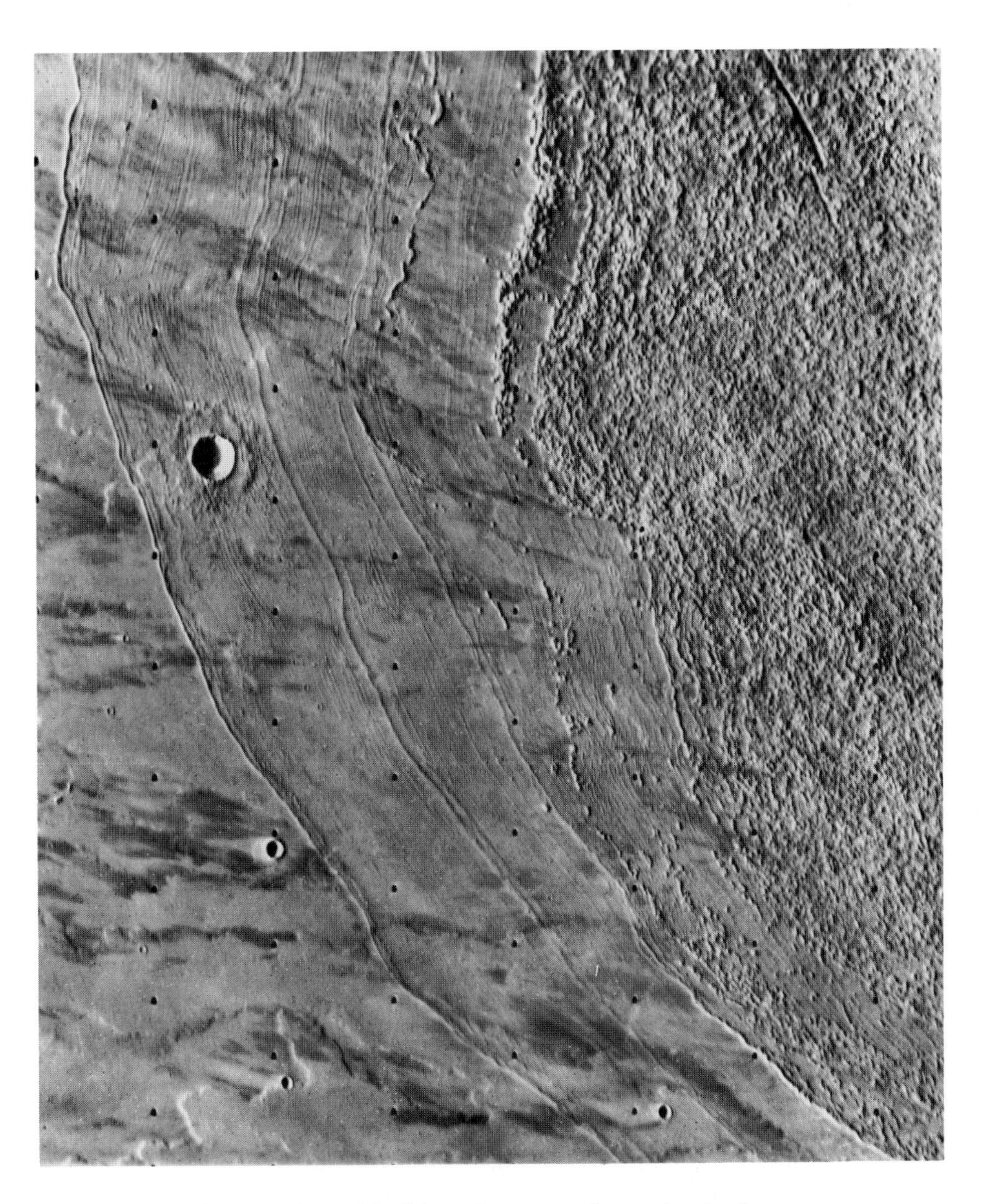

Figure 7.8. Detail of the margin of the lobe in the previous figure, showing fine striations cutting across lava flows and craters. The striations cut across, and are undeflected by, a crater at left center, which suggests that the linear texture is not the direct result of outward flow across the surface. Superposition from above, such as from ice, is a possibility. The frame is 200 km across. (42B35)

Figure 7.9. Flows, some with central channels, on the plains southeast of Arsia Mons. The flows originate mainly from the southwest embayment of the Arsia shield. They are mostly 5–10 km across, and many can be traced for hundreds of kilometers. The frame is 250 km across. (56A26)

Figure 7.10. (*below*) Flows on the flanks of Mauna Loa, Hawaii, shown for comparison with the previous picture. Many of the flows have central channels, some with distinct levees. Here the youngest flows are the darkest, in contrast with the previous figure. Note the enormous difference in scale between the two pictures. (U.S. Dept. of Agriculture, EKL 14 CC-35.)

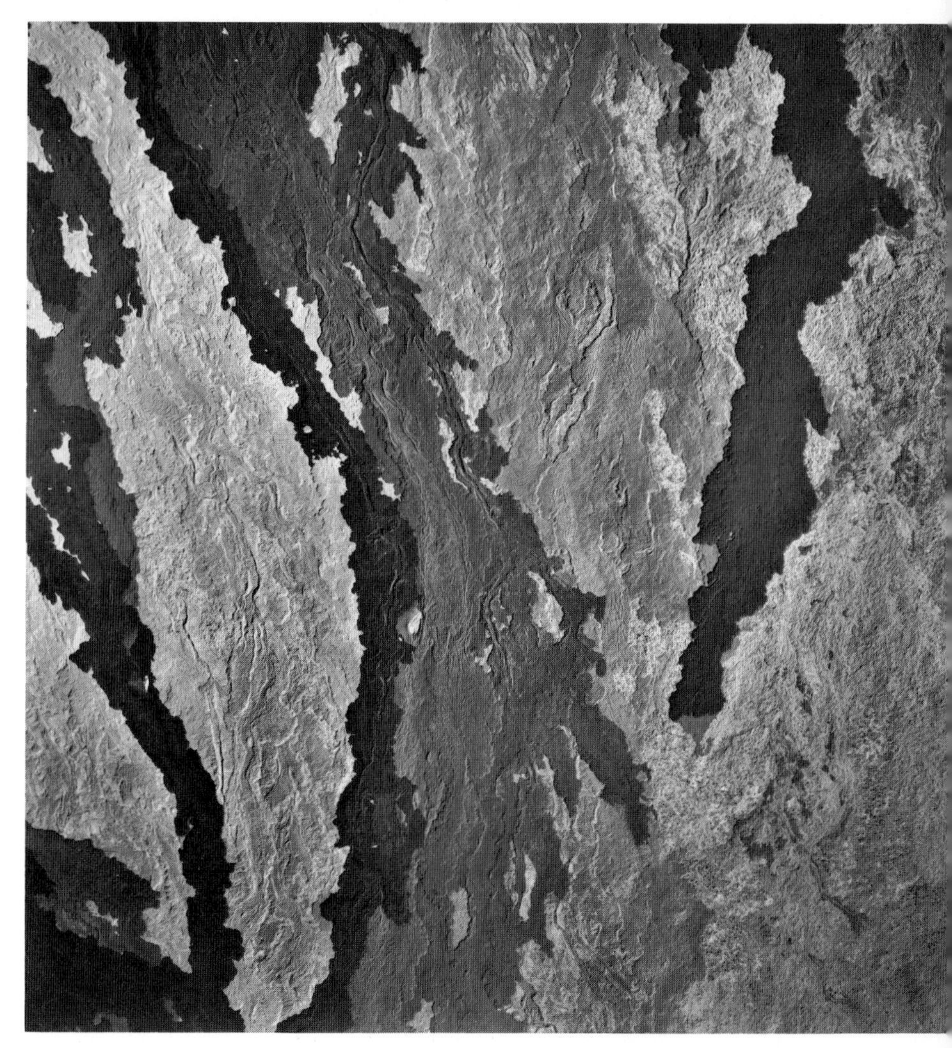

TABLE 7.1. VOLCANO AGES

		Crater Age (billions of years)		
	No. of Craters $>1\ km/10^6\ km^2$	*Minimum Likely*	*Best Estimate*	*Maximum Likely*
Olympus Mons	27	<0.1	<0.1	0.3
Arsia Mons	78	<0.1	0.1	0.5
Arsia Mons summit	150	<0.1	0.2	0.9
Arsia Mons flanks	390	0.2	0.6	1.6
Pavonis Mons	350	0.1	0.5	1.4
Biblis Patera	1,400	0.7	2.0	3.4
Alba Patera	1,850	1.0	2.7	3.6
Jovis Tholus	2,100	1.3	3.0	3.7
Uranius Patera	2,480	1.7	3.6	3.9
Apollinaris Patera	990	0.4	1.4	3.0
Tharsis Tholus	1,480	0.7	2.1	3.4
Albor Tholus	1,500	0.7	2.2	3.4
Hecates Tholus	1,800	1.0	2.6	3.6
Elysium Mons	2,350	1.6	3.4	3.8
Uranius Tholus	2,480	1.7	3.6	3.9
Ceraunius Tholus	2,600	1.8	3.8	3.9
Ulysses Patera	3,200	2.6	3.9	4.0
Hadriaca Patera	2,100	1.2	3.0	3.7
Tyrrhena Patera	2,400	1.7	3.5	3.9

NOTE: Counts are from Plescia and Saunders (1979) and Crumpler and Aubele (1978). Ages are based on the model of Hartmann et al. (1981), adapted as described in chapter 5.

of all the volcanoes continued into the recent geologic past and may still be continuing (Carr et al. 1977a).

Olympus Mons

Olympus Mons (figure 7.11) is a huge shield volcano around 700 km across, with a summit 27 km above the Mars datum and 25 km above the surrounding plains. The volcano is truly enormous, with a volume 50 to 100 times that of the largest volcano on Earth, Mauna Loa. Around the volcano is an escarpment, which in places is as high as 6 km. Extending for 300 km to 700 km beyond the escarpment are blocks of coarsely ridged terrain that form an incomplete ring, or aureole, around the volcano (figure 7.12). The shield has a complex summit caldera 80 km in diameter, which consists of several coalesced collapse craters with wrinkle ridges on their floors. Superposition relations suggest that the most recent areas of collapse are in the northeast and southwest. The flanks slope away from the summit at an average angle of 4°, but the volcano has a slightly sinusoidal

Figure 7.11. Oblique view of the summit of Olympus Mons, looking northwest. The caldera is 90 km across. Bright clouds beyond the caldera are orographic, but some standing-wave clouds are visible in the far distance. (41A11)

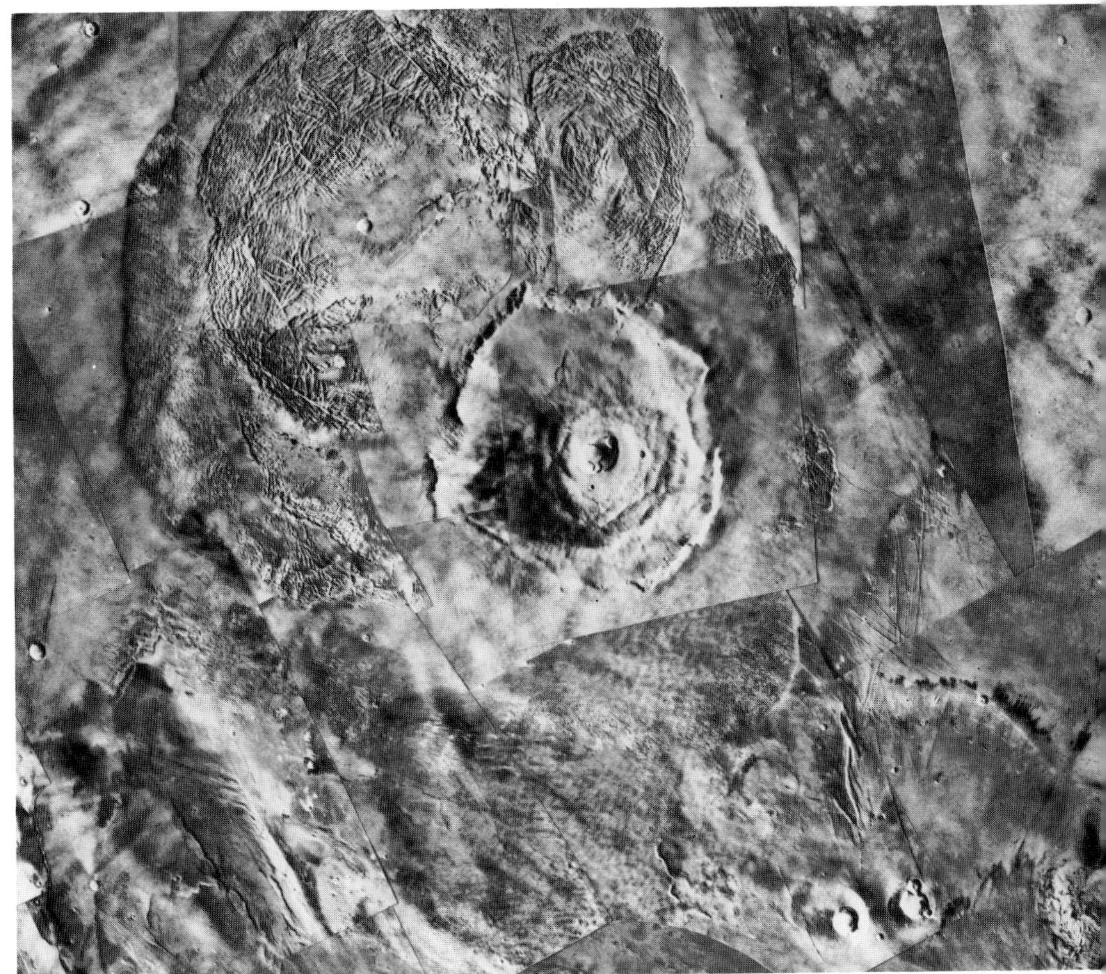

Figure 7.12. (*right*) Synoptic view of Olympus Mons and its surroundings. The main shield, as defined by the outward-facing cliff, is 550 km across. Lobes of highly textured terrain that extend far beyond the cliff, especially to the northwest, constitute the Olympus Mons "aureole."

profile, being flatter near the summit and the outer escarpment. The flanks are also terraced in a roughly concentric pattern, much like the Tharsis shields, with a spacing typically in the 15–50 km range. A fine radial fabric is caused by narrow flows, levees, sinuous channels, and an occasional line of craters suggesting a lava tube. Long, low, radial ridges with narrow (< 100 m) channels along their crests are common (figure 7.13). Again all the features indicate fluid lava. Nowhere in the volcano are there observable cinder cones, such as form in the late stages of growth of the Hawaiian shields, or any other indication of explosive activity. At the north and south margins of the shield, lava flows drape over the bounding escarpment and extend far beyond the edge of the main edifice (figure 7.14). Elsewhere the escarpment is a steep cliff, which truncates flows on the shield. Just inside the escarpment several small mesalike features stand above the surrounding flows, as though parts of the margin of the shield have been upfaulted. At the escarpment base are numerous lobate features suggestive of slumps, landslides, and other mass-wasting features (Blasius 1976a). Individual lobes may be as long as 100 km and, like the lobe to the west of Arsia Mons, may consist of a marginal zone of conformal ridges and a hummocky inner zone (Carr et al. 1977a).

All the features associated with Olympus Mons are young, although there is some discrepancy in the available crater numbers. Many of the flows on the plains surrounding Olympus Mons clearly cut and overlie the radial flows from the shield (Carr et al. 1977a, figure 18), yet published crater counts suggest the reverse relation. Plescia and Saunders (1979) gave a value of 27 craters larger than 1 km per million km^2 for the density of impact craters on Olympus Mons, which falls well below the range of 90–350 given by Schaber and co-workers (1978) for the immediately surrounding plains. A possible explanation is in the averaging. The groupings of Schaber and colleagues (1978) were large and may well have included flows younger than the range given. My colleagues and I (Carr et al. 1977a) also gave low crater numbers for Olympus Mons, with numbers ranging from 15 to 60 in the same units. In general the number of impact craters increases with elevation on the volcano, suggesting that eruption rates decline with altitude.

Figure 7.13. Detail of the Olympus Mons cliff, at 32°N, 129°W, showing flows draped over the bounding scarp of the shield. A perched lava channel is at the middle left of the picture, while numerous leveed channels can be seen in the upper right. The frame is 120 km across. (47B43)

Comparison with Hawaiian volcanoes

Olympus Mons and the three main Tharsis volcanoes resemble terrestrial shield volcanoes, so called because of their convex-upward, or inverted saucer, shape. The most intensely studied of the terrestrial shields are those in Hawaii, which grow to be as large as 120 km across and to a height of 9 km above the ocean floor. The active shields have a summit crater or caldera, which is not a vent, as in the case of the familiar stratovolcanoes, but a depression formed by downdropping of the summit region as a result of withdrawal of magma from below. The caldera generally includes a vent, but vents also occur on the volcano flanks, mainly along rift zones which, in the case of the Hawaiian volcanoes, are radial. On Fernandina in the Galapagos Islands, rift zones are concentric, more like the rift zones around the martian shield volcanoes. The Hawaiian rift zones are characterized by lines of collapse pits, spatter cones, and spatter ramparts, and eruptions from the rifts build broad ridges on the volcanoes, with the rift zone at the crest and lava flows diverging downslope. The Hawaiian shields all go through a similar eruptive cycle. During the early, main shield-building phase, as exemplified by Kilauea and Mauna Loa, mostly tholeitic basalt is erupted, and relatively high eruption rates are sustained. After a few hundred thousand years the style of eruption changes. The lavas become more silica-poor and alkali-rich, and the activity more explosive. The result is that at the end of the growth cycle most of the volcano—and the summit area in particular—becomes covered with cinder cones and ash deposits. Mauna Kea and Hualalai are in this stage.

Figure 7.14. Section of the cliff around Olympus Mons, at 15°N, 132°W. Massifs along the cliff edge are suggestive of upfaulting. Vague conical features on the shield are accumulations of lava at the mouths of channels. The frame is 110 km across. (222A64)

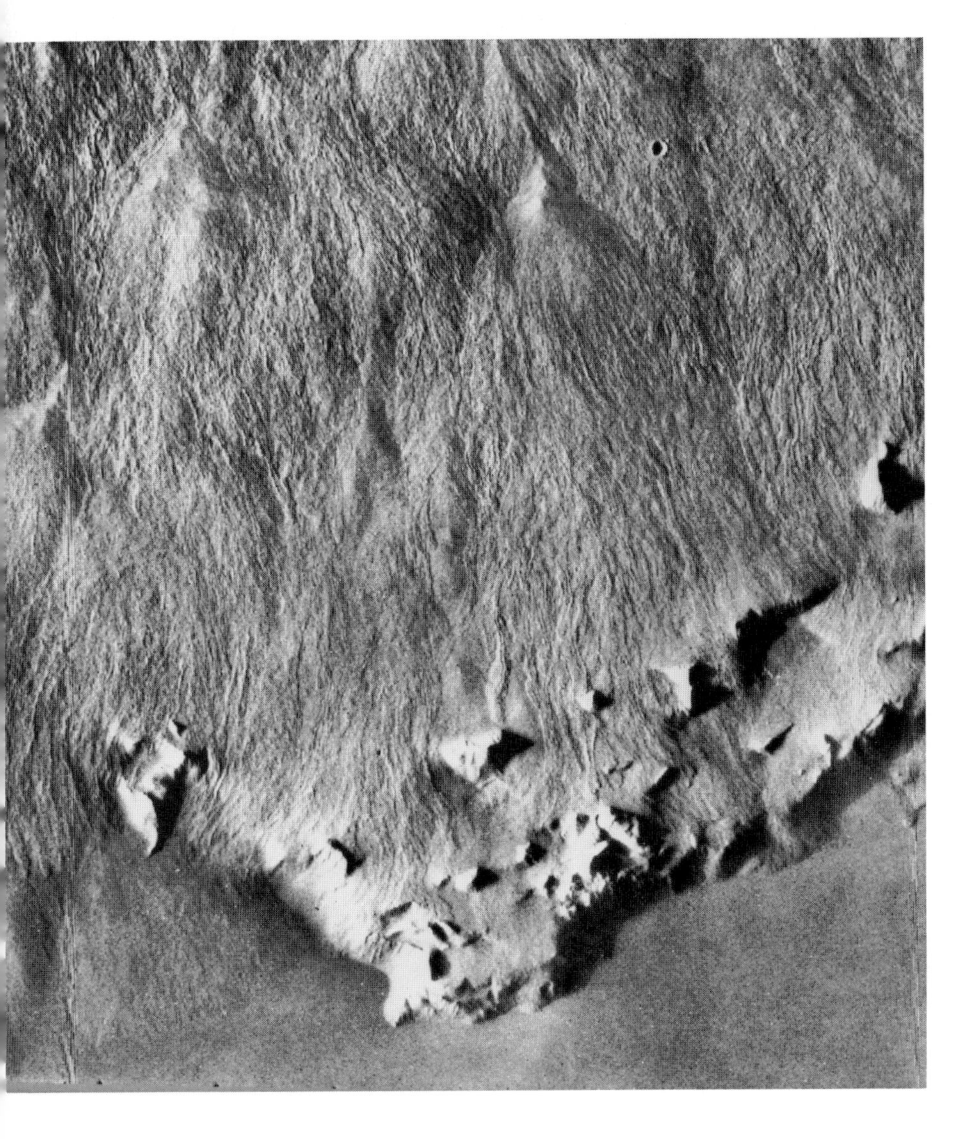

The martian shield volcanoes apparently only go through the first shield-building stage.

Seismic evidence suggests that the active Hawaiian volcanoes incorporate or overlie a relatively shallow magma chamber, which is supplied with magma from a source region at much greater depths. The source region is below the Pacific plate on which the volcanoes sit, so as the volcanoes grow, they are carried away from the magma source and ultimately become extinct (see below). At Kilauea, the chamber is approximately 2 km across and is 3–7 km below the summit. Eruptions are generally preceded by a swelling of the summit region. If eruption is on the flanks, then the summit normally deflates after the eruption; however, deflation does not generally follow summit eruptions. The caldera probably forms partly as a result of this inflation-deflation cycle.

Eruptions usually start from a fissure, to form a "curtain of fire," but in most cases activity rapidly becomes confined to one vent, with formation of a single lava fountain. Spatter from the fountain generally accumulates around the vent to form a cone. Lava is commonly carried away from the vent in channels, which may repeatedly overflow as a result of surges in the eruption rate to form levees. A channel may thus become situated on a low rise. During sustained eruptions the flowing lava crusts over, and the lava continues to flow below the surface in lava tubes. During the Mauna Ulu eruptions on Kilauea between 1969 to 1971, an elaborate system of tubes transported lava underground from the central vent to a subsurface reservoir in a former crater, then for many kilometers to the active flow fronts. Generally a tube leaves no trace at the surface, but occasionally lines of openings, or skylights, mark its course. Sometimes a lava tube may become blocked, and the lava creates a pressure head within the tube above the blockage. This can lead to breakout of lava at the surface and formation of a cone around a pseudo-vent. The tubes provide an efficient means for transporting lava over large distances. Heat loss from flowing lava on the surface is largely by radiation. If movement is through a tube, then radiation loss is greatly reduced, and flow can be sustained even when eruption rates are rather modest. Swanson (1973) estimated that Mauna Ulu lava in tubes cooled at a rate of only 1°C for every kilometer traveled. This figure almost certainly depends on the effusion rate, and at the larger effusion rates suspected for Mars the cooling rate could be considerably lower. The fluidity of the lava, combined with the heat-conserving flow through tubes, thus enables the lava to move long distances over shallow slopes and results in the characteristic low profile of the shield volcanoes.

Olympus Mons and the Tharsis shields have characteristics that suggest similar eruptive styles, although there are differences. The overall shapes of the edifices, with their low slopes and summit calderas, suggest eruption primarily of fluid lava with little ash. The presence of complex summit calderas indicates that the summits have undergone cycles of inflation and deflation like the Hawaiian volcanoes. The significance of the much larger size of the martian calderas is unclear, although a larger magma chamber may be indicated. The outlines of individual flows on the Hawaiian and martian volcanoes are similar, and the presence of leveed channels and channels atop ridges suggests similar processes whereby the magma is distributed from the source vents. Although lava channels are common, lines of pits indicative of lava tubes are rare. Whether this is an observational artifact or a real difference is not known. Another apparent difference between the Hawaiian and martian shields is the relative scarcity of observed vents. Rimless depressions, which could be vents, are common adjacent to the embayments on the northeast and southwest flanks of the Tharsis shields, but they are rare elsewhere on the shields and on Olympus Mons. Moreover, evidence of buildup of spatter around vents is almost completely lacking. Again the cause of the difference, supposing the difference to be real, is not known.

The most obvious difference between the Hawaiian and martian shields is the size of the shields themselves and the constituent features. The size of the flow features may be the result of high eruption rates. Flows on the martian shields commonly can be traced for tens of kilometers, and in some areas adjacent to the shields for hundreds of kilometers, and central channels may be hundreds of meters across. By comparison, flows on Hawaii are rarely more than 30 km long, and channels rarely more than a few tens of meters wide. Walker (1973) suggested that the effusion rate is the most important factor determining the length of a lava flow, with high effusion rates resulting in long flows. If true, then high effusion rates are implied for the martian shields (Carr et al. 1977a), which is consistent with the large size of the lava channels.

By the same reasoning as given in the previous chapter, the shields are probably built mostly of iron-rich basic magmas. Such a composition is consistent with the inferred high fluidities and with the estimated yield strengths. The method of estimating yield strengths was pioneered by Hulme (1974) and extended by Moore and by Schaber and colleagues (Schaber 1975; Schaber et al. 1978). The principle of the technique is that lava can be treated as a fluid that flows only when the basal stress at the margin of the flow exceeds some critical value. On level ground the higher the yield strength of the flow, the thicker the flow must be before it can move. On sloping ground a lava with a higher yield strength will tend to form a wider, thicker flow than one with a smaller yield strength. Estimates of yield strength can thus be made from the shape and thickness of flows, if the slope of the ground is known. Hulme (1974) suggested that the yield strength depends partly on the silica content of the lava, thereby providing a basis for estimating composition. Moore and co-workers (1978), however, calculated the yield strengths of many terrestrial lavas and found that the yield strength, while dependent on composition, also depends on the slope of the ground over which the lava flows, so is not a reliable index of composition. They (1978) estimated the yield strengths of many martian lava flows, mainly around Tharsis, and found values that range from 0.12 to 2.7×10^3 N/m^3, with an average of 1×10^3 N/m^3. These values, while not uniquely specifying composition, are more indicative of the compositions of basalts than of the more silicon-rich andesites, trachytes, or rhyolites.

The martian shields probably accumulated far more slowly than the Hawaiian shields. My colleagues and I (Carr et al. 1977a) showed that the flows to the southwest of Arsia Mons, which appear to be mostly derived from the embayment in the southwest flank of the shield, have a wide range in the number of superimposed impact craters, from less than 200 craters larger than 1 km per million km^2 close in to more than 2,000 further out. Such a spread implies that Arsia has been active for billions of years (table 7.1). We have no measure of how old Olympus Mons is. The relatively small number of impact craters suggests a relatively young age (10^7–10^8 years) for the present surface, but this may be more a measure of recent resurfacing rates than of the age of the volcano itself. Support for an older age for Olympus Mons is the pattern of flows around the edifice (Schaber et al. 1978), which indicates that the shield has been there for considerably longer than is implied by the age for the present surface, perhaps for billions of years. Shaw (1973) and Swanson (1973) estimated that accumulation rates for the Hawaiian volcanoes are in the range of 0.01 km^3/year to 0.02 km^3/year. Assuming a volume for Arsia Mons of 10^6 km^3 (Blasius and Cutts 1976) and an age of 2 billion years (Carr et al. 1977a), we derive an accumulation rate of 0.0005 km^3/year, which is considerably lower than that estimated for Hawaii. While the actual numbers are suspect, the general conclusion that the martian shields accumulated more slowly than the Hawaiian shields, is probably valid. The martian shields appear to have grown slowly over billions of years, by widely spaced eruptions of large volume.

Volcano size and lithosphere thickness

The most striking characteristic of Olympus Mons and the Tharsis volcanoes is their extreme size. This almost certainly results from a more stable and thicker lithosphere on Mars, as compared with Earth. The Hawaiian volcanoes are relatively short-lived because of plate motion. Hawaii lies at the southeast end of a long chain of extinct volcanoes, the Hawaiian Emperor chain, which stretches for several thousand kilometers across the Pacific. Within the Hawaiian Islands the volcanoes become progressively older to the northwest, and this same trend continues up the Hawaiian Emperor chain, so that the Koko seamount 5,000 km northwest of Hawaii is 46 million years old (Clague and Dalrymple 1973). The chain is believed to have formed as a result of the northwest movement of the Pacific plate on which Hawaii is located. It moves over a fixed magma source, or hot spot, that is currently under the southeast edge of the main island of Hawaii. Volcanoes initially start to grow over the magma source, but as they grow, they are slowly carried northwestward. Ultimately, they are removed so far from the source that they are cut off from the supply of magma and become extinct. As old volcanoes move northwestward and die, new volcanoes form to the southeast. Thus, the youngest and most active volcano in Hawaii, Kilauea, is at the southeast extremity of the Islands, and the other volcanoes become progressively older to the northwest. Hawaiian volcanoes are believed to remain active for a few hundred thousand years before the plate motion cuts off their magma supply and they become extinct. On Mars, however, there is no plate motion, so a volcano remains stationary over its source of magma and continues to grow as long as magma is available. As a consequence volcanoes on Mars can grow to much larger sizes than those on Earth (Carr 1973). The large martian shield volcanoes should, in effect, be compared to the entire Hawaiian Emperor chain, not just to individual volcanoes.

The height to which a volcano can grow depends on the depth of origin of the magma and the contrast in density between the lava and the rocks through which it passes on its way to the surface. Lava is pumped to the top of a shield volcano by the lithostatic pressure at the magma source (Eaton and Murata 1960). A volcano can continue to grow in elevation as long as the pressure exerted by the column of magma between the volcano summit and the magma source is less than the lithostatic pressure at the source depths. When the pressure at the base of the column equals the lithostatic pressure, then the lava can be pumped no higher, and any further growth of the volcano must be lateral. Eaton and Murata (1960) suggested that Mauna Loa is at its height limit, because it is almost at the same height as the nearby volcanoes Mauna Kea and Haleakala. Taking plausible values for the densities of the crust, mantle, and lava, they calculated that the Hawaiian lavas originate at depths of around 60 km.

Such calculations are extremely uncertain for Mars, since estimates of the densities are largely guesswork. Blasius and Cutts (1976), however, showed that irrespective of what densities are assumed, the larger elevation of Olympus Mons with respect to the surrounding plains, as compared with the Tharsis shields, implies a deeper source for its magma. Taking terrestrial values for the density of the lava and the mantle, and ignoring the effect of crustal rocks, they estimated that the magma that forms Olympus Mons is derived from depths that are 36 km deeper than those for the Tharsis shields. This is geologically plausible. Steeper temperature gradients and shallower depths to melting are expected under the Tharsis shields, which are closer to the center of the volcanic province, than under Olympus Mons, which is toward the edge of the volcanically active region. The absolute depth is more difficult to estimate. Taking 2.8 gm/cm^3 for the density of lava and 3.2 gm/cm^3 for the average density of the rocks above the magma source, including both crust and mantle, then a 23-km volcano height gives a 160-km depth to melting. The presence of a crust with a density close to that of the lava could significantly increase this figure. The implication is that the martian lithosphere is considerably thicker than Earth's, which ranges up to about 100 km at the thickest parts of the continents. This question of the thickness of the martian lithosphere is dealt with more fully in the next chapter.

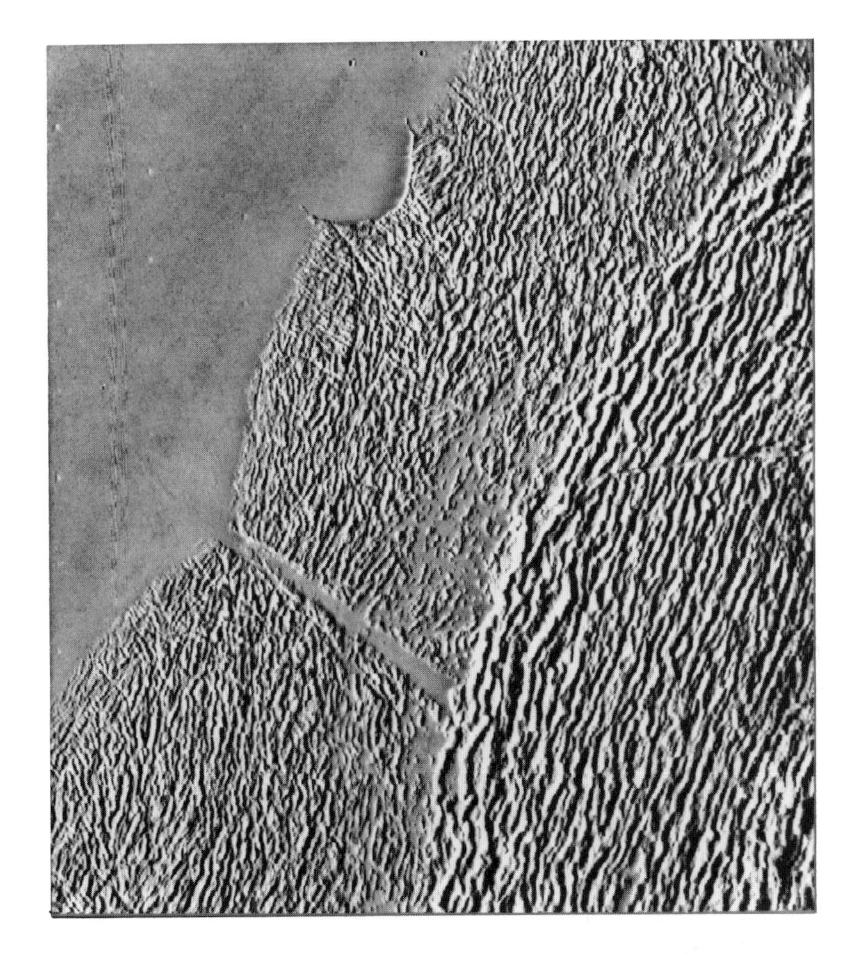

Figure 7.15. Detail of the northwest margin of the Olympus Mons aureole, showing the characteristic ridge and valley topography. The area covered is that in the upper left corner of figure 7.12. A younger, coarsely ridged lobe to the right appears to overlie a more finely ridged older lobe in the center. The outer lobe is partly embayed by smooth plains to the upper right. The origin of the ridged units is not known (see text). The frame is 220 km across. (512A30)

Olympus Mons aureole

The aureole of Olympus Mons has generated considerable controversy, and its origin is still a mystery. The aureole consists of several huge lobes of distinctively ridged terrain that extend out from the base of Olympus Mons for several hundred kilometers (see figure 7.12). The terrain within the lobes consists of closely spaced arcuate ridges that run roughly parallel to the boundary of the lobes (figure 7.15). Cutting across the parallel ridges are grabenlike troughs, oriented in various directions. The surface of each lobe appears to dip gently inward toward Olympus Mons, so that the inner parts are embayed by younger plains, and the outer margin of the lobes is an outward-facing escarpment. Only the outer margins of the older lobes are visible, mainly because of burial by younger lobes closer in. The aureole covers an enormous area, extending for 700 km to the northwest of the Olympus Mons scarp and for 300 km to 400 km from the scarp around the rest of the volcano. A positive free air gravity anomaly of several tens of milligals* appears to coincide with the aureole, at least to the northwest of Olympus Mons (Sjogren 1979).

The age of the aureole cannot be determined precisely. Few impact craters occur on the surface. Morris (1979) estimated that there are only 30 craters larger than 1 km per million km^2, which is comparable to the number on the surface of Olympus Mons. The extreme roughness of the terrain, however, could cause relatively rapid erosion of small impact craters by surface creep, as in the case of the lunar highlands, so this number may have little meaning with respect to age. However, the scarcity of craters larger than 5 km in diameter, which are less likely to be eroded away, as compared with the cratered plains just to the northwest of Olympus Mons (see figure 6.14) suggests that the aureole is considerably younger than these plains, which have 1,800–3,200 craters larger than 1 km per million km^2 (Schaber et al. 1978). The evidence from superposition is no better. Southeast of Olympus Mons the aureole overlaps fractured plains with a crater density of around 2,000 craters larger than 1 km per million km^2 and is itself overlapped by sparsely cratered plains that have a crater density of less than 100 in the same units.

Several suggestions have been made as to the origin of the aureole. King and Riehle (1974) proposed that Olympus Mons is a composite volcano that has erupted large volumes of ash and lava. They suggested that ash erupted from the central vent was transported by nuées ardentes—clouds of ash and hot gases—well beyond the present shield. They argued that the resulting ash deposits would decrease outward in their degree of compaction and coherency because of cooling of the ash during its transport. Erosion would therefore preferentially remove the outer, more friable parts of the ash sheets but leave the inner part of the volcano, comprised of lava and indurated ash, relatively unmodified. They suggested that the scarp of Olympus Mons is erosional and that it formed at the zone of transition between predominantly lava and indurated ash deposits close in and dominantly friable ash deposits farther out. By implication the aureole deposits are eroded remnants of the distal parts of the tuff sheets. A major objection to this hypothesis is the lack of evidence of pyroclastic activity of any kind on the main Olympus shield. In a related hypothesis, Morris (1979) suggested that the aureole deposits are vast tuff sheets, but that they were erupted from several volcanic vents around the Olympus shield. Here again wind is the dominant cause of the unique topography.

Harris (1977) proposed that the lobes are vast gravity-assisted thrust sheets, caused by a squeezing out of material from beneath Olympus Mons along planes of weakness. He suggested that the aureole materials are concentrated to the northwest, because this is down the regional slope caused by the Tharsis bulge. He pointed out a similarity between the Olympus Mons aureole and thrust sheets around the Bearpaw Mountains in Montana, which have also been attributed to volcanic loading. Another possibility is that the periphery of Olympus Mons fails when shear surfaces, such as those at the base of the observed landslides around the cliff, intersect a zone of weakness below the surface of the surrounding plains. The result would be movement outward of both the outer part of the shield and the plains materials above the failure surface. Movement could be aided by water under high pore pressure at the zone of failure. The main problems with both these hypotheses is that they do not explain the positive gravity anomaly over the aureole.

Hodges and Moore (1979) suggested that Olympus Mons is analogous to Icelandic table mountains, volcanoes that form partly under ice. The slopes of these volcanoes near the summit are gentle, like those of normal shield volcanoes, but around the outer margins are steep cliffs. The tops of the cliffs mark the height of the former ice cover, and the gently sloping parts above were formed subaerially. Hodges and Moore proposed that there was formerly a thick ice sheet around Olympus Mons, and that the height of the cliff, 4–6 km, indicates the thickness of the ice. The aureole, according to this proposal, is analogous to icelandic moberg, a friable, easily erodible rock that forms by intrusion of basaltic magma into ice. The difficulties with this hypothesis are numerous, including why no subaerial eruptions occurred over the vast area of the aureole, why the ice left no trace other than the aureole and the cliff, and why the thick ice-sheet should be preferentially located at Olympus Mons. Other suggestions are that the aureole is the erosional remnant of an ancient shield (Carr 1973) or an unroofed pluton (Blasius 1976a). Both of these suggestions appear to require higher erosion rates than are found elsewhere on the planet and so also appear implausible. Thus we are left with no satisfactory explanation of the features, despite several proposals, some quite imaginative.

Small Tharsis volcanoes

Several relatively small volcanoes occur within Tharsis. Some have features on their flanks that resemble those on the large shields; others are quite different. All have calderas that are large with respect to their overall dimensions. To the west of Pavonis Mons are two small shields, the 175 km by 105 km Biblis Patera (figure 7.16) and the 90-km–diameter Ulysses Patera. Both have indistinct flows and faint radial striae on their flanks, and both are cut by northwest-southeast–trending fractures. Loops of cratered mounds similar to those in Isidis (see below) are also present in Ulysses Patera. Uranius Patera, a 200-km–diameter volcano 750 km northeast of Ascraeus Mons, is similar to the Biblis and Ulysses Paterae in having a large central caldera (100 km in diameter) and a distinct radial texture; but on

*Accelerations due to gravity at the surface of the planet are measured in gals (1 gal = 1 cm/sec^2). A positive anomaly indicates a higher-than-average acceleration at the surface. Usual causes of a positive free air anomaly are higher-than-average elevations or higher-than-average densities of the near-surface rocks.

Figure 7.16. (*below*) Biblis Patera, a lesser shield volcano at 3°N, 124°W, between Olympus Mons and Arsia Mons. The shield and the surrounding plains are cut by fractures radial to the center of the Tharsis bulge to the southeast. The caldera is 50 km across. (44B50)

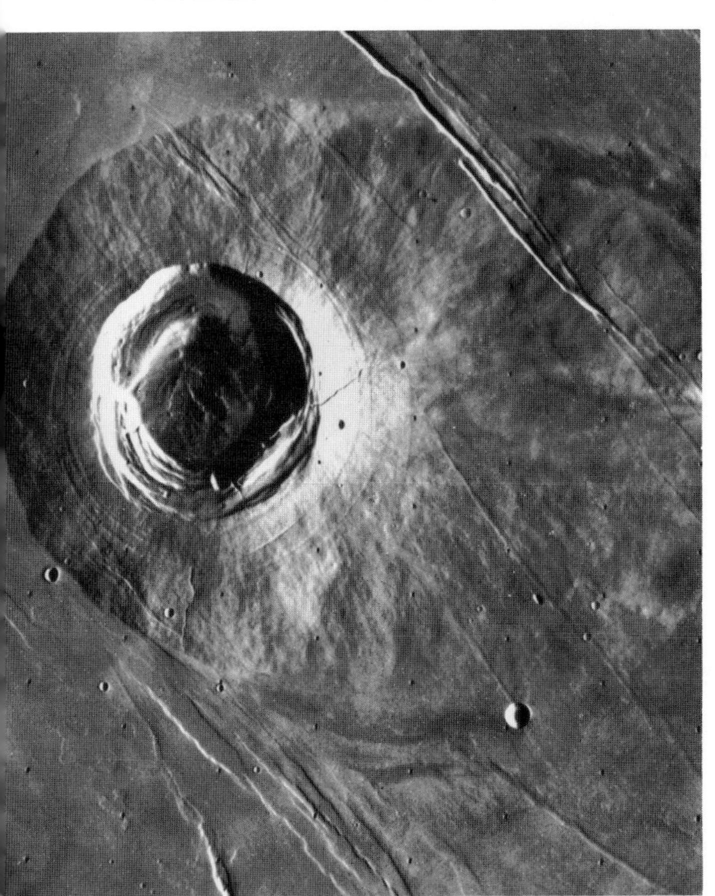

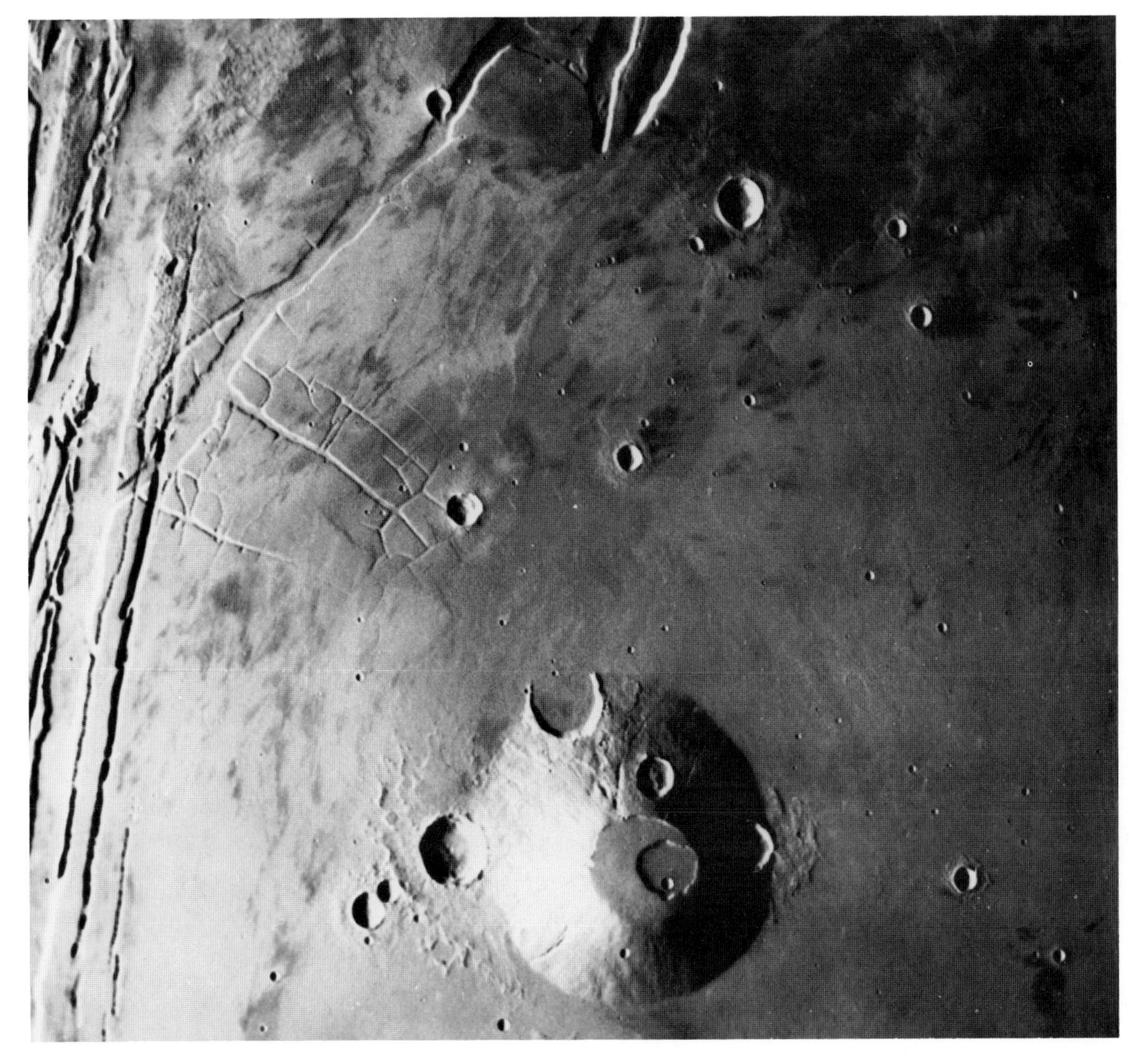

Figure 7.17. The 57-km–diameter Uranius Patera in northeast Tharsis. Several large superposed impact craters with well-developed fluidized ejecta patterns suggest a relatively old age. Fractures at the left are part of the set radial to Tharsis. (516A23)

Uranius Patera long thin flows are clearly visible. Two other volcanoes, the 50-km–diameter Jovis Tholus and the 120-km–diameter Tharsis Tholus, have relatively smooth flanks. Tharsis Tholus is dome-shaped and appears to have relatively steep peripheral slopes, although no measurements are available. The flanks are also cut by several roughly radial faults which form distinct scarps. The lower parts of all these volcanoes are buried by flows of the surrounding plains, so that their true areal extent and vertical dimensions are not known.

All the volcanoes just described have characteristics that suggest a type of volcanic activity similar to that which formed the large Tharsis shields. A possible exception is Tharsis Tholus, which has steeper slopes, which might suggest more viscous magmas; but lack of detail on the flanks prevents more informed interpretation. Two other volcanoes, Uranius Tholus (figure 7.17) and Ceraunius Tholus (figure 7.18) have features not found on the large Tharsis shields, although they are similar to Hecates Tholus in Elysium. Both have relatively steep slopes, and both have numerous fine radial channels on their flanks. Ceraunius Tholus is 120 km across, has a 25-km–diameter central caldera, and is 6 km high, giving it slopes that average 9° (Blasius 1979). Most of the channels lack the levees that charac-

Figure 7.18. Ceraunius Tholus in northeast Tharsis. A 2-km–wide channel extends from the 22-km–diameter summit caldera down the flanks of the volcano into an impact crater in the adjacent plains. Smaller channels are just visible elsewhere on the flanks. (516A24)

terize many of the channels on the large Tharsis shields and the lava plains; nor are the channels atop ridges, as they are on Alba and Olympus Mons. The Ceraunius channels are simply set into the volcanic surface. Most are a few hundred meters across or less, but a particularly large channel 2 km across starts at the summit caldera, extends down the northern flank of the volcano, then into a large impact crater in the surrounding plain. No large deposits occur within the crater at the end of the channel, so that most of the material eroded to form the channel must have been removed. Several smaller channels on the volcano flanks have conical accumulations of debris at their lower ends. The 60-km–diameter Uranius Tholus, just north of Ceraunius, is similar in also having finely channeled flanks and relatively steep slopes.

Reimers and Komar (1979) suggested that the channels on Ceraunius and Uranius Patera were cut by volcanic density currents, mixtures of gas and particulate debris that move radially outward from a volcanic center, either as a direct result of explosive activity at the vent or as a result of the collapse of a gas/ash column over the volcano. The currents, also known as base surges and nuées ardentes, can reach velocities as high as 50 m/sec and can travel for tens of kilometers. Their mobility is believed to be sustained by continued exsolution of gases from the entrained particles. Numerous examples of channels eroded by density currents exist. The largest is a channel on Asama volcano in Japan, which is 1–2 km wide and over 8 km long (Aramaki 1956). More commonly the channels are smaller and may form radial networks. Reimers and Komar (1979) remarked on the similarity between the array of channels on the flanks of Ceraunius Tholus and those on the flanks of Barcena Volcano, Mexico. They pointed out that the large channel on Ceraunius cannot be formed by lava, as I once suggested (Carr 1974b), since there are no volcanic constructs within the crater in which it terminates. Nor is it likely to have been carved by water, as suggested by Sharp and Malin (1975), because there is no catchment area, and erosion by seepage at a volcano summit is unlikely. They concluded that density currents, which could form either by rapid exsolution of juvenile water or interaction of magma with groundwater or ground-ice, cut this and the other channels on Ceraunius Tholus and Uranius Tholus. Even if Reimer and Komar (1979) are correct, a distinctive composition for Uranius and Ceraunius Tholus is not necessarily implied. Although nuées ardentes on Earth are more common in nonbasaltic lavas, such as andesites and rhyolites, they can form with basaltic magmas, particularly if there is strong interaction of the magma with groundwater.

Alba Patera

The 1,600-km–diameter Alba Patera far exceeds any other volcano in areal extent, covering eight times the area of Olympus Mons and 180 times the area of Mauna Loa. The height of the volcano is not known, although it is probably less than 6 km (Carr 1976). The slopes are shallow, therefore, less than half a degree. At the center is a 100-km–diameter caldera, around which splay numerous north-south– and northeast-southwest–trending fractures to form a fracture ring 600 km in diameter around the volcanic center (figure 7.19). A striking characteristic of the volcano is the presence of large, beautifully preserved lava flow features on the flanks.

Figure 7.19. Northwest summit region of Alba Patera. The partly buried central caldera is visible in the bottom right corner of the mosaic. Surrounding the summit, and seen in the center of the mosaic, is a ring of fractures approximately 600 km across. The fractures are part of the set radial to Tharsis but are deflected around the volcano. Most of them postdate and cut the flows on the flanks. The volcano appears to have little vertical relief, with flank slopes less than 0.5°. (211-5065)

Several types of flows are recognized (figure 7.20). Most common are sharp-crested radial ridges that maintain a uniform width over long distances. They average 8 km across and may be 300–400 km long. At the ridge crest is generally a channel, usually less than 500 m across, or a line of pits, suggesting the presence of a lava tube. The ridges are interpreted as having been built slowly by lava carried down the central channel. Occasionally low mounds and larger conical constructs are superposed on the ridge, or the ridge breaks up into a line of cones up to 10 km across. The mounds and cones resemble pseudovents on terrestrial volcanoes, which form where lava breaks out of a tube to the surface as a result of blockage of the tube downstream (Greeley and Hyde 1972). On the flanks of the ridges are numerous gullies, which give the ridges a striation roughly at right

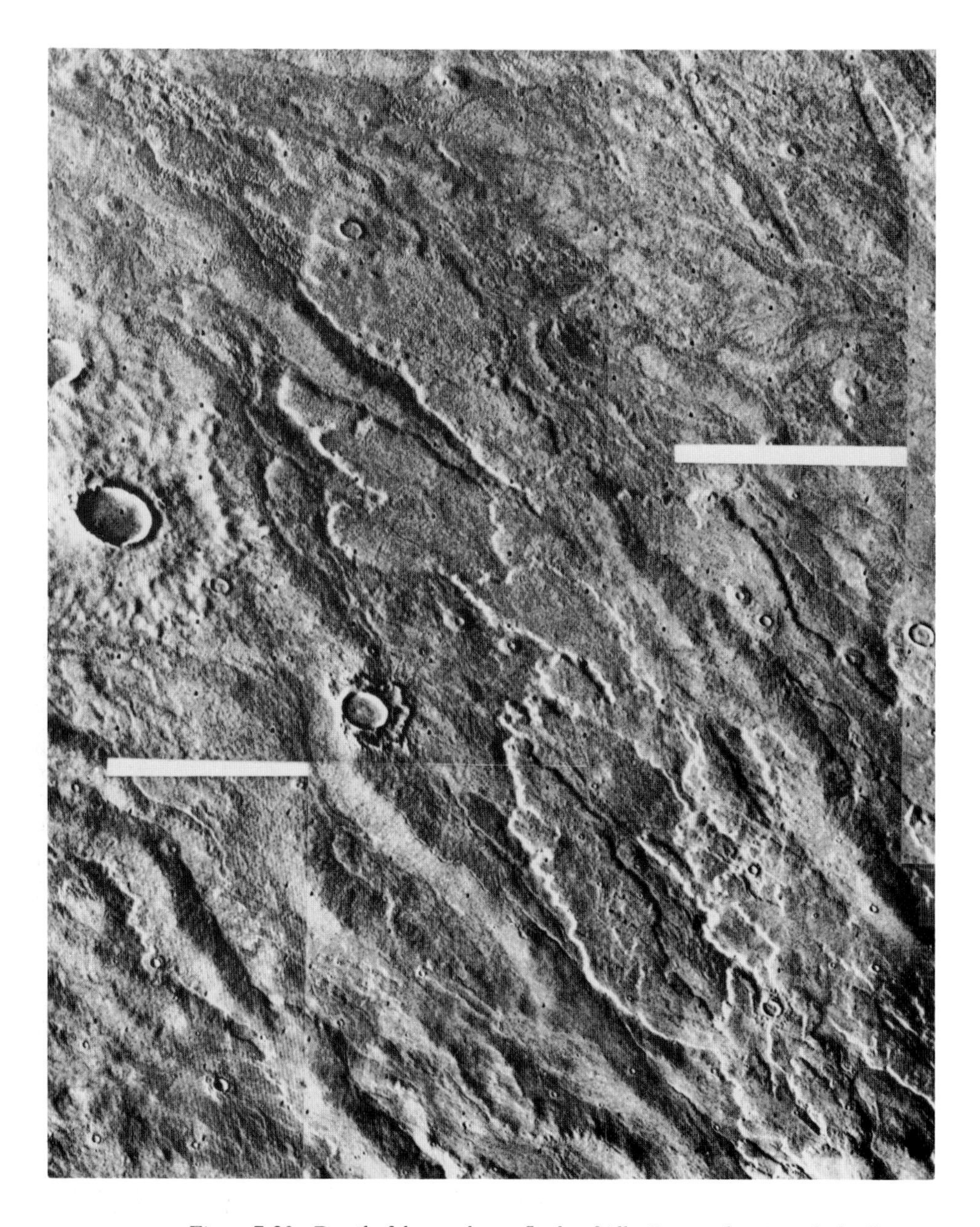

Figure 7.20. Detail of the northwest flanks of Alba Patera, showing tabular flows in the upper part of the mosaic and tube- and channel-fed flows in the lower left. Many of the channels run along the crests of ridges that are about 15 km across. It is predominantly these ridges that give the volcano its radial texture. The mosaic is approximately 150 km across. (211-5065)

Figure 7.21. (*right*) Index map of the Elysium region, showing the three volcanoes on the Elysium bulge and Apollinaris Patera in the lower right. Contour interval is 1 km. (Courtesy of U.S. Geological Survey.)

angles to its length. These are almost certainly formed by overflow of lava from the main channel which may have occurred relatively often as a consequence of the shallow longitudinal profile of the central channel. Occasionally channels leave the crest of the ridge and continue between the ridges. Presumably, lava overflowed the channel on the ridge crest, and subsequent flow was between the ridges. In this manner new ridges probably formed between the old ones as the volcano grew. Also present are tabular sheet flows with well-defined steep flow fronts. They resemble the flows on the lava plains and may be indicative of higher eruption rates than those associated with the flows that formed the long ridges. A third type of flow occurs mainly within the fracture ring close to the caldera. These are relatively short, narrow flows, less than 2 km across, with a central channel. The channels commonly have levees but are not atop ridges. The flows are similar to those in Olympus Mons and the large Tharsis shield volcanoes and may indicate steeper slopes toward the volcano summit. All the flow features just described have close terrestrial analogs. The Alba

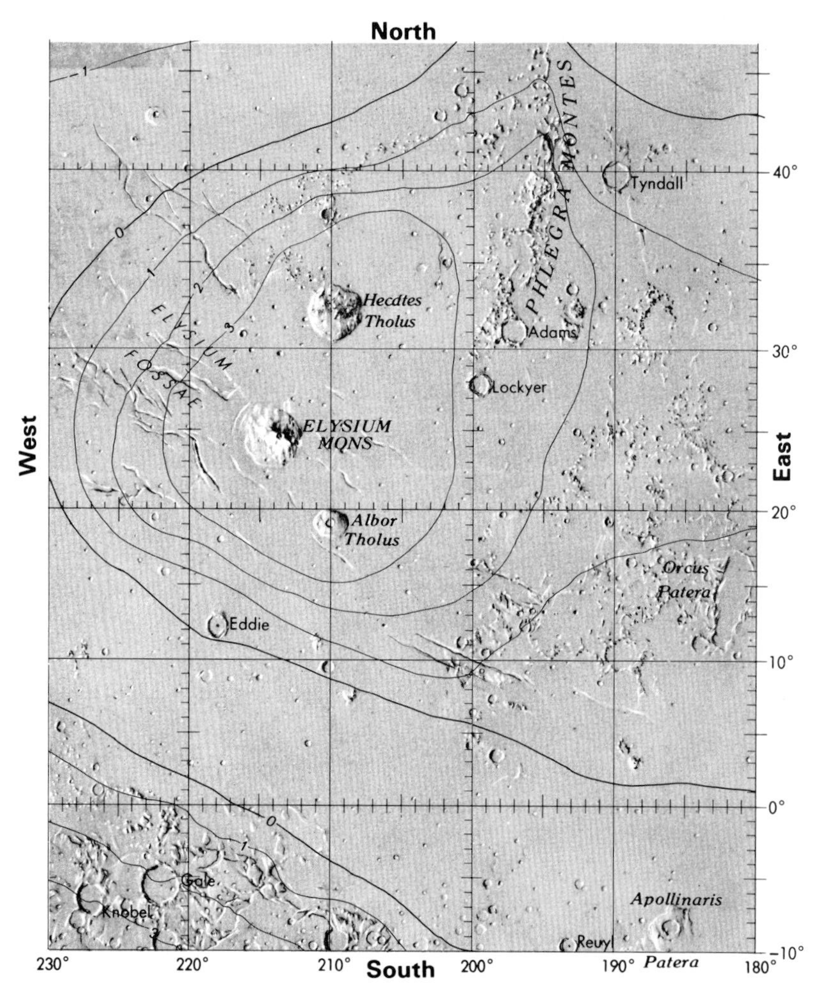

Figure 7.22. Southeast flank of Hecates Tholus, showing its highly channeled and pitted surface. Conical accumulations of lava or ash can be seen at the ends of some channels, particularly around the periphery of the volcano. The summit of the volcano is off the top of the picture. The frame width is 52 km. (651A20)

Patera features, however, are mostly an order of magnitude larger than any of their terrestrial counterparts.

ELYSIUM

The three Elysium volcanoes, Albor Tholus, Hecates Tholus, and Elysium Mons, lie on a broad arch 2,000 km across and 5 km high (figure 7.21). Little is known of the 30-km–diameter Albor Tholus, which will not be discussed further. The northernmost volcano, Hecates Tholus, is 180 km in diameter, and the Mariner 9 UVS gave 6 km for its height (Hord et al. 1974). The volcano appears dome-shaped, with a relatively flat top and steep outer slopes that terminate abruptly against the surrounding plains (figure 7.22). At the center is a complex caldera 11 km in diameter, from which several lines of depressions and graben radiate (figure 7.23). Long narrow channels similar to those on Ceraunius Tholus are common on the flanks, particularly on the steeper outer slopes. Conical piles of debris at the lower ends of the channels adjacent to the plains are common. Similar conical piles on the lower slopes of the volcano give it a hummocky

Figure 7.23. Summit of Hecates Tholus, showing the caldera and radial grabens and lines of craters. The surface appears to have a more subdued appearance immediately around the summit region. The large number of small pits, most of which are probably volcanic, indicates the difficulty in getting precise counts of the number of superposed impact craters on volcanoes. The frame is 52 km across. (651A19)

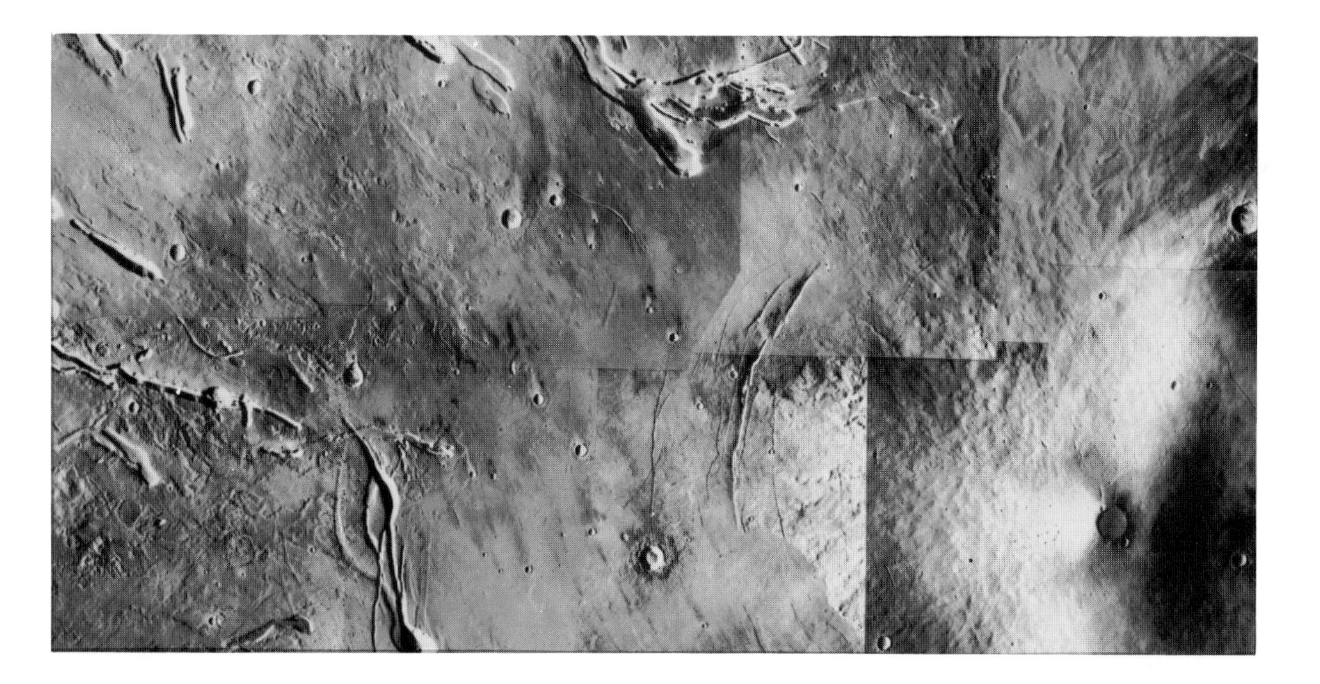

Figure 7.24. Mosaic showing Elysium Mons and the region to the northwest. Elysium Mons, with its 14-km-diameter central caldera, is to the right. Around the volcano flanks are concentric fractures, and out beyond these are some radial, grabenlike troughs, which give rise to several channels that extend for hundreds of kilometers to the northwest. See figure 10.16. (211-5043)

appearance. The flanks are densely pitted with craters smaller than 1 km in diameter, but around the summit caldera the surface is smoother, with fewer craters and fewer channels. On the west side a deep embayment is cut into the volcano. This does not appear to have been the source of flows like the embayments in the Tharsis shields, and its origin—whether by collapse or erosion—is unclear. The outer flanks of Hecates Tholus resemble Ceraunius Tholus, and the same arguments may be made for density currents playing a role in sculpting the surface. The muting of the fine-scale topography around the summit also supports a more pyroclastic type of volcanism than on the major shields, as does the lack of visible flows on the flanks, although the apparent lack of cinder cones is puzzling.

Elysium Mons has a significantly different appearance from other martian volcanoes (figure 7.24). Malin (1976) recognized two components: a main edifice, which is roughly circular and asymmetric about the central caldera, and a broad ridge approximately 200 km across and 2 km high that extends northward from the volcano center. On the ridge are flows and sinuous ridges, much like those on Alba. However, no flows are visible on the main edifice. The flanks appear hummocky, with individual hummocks up to 5 km across, but otherwise there is little surface detail apart from craters, even at high resolution. The summit of the volcano is estimated to be 9 km above the surrounding plain, giving an average slope of 3.5° for the flanks (Blasius 1979). Malin (1976) compared Elysium Mons to the terrestrial volcano Emi Koussi in the Tibetsi region of the Sahara. He pointed out the similarity in size and shape of the central caldera and the similar channel-like features near the summit, as well as a similarity in the slope of the entire edifice. The main difference between the two volcanoes is the presence of numerous flows and cinder cones on the flanks of Emi Koussi, in contrast to the smooth hummocks of Elysium Mons. Despite these differences, Malin suggested that Elysium Mons, like Emi Koussi, is a composite volcano built of both ash and lava and further suggested that andesitic and rhyolitic lavas may have been involved in its formation. The latter suggestion now appears less likely, in view of the chemistry of the soils at the Viking landing sites (see chapter 13).

The most puzzling features of the Elysium volcanoes occur beyond their edges. A fracture ring approximately 400 km in diameter and centered on the summit caldera almost completely surrounds Elysium Mons. Presumably the fracture ring is caused by flexure of the crust under loading of the volcano, as suggested by Thurber and Toksoz (1978) for Olympus Mons. Outside the ring are numerous west-northwest– to east-southeast–trending, flat-floored troughs with straight walls, much like graben. The troughs are roughly parallel to some much larger fractures, the Cerberus Rupes, southeast of Elysium. The most startling aspect of these troughs is that they change northwestward into fluvial-like channels, which branch, rejoin, and meander for several hundred kilometers across the plains to the northwest. Similar, but smaller, channels also arise close to the northwest flank of Hecates Tholus. The channels are discussed more fully in chapter 10. Their origin is controversial. The two main possibilities are that they were cut by very fluid lava that erupted around the periphery of the Elysium volcanoes or that they were cut by water somehow released by the volcanic activity at Elysium, perhaps by melting of ground-ice.

HELLAS REGION

Around the rim of Hellas are several old, sprawling, volcanic features. Hadriaca Patera, at 264°W, 30°S, on the northeast rim has at its center a smooth-floored caldera 60 km in diameter (figure 7.25). Outside the caldera are numerous low ridges which, to the south, toward the Hellas basin, extend as far as 300 km from the caldera rim. Although no photogrammetric measurements have been made, the volcano appears to have little vertical relief, less than 2 km judging from the photometry (Carr 1976). The radial ridges resemble those around Alba, which have a lava channel at their crest. Indeed, the whole volcano, with its low profile, radial ridges, and inner ring, resembles Alba, although considerably smaller in size. Several large impact craters up to 40 km in diameter suggest that the volcano is old. A large channel starts on the southeast flank of the volcano in a steep-walled spatulate depression 40 km across. The channel, Harmahkis Vallis, extends southwestward for 800 km before merging with the floor of Hellas. Hadriaca thus provides us with a third example, in addition to Elysium Mons and Hecates Tholus, of a volcano around which large channels start.

Figure 7.25. Hadriaca Patera (30°S, 267°W) on the northeast rim of Hellas. The volcano center is at the top of the frame, and the low ridges extending outward are probably analogous to the ridges in the lower left of figure 7.20. The origin of the steep-walled depressions is not known. They give rise to channels that extend southwest, down into the Hellas basin, to the right of the picture. Picture width is 275 km. (106A09)

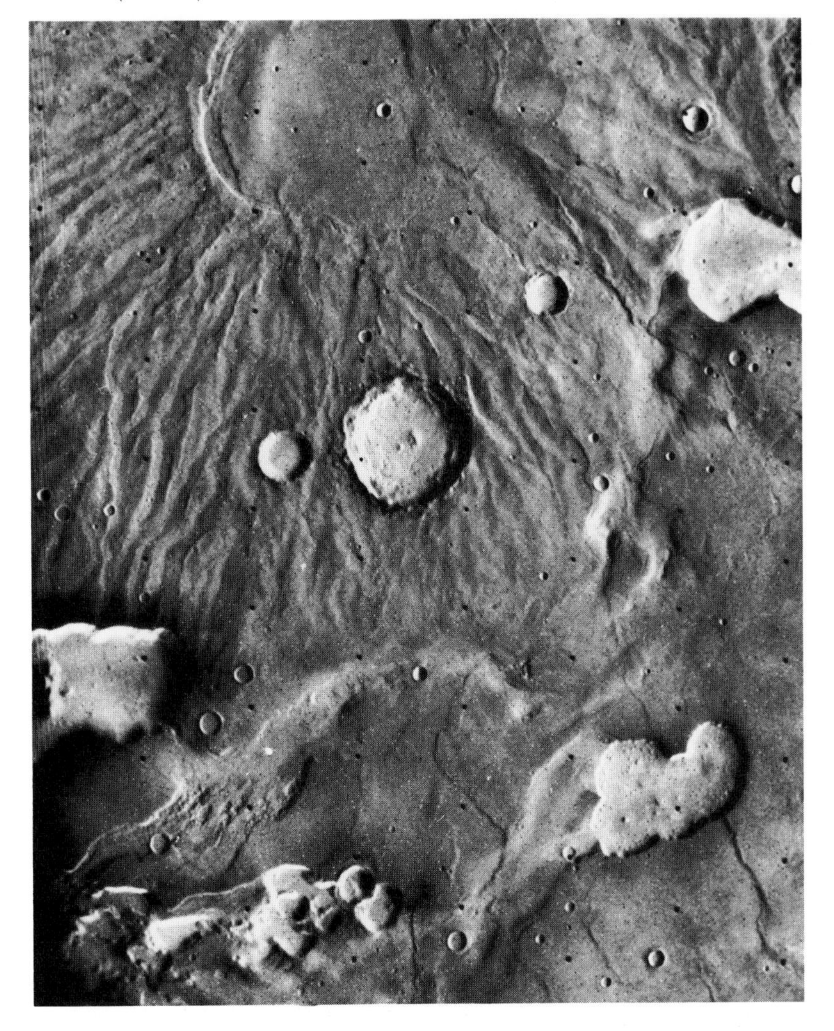

On the south rim of Hellas are several indistinct ring structures from which radiate numerous ridges, similar to those around Hadriaca Patera (Peterson 1978). The 110-km–diameter Amphitrites Patera, at 58°S, 298°W, is the only such volcano formally named (figure 7.26). The characteristic ridged pattern around these centers can be traced northward for 400 km, well into the Hellas basin. The centers have little relief. They merge outward with the ridged plains that occur over much of the south rim of Hellas and appear to be the source of the lava flows that constitute the plains.

Another low-profile volcanic center, Tyrrhena Patera (figure 7.27), occurs in Hesperia Planum, 800 km northeast of Hadriaca Patera. A 12-km–diameter caldera at the center is surrounded by a 45-km–diameter fracture ring. Around the fracture ring, the terrain appears highly eroded, forming ragged outward-facing escarpments, as though successive flat-lying layers had been eroded back. The escarpments are not primary or tectonic features, since numerous outliers occur as mesas beyond the escarpment edge. Incised into the sequence are several flat-floored channels that extend outward as far as 200 km from the volcano center. The extreme dissection of the volcano, as compared with Hadriaca and Amphitrites Paterae or the surrounding ridged plains, which are all of comparable age, suggests that Tyrrhena Patera is composed of easily erodible materials. One possibility is that the volcano is composed largely of ash, although it must have been dispersed over large distances, since the escarpments occur as far as 300 km from the volcano center. The profile is very similar to that given by Pike (1978) for alkalic and calc-alkalic ash flow plains with central calderas. The origin of the channels is uncertain. They more closely resemble those that start near Hadriaca Patera and Elysium Mons than the narrow, deeply incised channels of Hecates and Ceraunius Tholus that Reimers and Komar (1979) attributed to volcanic density currents. Again, a fluvial origin is possible, with the release of water being triggered by volcanic activity, as we shall see in chapter 10. It is the oldest of the central constructs, the only large volcano to be extensively dissected, and the only one that appears to be comprised of horizontal sheets.

Apollinaris Patera (figure 7.28), an isolated volcano at 9°S, 186°W, is 400 km across, has a central, roughly circular caldera 70 km in diameter, and appears to have a low profile, although no accurate elevation data are available. The main edifice is surrounded by a low escarpment, except to the south, where a broad spur extends from close to the caldera rim south-

Figure 7.26. Edge of Amphitrites Patera, on the south rim of Hellas. The vague circular ridges at the bottom of the frame mark the edge of the central ring. Roughly radial ridges in the upper half of the frame extend northward into the Hellas basin (see figure 6.25). The scabby appearance is typical of many high-latitude areas. Compare with figure 11.10, for example. The frame is 80 km across. (578B01)

ward for about 350 km. On the spur is a fan-shaped array of low ridges. Numerous fine radial channels cut the flanks around the rest of the volcano. Apollinaris appears to have had a similar history to Arsia Mons, in that after the main edifice was built, eruption continued on the volcano flank, to form a broad ridge adjacent to and overlapping the original shield.

SMALL VOLCANIC FEATURES

Most of the literature on martian volcanism has focused on the large central constructs and the lava plains. These are large enough that secondary characteristics, such as calderas, lava flows, levees, flow fronts, and so forth, can be readily identified, and a volcanic origin can hardly be doubted. In many areas, however, there are numerous small (< 5 km) structures that can also plausibly be interpreted as volcanic (West 1974; Peterson 1978; Hodges 1979). Generally the features are so small that the only clue to their volcanic origin is their overall shape. Subsidiary features, such as flows and lava channels, cannot be seen. Nevertheless, the shape alone may be sufficiently diagnostic that a volcanic origin is more plausible than any other possibility. The most common of these small putative volcanoes are cratered domes. They are particularly numerous in Isidis Planitia (figure 7.29) but also are found in Mare Acidalium, Utopia Planitia, and east of Hellas, around 45°S, 240°W (Hodges 1979). In all

Figure 7.27. Tyrrhena Patera, at 20°S, 252°W. A central circular depression at the left of the picture is 45 km across and is surrounded by circular fractures and highly dissected horizontal sheets. The structure may be comprised largely of ashflows, in contrast to the shield volcanoes, which are mostly lava. (211-5730)

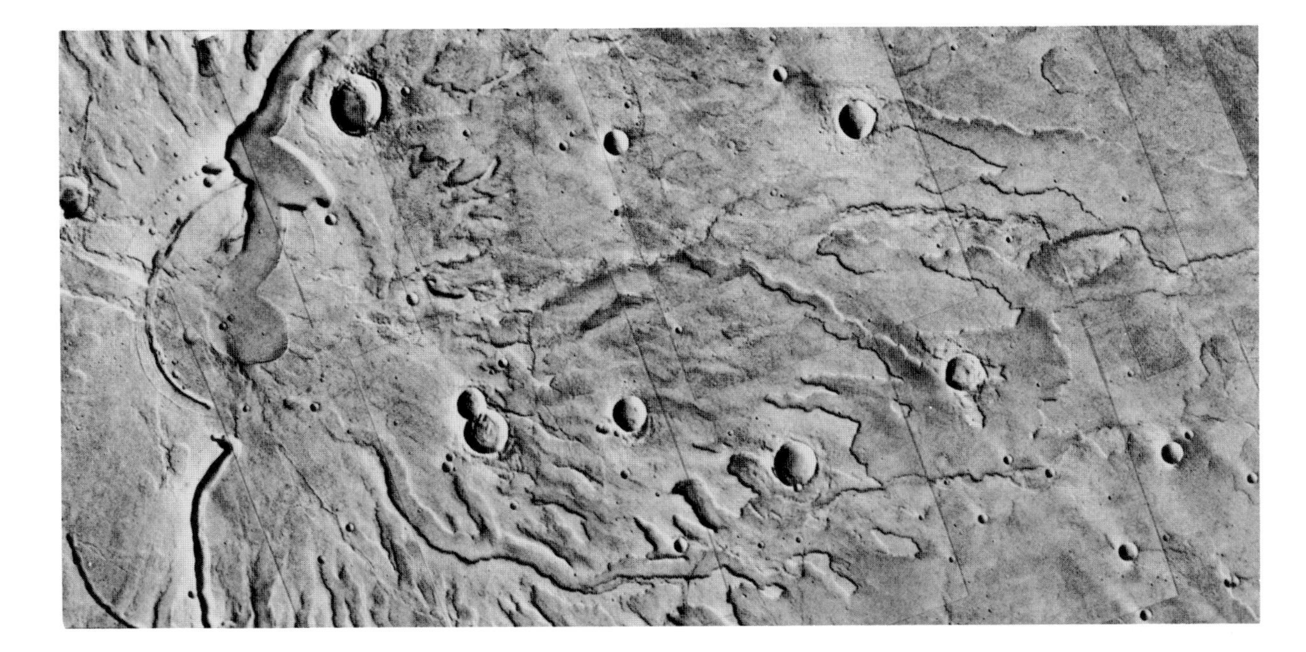

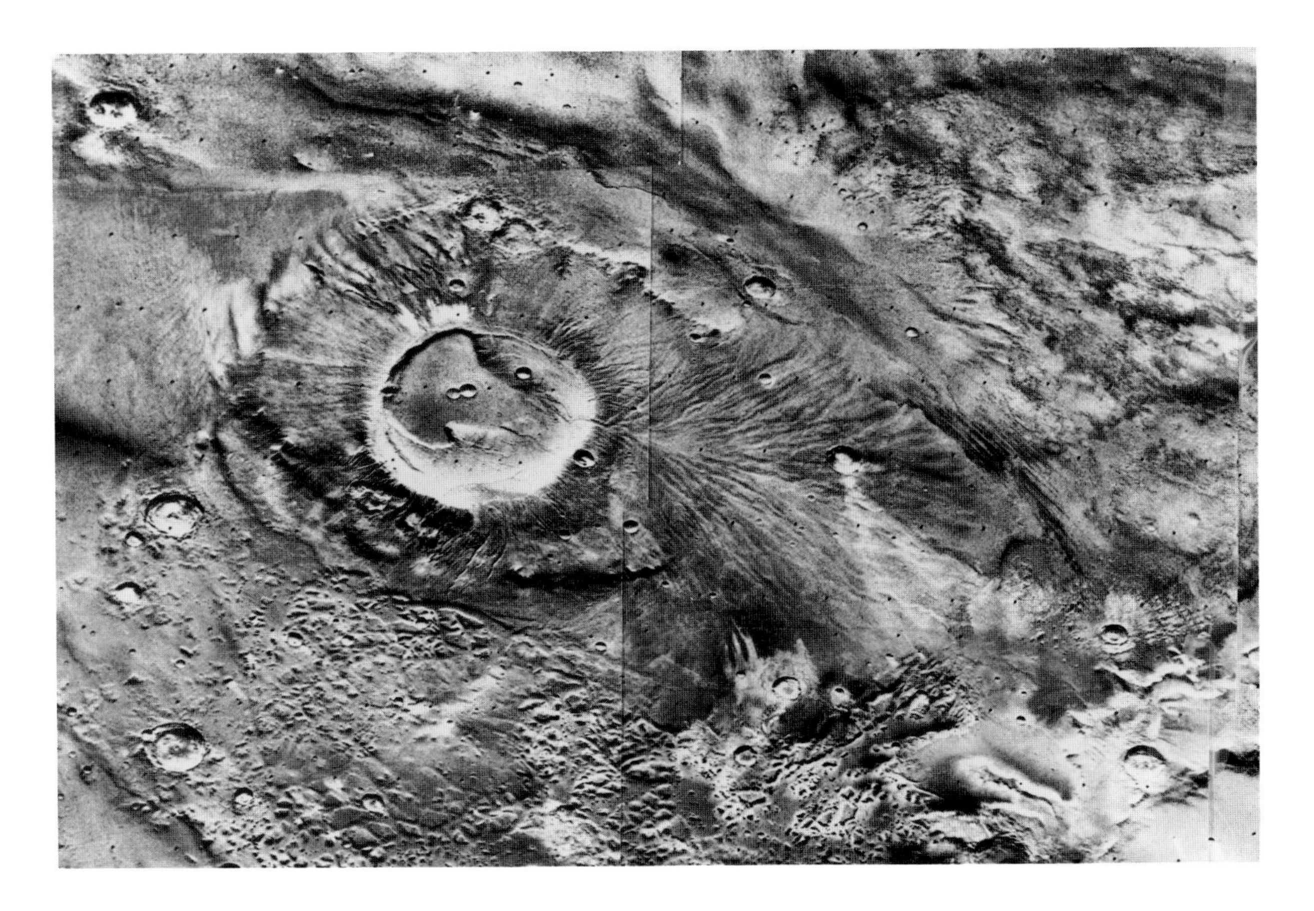

Figure 7.28. Apollinaris Patera, with its 100-km–wide central caldera. The plumelike array of flows to the southeast is somewhat similar to that associated with rift zones on the Hawaiian shield volcanoes. (P-18143)

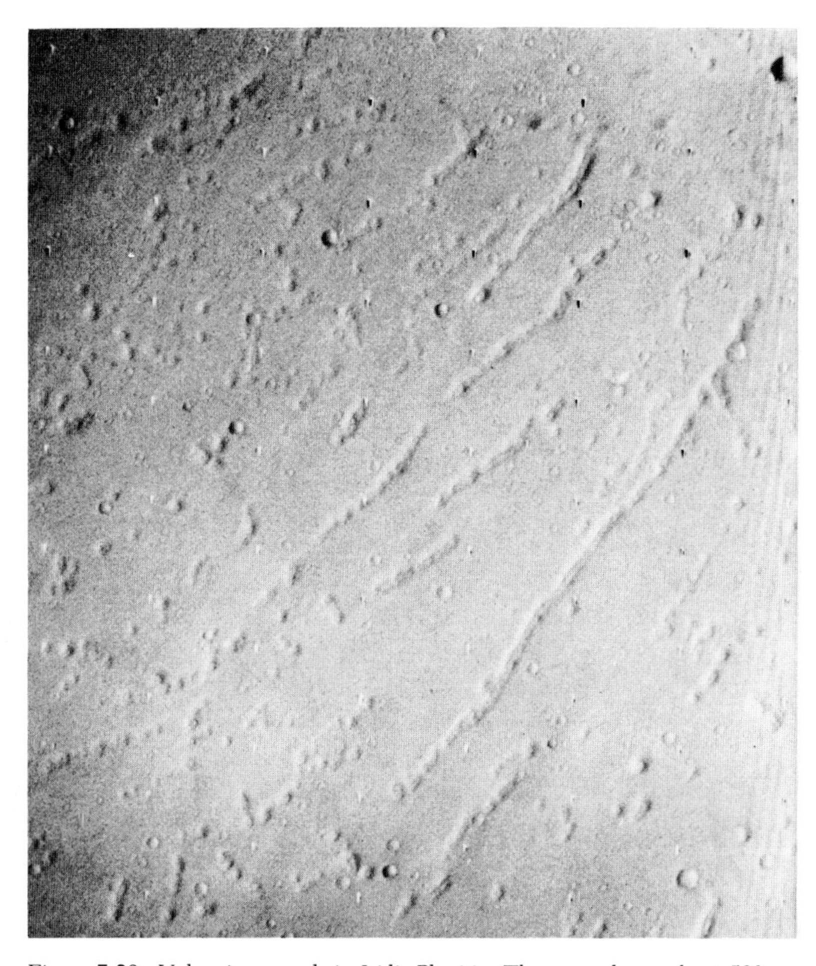

Figure 7.29. Volcanic mounds in Isidis Planitia. The mounds are about 500 m across and are mostly topped by a small crater and arranged in lines. They may be spatter cones that formed along the fissures from which the lava that formed the surrounding plains erupted. (146S23)

Figure 7.30. Low volcanic domes at 43°S, 239°W, east of Hellas. Most are from 400 m to 1 km across and some are slotlike summit vents. The frame is 38 km across. (586B34)

these areas are numerous low, convex-upward domes with central pits. East of Hellas many of the domes are elongate and darker than their surroundings and have slotlike vents (figure 7.30). In Isidis, the domes are arrayed in long strings, apparently along faults. All occur on what appear to be lava plains. They are generally several hundred meters across and strongly resemble spatter cones and ramparts, both in shape and size. They may well mark the vents from which the lavas that formed the surrounding plains erupted. Similar domical features have also been interpreted as resulting from eolian modification of impact craters (see chapter 4). The evidence for modification of impact craters to produce pedestal craters, and in the case of small craters, cratered domes, is strong. However, it is unlikely that impact could produce the lines of domes that occur throughout Isidis or the domes with summit slits east of Hellas. It appears that two processes, impact and volcanism, can produce similarly shaped cratered domes, and in many areas, it cannot be determined which of the two processes was involved.

Also of ambiguous origin are the cratered cones described by West (1974). Most of West's examples are from the knobby terrain at the southeastern edge of Elysium Planitia. The formation of the knobby terrain by breakup of the old cratered terrain is well established (see previous chapter). Remnants of old craters, shoulder to shoulder, are common, and the transition from unaltered cratered terrain to knobby terrain can be traced in detail. The presence of occasional knobs with summit craters should not be surprising, since the original surface that broke up was heavily cratered.

The presence of craters may well have controlled the breakup and left the craters on top of plateau remnants. Nevertheless, some of the cratered knobs do resemble stratocones, and the possibility of a volcanic origin should be left open.

Less ambiguous are many small volcanoes in the Tempe region. These are low conical constructs that merge with the surrounding plains (figure 7.31). Many lie on faults and have slotlike vents. They are generally a few hundred meters across. Many of the graben on the plains west of the Tempe plateau have on their floors lines of craters that may also be volcanic. In addition, in southwest Utopia and southwest of Olympus Mons are several flat-topped pitted mountains that show a remarkable resemblance to table mountains (Hodges and Moore 1979; Allen 1979a,b) which on Earth result from subglacial eruptions. Finally, Greeley and co-workers (1977) noted several low circular shields less than 10 km across in south Chryse Planitia. Some lie astride a long linear feature that is interpreted as a dike. Thus, examples of small volcanic features are numerous, and no doubt many more will be found as the Viking pictures are more closely examined.

COMPARISON OF MARTIAN AND TERRESTRIAL VOLCANISM

Terrestrial volcanoes range widely in shape, from the gently sloping shield volcanoes composed mostly of lava, through the classic stratocones built of both ash and lava, to cinder cones and tuff rings composed almost entirely of fragmental debris. The style of volcanic activity and the shape of the resulting volcano depend strongly on the fluidity of the lava. Fluidity can affect the shape of a volcano simply because for more viscous lava to flow requires steeper slopes, thereby producing more steeply sloping features. However, a more important effect of fluidity is that it controls the rate of outgassing. Lava is brought to the surface from great depths. As the lava ascends, gases that were dissolved in the magma under large confining pressures start to exsolve and may help in driving the magma to the surface. If the lava is fluid, it is easily forced out of the vent and once out, it will continue to outgas relatively quietly. Observations of lava lakes in Hawaii indicate substantial volume reductions as a result of outgassing. If the lava is viscous it may block the vent. Gas pressures can then build up, leading ultimately to explosive eruptions. The explosive activity may be limited in scale and merely result in explosive ejection and disruption of clots of lava. It can also occur on a grand scale and cause destruction of much of the volcano, as with Mt. St. Helens in 1980 and Krakatoa in 1883. When the magma is viscous, the exsolved gases tend to blow the cooling lava apart and produce ash and cinders (pyroclastic debris) rather than coherent lava flows. The fragmental debris may be spread over large areas to produce bedded tuff deposits or fall close to the vent to produce a steeply sloping cone. Thus at one extreme, with fluid lavas, we have quiet effusion of lava to form lava plains or shield volcanoes, and at the other extreme, with viscous lava, we have explosive eruptions with production of large quantities of pyroclastic debris and no lava. More commonly, both ash and lava are erupted, and a stratovolcano or stratocone forms. Usually basaltic lavas are fluid and produce constructs with low profiles, whereas nonbasaltic lavas, such as andesites, trachytes, and rhyolites, produce steeper-profile volcanoes with high proportions of ash and cinders. But this is only a generality, for basaltic eruptions can result in steeply sloping features,

Figure 7.31. A small volcano in the Tempe region (36°N, 86°W), to the northeast of Tharsis. An old surface, faulted by Tharsis radials, on the right is embayed by younger plains on the left. The volcano, near the center of the picture, is 4 km across, and has an elongate summit vent. (627A28)

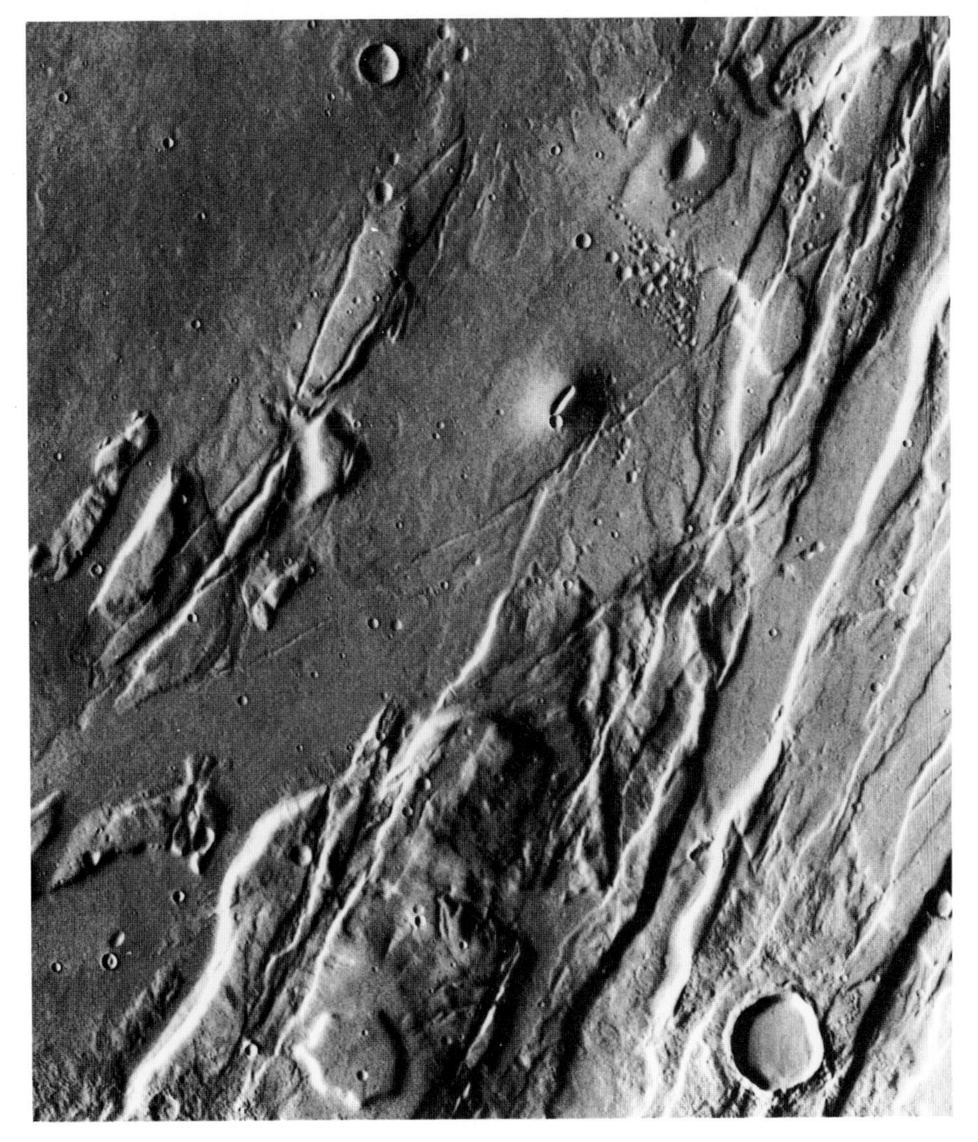

especially if eruption rates are low, or large amounts of ash are produced as a result of interaction of the lava with groundwater or seawater.

The type and location of terrestrial volcanoes are governed largely by plate motion (figure 7.32). Most, but not all, volcanism is concentrated along the plate margins, and the type of volcanism depends on the type of plate junction. Along midoceanic ridges volcanic activity is almost exclusively basaltic. The only exceptions are where major mountains form, such as in Iceland, where small amounts of more silicic magmas may also be erupted. Along subduction zones volcanism is more varied. Where a continental plate overrides a downgoing oceanic plate, as along the western coast of South America, the continental plate margin may be zoned, with andesites and basalts near the junction and more potassic rocks inland. At the junction of two oceanic plates, as in the Philippines and the Marianas, the volcanism is predominantly andesitic. Thus, along regions of plate convergence the volcanism tends to be nonbasaltic, and the typical volcano is a stratified cone. Several types of volcanic features are difficult to relate to plate margins, although they may have been affected by plate motion. Included are some shield volcanoes, such as those in Hawaii, and the plateau basalts of northwestern United States, India, and South America.

On Mars there is no evidence of plate motion. The arcuate chains of mountains that characterize terrestrial subduction zones are totally absent. Indeed compressional features of any type are rare. Where layering is visible, it is almost always horizontal, with no indication of folding. Evidence of strike slip movement, which on Earth occurs as a result of plates moving laterally with respect to one another, is also absent on Mars, despite the survival of large areas of ancient cratered terrain in which such movements could easily be detected. It appears that the crust of Mars is stable and is not broken into moving plates like that of Earth. The type of volcanism on Mars is correspondingly different. Volcanic cones, which typically occur along subduction zones on Earth, are relatively rare, and it is not unreasonable to assume that the andesitic and alkalic magmas that typify subduction zone volcanism are also rare. Such an assumption is consistent with the limited compositional data we have for Mars, which suggests that most of the primary rocks are iron-rich and basic or ultrabasic. The most prominent volcanic features on Mars are shield volcanoes and flood basalts, both of which occur within plates on Earth and are difficult to relate to plate motion. The martian shield volcanoes, however, grow to much greater sizes than those on Earth, probably because of the greater stability and thickness of the martian crust.

HISTORY OF VOLCANIC ACTIVITY

The history of volcanic activity on Mars is dominated by the formation of lava plains, discussed in the previous chapter, and large shield volcanoes. The history can be reconstructed reasonably well from crater ages on different volcanic features. The oldest recognizable volcanics are the ridged plateau plains which, according to Greeley and Spudis (1978), cover 36 percent of the old cratered terrain. Flow fronts are rare on these plains, but their vast extent, smooth surface, spectral signature, and wrinkle ridges all support a volcanic origin. Most of the plateau plains are ancient, having more than 2,800 craters greater than 1 km per million km^2, which is as many as any other part of the planet except for the rough plateau areas of the densely cratered terrain. Despite an age that is probably in excess of 3.5 billion years, most of the craters on the ridged plains are relatively fresh-appearing, indicating that they formed after the early era of intense crater obliteration was over, and after the early high cratering rates had declined. It is unlikely that the start of intense volcanism coincided with the decline in cratering and obliteration. It is more probable that volcanic activity was continuous, but that we can only recognize as volcanic those surfaces that formed after the landscape stabilized. All evidence of the volcanism that occurred before 3.9 billion years ago has been either lost or is unrecognized.

The next oldest recognizable volcanic units are the floors of Hellas, Hesperia Planum, and Tyrrhena Patera, all of which have crater numbers in excess of 2,600. These features are still within the cratered hemisphere. Slightly younger, with crater numbers between 2,000 and 2,600, are most of the ridged plains of the sparsely cratered hemisphere, including Lunae Planum, Solis Planum, Isidis Planitia, and Syrtis Major Planitia. Some of the more intensely cratered plains peripheral to Tharsis and most of the small Tharsis shields also formed at this time, which, in absolute terms, is probably close to 3 billion years ago. At a slightly younger age are the Elysium, Amazonis, Syrtis, and Noachis plains, Alba Patera, and the Elysium volcanoes, all of which have crater numbers close to 2,000 and probably date from before 2.5 billion years ago.

Since this time volcanic activity has been almost completely restricted to Tharsis. The only exceptions are the plains of Mare Acidalium and Utopia, which have crater numbers between 800 and 1,500, and possibly Apollinaris Patera, with a crater number of 990 (see table 7.1). Within Tharsis volcanic plains date from in excess of 3 billion years ago to the present. Volcanic activity has thus been continuous for at least two-thirds of the planet's history.

The surfaces of the large Tharsis shields and Olympus Mons are very young, with average ages of a few hundred million years or less. The ages of the edifices, however, are unclear. As we saw above, the edifices could have been accumulating for a considerable time. For example, the range in age of flows south of Arsia Mons, which appear to be derived from the Arsia center, is around 2.5 billion years. Comparable accumulation times for Ascraeus, Pavonis, and Olympus Mons seem reasonable. Volcanic activity on the southeast flanks of the Tharsis bulge terminated earlier than that in Tharsis proper. Counts in the Sinai-Solis plains range from approximately 2,300 down to 1,000, indicating an approximate age range of 3 to 1.5 billion years.

Figure 7.32. A generalized map of the plate system of Earth. The plates move in the directions shown at rates of a few centimeters per year. New lithosphere forms along the midoceanic ridges; old lithosphere is subducted into the underlying asthenosphere at zones of subduction, as shown in the idealized cross section. Most of Earth's seismic and volcanic activity is concentrated along the subduction zones, which are also sites of enhanced sedimentation, since their traces often coincide with ocean deeps adjacent to mountain chains. The thick sedimentary and volcanic sequences that accumulate along the subduction zones commonly become folded and metamorphosed and may ultimately form mountain chains. This tectonic framework provides a striking contrast to that on Mars, where there are no plate tectonics. (Adapted from Press and Siever 1978.)

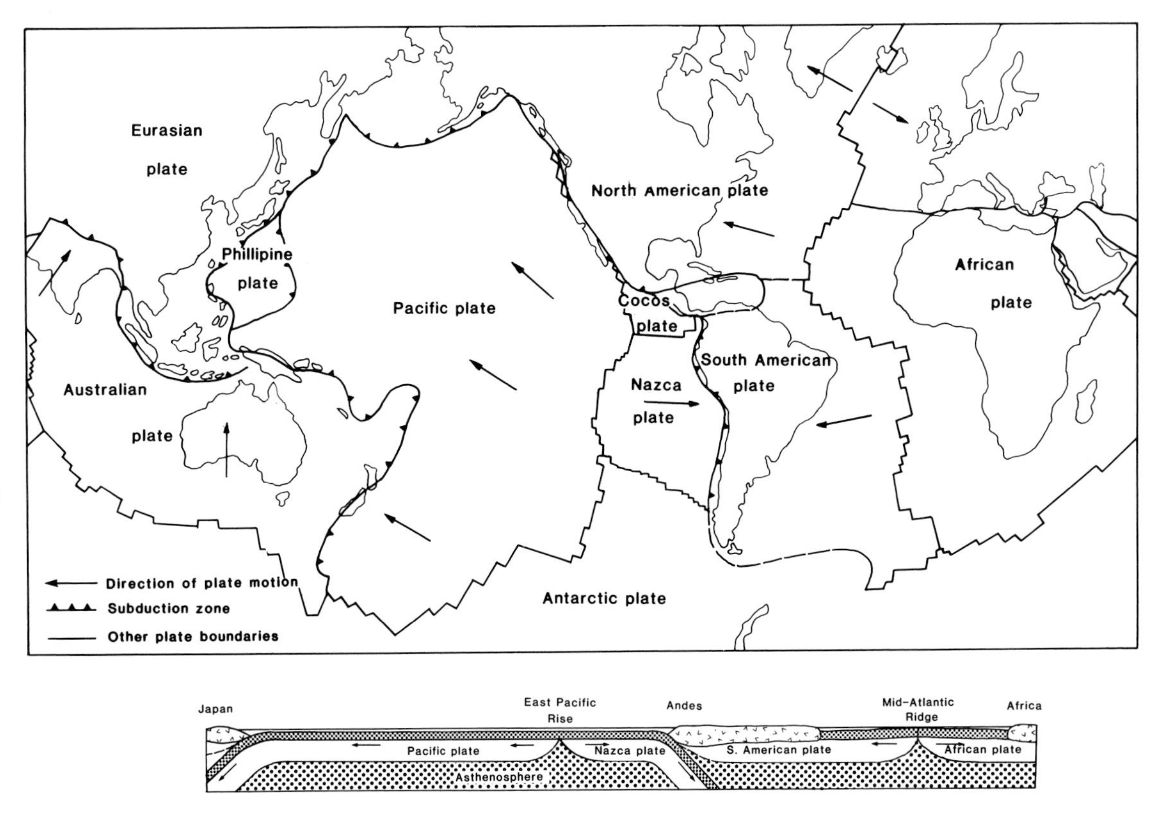

Thus the history of volcanic activity on Mars is one in which activity became progressively more restricted with time. Between 2.5 and 4 billion years ago extensive volcanic activity resulted in the formation of most of the sparsely cratered plains of the northern hemisphere and the ridged plains of the densely cratered terrain. After 2.5 billion years ago, activity was largely restricted to around the Tharsis bulge and to some of the northern plains, while in the last billion years, volcanism occurred exclusively near the crest and on the northwest flank of the bulge.

8 TECTONICS AND THE THARSIS BULGE

The tectonic history of Mars is vastly different from that of Earth. Deformation of Earth's crust is controlled by the constant motion of the lithospheric plates with respect to one another (see figure 7.32). At subduction zones thick sedimentary sequences accumulate in linear troughs, then become squeezed, folded, and metamorphosed as they are caught between the converging plates. Enormous transcurrent fault zones develop where one plate slides by another, and rift zones form where plates diverge. The whole system is dynamic, with stresses in the lithosphere constantly changing as the configuration of the plates alters. Despite arguments that the equatorial canyons may be analogous to terrestrial midoceanic ridges (see next chapter), the surface of Mars appears to have been quite stable throughout its history. There are no linear depositional trenches, no mountain chains, few, if any, folded rock sequences, and no large transcurrent fault zones. Moreover, as we shall see below, the stresses within the crust appear to have maintained one configuration for billions of years, in striking contrast to the continuously shifting patterns of Earth.

The tectonics of Mars are dominated by the Tharsis bulge. Away from the bulge deformation features are rare. Several discontinuous escarpments that surround the Hellas, Argyre, and Isidis basins probably result from relaxation of the crust following basin formation. Circular fractures around some of the large shield volcanoes, particularly Ascraeus Mons and Elysium Mons, appear to have formed by loading of the lithosphere by the volcanoes. Some crevasselike features that run parallel to the plains/uplands boundary in places may be caused by faults that are in some way connected with the destruction of the old cratered terrain in the plains hemisphere, and polygonal ground (see chapter 6) may be tectonic in origin and result from stresses caused by regional warping. Almost all other deformation features appear to be related to Tharsis. One possible exception is a series of fractures on the flanks of the Elysium bulge, but even here the fractures are best developed where the Elysium and Tharsis radial directions are roughly coincidental.

The Tharsis bulge has clearly played a major role in the evolution of the planet. Radial fractures around the rise affect almost an entire hemisphere, and wrinkle ridges which resemble those on the lunar maria, occur around much of the bulge. The largest and youngest volcanoes are on its flanks, as is the vast system of equatorial canyons. The bulge is so large that its formation may have affected precession of the planet's rotation axis, thereby affecting climates (Ward et al. 1979). It has even influenced surficial processes by perturbing the wind regime and possibly by controlling migration of groundwater. But the bulge is of interest not only for its effects on surface geology but also for the clues it provides concerning the nature of the planet's interior. Coincident with the bulge is a large gravity anomaly, which provides a means of assessing the structure of the crust and the density distribution down to depths of several hundred kilometers. The anomaly, in addition, gives some indication as to whether the bulge is being actively supported from below, such as by convection, or whether it is supported passively by the strength of the crust. Somewhat less definitively the bulge provides clues concerning events that took place early in the history of the planet, since formation of the rise must have been related in some way to dynamical and chemical anisotropies, such as might be associated with mantle convection, core formation, or an uneven distribution of radioactive elements. Discussion of the Tharsis bulge will thus touch on some of the broadest issues concerning the evolution of the planet.

TOPOGRAPHY AND PHYSIOGRAPHY

The crest of the Tharsis bulge is close to the west end of Noctis Labyrinthus, at approximately 0° latitude and 105°W (Christensen 1975; Wu 1978). The exact location cannot be specified because of large uncertainties in the surface elevations. The crest region is somewhat diamond-shaped, with a broad, relatively flat summit and with flanks that slope away, mostly in directions diagonal to the latitude-longitude grid. To the northeast and northwest the slopes are steeper, being mostly in the 0.2–0.4° range; the southeast and southwest slopes are about half these values. Superimposed on this broad regional high are local topographic features, such as volcanoes, canyons, and fracture zones. To the north the regional slopes continue down below the −1-km level into the low-lying plains of Chryse and Amazonis Planitiae. To the south the outward-facing slopes are maintained only down to about the 4-km level, beyond which the bulge merges with the relatively high southern uplands. The bulge is therefore asymmetric, being steeper and more extensive to the north, probably as a result of straddling the plains/upland boundary. If the 4-km contour is taken as the base of the bulge, it is 5,500 km across and 7 km high, but to the north, of course, it is far more extensive.

The bulge encompasses the most intensely and most recently active volcanic region of the planet. As we saw in the previous chapter, almost all the volcanoes are on the northwest flank. The cause of the preferred location is not known, but, like the topographic asymmetry, it may be related to the plains/upland boundary. On the southeast flank the cratered uplands are almost everywhere close to the surface, and volcanoes are absent. To the northwest ancient cratered terrain is rarely exposed, and it is presumably buried much more deeply. The distribution of volcanic plains is less asymmetric. Large lava flow fields occur on the southeast flanks, in Solis and Syria Planum, for example, as well as to the northwest.

The extent to which the topographic high results from upwarping, and the extent to which it results from accumulation of volcanic materials at the surface are unclear. Remnants of the old cratered terrain occur at an elevation of 7 km above the datum south of Arsia Mons and the Solis-Sinai region. Old fractured craters are also visible in the Claritas Fossae at elevations as high as 8 km. The relations could be explained in two ways. The old cratered terrain may have been uplifted to form the bulge. Since the terrain around the bulge is 3–4 km above the datum, uplift of at least 3–5 km is implied. Alternatively, no uplift occurred, but volcanics started accumulating on the bulge before the decline in the cratering rate 3.9 billion years ago. The predecline volcanics are thus heavily cratered and are indistinguishable from the rest of the cratered terrain. This explanation implies that at least 3.5 km of volcanics had accumulated by the time of the decline.

One of the most striking features of the bulge is its extensive array of roughly radial fractures (figure 8.1). The fractures appear to be almost all normal faults. No strike-slip faults have been positively identified, despite numerous transections of craters where strike-slip movement would be very obvious. Radial compressional features are also rare, although Wise and colleagues (1979) suggested that the Claritas Fossae, south of the Tharsis summit, may be in part an anticlinorium. Single isolated fractures are uncommon; most form graben. In some places, such as Memnonia, southwest of the bulge, the graben maintain a continuity and parallelism for several hundred kilometers, indicating a uniform stress over large areas. In other places, such as north of Tharsis, several fault sets, each with a slightly different orientation, intersect one another, creating complex patterns that suggest episodic shifts in the stress orientations (figures 8.2 and 8.3).

Several attempts have been made to reconstruct the sequence and timing of faulting around the Tharsis bulge (Wise et al. 1979; McGill 1978; Frey 1979; Plescia and Saunders 1980). Three kinds of evidence are used: (1) intersections between fractures with different orientations; (2) crater densities on fractured and unfractured surfaces; (3) transection relations between surface units and fractures. The information is often ambiguous. A surface may be partly fractured and partly unfractured, for example. This could be explained by localization of the fracturing or by partial burial of a fractured surface by an unfractured unit. Nevertheless, the broad outlines of the deformational history around Tharsis are relatively clear. The period of most intense fracturing postdates the surface of Lunae Planum and features of equivalent age. These surfaces have about 120 craters in the 4–10 km size range per million km^2 (Condit 1978), or 2,400 craters larger than 1 km per million km^2 (Gregory 1979), suggesting an age of about 3.5 billion years. Most fractures also cut the cratered plains of Scott and Carr (1978), which have corresponding crater densities of 89 (Condit 1978) and 1,800–2,400 (Schaber et al. 1978), respectively, suggesting ages in the 2.5–3.5 billion years range. Surfaces younger than the cratered plains unit are only sparsely fractured. The fracturing thus appears to have tapered off rapidly after deposition of the cratered plains, and while there is considerable uncertainty concerning absolute dates, the falloff probably occurred over 2.5 billion years ago.

Fracturing continued after this time, but with considerably less intensity. Occasional fractures are visible, for example, among the lava flows of central Tharsis, where many flows that clearly transect fractures are themselves cut by fractures with similar orientations. The simplest explanation is that volcanic activity and faulting took place simultaneously. The faulting apparently continued into the geologically recent past, since many of the flows are young. The general picture, therefore, is that the deformation took place mostly in the first half of the planet's history but continued to the present at a low level. The long period of deformation implies a stress system of astonishing stability, which has been sustained for much of the lifetime of the planet.

Although the general region of Tharsis remained at the center of the fracture system, the precise location of the center may have shifted slightly on occasion, as is evident from the different orientations of different aged fractures. Some of the earliest faulting may have been caused by a local high in the Thaumasia region on the southern periphery of the bulge (Frey 1979). If such faulting occurred, it terminated early, for all the possible Thaumasia radial faults are cut by Tharsis radials. Subsequent faulting was all centered on Tharsis, mostly on the Pavonis–Noctis Labyrinthus region. Plescia and Saunders (1980) suggested that the center was originally at 8°S, 100°W, near the west end of Noctis Labyrinthus. Faults that formed at this time are mainly exposed in the Ceraunius, Claritas, and Thaumasia Fossae and just north of Noctis Labyrinthus. Plescia and Saunders suggested that the center of faulting then shifted to 4°S, 110°W, close to Pavonis Mons. Relatively early, intense faulting about this center resulted in the Memnonia, Sirenum, and Mareotis Fossae and the faults within the canyon. After this early intense episode faulting continued about the same center at a relatively low level up to the present time, resulting in the fractures among the young flows of central Tharsis. Wise and co-workers (1979) deduced a similar sequence of faulting, although they gave a somewhat different location, 14°S, 101°W, for the center of the fracture system.

A plausible explanation for the stability of the stresses is that they are caused by the bulge itself, and that the long-sustained deformation merely reflects the long persistence of the topographic high. Phillips and col-

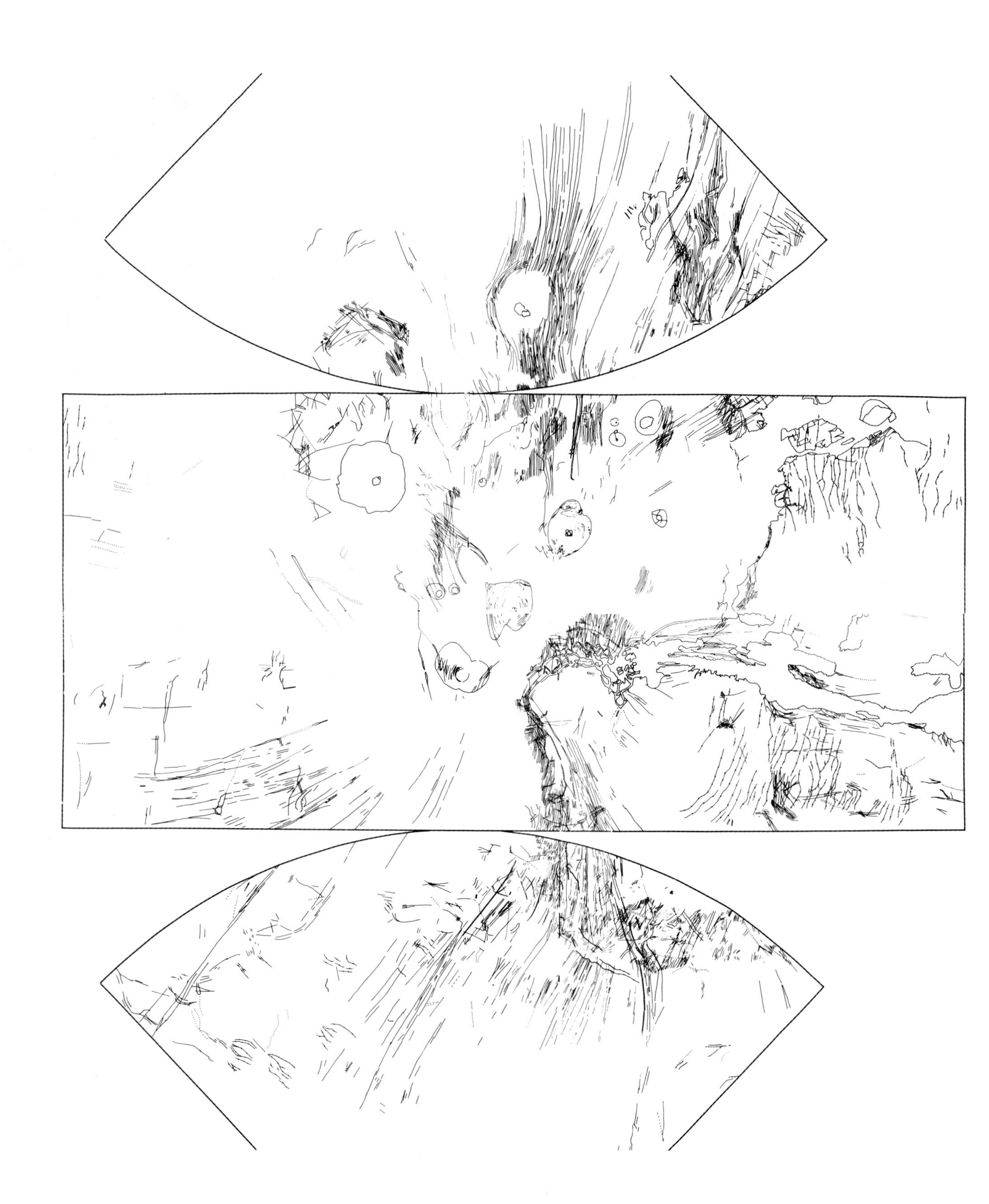

Figure 8.1. Deformational features around the Tharsis bulge. The central part of the map covers the area that extends from 45°W to 180°W, between latitudes 30°N and 30°S. For feature names, see figure 7.1. Faults are shown as simple lines, mare ridges as beaded lines. Volcanoes, canyons, and steep escarpments are also outlined. The center of the Tharsis bulge is close to the center of the figure, near Noctis Labyrinthus and Pavonis Mons. The roughly radial orientation of the fractures is obvious. Areas around the center of the figure that lack fractures are mostly covered by relatively young lava plains. Fractures may be present in these areas but buried below the surface. Most mare ridges are to the east of the bulge and roughly concentric with it. (Courtesy J. Plescia, Jet Propulsion Laboratory.)

Figure 8.2. Fractures and volcanoes in northwest Tharsis. The larger volcano to the right is the 115-km–diameter Ceraunius Tholus, at 24°N, 97°W. Fractures of slightly different orientation cut each other and imply changes in the stress pattern within the region. Older, intensely fractured terrain appears to be partly covered by younger plains, which are only sparsely fractured. (211-5639)

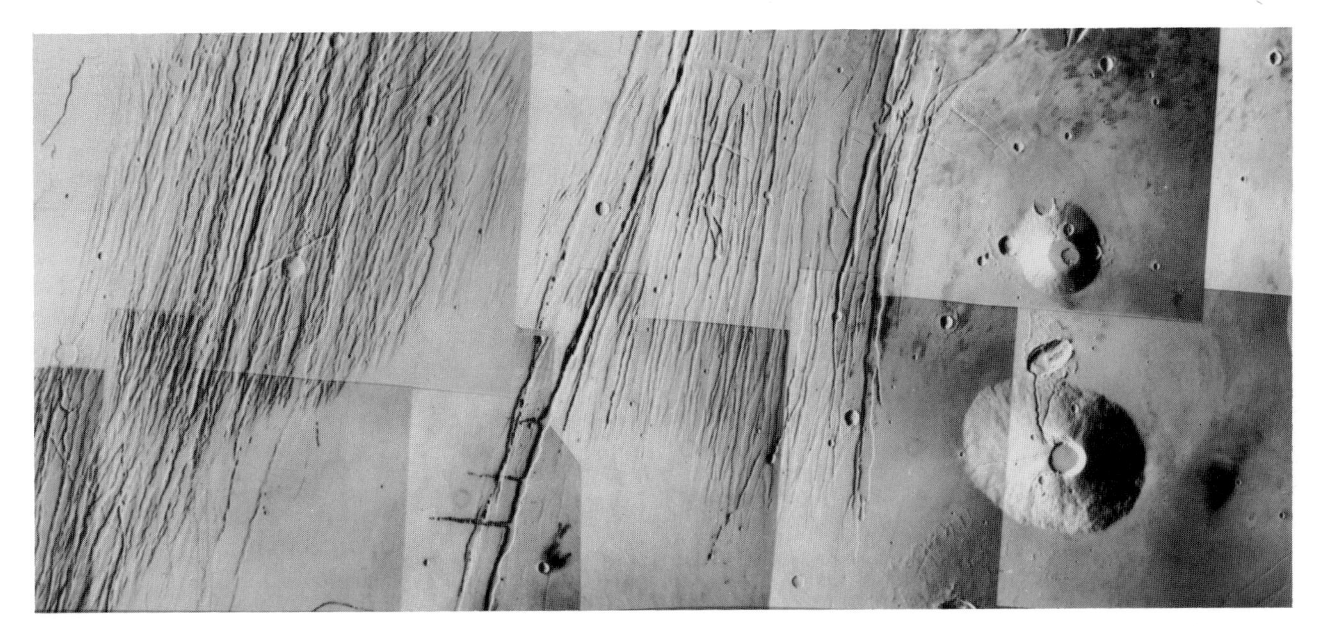

Figure 8.3. (*below*) Detail of fractures in northeast Tharsis shown in the previous figure. The fractures commonly persist with uniform spacing and orientation over long distances. Graben may have cuspate walls or include a line of central pits, as in the left of the frame. Picture width is 150 km. (39B59)

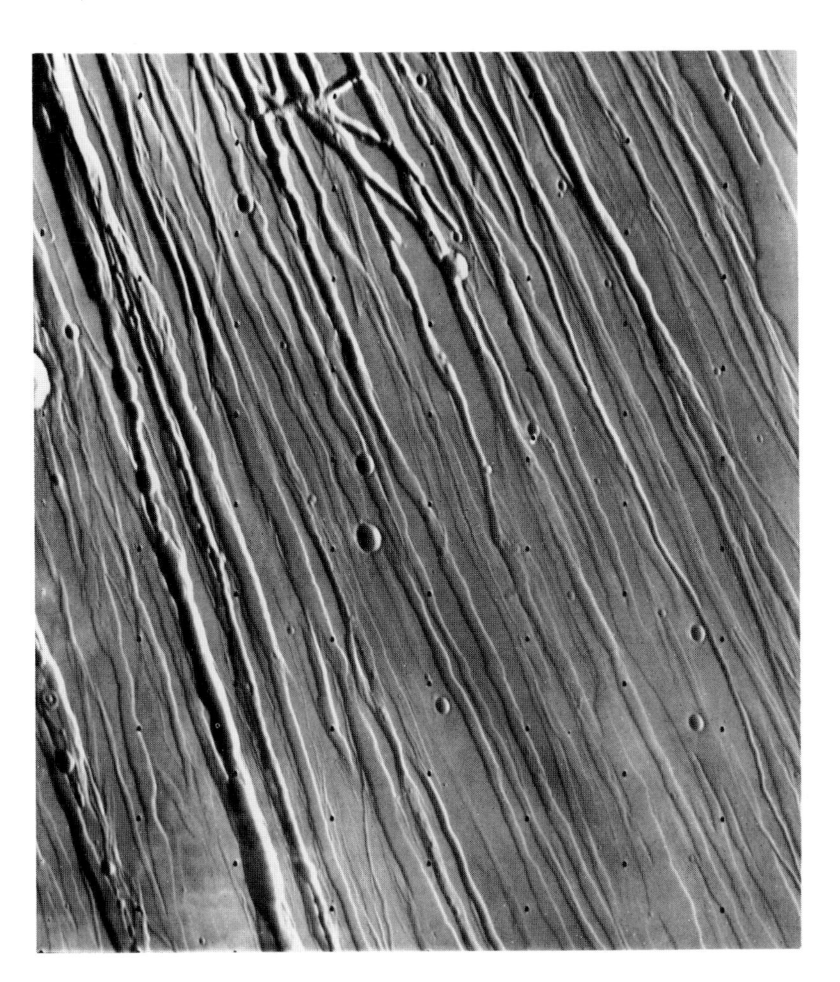

leagues (Phillips and Ivins 1979; Phillips et al. 1981) calculated the stresses in the martian lithosphere that are caused by the present areal variations in surface gravity and topography. They showed that the stress directions are strongly correlated with the orientations of deformational features such as fractures and ridges (figure 8.4). As expected, the largest stresses are around Tharsis. The directions of maximum tensile stress are mostly circumferential to the bulge, at right angles to the trend of the radial fractures. The direction of maximum compressional stress is radial, at right angles to the trend of most wrinkle ridges. The latter are particularly numerous east of Tharsis, in the Lunae, Syria, and Sinai Plana. The correlation is remarkably consistent; indeed, it is very rare in geology that such a consistent relation between theory and observation is encountered. The conclusion is almost inescapable that the fractures and ridges around Tharsis are caused by the topographic and gravity high that are now centered on it. The important point here is that the fractures are caused by the presence of the high and not by the process of its formation, as I suggested earlier (Carr 1974a). Since the fractures appear to date back at least 2.5 billion years, so does the bulge.

GRAVITY

Associated with the Tharsis bulge is a large free-air gravity high (Lorell et al. 1972; Gapcynski et al. 1977; Sjogren et al. 1975). Local highs also occur over the large volcanoes (Sjogren 1979). The gravity data are of interest in that they give an indication of the structure of the planet beneath Tharsis and of whether or not the bulge is being actively supported, such as by convection.

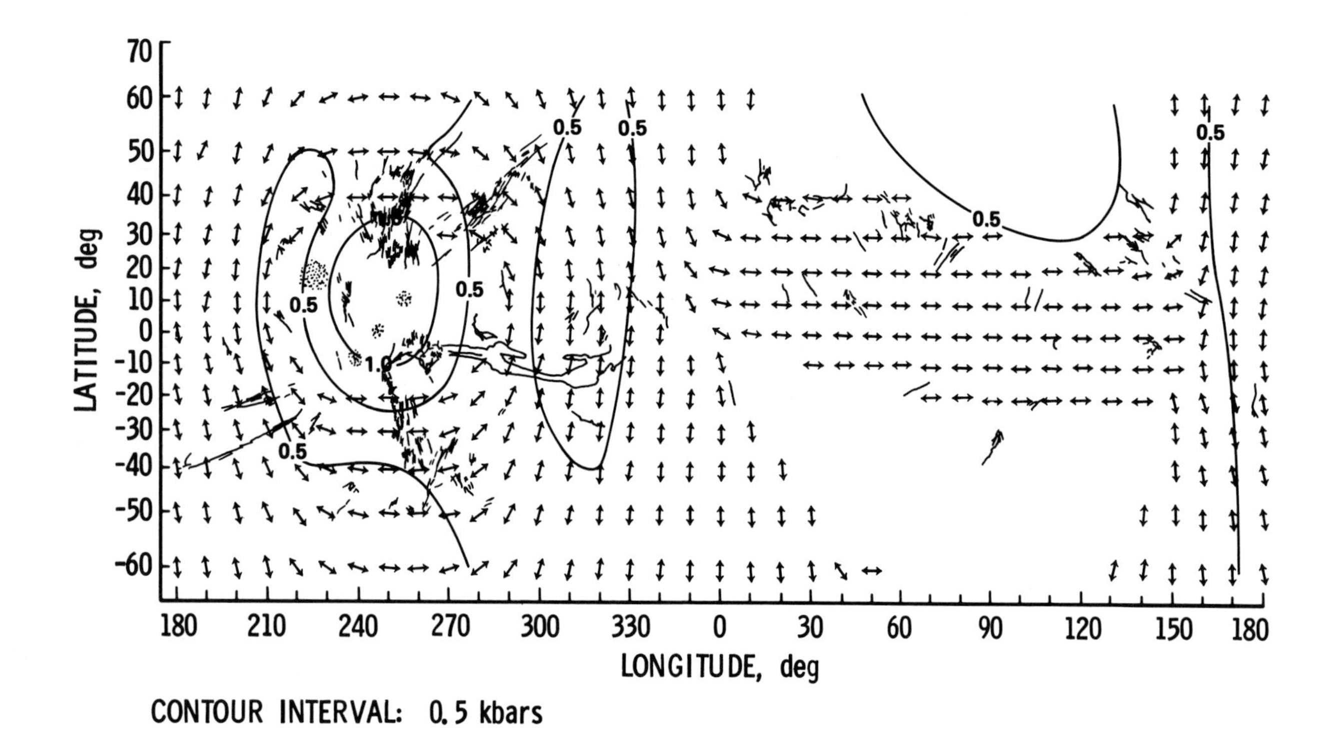

Figure 8.4. Directions of maximum tensile stresses within the martian lithosphere, which result from the present topography and gravity. The stress directions show a strong correlation with deformational features. Most are at right angles to the observed fractures and parallel to mare ridges, where present. (Published with permission from Phillips and Ivins 1979.)

If the planet were radially symmetrical, then the acceleration due to gravity at the surface would be everywhere the same. Differences arise because of topography and areal variations in the density of the subsurface materials. A free-air gravity anomaly is that which is actually measured, referenced to some mean. The free-air gravity anomaly can be corrected for the observed topography by assuming some density for the near-surface rocks and calculating what their effect would be on the gravity. What remains is termed a Bouguer anomaly and is useful for detecting density variations within the crust. On Earth free-air anomalies are normally small; mountains tend to be compensated for at depth by low-density roots, so do not greatly perturb the surface gravity. Correction for positive relief, therefore, generally results in a large negative Bouguer anomaly, indicative of less dense material below the topographic high. On most parts of Earth, because of this isostatic compensation, there exists a surface at some relatively shallow depth—a few tens of kilometers—where the topography is fully compensated; the free-air gravity anomaly is completely accounted for by topography and density variations above that surface. On Mars, as we shall see, gravity anomalies are much larger than on Earth, and so no such surface exists, at least at shallow depths, above which there is complete compensation for the topography.

Planetwide variations in gravity are commonly depicted in terms of spherical harmonics. The shape of an irregular object can be described mathematically in terms of a number of components with different frequencies. Associated with each component is a coefficient (J) that is a measure of the importance of that particular frequency in the overall shape. Low-order, long-wavelength components describe broad planetwide variations, such as a systematic variation from equator to pole. Higher-order, or shorter-wavelength components describe more local variations, such as the anomaly associated with the Tharsis bulge, or, at even higher orders, those associated with individual volcanoes. The representation is only an approximation, and in practice large numbers of terms are required to describe adequately a shape of only modest complexity.

The gravity field of Mars has been reconstructed in two ways, both involving Doppler tracking of spacecraft. In the first method (Lorell et al. 1972) samples were taken over many orbits of the position and velocity of the Mariner 9 spacecraft, then a mathematical model of the gravity field was constructed to explain the observed motions. This model provided a good description of the low-frequency variations in the gravity field but did not provide reliable high-frequency information, partly because of the relatively high periapsis altitude of the Mariner 9 spacecraft (1,650 km) and partly because of the inability of the mathematical modeling to describe accurately the high frequencies. In the second method, which was applied to the second Viking orbiter (Sjogren 1979), the gravity field was reconstructed directly, by continuously monitoring the changing velocity of the spacecraft as it passed through the nonuniform gravity field. By virtue of the low altitude of the Viking spacecraft at periapsis (300 km) and the continuous velocity readings, good high-frequency information was obtained.

Both techniques indicate that a long-wavelength, free-air gravity anomaly of approximately 500 mgals exists over the Tharsis province, and that gravity lows of around 200 mgals occur over the adjacent lowlands of Chryse and Amazonis (figure 8.5). Other lows exist over Hellas and Isidis (Sjogren et al. 1975). High-resolution data, derived by direct measurements of the accelerations of the second Viking orbiter (Sjogren 1979), show that there are also large local positive anomalies over individual volcanoes (figure 8.6). The largest is over Olympus Mons, which has a 344-mgal anomaly, as measured at an altitude of 275 km. Alba Patera has a positive anomaly of close to 70 mgals, and somewhat smaller anomalies are associated with the other large volcanoes. Local negative gravity lows occur over Isidis and the canyons.

The extent to which gravity and topography are correlated is a measure of the degree to which the topography is isostatically compensated and of the depth of the zone of compensation. In figure 8.7 the observed gravity deviations are plotted against those that are produced directly by the observed topography. If there is no correlation between gravity and topography, the points form a horizontal line; if there is a perfect correlation—that is, the entire gravity anomaly is the result of the observed topography—then the plot has a slope of 1. On Earth gravity and topography are largely uncorrelated because of isostatic compensation. The light rigid outer parts of Earth are essentially floating on a more fluid, denser interior, so that at a relatively shallow depth (≃100 km) the pressure is everywhere the same, and the variations of topography at the surface are fully compensated. High mountains, for example, have deep, low-density roots, and so do not cause large gravity anomalies. On Mars the situation is

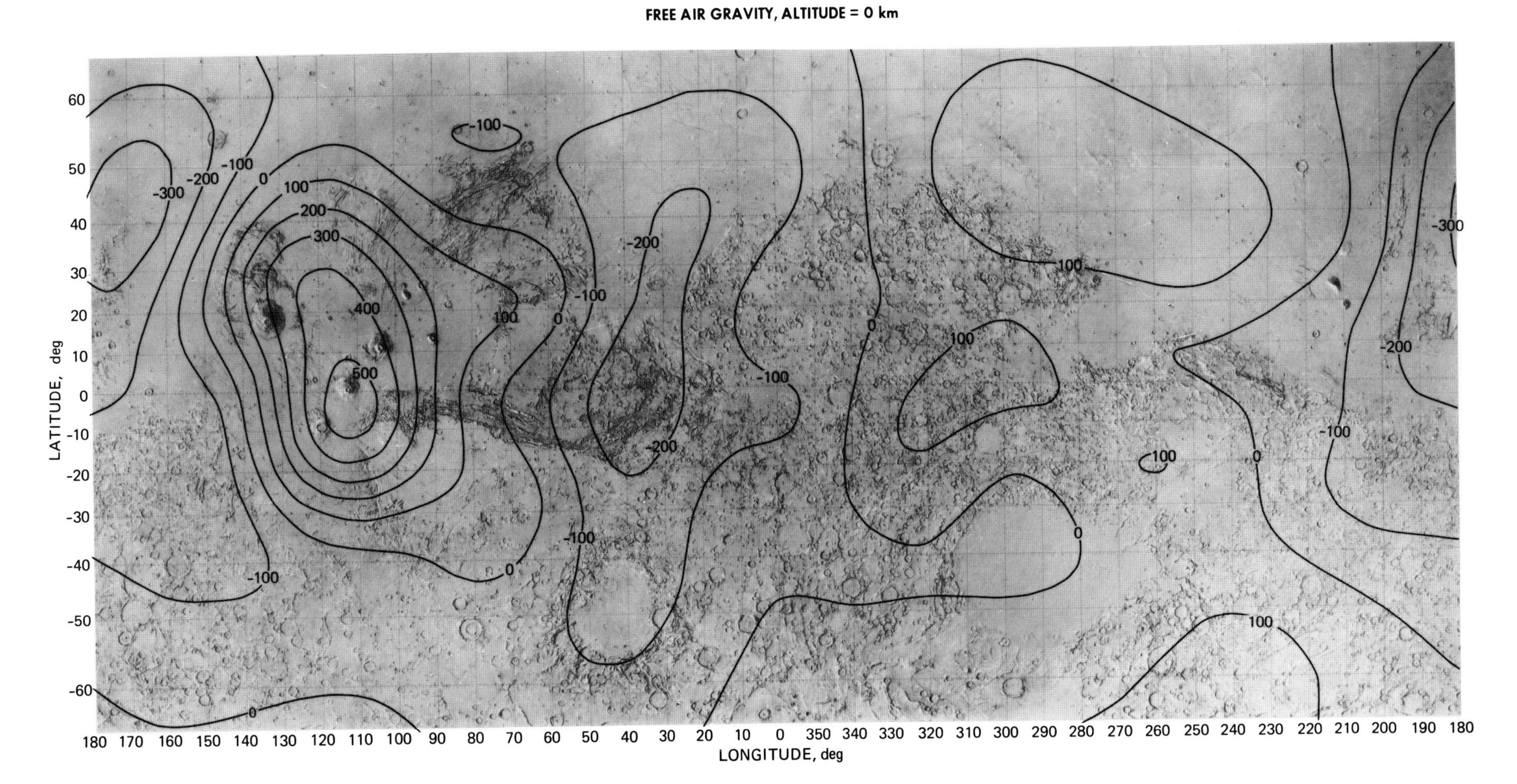

Figure 8.5. Eighth-order model of the gravity at the surface of Mars. The model is dominated by a broad high over Tharsis and complementary lows over the adjacent Chryse and Amazonis lowlands. Gravity is given in mgals. (Courtesy R. J. Phillips, Lunar and Planetary Institute.)

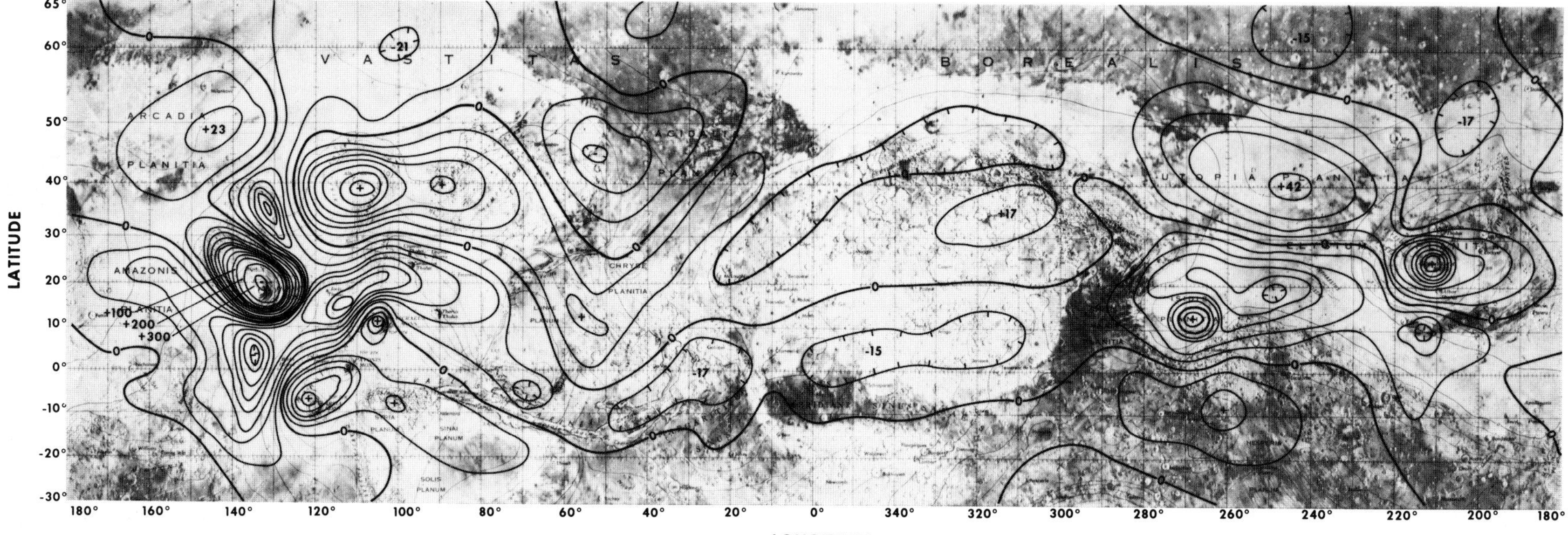

Figure 8.6. High-frequency component of the gravity field, derived by direct line-of-site accelerations. The values are deviations from a fourth-order spherical harmonic representation of the gravity field, so have had their low-frequency component—such as that created by the Tharsis bulge—removed. The main features are large positive gravity anomalies over the Tharsis volcanoes, particularly Olympus Mons; the Elysium volcanoes; and Isidis Planitia. The contour interval is 10 mgals, and the reference altitude is 1,000 km. (Reproduced with permission from Sjogren 1979, copyright 1979, American Association for the Advancement of Science.)

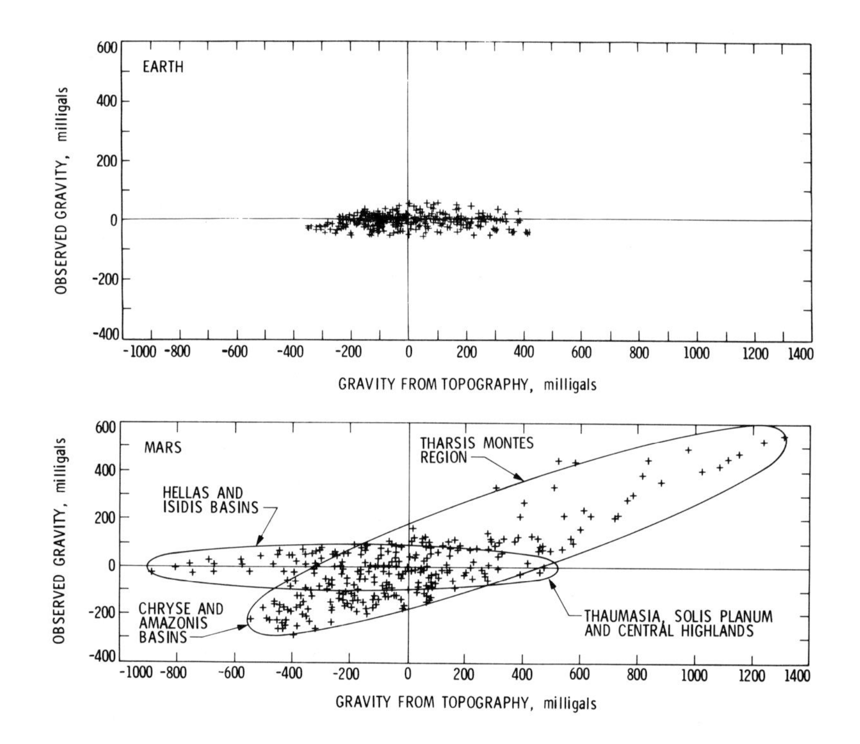

more complicated. In figure 8.7 the points from outside Tharsis form a roughly horizontal line like those from Earth, indicating poor correlation between gravity and topography and thus compensation at a relatively shallow depth. The data from Tharsis, however, form a line with a positive slope, showing that the Tharsis topography is only partly compensated, at least at a shallow depth (Phillips and Saunders 1975). Tharsis could still be fully compensated at a depth of at least several hundred kilometers, for in this case the deep compensating elements would exert only a small effect on the surface gravity, and an anomaly would still remain. The data from the Amazonis and Chryse basins follow the same trend as the Tharsis data, indicating that their formation was somehow coupled to the formation of the bulge. This is of interest in that it suggests that a significant amount of the topography is due to warping of the surface, as opposed to accumulation of volcanics. As seen in the previous section, upwarping is also consistent with the distribution of remnants of old terrain in Tharsis.

Two simple models have been used to explain isostatic compensation on Earth. In the Airy model topographic variations are compensated for by variations in the thickness of the crust. High mountains, for example, are underlain by a thicker crust than the adjacent lowlands. With the Pratt

Figure 8.7. Plots of observed gravity against surface gravity, calculated from the observed topography. Topographic densities are assumed to be 2.67 gm/cm³ for Earth and 3.0 gm/cm³ for Mars. On Earth correlation between the computed and observed values is poor because of isostatic compensation. On Mars it is similarly poor for the old cratered terrain, indicating that the topography there is compensated at a relatively shallow depth. Good correlation between observed and computed gravities for Tharsis and the adjacent lowlands indicates poor compensation of the topography at shallow depths. (Reproduced with permission from Phillips and Saunders 1975, copyrighted by the American Geophysical Union.)

model the compensating layer has a uniform thickness but varies laterally in density, being less dense under mountains. In most terrestrial situations the Airy model is more applicable. On Mars, if isostatic compensation exists, it cannot be by either a simple Pratt or Airy mechanism, for no single depth exists at which all frequencies in the gravity field can be compensated. The high-frequency components require shallow (≃100 km) depths for compensation, whereas the low-frequency components, which are due mostly to Tharsis, require very great (≃1,000 km) depths for compensation (Phillips and Lambeck 1980). One possibility is that the topography of the old cratered terrain was compensated for very early in the history of the planet, when the rigid outer rind of the planet—its lithosphere—was relatively thin, whereas the Tharsis anomaly was created at a later time, after the lithosphere had thickened considerably.

Phillips and co-workers (Phillips et al. 1973; Phillips and Saunders 1975) examined the Mars gravity data in terms of a simple Airy model. They calculated the gravitational effects of the observed topography and subtracted these from the measured gravity to get a Bouguer anomaly. This had large negative values for Tharsis and positive values for some of the surrounding lowlands, indicating that the topography is at least partly compensated. They suggested that the anomaly could be the result of a crust of variable thickness. Taking 50 km for the thickness of the crust where the Bouguer anomaly is zero and 3.0 gm/cm^3 and 3.3 gm/cm^3 for the crust and mantle densities, respectively, they calculated crustal thicknesses as high as 130 km under Tharsis (figure 8.8). They recognized

Figure 8.8. Profile along the 15°S latitude showing the eighth-order representation of the Bouguer gravity and the crustal thickness according to different models. The Airy isostatic model shows the crustal thickness needed to compensate for the observed topography. The gravity model shows the variation in crustal thickness needed to account for the observed gravity. Coincidence of the two lines would indicate perfect Airy isostatic compensation. (Reproduced with permission from Phillips et al. 1973, copyrighted by the American Geophysical Union.)

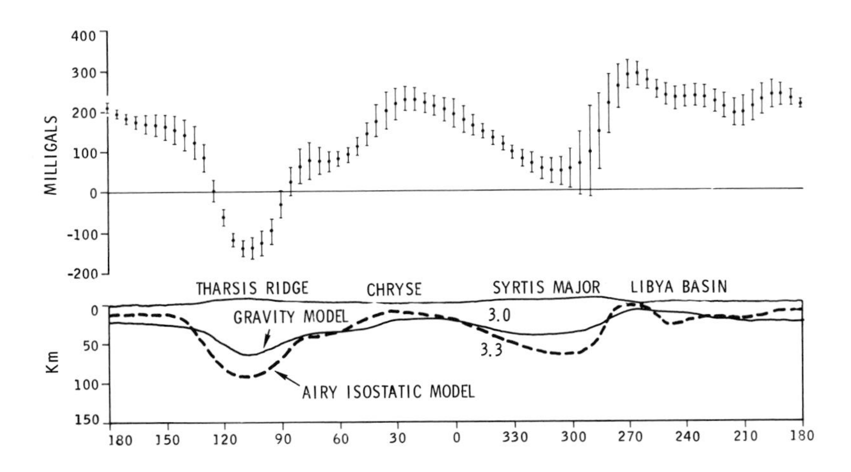

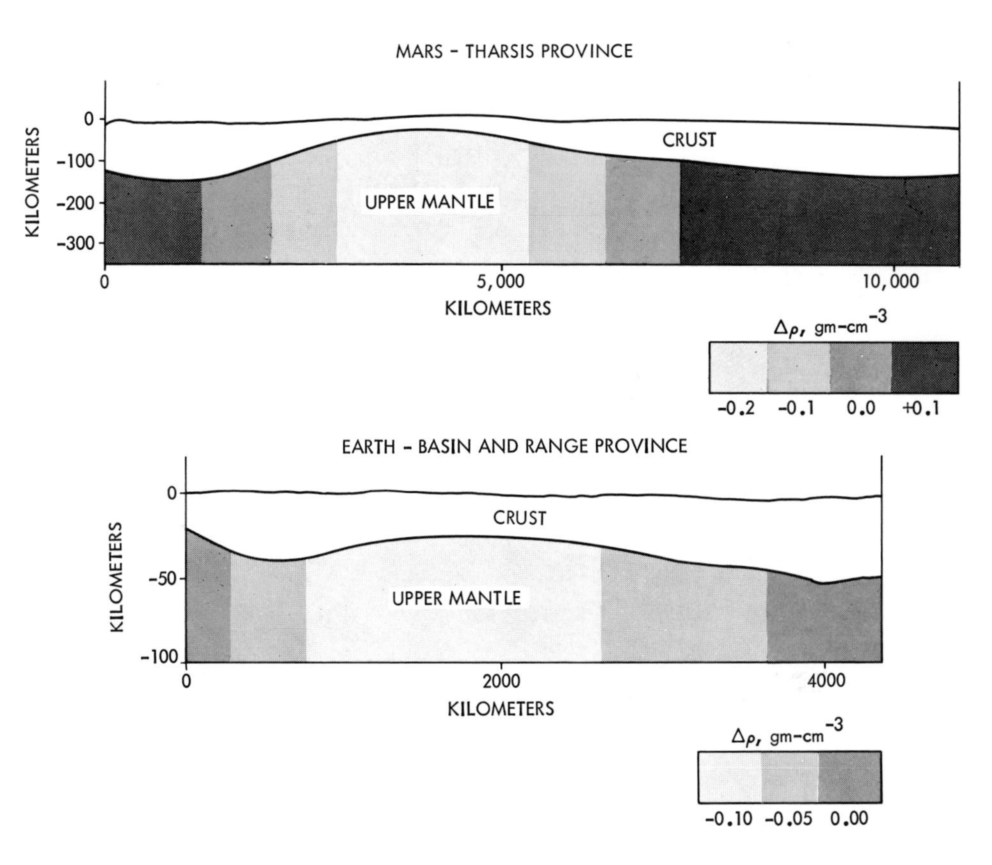

Figure 8.9. Combined Pratt and Airy model for Tharsis. The gravity anomaly is assumed to be caused by variations in both mantle density and crustal thickness. A similar model has been proposed for the Basin and Range province in the western United States. (Reproduced with permission from Sleep and Phillips 1979.)

that compensation is incomplete and calculated the "isostatic deviation," which is that fraction of the gravity anomaly that remains after correcting for isostatic compensation at some relatively shallow depth. The deviation has significant values only for the Tharsis region and its surroundings, as anticipated from figure 8.7.

As we saw above, compensation of both the high and the low frequencies of the topography cannot be achieved with either simple Pratt or Airy models. As an alternative Sleep and Phillips (1979) proposed a compensating mechanism that combines the Pratt and Airy concepts (figure 8.9). They assumed that a crust of variable thickness (Airy model) overlies an upper mantle with lateral variations in density (Pratt model). In their model the crust was thinner under Tharsis than elsewhere, and the upper mantle was less dense. They pointed out that low mantle densities under Tharsis are plausible because of the higher temperatures and different compositions expected under a volcanic region. The model was similar to that proposed by Thompson and Burke (1974) for the Basin and Range province of the United States. Sleep and Phillips were able to achieve compensation under Tharsis at depths as shallow as 300 km. Shallower depths were not possible with their model, for this is where the crust thins to zero. It is interesting to note that in this model the crust thinned to zero under Tharsis, whereas in the simple Airy model discussed above, the crust under Tharsis was 130 km thick. Clearly, a wide range of configurations is possible.

One reason for the interest in isostatic models is that they minimize stresses in the lithosphere. If it can be shown that the stresses within the lithosphere exceed the strength of the rocks for the isostatic case, then it will be true for all other cases, and dynamic support of the lithosphere will be required (Phillips and Ivins 1979). Kuckes (1977) suggested that the lithosphere can be thought of in either of two ways: a thermal lithosphere, in which the dominant mode of heat transport is conduction rather than convection; or an elastic lithosphere, which can maintain elastic stresses for long times and mechanically support long-term surface loads. The elastic lithosphere is generally two or three times thinner than the thermal lithosphere. The heights of the volcanoes suggest depths to melting of 150–300 km (Carr 1973), which should crudely approximate the thickness of the thermal lithosphere. Solomon and colleagues (1979) have suggested thicknesses of 25–150 km for the elastic lithosphere, on the basis of flexure under loading of the volcanoes, which is roughly consistent with 150–300 km for the thermal lithosphere.

Generalized isostatic models for Tharsis yield stresses that are typically in excess of 1 kbar, which must be sustained in the outer few hundred kilometers of the planet because of high temperatures and low yield strengths at greater depths (Phillips et al. 1981). Sleep and Phillips (1979) calculated that for the limiting case of a 300-km–thick elastic lithosphere, stresses are close to 750 bars. A 300-km thickness is unlikely, however, for the elastic lithosphere in this volcanically active region. The elastic lithosphere is almost certainly thinner, and the stresses within it correspondingly higher. Thus the question arises as to whether stresses in excess of 750 bars can be maintained passively for billions of years by the finite strength of the rocks at depths close to 300 km. The question cannot be answered with any assurance because of uncertainties in the finite strength of the rocks and the temperature gradient under Tharsis, but some dynamic support, such as would be provided by convection below or lateral compression, does not appear unreasonable.

FORMATION OF THE THARSIS BULGE

Phillips and Saunders (Phillips 1978; Phillips and Saunders 1979) outlined several possible ways in which the bulge and all its associated features could have formed. One possibility is that Tharsis and the adjacent lowlands result from lateral compression, caused by global shrinking following overall cooling. This seems improbable because of the general scarcity of compressional features on the planet and the difficulty of maintaining compressional stresses over the long periods of time required. Another possibility is that the unique character of Tharsis is somehow related to its roughly antipodal position to Hellas, the planet's largest impact basin. Theoretical calculations and experimental data show that for a large impact into a sphere, a considerable amount of shock energy is focused in the antipodal region. Dissipation of this energy could cause heating and fracturing. Tharsis, however, is not truly antipodal to Hellas, and long-sustained volcanism in Tharsis is unlikely, since the impact-induced heat anomaly should decay relatively quickly (< 1 billion years). Furthermore, orders of magnitude more energy would be dissipated locally around Hellas, yet this region has not had a volcanic history as long-lasting or intense as that of Tharsis.

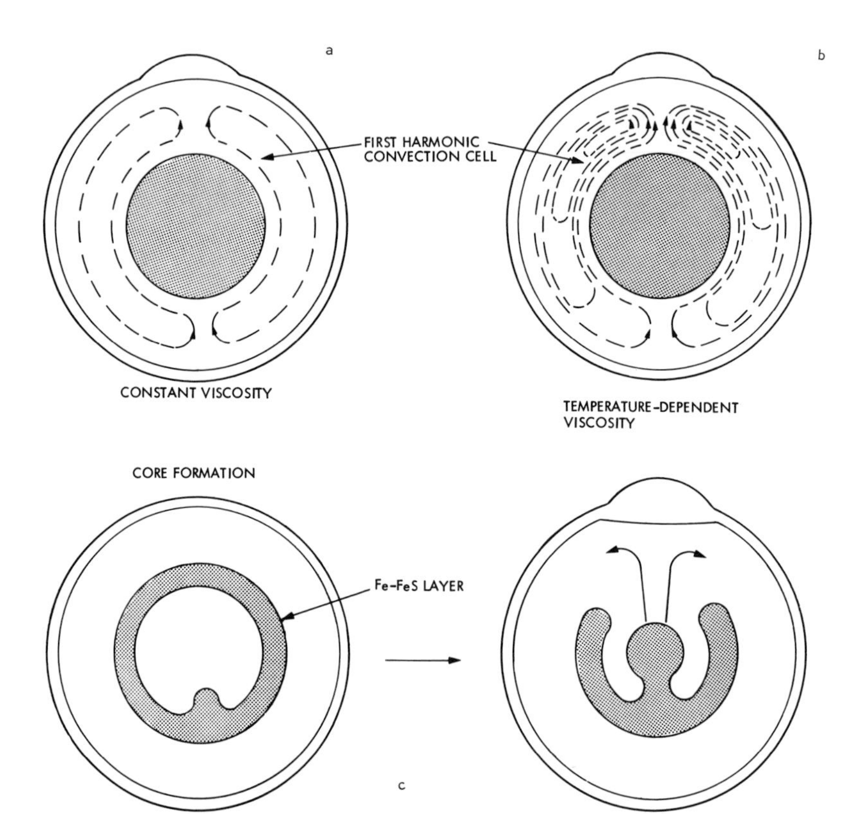

Figure 8.10. Three possible models to explain the Tharsis bulge. In model (a), a single convection cell in the martian mantle upwells under Tharsis. Model (b) is similar but takes into account the temperature dependence of viscosity, so there is no region of strong downwelling as in model (a). In model (c) the Tharsis bulge forms as a consequence of upwelling that is triggered by core formation. (From Phillips 1978.)

Mechanisms involving mantle convection appear to be the more attractive. The bulge may be over the region of upwelling of a single-cell, convective system within the mantle (figure 8.10). However, no corresponding downwelling effect is observed at the antipode, although such a downwelling should occur if viscosities are constant throughout the mantle, so this configuration is unlikely. A more probable situation is that viscosity depends on temperature, in which case the downwelling would be diffuse, and no antipodal effect would occur. Several mechanisms could trigger the convection and localize upwelling under Tharsis. Convection could result simply from radiogenic heating of the mantle and the establishment of a supercritical vertical temperature gradient. Alternatively, inhomogeneous accretion of the planet could lead to a lower mantle that is less dense than the upper mantle, as a result of which a "chemical plume" forms, which carries radiogenically rich materials upward under Tharsis. Convection may also have been triggered by formation of the core. Elsasser and colleagues (1979) have proposed that as Earth heated up early in its history, an Fe/FeS-rich layer separated out at intermediate depths. Further heating caused melting at the center, allowing the Fe/FeS to fall inward and form the core. Such collapse would be accompanied by convective overturn of the mantle. As applied to Mars, the core-forming process could cause major upwelling and heat release under the Tharsis region (see figure 8.10). Whatever the mechanism, it must be capable of sustaining the Tharsis anomaly for 2.5 to 4 billion years, for that appears to be the lifetime of volcanism and lithospheric support within the region.

In summary the Tharsis bulge formed early in the history of Mars. The topographic and gravity high centered on the bulge caused stresses in the crust, which led to extensive radial fractures. The fracturing occurred mostly in the first half of the planet's history but continued into the geologically recent past. Formation of the bulge caused localization of much of the planet's volcanism in the Tharsis region, and preferential positioning of the large volcanoes on the northwest flanks may be in some way related to the bulge straddling the plains/upland boundary. A large gravity anomaly associated with the bulge cannot be accounted for by simple isostatic modeling of the upper few hundred kilometers of the crust. Either the lithosphere supports large stresses, or it is supported from below, such as by convection. The origin of the bulge is uncertain, but a plausible model is that the bulge results from mantle convection that was triggered by formation of the core.

Figure 9.1. Synoptic view of the canyons and the region to the north. For feature names, see the maps at the back of the book and figure 10.7. Noctis Labyrinthus to the left is close to the summit of the Tharsis bulge, at an elevation of 10 km. Melas Chasma, the broad canyon with a chaotic floor to the right of the picture, is at the edge of the bulge, at an elevation below 3 km. The vague north-south depression that gives rise to Kasei Vallis is visible in the left center, and Juventae Chasma, which gives rise to a large channel system along the east side of Lunae Planum, is seen in the right center. Layered deposits within the Ophir and Candor Chasmas can just be seen as bright patches. The scene is 4,100 km across, about one-fifth of the circumference of the planet.

9 CANYONS

Just south of the equator, between longitudes 30°W and 110°W, are several enormous, interconnected canyons, which have been collectively called Valles Marineris. They extend roughly east-west for over 4,000 km, from the summit of the Tharsis bulge at Noctis Labyrinthus, down the crest of a broad rise on the eastern flank of the bulge, to some low-lying areas of chaotic terrain between Chryse Planitia and Margaritifer Sinus (figure 9.1). Individual canyons may be over 200 km wide and 7 km deep. In the central section, where there are three partly connected parallel canyons, the entire system is over 700 km across. By comparison, the Grand Canyon, Arizona, is 450 km long, 30 km across, at its widest point, and 2 km deep. A distinction is made between channels and canyons. The canyons are poorly graded and are commonly choked with landslide debris, and most of them lack indications of fluvial action on their floors. Many sections are segmented and consist of strings of closed depressions rather than a continuous canyon. In contrast the channels have numerous indications of fluvial activity, such as graded floors, teardrop-shaped islands, longitudinal scour, and hydrodynamically shaped walls. Because of the lack of fluvial features, Sharp (1973b) suggested that the use of the term "canyon," which has connotations of water erosion, was inappropriate. However, the term is now well entrenched in martian literature, and we will continue to use it here. Much of the relief of the canyons appears to result from faulting along roughly east-west faults radial to the center of the Tharsis bulge, but other processes, such as undermining, landsliding, and gullying, have also clearly played a role, and precisely how the canyons formed is unclear. The main issue is the relative roles of deep-seated processes such as tectonism, as opposed to surficial processes such as erosion. Puzzling also is the presence within some canyons of thick sequences of layered deposits, suggestive of cyclic sedimentation on a large scale.

Although the canyons show little direct evidence of fluvial activity, their origin appears to be related in some way to the formation of channels. The main canyons merge eastward with chaotic terrain as the canyon floors become rubbly and chaotic and the depths of the canyons decrease. The term "chaotic terrain" refers to areas that have seemingly collapsed to form a haphazard jumble of large angular blocks at a lower elevation than the surrounding undisturbed terrain (Sharp 1973a). The chaos may be restricted to discrete areas, clearly defined by an inward-facing escarpment, or may form ill-defined areas that merge with the surrounding terrain as breakup and collapse of the original surface becomes less severe. Most chaotic terrain is just east of the canyons and south of Chryse Planitia, but small areas occur elsewhere, such as north of Hecates Tholus in Elysium. Many of the largest channels on Mars start in the chaotic terrain. They start full size and have numerous attributes suggesting that they were formed by catastrophic floods. There is, therefore, a continuous transition eastward, from canyons through chaos to channels. The transition is repeated north of the main canyons, where two large channel systems start in box canyons containing chaos. The cause of the connection between the canyons, which appear to have formed largely by faulting and wall recession, and the channels, which have every indication of having been formed by floods of water, is not known. A possibility is that extensive reservoirs of groundwater once existed in the low areas east and north of the canyons, and that the reservoirs gave rise to the large channels and were occasionally disrupted by canyon faults. In this chapter the focus is on canyons; the large flood features will be discussed in chapter 10, along with other types of channels.

CANYON PHYSIOGRAPHY

The canyons can be divided into three sections: a complex maze of interconnected canyons—the Noctis Labyrinthus—at the west; the main section of roughly east-west–trending canyons in the center; and some irregular depressions which merge with the chaotic terrain further east (see map, figure 10.7). The canyons that comprise Noctis Labyrinthus are relatively short and narrow and have different orientations, so that they divide the upland into a complex mosaic of blocks (figure 9.2). The labyrinth is at the summit of the Tharsis bulge, close to the point of convergence of several sets of fractures that are part of the radial fracture system around the bulge. To the north of the labyrinth the direction of the fractures is predominantly north-northeast–south-southwest, to the east it is mainly east-west, and to the south it is mostly northwest-southeast. All these trends are represented in the canyon segments that make up the labyrinth. To the west the different-trending fractures merge and turn southward to form a complex fracture zone, the Claritas Fossae. Most of the depressions that constitute the western part of the labyrinth are unconnected and consist mostly of lines of coalesced pits, with the floors of each pit at a different level. They

Figure 9.2. Part of Noctis Labyrinthus, centered at 5°S, 99°W. The "labyrinth" is composed of numerous intersecting, closed depressions. Within many of the depressions, faults in the surrounding plateau can be seen in section to have near vertical dips. The frame is 120 km across. (48A27)

appear to have formed by localized collapse, rather than through any mechanism involving longitudinal transport. Further east, connections between the depressions become more common, and ultimately the labyrinth merges with the more or less continuous canyons to the east.

The main canyon extends from 52°W to 90°W, a distance of 2,400 km. Along this entire length the canyon is multiple, consisting of parallel canyons and chains of craters and graben, nearly all trending east-southeast–west-northwest. There appears to be a continuous sequence, from flat-floored graben to graben that enclose small pits to chains of mutually intersecting craters and finally to continuous canyons. The canyons tend to be better integrated to the east and more segmented to the west, although discontinuous troughs parallel to the main canyons occur in all sections (figure 9.3). In the part just east of Noctis Labyrinthus, the two largest canyons are Tithonium Chasma to the north and Ius Chasma to the south. Tithonium Chasma narrows eastward and ultimately degenerates into a line of closed depressions (see left side of figure 9.6). Ius Chasma consists of two parallel troughs separated by a longitudinal ridge. Both troughs broaden eastward and finally merge with the broad Melas Chasma (see figure 9.1).

The canyons are widest and deepest in the central section, between 65°W and 77°W, where three huge parallel troughs are each close to 200 km across. Separating the three are partly breached ridges, which appear to be remnants of the former plateau. The two northernmost troughs, the Ophir and Candor Chasmas, are somewhat spatulate in outline, having blunt east and west ends, although crater chains continue along the main east-west trend of the canyons beyond their terminations (see figure 9.6). It is in this broad central section that layered sequences, suggestive of cyclic sedimentation, are most common. The southernmost canyon of the central section, Melas Chasma, continues eastward into Coprates Chasma, which, like the western canyons, is composite, consisting of several parallel troughs, crater chains, and graben.

The characteristics of the canyons change greatly at about 52°W. West of this longitude the canyon walls are linear and follow a roughly east-southeast–west-northwest trend, as do the crater chains, graben, and subsidiary troughs. East of 52°W, however, the canyons are irregular in outline and follow no consistent trend. The floors also change from being mainly smooth to the west to being mainly rubbly to the east. East of 52°W the canyons consist of two main depressions, the relatively narrow Ganges Chasma to the north and a broad depression including the Eos and Capri Chasmas to the south. Both depressions continue eastward into irregular areas of chaotic terrain.

North of the main canyons and completely unconnected with them are three additional canyons. The westernmost canyon, Echus Chasma, broadens northward into a vague north-south depression along the west edge of Lunae Planum. At its widest this depression is almost 400 km across, and from it emerges the huge channel Kasei Vallis, which cuts through Lunae Planum eastward to Chryse Planitia. Immediately to the west of Echus Chasma is the completely enclosed Hebes Chasma, and still further east is Juventae Chasma, the point of origin of another large channel system, which extends northward along the eastern boundary of Lunae Planum and ultimately reaches Chryse Planitia also (Carr 1979). Thus, both to the north and the east the canyons are continuous, with outflow channels that appear to be fluvial in origin, and a genetic connection appears inescapable.

Large areas of chaotic terrain occur east of the canyon, between longitudes 10°W and 50°W and latitudes 20°S and 10°N, and numerous smaller areas occur elsewhere, as in Juventae Chasma, in the old cratered terrain between Lunae Planum and Chryse Planitia, and in the vague north-south depression just west of Lunae Planum. Much of the chaotic terrain east of the canyon is interconnected, as though part of extensive drainage systems. Valles Marineris is continuous with several irregular chaos areas, the largest being the Aureum Chaos and the Hydroates Chaos (see map, figure 10.7). These in turn connect northward with two large channels, the Simud and Tiu Valles, which drain into the Chryse basin. Other interconnected, irregular, and seemingly more shallow, areas of chaos further east, the Margaritifer Chaos and the Aram Chaos, drain northward into Ares Vallis and appear to have no direct connection with the canyons.

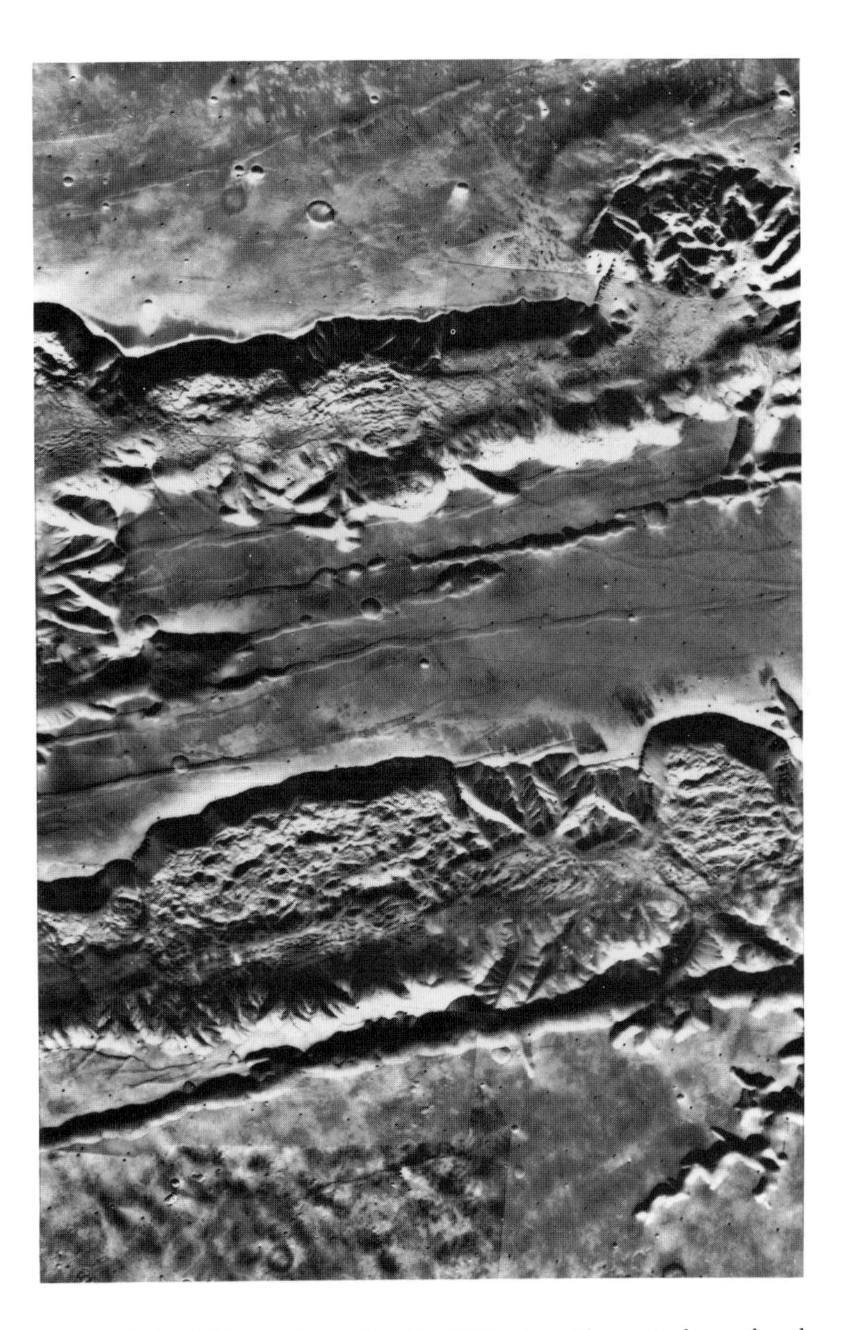

Figure 9.3. Landslides on the north walls of Tithonium Chasma to the north and Ius Chasma to the south. The alcove left in the wall by the landslide in the lower right is 40 km across. The huge landslide to the west is 100 km across. The canyon walls at this point are 2–3 km high. Most of the larger landslides extend all the way across the canyon and ride part way up the opposite wall. (P-17708)

FAULTING WITHIN THE CANYONS

Faulting appears to have played a key role in formation of the canyons, both by providing sites for preferential erosion and by providing much of the vertical relief necessary for erosion to occur. Evidence for faulting is from several sources: (1) The orientation of the canyons radial to the center of the Tharsis bulge and parallel to the radial fractures around the bulge indicates strong structural control over orientation and location. (2) Some faults in the surrounding plateau can be traced directly into the canyon walls, particularly in Noctis Labyrinthus. (3) Craters on parts of the canyon floor suggest that they are downfaulted segments of the preexisting plateau surface (Blasius et al. 1977); an especially large crater occurs, for example, on the floor of Coprates Chasma, at 65°W. (4) Linear scarps, resembling fault scarps, truncate the bases of ridges and gullies on the canyon walls, leaving hanging gullies and triangular faceted spurs; this is particularly evident on the north walls of the Coprates and Ius Chasmas; similarly, side canyons on the south wall of Ius Chasma are in places left hanging above inward-facing scarps. The relations suggest that faulting and erosion proceeded simultaneously rather than succeeding each other, because erosional features, which could not have formed until after faulting started, are themselves cut by faults. Many of the faults can be seen in section in the canyon walls, particularly in Noctis Labyrinthus and at the west end of Candor Chasma. Somewhat surprisingly the faults are mostly near vertical, rather than being close to 60°, as expected of normal faults.

As we saw in the previous chapter, the faults appear to result mainly from the presence of the Tharsis bulge. The cause of the unique location of the canyons, on the eastern flank of the bulge, rather than in some other sector of the radial fracture system, is unclear, although "keystoning" along the crest of the low ridge on this flank is a possibility (Blasius et al. 1977). In an alternative hypothesis, Allegre and colleagues (1974), Sengor and Jones (1975), Courtillot and co-workers (1975), and Masson (1977) have suggested that the Valles Marineris formed by faulting and subsidence, as a result of the southward movement of the Syria-Solis Planum block, which lies to the south of Valles Marineris. These authors proposed movements of several tens of kilometers, implying that most of the volume of the canyons is of tectonic origin. Wise and colleagues (1979) argued, however, that much smaller movements, possibly less than 3 km, are sufficient to explain the geometry of the faults actually observed, and that the opposing walls of the canyons are unlikely ever to have been in contact.

Given the strong evidence for faulting, it seems likely that faulting played a major role in the formation of the canyons. A question arises, however, as to what extent the dimensions of the canyons are due to tectonic subsidence, and to what extent they are due to erosion, with faulting merely acting as a trigger and creating escarpments along which erosion could occur. For some of the canyons, such as the Ius and Tithonium Chasmas, a case can be made for their present volume being largely due to faulting; in the case of others, such as the Hebes Chasma and the broad central section containing the Ophir, Candor, and Melas Chasmas, erosion appears to have predominated.

CANYON WALLS

The canyon walls show abundant evidence of enlargement of the canyons following, and contemporaneous with, creation of relief by faulting. The erosional style varies, suggesting different erosional mechanisms in different places. Mass wasting, perhaps aided by melting and sublimation of ground-ice, and groundwater sapping appear to be the dominant processes. Most walls are characterized either by ridge and gully topography, landslides, or tributary canyons (McCauley et al. 1972; McCauley 1978; Sharp 1973b; Lucchitta 1978b; Blasius et al. 1977). Ridge and gully topography, the most commonly occurring wall type, consists of sharp-crested, downward-bifurcating ridges with steep talus chutes between. Along the ridge crests is commonly a rough spine, below which the smooth talus slopes begin. The spurs have slopes of 15–20° (Wu et al. 1973) and are usually oriented at right angles to the length of the canyons, although they may be oblique where faults intersect the walls. Ridge and gully topography is visible between the Ius landslides in figure 9.3 and on the central spine in figure 9.5. At the base of some gullies are debris fans, which can be traced upslope into leveed channels (Lucchitta 1978b). Similar fans with channels in the Wright Valley of Antarctica (Gunn and Warren 1962) and in Spitzbergen (Rapp 1975) form by mass wasting. The general character of the ridges and gullies themselves is also similar to that which develops on terrestrial scarps in relatively homogeneous deposits in alpine and arctic environments (Sharp 1973b; Lucchitta 1978b), where erosion is largely by dry mass wasting.

Spectacular landslides, some of enormous size, occur along most of the walls of the canyon, indicating catastrophic failure in addition to slow, steady erosion. One on the north wall of Ius Chasma, for example, is over 100 km across (figure 9.3). The landslides can generally be divided into two parts: an upper layer, consisting of coarsely textured debris with a crude layering parallel to the length of the canyon; and below this a debris fan with fine striae oriented parallel to the direction of movement of the slide (figure 9.4). The coarse debris is probably made up of large rotated blocks of the former surface, possibly lava flows, whereas the lower layer probably consists of highly brecciated, much finer-grained materials of the old cratered terrain. The coarse debris is generally visible adjacent to the canyon wall, and the more finely textured debris further out into the canyon. However, with the larger landslides, or where the canyon is narrow, the upper coarse debris may extend all the way across the canyon and ride up the opposite wall, so that the fine texture of the lower portion is hidden by

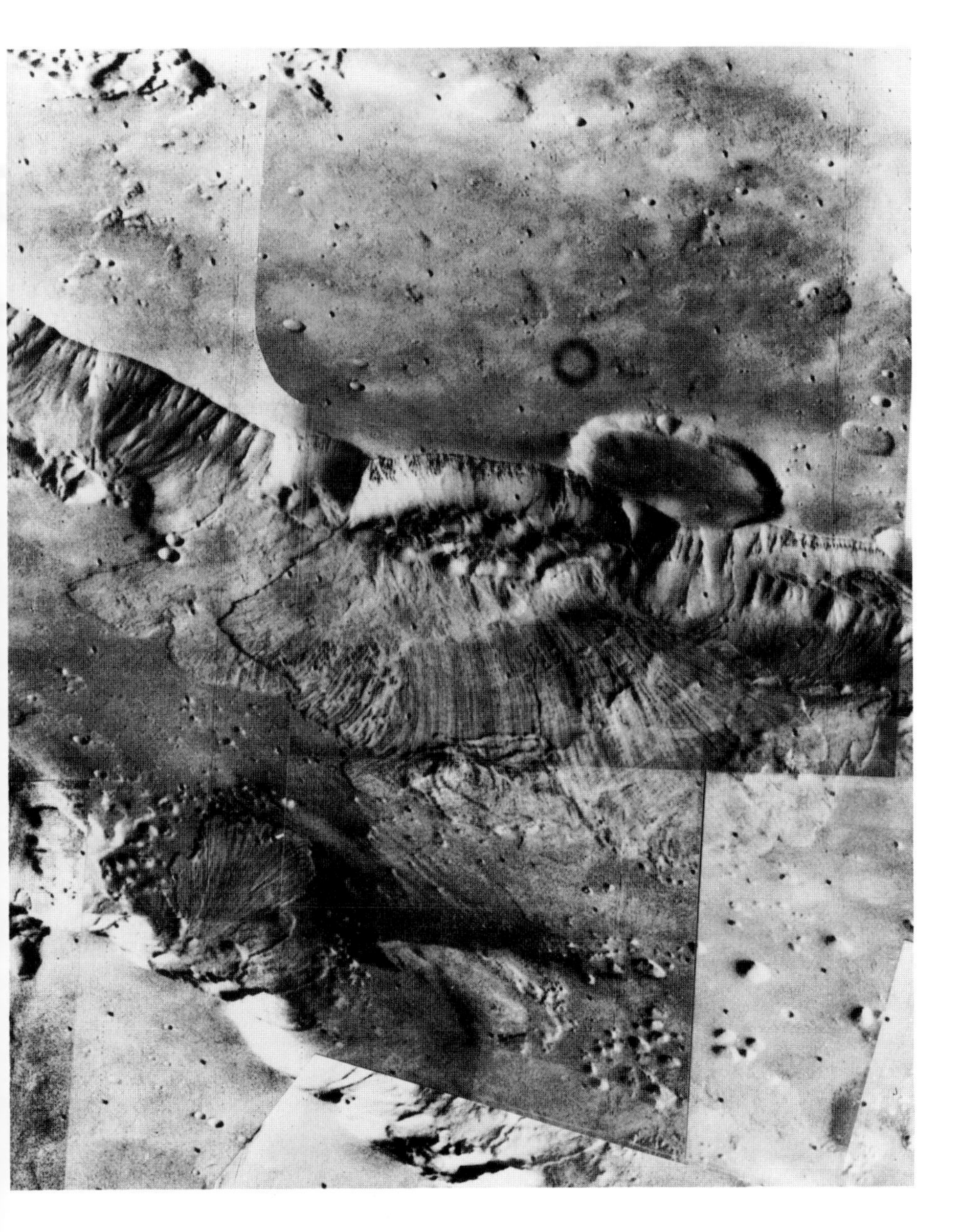

Figure 9.4. Landslides in Ganges Chasma. The landslide on the far wall has two components: an upper, blocky layer, which is probably disrupted cap rock; and a finely striated lower layer, which is probably mostly debris from the old brecciated cratered terrain. Similar striations are seen on some large terrestrial landslides. Fan-shaped slides are also visible on the near wall. The transected crater on the far wall is 16 km across. (P16952)

the upper facies or is destroyed by interaction with the opposing cliff. The landslides produce smooth-walled alcoves in the canyon walls, which contrast sharply with the ridge and gully morphology of the undisturbed parts.

Lucchitta (1978a) compared the mechanics of emplacement of the large landslides of the canyons with some large terrestrial landslides. Since many landslides ride part way up the opposite wall of the canyon or over obstructions on the canyon floor, they must form by catastrophic collapse rather than by slow creep. From the dimensions of some of the overridden obstacles, Lucchitta estimated that emplacement velocities were in the 100–140 km/hr range, which is comparable to the range of velocities for large terrestrial landslides. She also showed that the effective coefficients of friction of the martian landslides are low (< 0.1) compared with most terrestrial examples, which are in the 0.1–0.5 range. Some large Alaskan slides, however, have values that are almost as low as the martian ones. To explain the low coefficient of friction of the Sherman landslide in Alaska, Shreve (1966) proposed that it was emplaced on a cushion of air. However, Heim (1932), Varnes (1958), and Kent (1966), writing of landslides in general, contended that partial fluidization of landslide debris by interaction of the constituent debris and entrapped air can result in high emplacement efficiencies, and that an air cushion is not required. Flow of wet debris by incorporation of water or ice has also been proposed (Plafker and Erickson 1978) to explain the very fluid properties. The precise mechanism of emplacement of large terrestrial landslides, and by extension martian landslides, is thus poorly understood. Water or water-ice may have played a role in the formation of the landslides within Valles Marineris. The depths of the canyons considerably exceed the probable depths of the permafrost (Fanale 1976). Groundwater could therefore have existed beneath the permafrost behind the canyon walls, lowering the shear strength of the walls and making them susceptible to collapse, which may have been triggered by earthquakes along the longitudinal faults.

CANYON FLOORS

The canyon floors vary considerably from place to place. Where the canyons are narrow, such as along the western part of Tithonium Chasma and Noctis Labyrinthus, the floors tend to be segmented, consisting of partly coalesced elliptical pits separated by transverse septa, with the floors of adjacent pits at slightly different levels. Where the troughs are wide, the floors are flatter and better integrated and may include landslide deposits, hilly plateau remnants, downfaulted sections of the former surface, and, more rarely, fields of dunes. Landslides are present in most of the canyons, but, as indicated above, they are particularly common on the floor of Ius Chasma, where several extend across the canyon, covering the floor with blocky debris or fans with fine striae transverse to the canyon. The plateau remnants consist mostly of longitudinal ridges or spines, whose slopes have been eroded in the same gullied style as the canyon walls. Where they are

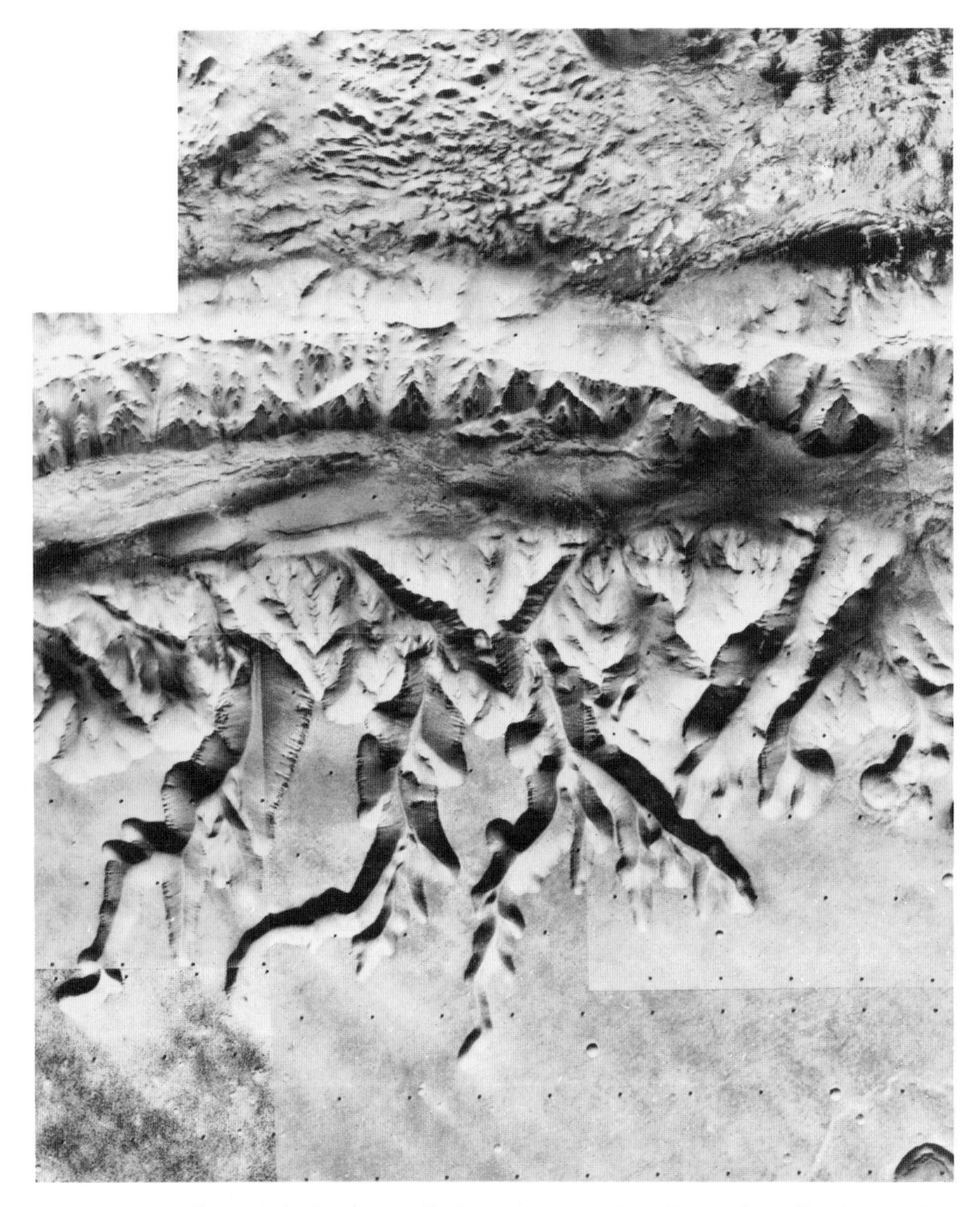

Figure 9.5. Southern wall of Ius Chasma, at 7°S, 82°W. The wall is dissected by deep valleys with V-shaped cross sections and alcovelike terminations. Some of the valleys are offset where they enter the canyon proper (*center*), as though by faulting. The east-west spine at the upper center of the frame runs along the middle of the canyon and is eroded into the ridge and gully topography that typifies many of the canyon walls. Many of the spurs are truncated where they meet the floor, another suggestion of faults. The mosaic is 110 km across. (211-5158)

eroded below the level of the plateau, the ridges have sharp crests; otherwise, they have flat tops continuous with the plateau. East of the 52°W longitude, seemingly undisturbed longitudinal ridges are not found, but instead the floors have a rubbly appearance. The rubble is more like the chaotic terrain further east, which appears to have formed by collapse, than the longitudinal ridges to the west. Unlike the longitudinal ridges, which appear to be in place, therefore, the rubble may be below its original stratigraphic position. Most of the level section of the canyon floor is sparsely cratered. The density of the craters, however, is uneven, and locally numerous poorly preserved craters are present, which suggested to Blasius and co-workers (1977) that at least some parts of the floor are sections of the former surface that have been downfaulted.

The third erosional style, involving formation of tributary canyons, is best seen on the south wall of Ius Chasma (figure 9.5). Most side canyons have rounded cirquelike terminations, V-shaped cross sections, and very narrow flat floors. The branches are usually concordant, but intersections with the main canyon may be slightly discordant because of the presence of a fault scarp. No obvious accumulations of debris are found on the canyon floor at the mouths of the tributaries. What happened to the eroded materials is uncertain, although, as suggested above, some may simply accumulate within the main canyon, as its floor is lowered by faulting. Sharp (1973b) suggested that the side canyons originated by sapping, as a result of groundwater seepage or by dry mass wasting, possibly after sublimation of ground-ice caused disaggregation of the wall materials. Some of the tributaries have a crude orthogonal pattern, suggesting structural control, such as could occur if the rocks were more erodible along lines of structural weakness, or if preferential movement of water took place along these zones. Sharp and Malin (1975) proposed that the better development of the tributaries on the south wall of Ius Chasma, as opposed to the north wall, is due to a north regional dip. Tributaries of the Grand Canyon, Arizona, for example, are better developed on the north rim, where they extend up the regional dip, than on the south rim, because of migration of water down the bedding planes. The relative roles of mass wasting and water transport in the formation of the side canyons of Valles Marineris, however, are poorly understood. The longitudinal slopes are relatively steep (1–3°), and mass wasting processes alone, with no involvement of liquid water, cannot be ruled out. If cut by water, the side canyons represent some of the youngest waterworn features on the planet.

There is thus abundant evidence of erosion on the canyon walls, but the role of water in the erosion is uncertain. The ridge and gully topography and the landslides are both readily explicable in terms of dry mass wasting, although water would facilitate the formation of both these features. A more plausible case can be made for a fluvial origin of the side canyons, but here also, dry mass wasting, possibly involving ice, could be responsible. Whatever the erosive process, it is clear that erosion has extensively modified the canyon walls and significantly added to the canyon's size.

One of the most puzzling features of the canyons is the presence in some sections of rhythmically layered sediments (Masursky 1973; Sharp and Malin 1975; McCauley 1978; Blasius et al. 1977), which suggests a quiet depositional environment, such as is provided by standing water. The best examples are in the Ophir and Candor Chasmas (figures 9.6 and 9.7)

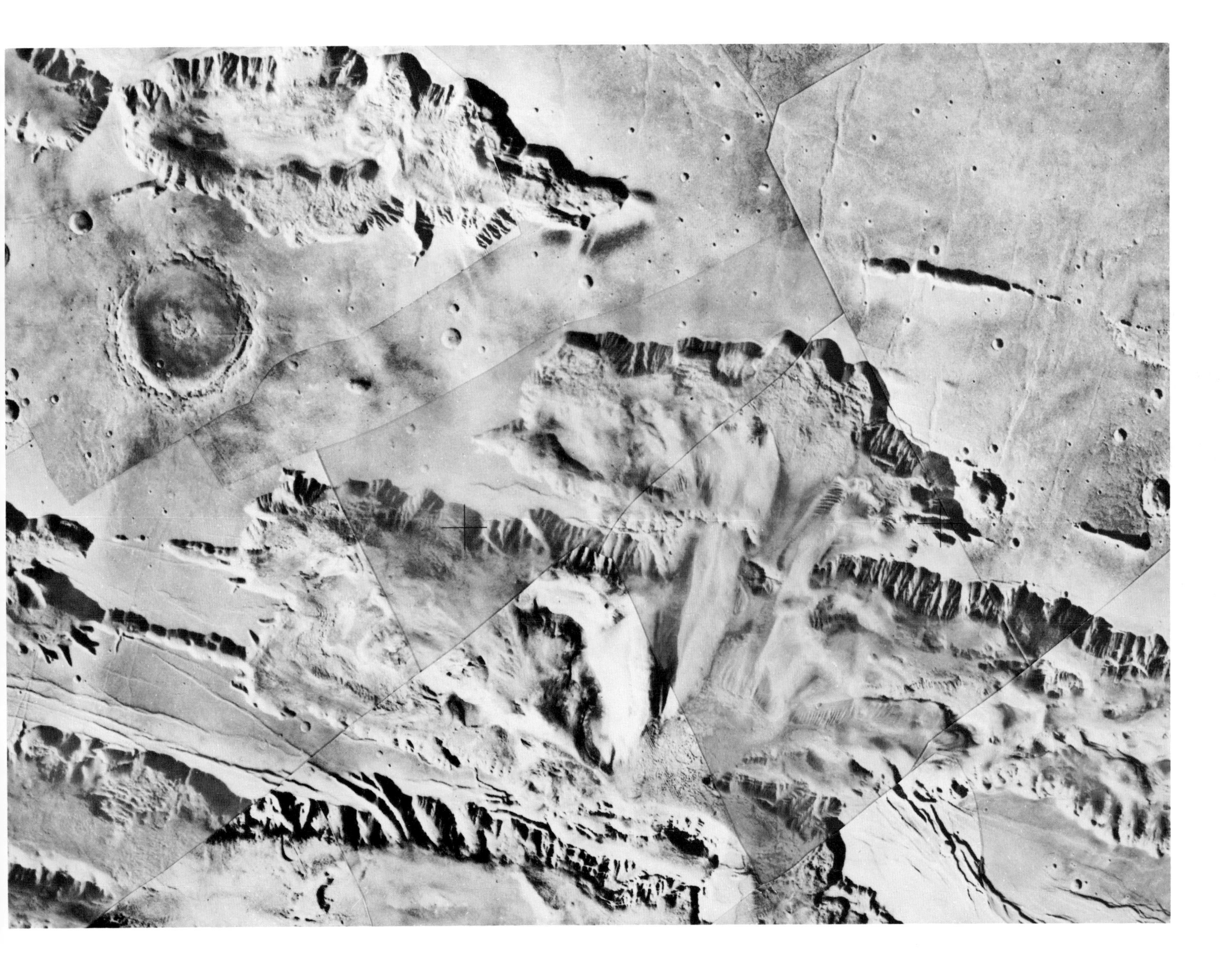

Figure 9.6. The Hebes (*top*), Ophir (*middle*), and Candor (*bottom*) Chasmas in the central section of Valles Marineris. Layered deposits are present in each of the canyons but are best seen in the completely enclosed Hebes Chasma. Erosion of deposits in the Ophir and Candor Chasmas has imposed a coarse north-south texture near the divide between the two canyons. A gap in the south wall of Candor Chasma may indicate drainage of material southward into the main east-west throughgoing canyon to the south. The north wall of Candor Chasma has typical ridge and gully topography. Numerous landslides are visible in Hebes Chasma and on the north wall of Ophir Chasma. The west end of Ophir Chasma is 140 km across. (From U.S. Geological Survey Map MC-19 NW.)

but similar sequences occur in other canyons. The deposits form irregular hills or flat-topped mesas with a fine and regular horizontal layering. They are mostly restricted to the centers of the canyons, so that contacts between the sediments and the canyon walls are rarely seen. One place where the two are in contact is along the divide between the Ophir and Candor Chasmas. Here, layered sediments in the form of flat-topped hills with finely fluted slopes clearly transect the coarse ridge and gully topography of the divide (figure 9.7).

The origin of the layered deposits has been the subject of considerable speculation. Malin (1976) suggested that they are simply erosional remnants of the materials into which the canyons are cut. However, this interpretation appears inconsistent with the relations at the Ophir-Candor divide, where the contrasting erosional styles and discordant layering between sediments and canyon wall are striking. Sharp (1973b) and McCauley (1978) both agreed that the layered deposits formed after the canyons and have since been partly eroded themselves. McCauley (1978) pointed out that there are relatively few ways of depositing a rhythmically layered rock sequence. Another place on Mars where such regular sequences occur is at the poles. These layered deposits at the poles are believed to form by cyclic deposition of frost and dust, the relative proportions perhaps modulated in some way by climatic changes. Such an origin seems unlikely for the canyon deposits because of the restriction of the deposits to the canyon floors and the location of the canyon at the equator. McCauley alternatively proposed that the canyon sequences were deposited within lakes, asserting that no other process can explain the horizontal nature and continuity of the beds and the subtle differences in reflectance and competence of the various beds in the sequence. This implies that large, long-lasting, standing bodies of water must have once existed within the canyons. While initially this might seem a somewhat outrageous hypothesis, on closer examination it acquires some credibility. As we shall see from considerations of the general morphology of the canyons and the origin of outflow channels, there is considerable evidence for extensive eastward movement of groundwater within the general vicinity of the canyons. This groundwater could have supplied water for the hypothesized lakes.

If liquid water were stable with respect to the atmosphere—that is, if the atmosphere was once thicker and the temperatures higher than at present—then lakes would be likely to occur within the canyon, given an extensive groundwater system. Such clement climatic conditions, however, are unlikely. Most models of atmospheric evolution (Pollack 1979) do not allow for a thick atmosphere in the geologically recent past, postdating the relatively young canyons, as is required. Moreover, a thick atmosphere and warm temperatures late in the planet's history should have left a much

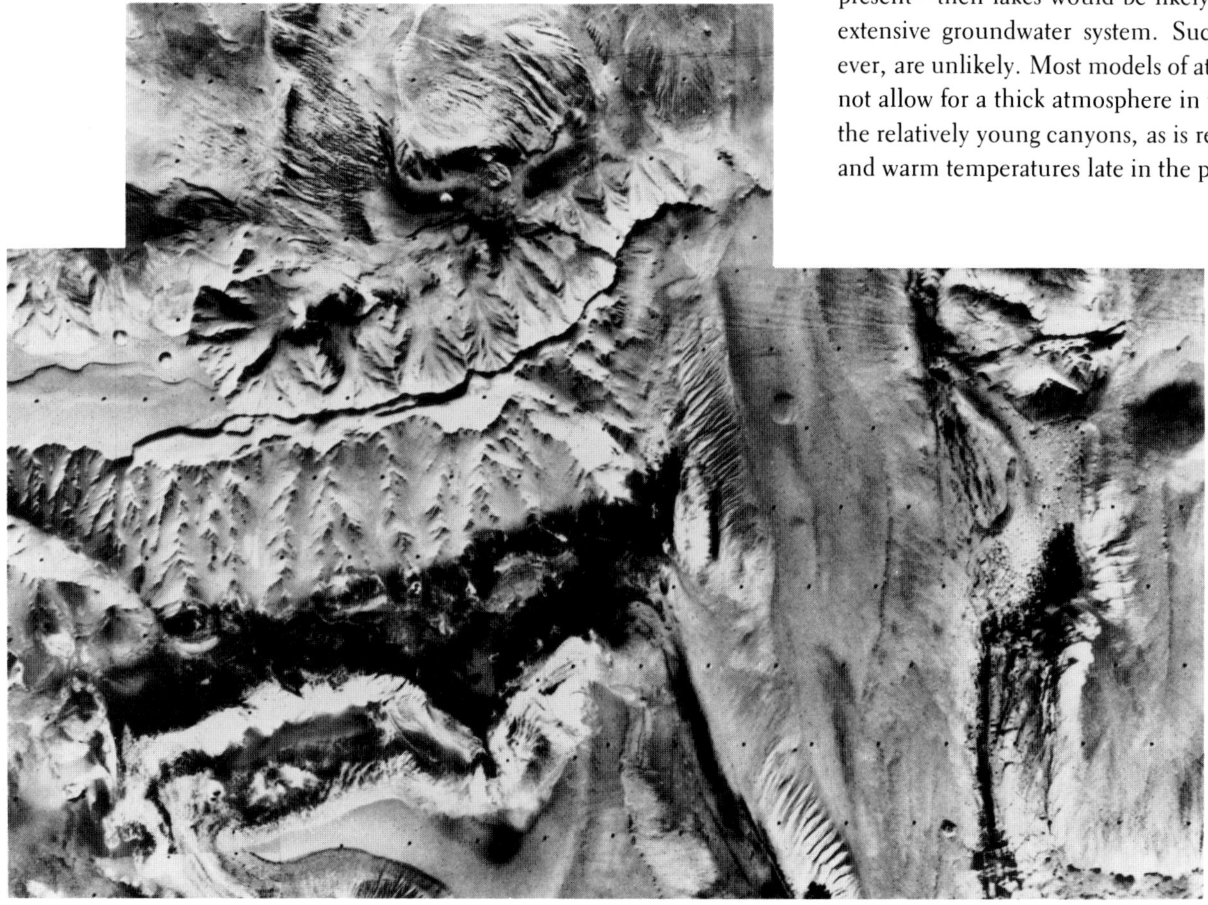

Figure 9.7. The divide between the Ophir and Candor Chasmas. The divide appears to be composed of the materials into which the canyons are cut. Superimposed on the divide (*right center*) is a flat-topped, mesalike formation with finely fluted slopes. A similar sequence in the lower left is clearly layered. The layered deposits appear to postdate formation of the canyon and lie unconformably on both the canyon floor and the divide. The mosaic is 150 km across and centered at 5°S, 83°W. (211-5157)

wider imprint on the general planetary surface than is evident. I (Carr 1979) proposed that massive breakout of water from underground could occur under the present climatic conditions. Such breakouts could result in the partial filling of closed canyons, but whether a lake so formed could survive long is unclear. As Wallace and Sagan (1979) pointed out, a lake on Mars will rapidly form an ice cover, and its lifetime would depend on the balance between seepage into the lake from groundwater and sublimation from the ice surface. Cyclic sedimentation within an ice-covered lake presents yet another problem. Conceivably, it could occur by periodic introduction of debris into the lake by groundwater. Subsurface erosion by movement of groundwater is not uncommon on Earth and has been variously referred to as "piping", "tunneling", and "sink-hole erosion" (Parker et al. 1964). It is particularly common in fine-grained deposits such as loess and such as might be expected in the canyon walls, since most of the walls are incised into the highly brecciated and gardened old cratered terrain. Such "tunneling" could provide sediment to canyon lakes but would have to occur on a grand scale to provide the necessary volume. Another potential source for the sediments are the canyon walls themselves, although how their erosion would produce rhythmic sedimentation is unclear.

McCauley (1978) took the lake hypothesis one step further, suggesting that the lakes drained eastward as the containing walls were eroded away. Thus the divides between the Ophir and Candor Chasmas and the Candor and Melas Chasmas formerly dammed lakes behind them. The lakes could have drained catastrophically or slowly, depending on the erosion rates at the outlet, but McCauley favored catastrophic drainage, for this explains the dissection of the layered deposits and some of the large flood features that are continuous with the canyons further east. Presumably, evidence of fluvial activity within the canyons themselves has been destroyed by subsequent faulting and mass wasting. All this, however, is extremely speculative, and the origin of the canyon layered deposits must remain one of the more puzzling of the canyons' unsolved problems.

FORMATION OF THE CANYONS

While several processes have been identified that contributed to the formation of the canyons, their origin remains poorly understood. The following discussion outlines some of the possibilities but, unfortunately, does not lead to a comprehensive model. Faulting has clearly played a key role. As pointed out by Sharp (1973b), however, the general geometry of the canyons is inconsistent with most of the volume having been created by tectonic subsidence. Especially difficult to explain are the large central troughs. If they were formed by southward movement of the Syria-Solis Planum block, as discussed above, then transform faults are required at the east and west ends of the troughs, as well as some means of accommodating to the extension beyond the ends of the canyons, such as between the Ophir and Ganges Chasmas. Neither of these two conditions is fulfilled, and an extension of several tens of kilometers, as has been suggested, appears unlikely. Alternatively, rectangular blocks could simply have been downfaulted to form the Ophir and Candor chasmas. This requires normal faults at the canyon ends and east-west extension, for which there is also no evidence. Indeed, the stress field created by the Tharsis bulge would cause compression in this direction rather than extension. Thus, formation of the central section of the canyons simply by extension appears unlikely.

Despite these arguments, some parts of Valles Marineris, such as the Ius and Tithonium Chasmas to the west and the Coprates Chasma to the east, may still have formed mainly by faulting. Parts of the floors of these canyons are suspected of being downfaulted sections of plateau, and erosion does not appear to have enlarged the canyons greatly, since fault scarps are still visible at the bases of the walls. Central spines suggest that some sections formed by merging of adjacent graben by erosion.

As an alternative to an origin in which tectonic subsidence dominated, Sharp (1973b) suggested a combination of faulting and scarp recession. He proposed that the first step was the formation of parallel graben, trending roughly east-southeast–west-southwest. Differential subsidence along the graben caused lines of pits to form within the graben. Continued subsidence, and perhaps wall recession, enlarged the pits until they merged longitudinally to form narrow troughs with scalloped walls. Lateral enlargement by sustained subsidence and further wall recession caused merger of adjacent troughs to form wider canyons, sometimes leaving longitudinal ridges along the former divides. Several such lateral mergers would be required to form the wide central troughs. Wall recession could have resulted from a combination of several processes, including dry mass wasting, groundwater sapping, and sublimation or melting of ground-ice.

The main problem with any mechanism involving large-scale erosion is elimination of the erosion products. If much of the volume removed was ice, then it could easily be eliminated by melting or sublimation. As pointed out by Sharp (1973b), over 10^6 km^3 would have to be removed. This is about 60 m of water over the whole planet and is close to the total amount of water outgassed from the planet according to many models, so removal of ice alone would imply large modeling errors. If the material were mostly rock debris, as appears likely, then its removal would require some form of mechanical transport, such as subsidence or movement by wind or water. We have already discussed subsidence as a possible mechanism. Tectonically induced subsidence could explain many of the crater chains, the lines of pits, and the relatively narrow canyons of Noctis Labyrinthus and the western canyons, but it appears inadequate to explain the broad central troughs. Eolian deflation, the other obvious means of moving material up and out, is constrained by the need to reduce all materials to fine grains and to prevent formation of protective lag concentrates, and Sharp (1973b) has concluded that eolian transport has played a relatively minor role. This position is supported by the rarity of obvious eolian landforms either in or adjacent to the canyons and to the apparent inefficiency of eolian processes in modifying the martian landscape in general.

Surface transport by water is unlikely to have played a major role in removal of erosional debris from Noctis Labyrinthus, Tithonium Chasma, and Hebes Chasma, since they are all completely closed or include closed sections. If movement of water were involved in formation of these canyons, it would have to take place beneath the surface. A better case can be made for water transport in the cases of the central and eastern troughs. We have already discussed possible lakes within the canyons and drainage of the lakes to the east. If drainage occurred, then large amounts of erosional debris would inevitably have been removed. However, we have seen that the lake hypothesis raises other questions that have not been adequately answered. Removal of debris by slow water transport, such as might occur as a result of seepage of groundwater from the canyon walls, would require large climatic and atmospheric perturbations relatively late in the history of the planet, for which there is little evidence.

Subsurface migration of water may have affected canyon morphology. I suggested that the chaotic terrain at the east end of the canyon resulted from collapse following release of artesian water (Carr 1979) (see next chapter). I proposed that water accumulated in the low regions east of the canyons as a result of subsurface migration down the flank of the Tharsis bulge. On Earth underground movement of water through fine-grained sediments can result in subsurface erosion and subsequent collapse of the surface. This is particularly common in loess regions of China, where tunnels, natural bridges, and circular pits form as a result of subsurface erosion (Fuller 1922). Lines of pits parallel to the martian canyons may form in a similar manner. More direct evidence of groundwater is the emergence of channels from local depressions. In a striking example at 11°S, 55°W, just west of Ganges Chasma, a channel emerges from one of the narrow linear troughs that runs parallel to Coprates Chasma, which suggests that the crater chains and lines of depressions adjacent and parallel to the canyons do indeed mark paths along which groundwater has moved in the past.

The presence of an extensive interconnected groundwater system would explain why the canyons, in which tectonic and mass-wasting features predominate, merge eastward into chaos and channels, which have abundant indications of fluvial activity (see figures 10.7 and 10.10). Because of the accumulation of water in the low regions at the east end of the canyons, faults could disrupt the aquifers there, allowing water to access the surface to form fluvial features. Possibly the lakes postulated for the middle reaches of the canyon were formed in a similar way, although here, because of the lower pore pressures, outflow may have been less violent than further east. In contrast the upper western parts of the canyons may have formed in terrain largely drained of water, so faulting failed to release any significant amounts of water. As a consequence the upper western parts of the canyon are poorly graded, have no fluvial features, and contain no evidence of rhythmic sedimentation.

In summary, many of the characteristics of the canyons can plausibly be explained by faulting, mass wasting, and the effects of groundwater. We are left, however, with two major unsolved problems: the origin of the layered sediments in the central section and the fate of the materials excavated to create the enormous void that the canyons represent.

10 CHANNELS

Of the many spectacular achievements of the Mariner 9 mission, none was more surprising than the discovery of channels on the martian surface. Particularly puzzling are several huge channels tens of kilometers wide and many hundreds of kilometers long with characteristics that suggest floods of enormous magnitude. The channels were unexpected, first, because there was no indication of their presence from earlier missions to the planet, and second, because water cannot exist as a liquid on the surface of Mars under present climatic conditions. Several authors had previously discussed the possibility of extensive ground-ice on Mars (Anderson et al. 1967; Lederberg and Sagan 1962; Salisbury 1966; Wade and de Wys 1968), simply on the basis of the observed annual variation in surface temperatures and an assumed planetary outgassing. However, liquid water appeared to be ruled out by the temperature and pressure conditions. With temperatures in the 150–290°K range (Leighton and Murray 1966; Neugebauer et al. 1971; Sinton and Strong 1960) and pressures, due largely to carbon dioxide, in the 6 mb range, liquid water will either evaporate or freeze, depending on local conditions. These constraints were well established by the time the Mariner 9 data was acquired. When the channels were first seen, therefore, there was considerable resistance to accepting the channels as waterworn, despite their close resemblance to terrestrial fluvial features, and alternative mechanisms for their formation were aggressively sought. Water, however, still appears to be a likely cause for most of the channels, and various proposals have been made to resolve the seeming incompatibility of the water hypothesis with current conditions. While this book was in its final stages of preparation, a new hypothesis was proposed (Lucchitta 1980), namely that the large channels were cut by artesian-fed glaciers, but the full implications of the model have yet to be explored.

The issue of whether or not the channels were cut by water (or ice) has a profound impact on our perception of the planet's geology. Such large volumes of water are required to cut the large flood features that either the more conservative estimates of the amounts of water outgassed from the planet are wrong or the channels were cut by some other fluid. If water is more abundant than the estimates, then so are other volatiles, and appropriate sinks must be present. Origin of the smaller channels by fluvial erosion, in addition, appears to require climatic conditions significantly different from those that prevail today, although this has been questioned recently (Wallace and Sagan 1979). Considerable attention has therefore been directed to determining mechanisms whereby the climate on Mars might have changed and to assessing the magnitude of the climatic changes. Clearly, if the channels were formed by wind or lava, as has been suggested, then the volatile and climatic implications would be quite different.

Viking has provided substantial supporting evidence for water. Pictures of the old cratered terrain have revealed numerous, well integrated tributary systems, strongly suggestive of surface flow of water. These systems could only vaguely be seen in the Mariner 9 data, and evidence of a fluvial origin was thus unconvincing. The clear drainage patterns now seen are persuasive, however, and few geologists argue against a fluvial origin for these particular channels. One argument raised against a fluvial origin for the larger channels, the nonavailability of water, has thus become considerably weaker. Viking has also provided better data on the relative ages of the channels. The ages are of special interest in that they put constraints on outgassing and climatic models. We will see that most of the small channels appear to be old and so require early outgassing. The required water, therefore, could not have been outgassed as a result of the Tharsis volcanism, which mostly postdates the channels. Similarly, models of the evolution of the atmosphere must allow for climate changes on a time scale that is consistent with the ages of the channels.

In this chapter I shall describe different types of channels and discuss different hypotheses for their formation. The general conclusions are that the small tributary networks in the old cratered terrain are formed almost entirely by running water, which means that more moderate climatic conditions must once have prevailed. Most of the large channels may also have been cut by water, although ice is a possibility, but at a later date and under climatic conditions that could be similar to those at present. Some large channels that start near volcanoes may have been cut by lava.

The term "channel" is used somewhat erroneously with respect to Mars. Strictly it should be applied only to conduits that are, at times, almost or completely full of liquid. Channels are therefore distinct from valleys, which are never filled with water but which, of course, usually contain a channel. Many of the small martian "channels" are probably not true channels but valleys. Nevertheless the term channel is widely used, and we will continue to use it here, acknowledging the error in doing so. Sharp and Malin (1975) identified three major classes of channels:

1. Runoff channels. These start small, increase in size downstream, and have tributaries. Most are a few tens of kilometers long, 1 km wide or less, and are probably valleys.

2. Outflow channels. These are mostly large features that start full-born from local sources. They rarely have tributaries, and most have features suggestive of large-volume flow, such as occurs in catastrophic floods or glaciers.

3. Fretted channels. These are wide, steep-walled channels with smooth flat floors. The walls may intersect craters, other channels, or low escarpments, to isolate buttes and mesalike features. They become narrower upstream and may have tributaries.

Runoff channels and outflow channels are discussed in detail below. Fretted channels are only briefly mentioned, because fretted terrain was covered at some length in chapter 6.

RUNOFF CHANNELS

Runoff channels are almost ubiquitous in the heavily cratered terrain and are almost totally absent on the plains. The most common type is a simple gully a few tens of kilometers long, generally on steep slopes such as crater walls and escarpments (figure 10.1). Most gullies on crater walls simply terminate against the crater floor; gullies on other slopes may similarly terminate abruptly, but many connect with other channels to form tributary networks (figures 10.2 and 10.3). Most networks are open, with wide, undissected areas between individual channels. Moreover, even in the more dissected parts of the old terrain, the channel systems are spaced widely apart, indicating that they are relatively immature, and that there has been little competition between adjacent drainage networks. Although such networks are common, they rarely exceed 300 km in length, from the upstream end of the most distant tributary to the termination of the main branch. Larger networks do occur. The channels Ma'adim and Al Qahira, for example, at 20°S, 182°W, and 18°S, 196°W, respectively, are 600 km and 800 km long, but even these relatively large drainage systems are considerably smaller than the larger terrestrial river systems. Junction angles between tributaries commonly show a wide variation, which gives the channel systems a more random directional pattern than typical terrestrial drainage networks (Pieri and Sagan 1978). The pattern may be so random that the direction of flow is difficult to determine; structural control is often obvious. The main trunk channels generally wind between the craters, but occasionally such a channel breaches a crater wall and ends within the crater. Many of the trunk channels terminate abruptly in low areas, with no indication of deposits at their mouths. Termination may result simply from burial by younger deposits. This is not the general case, however, and may not be true of the majority of cases. Many of the channels simply stop abruptly. The pattern is somewhat similar to that in karst regions of Earth, where flow over the surface ends suddenly but continues underground.

Figure 10.1. Dissected cratered terrain at 25°S, 358°W. The large crater in the center is 230 km across. Dissection is typically by parallel gullies, with little integration into larger networks. (211-5207)

Although most areas of the cratered terrain show abundant evidence of dissection, in some areas channels are rare. The region extending from the rim of Argyre 1,000 km to the east, between latitudes 40°S and 60°S, for example, has few channels. Runoff channels are also few between latitudes 0° and 30°N and longitudes 20°W and 350°W, just east of the Chryse basin. In both these areas the lack of channels may be due to burial by younger deposits. The intercrater areas are unusually smooth, crater rims appear buried, and rugged plateau areas are largely missing, as though the old densely cratered surface were at a shallow depth below the present surface. Channels are also less common where large proportions of ridged plains are present. They are almost completely absent, for example, on the ridged plains of Hesperia Planum, which clearly bury channels cut into the surrounding plateau areas. Runoff channels are also scarce in the old cratered terrain north of 30°N, although fretted channels are common. Here the cratered terrain has a different character than in most other areas.

The upper surface of the plateau, where not affected by the fretting process, is unusually smooth, and flow features are common within and outside craters. The general impression is that the high frequency relief has been smoothed out by some mass-wasting process. Elsewhere in the old cratered terrain channels are omnipresent but in dispersed networks. The ridged plains of the old cratered terrain have fewer channels than the plateau areas, but channels are present—Nirgal Vallis, for example (figures 10.4 and 10.5). The channels of the ridged plains tend to be crisp and well delineated, in contrast to those of the plateau areas, which show a wider range of preservation. Runoff channels are almost totally absent outside the old cratered terrain. The main exceptions are channels on the canyon walls, which clearly cut the surface of the adjacent plains.

Figure 10.2. Open channel networks and gullies at 24°S, 8°W. Within the networks are largely undissected interfluves, and at the upper reaches of the channels there is no competition from adjacent drainage basins, which suggests an immature drainage system. The picture is approximately 300 km across. (P-18115)

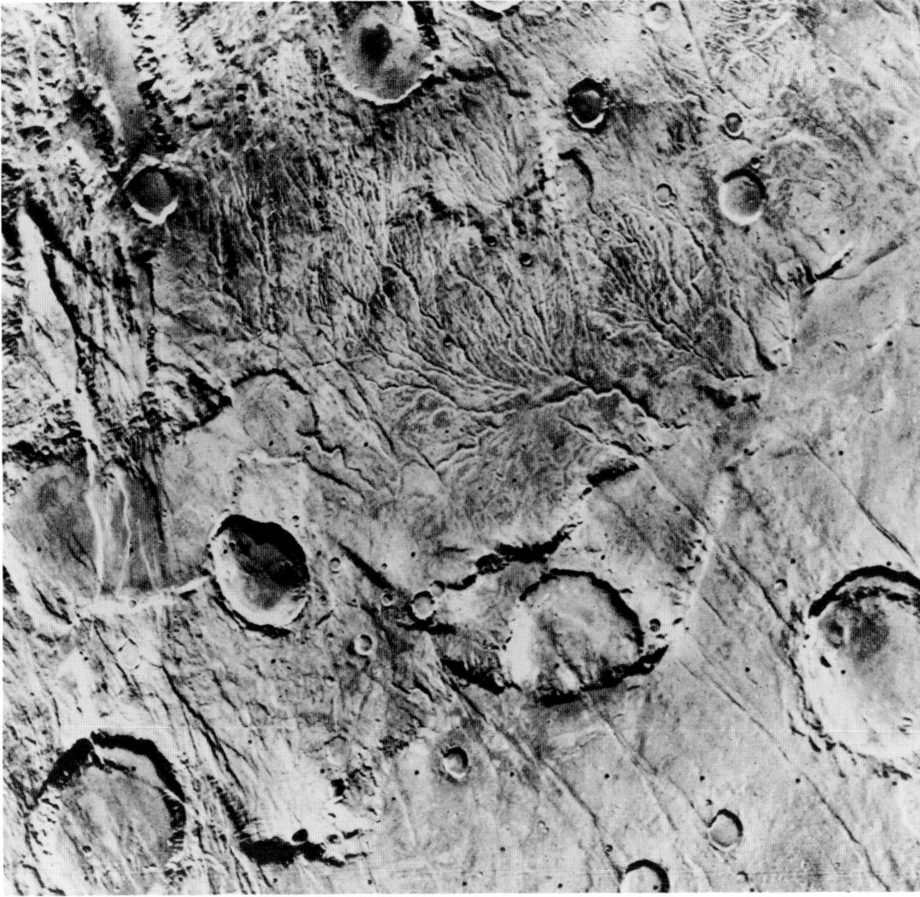

Figure 10.3. Highly fractured old terrain at 48°S, 98°W, is finely dissected by channels. This is one of the densest drainage networks observed on the planet. The picture is 250 km across. (63A09)

The almost complete restriction of runoff channels to old surfaces suggests that most of the runoff channels are old. Alternative explanations are possible, however. The channels may occur in the old terrain because they form by sapping (Pieri 1979; Sharp 1973a,b), and only in the old terrain is the groundwater or ground-ice available to sustain the process. Alternatively, the old terrain may be more erodible than the younger terrains. However, the simplest explanation of the scarcity of runoff channels on terrains younger than about 3.5 billion years is that the conditions necessary for their formation were not sustained after that time.

The open networks and lack of competition between adjacent drainage

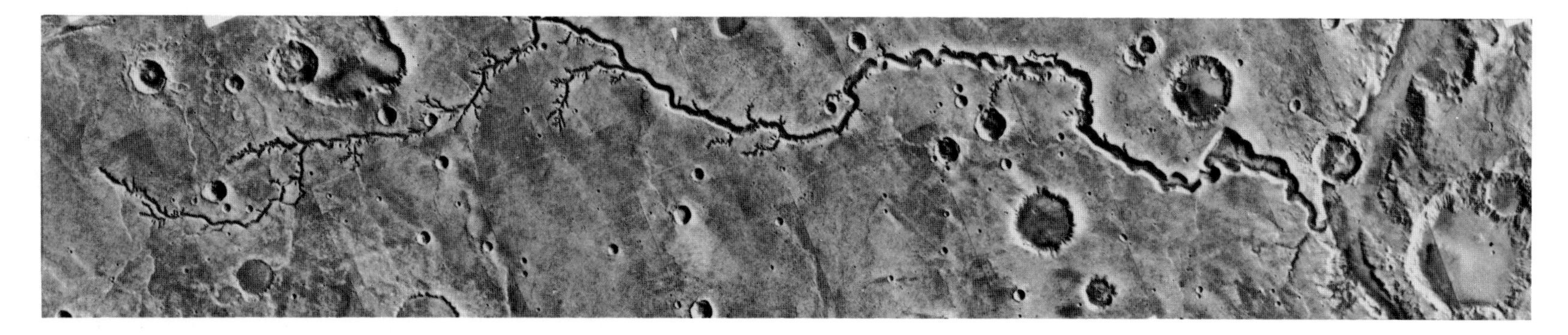

Figure 10.4. Nirgal Vallis, at 28°S, 40°W. The channel is 800 km long and has numerous tributaries incised into the ridged plains of the old cratered terrain. Drainage, however, is restricted to a narrow band adjacent to the trunk stream. The open nature of the network and the lack of a large catchment area suggest formation by groundwater sapping rather than by surface runoff. (211-5540)

basins suggest an immature drainage system, in which larger channels have not had time to capture adjacent streams and integrate drainage over large areas. This conclusion of poorly established drainage patterns and relative immaturity is somewhat at odds with the wide range in preservation of the small runoff channels, which suggests a range of ages, some of which must be quite old with respect to the present landscape. The seeming discrepancy may be the result of other processes. Very early in the history of the planet, when most of the runoff channels appear to have formed, crater formation and obliteration rates were high, so that the landscape was being continually re-formed. As crater obliteration and formation tapered off, the fluvial features became superimposed on a more stable surface, which still retained some evidence of more eroded channels. The more crisp-appearing channels are thus somewhat analogous to the fresh-appearing craters, in that they formed after the major obliteration event was over. Fluvial action could not have continued long after the surface stabilized, however, unless at extremely low rates, for large mature drainage basins never developed.

The runoff channels are difficult to reconcile with any origin other than erosion by running water. Several alternatives have been explored to explain channels on Mars in general. These include erosion by lava (Carr 1974b; Schonfeld 1977), erosion by liquid hydrocarbons (Yung and Pinto 1978), erosion by wind (Cutts et al. 1978a; Cutts and Blasius 1979), and faulting (Schumm 1974). While some of the alternatives may plausibly explain some of the large outflow channels, they conspicuously fail to explain the integrated tributary networks of the old cratered terrain. Volcanism would require lava sources at the head of each tributary, in areas where there is no indication of volcanic activity. Wind does not form complicated drainage networks, particularly with branches meeting at large angles, because of strong coupling between flow near the surface and regional flow patterns higher in the atmosphere. Faulting can simulate drainage patterns occasionally, but can hardly explain the repetitive mimicry from network to network. Finally, there is no direct evidence of hydrocarbons on the martian surface, despite attempts to detect them with the extremely sensitive gas chromatograph–mass spectrometers on the Viking landers, so that invoking hydrocarbons rather than water raises more problems than it solves. It thus seems almost inescapable that the runoff channels were formed by running water.

The water that cut the channels may have been derived directly from rainfall, as suggested by Masursky and co-workers (1977), or by seepage from the ground. Pieri (1979) examined several characteristics of runoff channels and concluded that in most cases seepage was more probable than rainfall. Seepage can occur by melting of ground-ice or release of groundwater. Channels that form by seepage extend themselves headward by undermining the terrain at the point where water comes to the surface. The valleys produced have somewhat different shapes from those produced by surface runoff. Seepage networks tend to be more open, with undissected interfluves and alcovelike terminations to individual valleys. In contrast, networks caused directly by rainfall tend to be dense, with dissected interfluves and tributaries that branch into ever finer valleys. In practice, the distinction between the two types of network is rarely clear, because rainfall nearly always involves the charging and discharging of a groundwater system. Nevertheless, in cases such as Nirgal Vallis and the channels that are deeply incised in the canyon walls, origin by sapping, as originally suggested by Sharp (1973b), is convincing. These deeply incised networks have no traces of drainage between the major branches, and all the tributaries have cirquelike terminations. The channels on the walls of the canyons have rectilinear patterns, indicating strong structural control, which is common with channels that form by sapping because of preferential flow of groundwater along lines of structural weakness. For smaller channels the case for sapping is not as strong, because the diagnostic characteristics are not as readily observable. One common relationship, however, that strongly supports sapping is the breaching of a crater wall by a trunk channel and the drainage of a network into a crater. This relationship is explained simply by sapping. The channels started on the walls within the crater, then cut through the rim by headward erosion, and extended themselves into the surrounding terrain. To breach the crater rim by surface drainage requires superposition from a former surface over the crater and coincidental termination of the channel in the crater, both of which are unlikely. The evidence for sapping in these cases is very convincing. An equally convincing case cannot be made for rainfall, but neither can rainfall be ruled out; it is certainly the simplest way of charging the groundwater system.

In most areas there is no consistent direction of drainage, and adjacent networks may drain in different directions. Two exceptions are the channels adjacent to the uplands/plains boundary, which consistently drain toward the plains to the north, and the area southeast of the main chaos region of the planet in Margaritifer Sinus, where the runoff channels drain northwest toward the chaos. As previously indicated, most channels terminate abruptly, with no indication of deposits at their mouths. One possible explanation is that the water that cut the channels disappeared underground. The old cratered terrain is likely to be porous as a result of brecciation by impact. We have seen that there is good evidence for seepage, which requires either groundwater or ground-ice. It is not unreasonable, therefore, to attribute the abrupt channel terminations to return of the water to the artesian system through rainfall. Alternatively, the water that cut the channels may simply have evaporated into the atmosphere, in which case evaporite deposits should occur at their mouths.

Figure 10.5. Detail of Nirgal Vallis, showing the open tributary network and the lack of dissection between branches. The alcovelike terminations are also suggestive of sapping. The frame is 80 km across. (466A54)

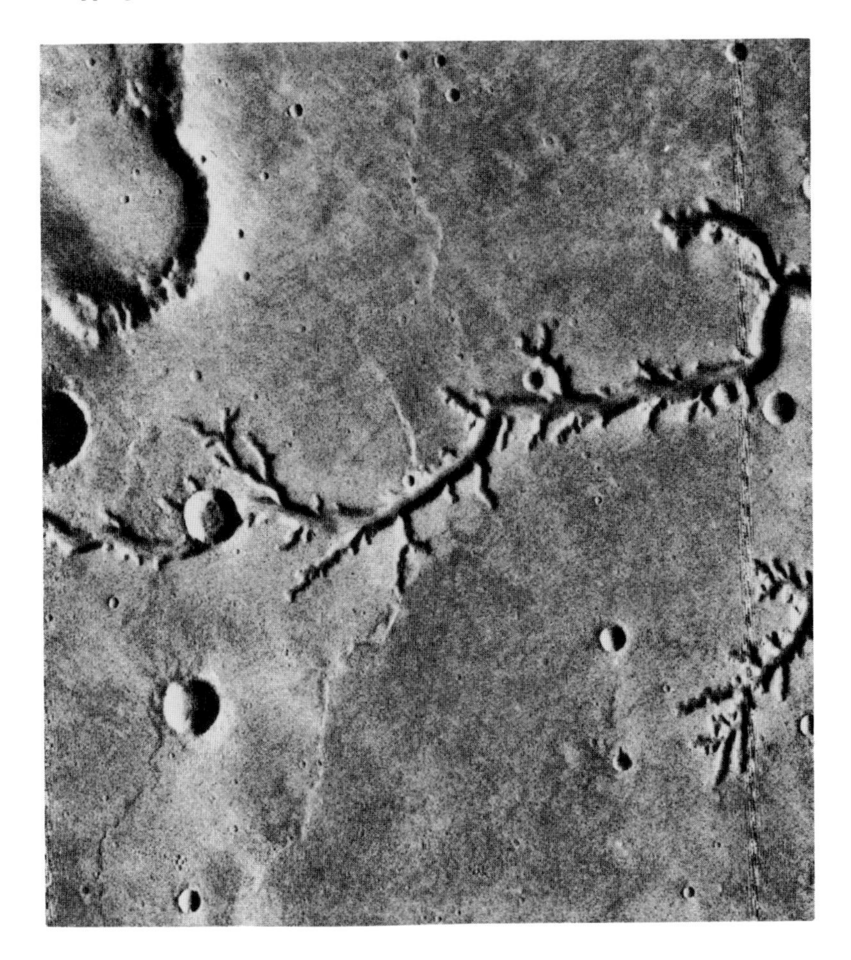

Formation of the small runoff channels almost certainly requires climatic conditions different from those that currently prevail on the planet. Wallace and Sagan (1979) examined this problem in some detail, elaborating upon earlier work by Lingenfelter and colleagues (1968), who explored the possible fluvial origin of lunar rilles. Lingenfelter and co-workers demonstrated that with large discharges, flow of water across the lunar surface could be sustained, because rapid evaporative cooling would cause the surface to freeze, thereby retarding further evaporation. Wallace and Sagan explored the implications of such a model for Mars, taking into account the temperature and pressure conditions there and additional effects such as wind. They concluded that a protective ice cover 10 m to 30 m thick could develop on a martian river, and that flow could continue under the ice layer at relatively low discharge rates. However, these results are probably only applicable to the large flood features. The trunk channels of the tributary networks are fed by small streams. Even in the unlikely event that the main trunk channel held water 10–30 m deep, the distal parts of the network are likely to be extremely shallow. They would therefore freeze and cut off the supply to the main stream, just as terrestrial rivers in the arctic do under much less extreme conditions than those that occur on Mars. It appears, therefore, that temperatures and pressures must have been higher when the runoff channels formed.

We can conclude that most of the runoff channels were formed by running water, some definitely by sapping, but some possibly by surface drainage following rainfall also. The majority probably formed early in the planet's history, before most of the plains formed, at a time when climatic conditions were more conducive to liquid water at the surface than they are now.

OUTFLOW CHANNELS

No other aspect of martian geology has generated as much controversy as the formation of the large outflow channels. Because of their close resemblance to large terrestrial flood features, some researchers hold that origin by catastrophic floods is inescapable. Others argue that the size and shape of the channels are more consistent with glacial erosion. Still others point

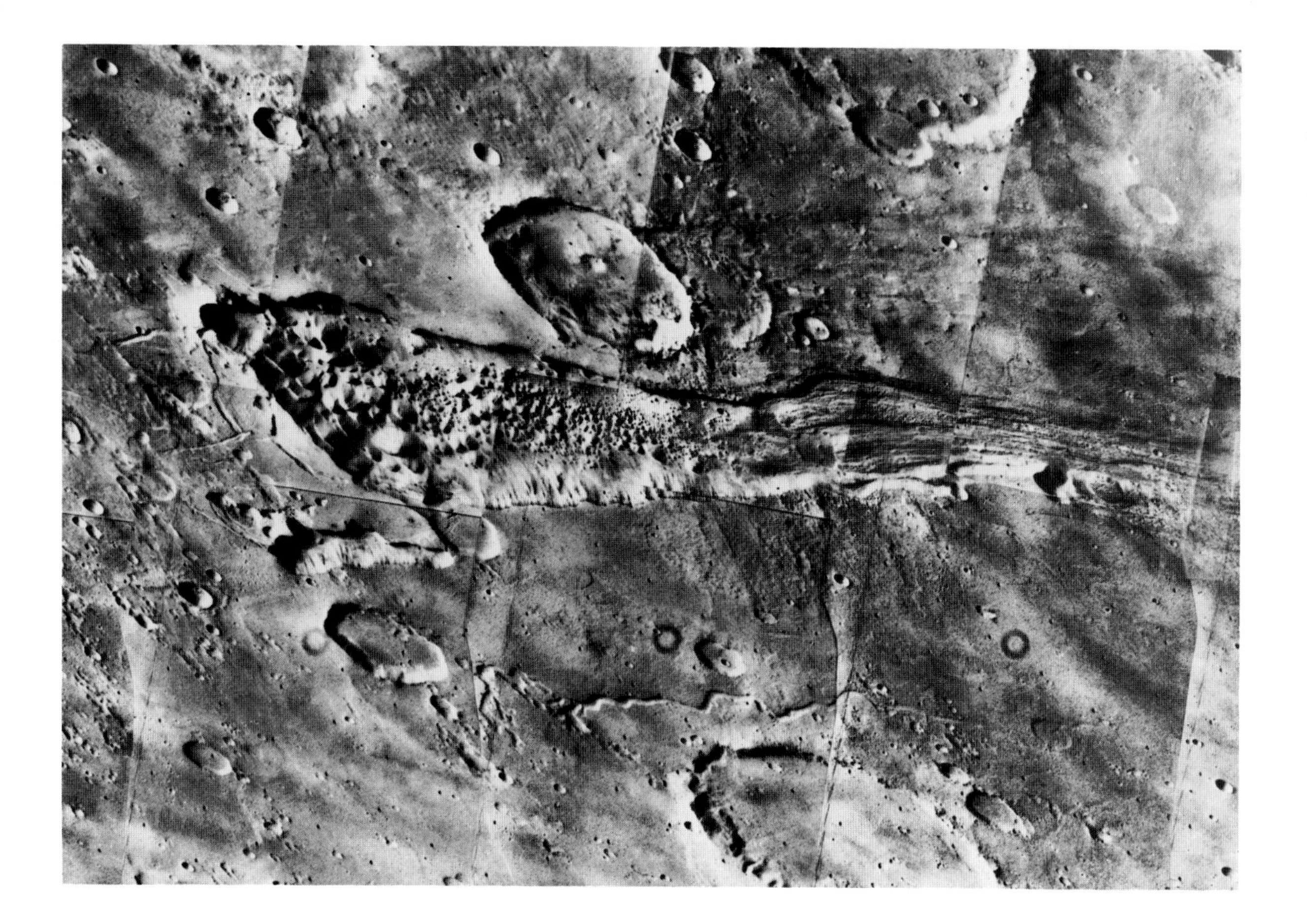

Figure 10.6. Chaos and channels at 1°S, 42°W. A 20-km–diameter channel emerges full-born from a 40-km–diameter depression enclosing chaotic terrain. The channel continues eastward off the picture to connect with Simud Vallis. (P-16983)

to the lack of sources and sinks for the vast quantities of water required to erode the outflow channels and argue for erosion by wind or lava. Outflow channels start full-scale from local sources (figure 10.6). Some are enormous. Parts of the upper reaches of Kasei Vallis, for example, are over 200 km across, and some of the fluvial features south of Chryse can be traced for over 2,000 km. The largest are around the Chryse Basin, but others occur in Memnonia, Amazonis Planitia, northwest of Elysium, and on the rim of Hellas. Outflow channels can be grouped into two main classes: unconfined types, in which flow features are found over broad, poorly delineated swaths of terrain; and confined types, in which flow appears to have been restricted to distinct, well-defined channels. Most of the unconfined types occur around the Chryse basin, where flow from the channels scoured much of the periphery of the basin. They have numerous characteristics that suggest erosion by catastrophic floods, or possibly ice. Several large channels northwest of Elysium and on the northeast rim of Hellas are of the confined type. They have some features in common with lunar rilles and can plausibly be interpreted as having formed by lava erosion, although water or ice still seem more probable.

Unconfined outflow channels

CIRCUM-CHRYSE CHANNELS. Several large channels converge on Chryse Planitia from the west, south, and east, then continue northward across the surface of the Chryse plains until they disappear between 30°N and 40°N (figure 10.7). All start full-born in box canyons or chaos regions that are continuous with, or adjacent to, the equatorial canyons. As we saw in the previous chapter, the main canyon system merges eastward into chaotic terrain, as the rectilinear, roughly east-west walls of the canyon become irregular in outline, and the smooth floor of the canyon assumes a rubbly, chaotic appearance. Further to the northeast, there emerge from the chaos smooth linear features, which, when traced downslope, display numerous characteristics suggestive of fluid erosion, such as teardrop-shaped islands, longitudinal grooves, and curvilinear banks.

The connection between canyon, chaos, and channels is repetitive at a range of scales. Juventae Chasma, at 2°S, 61°W, typifies the sequence. Its configuration is much the same as that of Hydraotes Chaos (see figure

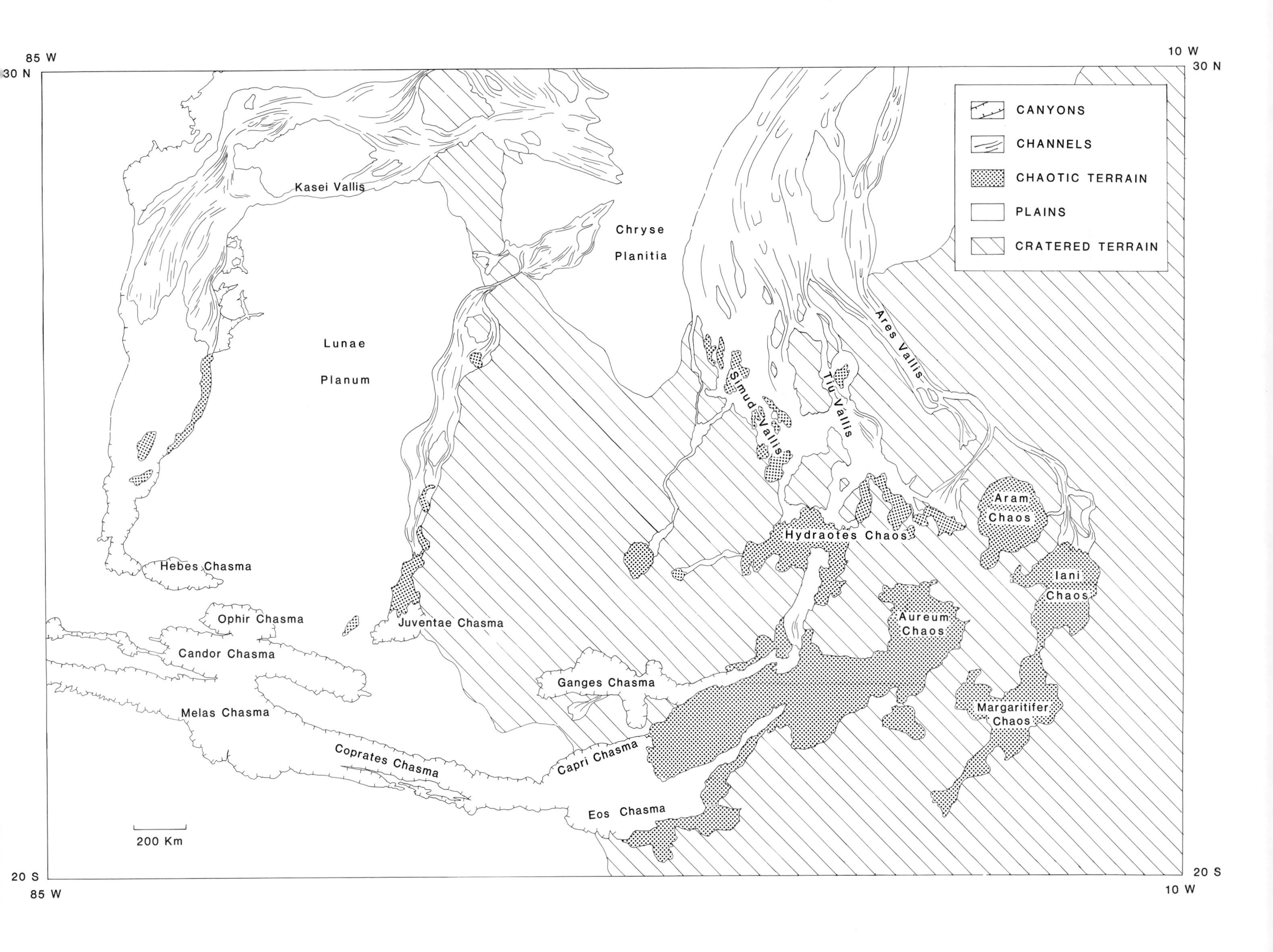

Figure 10.7. Map showing the relations between canyons, chaos, and outflow channels around the Chryse basin. The flow lines on the channels are largely schematic but roughly follow the flow, as indicated by streamlined islands and scour. The direction of flow is all northward into Chryse Planitia, and most of the flow features fade away just to the north of the map area. The region portrayed is 4,400 km across at the equator.

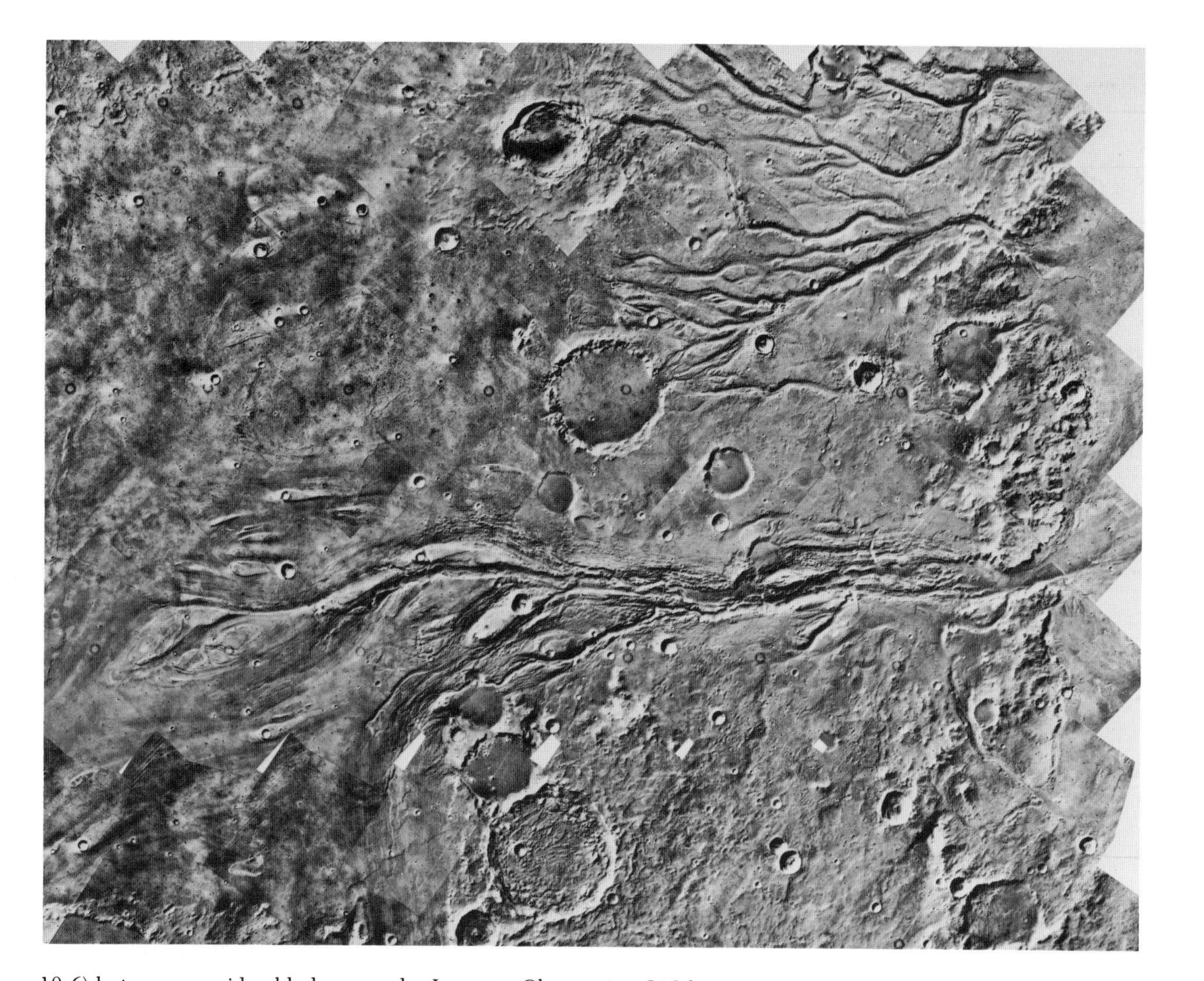

Figure 10.8. Maja Vallis to the south and Vedra Vallis to the north are deeply incised into the old cratered terrain between Lunae Planum to the left and Chryse Planitia to the right. Juventae Chasma, several hundred kilometers to the south of the area shown, gives rise to a large channel that extends northward along the eastern boundary of Lunae Planum into this region. Flow apparently converged on Maja Vallis to cut its gorge. The scene is 300 km across. (211-5190)

10.6) but on a considerably larger scale. Juventae Chasma is a 250 km × 100 km depression that is completely enclosed by an escarpment, except to the north where a large channel emerges. The southern half of the chasma, with its high escarpment and relatively flat floor, resembles many sections of Valles Marineris, whereas the northern half is filled with jostled blocks and is typical chaos. North of the chaotic area the boundary between the old cratered terrain to the east and the younger Lunae Planum to the west is extensively modified by branching channels, which emerge from the chaos and diverge and converge around obstacles such as low hills and craters. Longitudinal grooves, teardrop-shaped islands, and low horseshoe-shaped escarpments can be traced northward along the boundary for about 1,000 km. At 16°N the flow lines converge on several northeast-trending channels that are deeply incised into the old cratered terrain between Lunae Planum and Chryse Planitia (figure 10.8). Most of the flow apparently converged on Maja Vallis, a 10-km–wide, 180-km–long gorge, over 1 km deep. Beyond the mouth of Maja Vallis, flow diverged over the gently sloping plains of Chryse Planitia to form a pattern of gently sinuous, anastomosing channels, elongate hills, and longitudinal grooves. Where obstacles, such as craters and ridges, occur, the flow lines diverge around them or become concentrated at gaps (fig. 10.9). Further to the northeast the flow lines disappear, either because the flow ceased, or because the modified surface is covered by later deposits.

The large channels that enter the Chryse basin from the southeast start either in isolated areas of chaotic terrain or in areas of chaos that merge southwestward with the canyons. Ares Vallis, a 25-km–wide channel, emerges close to the equator at 18°W from several areas of chaos, seemingly unconnected with the canyons. The chaos areas are either circular in shape and completely contained within a former crater, as with the Aram Chaos at 2°N, 21°W, or irregular and ill-defined, as with the Iani and Margaritifer Chaos just south of the equator. The two large channels to the east of Ares, the Tiu and Simud Valles, also have local areas of chaos in their upper reaches (figures 10.10 and 10.11), but the walls that enclose these channels and their chaos are continuous to the south and west with the walls of the canyons. The Hydroates Chaos at 9°N, 35°W, for example, merges southward with Capri Chasma, and the Aureum Chaos merges to the west with Eos Chasma. In the chaos region north of approximately 5°S fluvial-like features are abundant; in the canyons west of 50°W, they are absent except for the gullies in the walls described previously. Between is a transitional section in which canyon, chaos, and channel features occur together. Ganges Chasma, for example, has a channel on its south rim, and another emerges from a closed depression to the west. Both channels flow to the canyon, where they are cut off by the rim. Chaos and smooth curvilinear features occur within the canyon. Canyons, chaos, and outflow channels are thus physically connected, and their origins may be in some way related.

At the mouths of the Ares, Simud, and Tiu Valles, are rich arrays of sculpted landforms (Carr et al. 1976; Masursky et al. 1977; Greeley et al. 1977). Remnants of the plateau in which the channels are incised have been shaped into streamlined islands, elongate in the direction of flow (figure 10.12). Many of the islands are teardrop-shaped or lemniscate in outline, with the blunter end pointing upstream and a long tail pointing downstream. Some islands have a pointed prowlike upstream termination and a shallow moat around them, as though the flow had separated and swept around the island to cause enhanced erosion around its periphery. Craters and low hills apparently acted as obstacles to the flow and commonly occupy the broader upstream part of the teardrop. The trailing sections are occasionally terraced, as though a layered sequence had been eroded back. Between the plateau remnants the surface has a longitudinal scour, which sweeps northwestward out of the channels into Chryse Planitia, then turns to the northeast to run along its eastern boundary (see

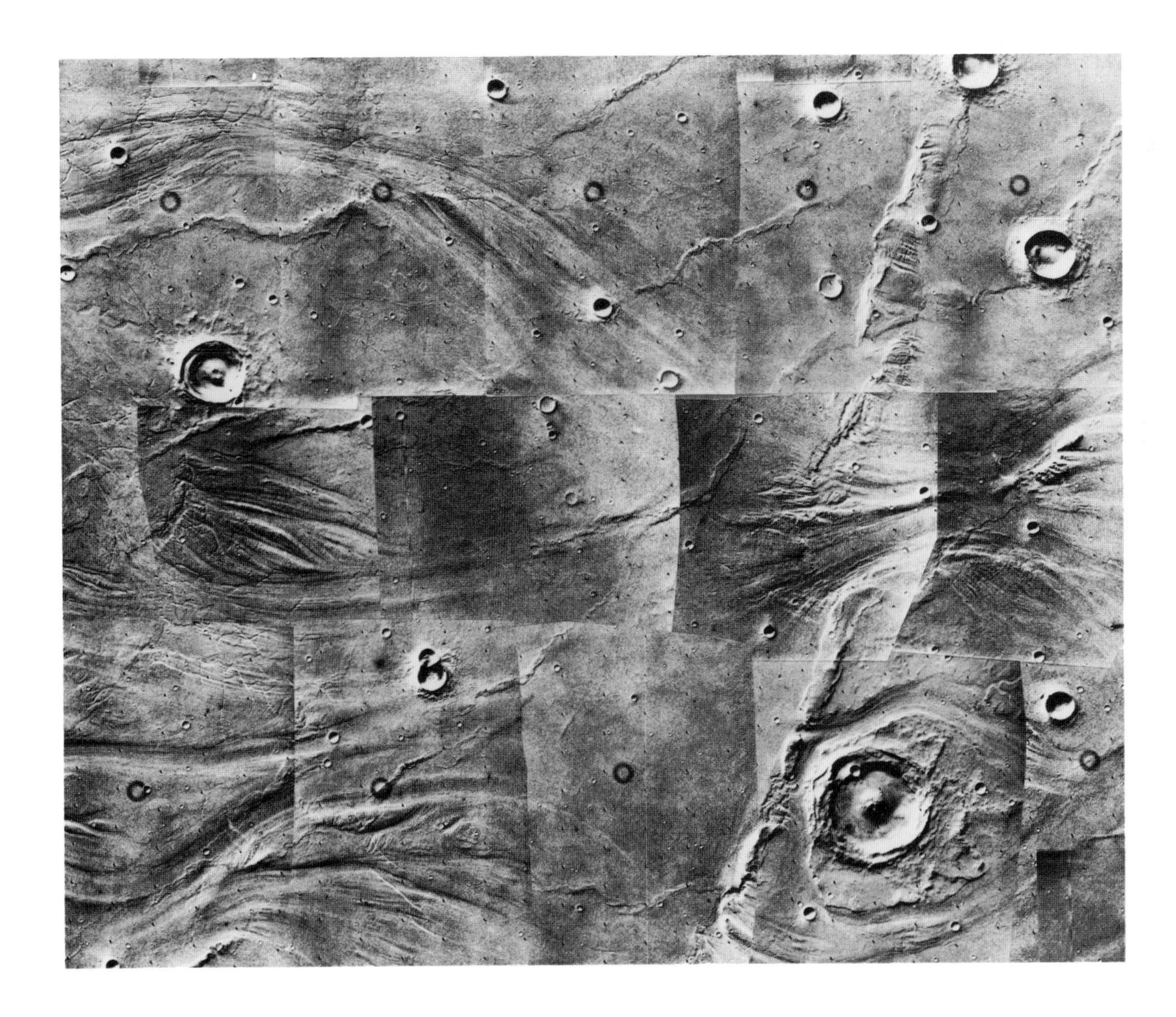

Figure 10.9. The plains of Chryse, just to the east of the area shown in the previous figure. Flow from Maja Vallis diverged across the plains and deeply scoured the surface. Ridges that resemble those on the lunar maria partly obstructed the flow, which was funneled through gaps and low points, where intense erosion occurred. The scene is 155 km across. (P-17002)

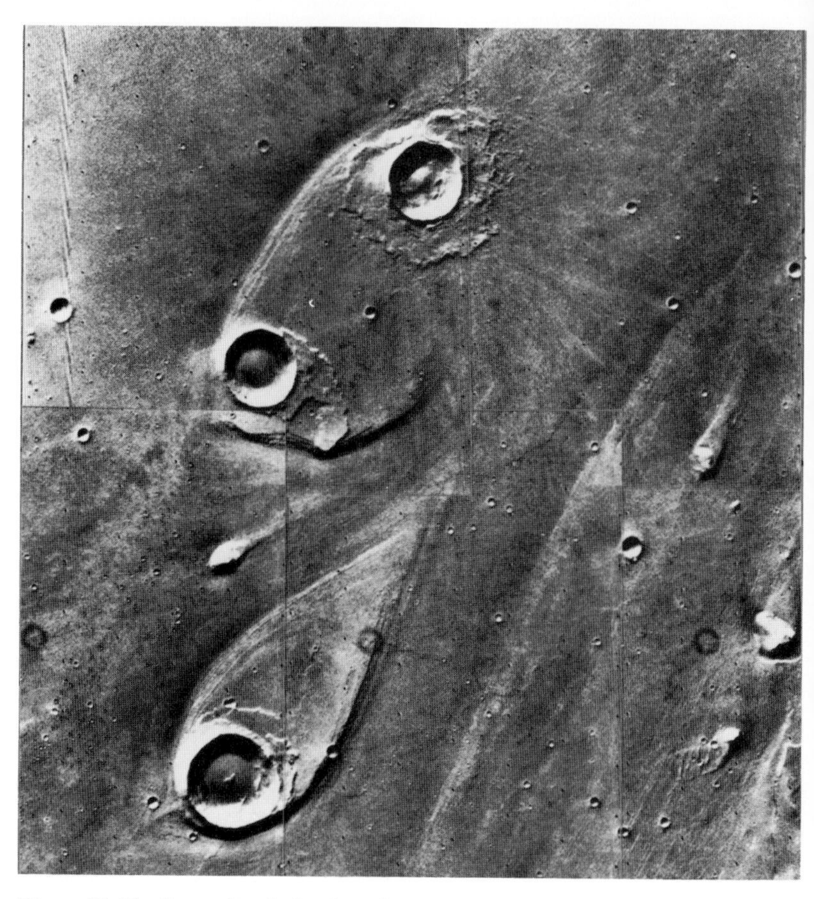

Figure 10.12. Streamlined islands at the mouth of Ares Vallis, at 20°N, 31°W. The islands appear to be eroded remnants of a former plateau. The flow from the lower left diverged around two craters to form islands, with a sharp-pointed prow upstream and a long tapering tail downstream. The adjacent channel floor has a faint longitudinal scour. The upper crater postdates formation of the channel. Each island is approximately 40 km long. (4A50-54)

Figure 10.10. (*above*) Mosaic of the chaos and channel source region to the southeast of Chryse Planitia. The scene covers most of the bottom right quadrant of the map in figure 10.7 and connects to the left with the synoptic view of the canyon in figure 9.1. The mosaic is 1,200 km across. (211-5821)

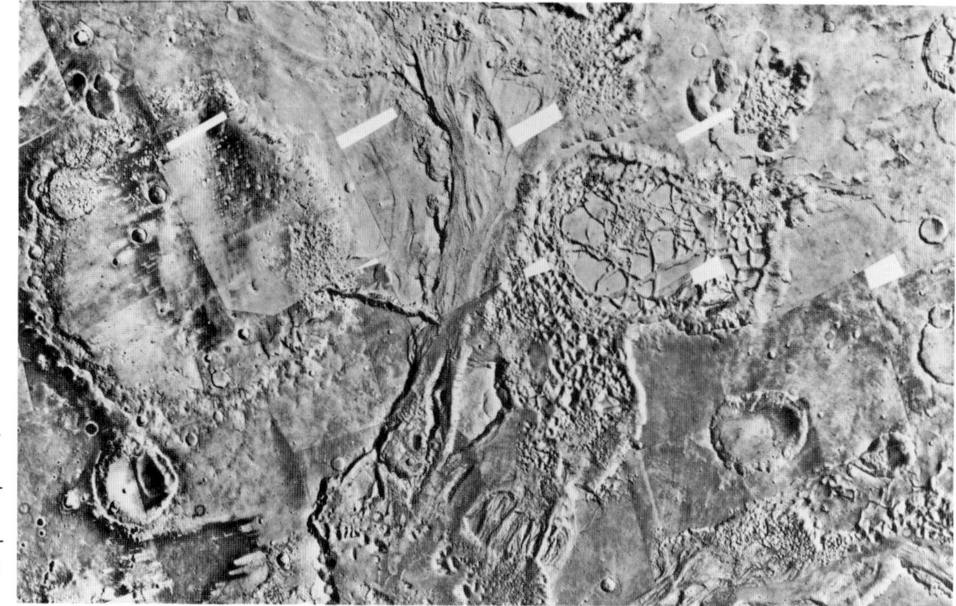

Figure 10.11. Detail of part of the previous figure, showing the source region of Tiu Vallis in chaotic terrain at 3°N, 34°W. The gorge connecting the large crater to the right with the main channel is a common relation in the chaotic regions. The roughly circular area of chaos at left center is 100 km across. (P-19131)

figure 10.7). Parts of the Chryse plains also appear etched, as though flow had plucked part of the surface to form arrays of jagged, irregularly shaped depressions, roughly elongate in the flow direction. The scour and sculpted islands die out at around 30°N.

The largest of the outflow channels, and the one in which outflow characteristics are best displayed, is Kasei Vallis, which enters Chryse Planitia from the west (figure 10.13; see also figure 9.1). The channel originates in a shallow, north-south–trending depression, 300 km wide and 1,500 km long, which merges southward with the box canyon Echus Chasma. The depression is well delineated to the east by the escarpment along the western boundary of Lunae Planum, but to the west the edge of the trough is indicated only by indistinct hills and breaks in slope, probably because of partial burial of the trough boundary by young flows from Tharsis. North of 15°N the floor of the depression is pervasively scoured with closely spaced longitudinal grooves. Numerous streamlined islands, similar to those near the mouths of the south Chryse channels, are also present. The scour pattern indicates flow to the north, but downstream the flow turns eastward, and the channel splits into two main east-west branches, which cut through the ridged plains of Lunae Planum. These join 700 km downstream, to isolate a large island of undissected plains 300 km across. Further east the pattern of flow is complex, as several branches separate and rejoin around obstacles. After the channel enters Chryse Planitia, traces of the flow are rapidly lost, and none can be found east of about 47°W.

The floor of Kasei Vallis is highly textured (figure 10.14). The northern branch is a scoured swath about 100 km across, which contains at its southern edge a deep, steep-walled, smooth-floored channel, 10–15 km across. Flow lines from the broad depression to the west tend to converge on this channel, but convergence is not achieved completely, and the flow lines sweep over the plateau remnant south of the channel, then beyond the channel to scour the area to the north, as though the flow had overshot the channel. In the longitudinally scoured sections are numerous rectilinear crevasselike features. These have mainly north-northwest–east-southeast and north-northeast–west-southwest orientations and form local, gridlike arrays of steep-walled depressions. These same orientations recur in the escarpments surrounding Lunae Planum and other plateau remnants and appear to reflect a regional joint or fault pattern. East of 53°W the north Kasei branch broadens, and several barlike features are present. Still further east the flow diverges around the 100-km crater Sharanov. Flow along the southern branch of Kasei appears to have been almost completely confined to a relatively deep, smooth-floored channel, except to the east where the channel broadens, and bars and several plateau remnants are present.

The escarpment at the western edge of Lunae Planum (see figure 9.1) has a jagged outline, with long, linear depressions reaching deep into Lunae Planum. The escarpments that bound the channels have similar

Figure 10.13. Synoptic view of the upper reaches of Kasei Vallis. Flow from a broad depression to the south changed its course eastward to form the deeply incised Kasei Vallis, which is mainly off the picture to the right. Flow is indicated by longitudinal scour and streamlined islands that occur in a swath that in places is over 200 km across. The mosaic is 400 km wide. (211-5642)

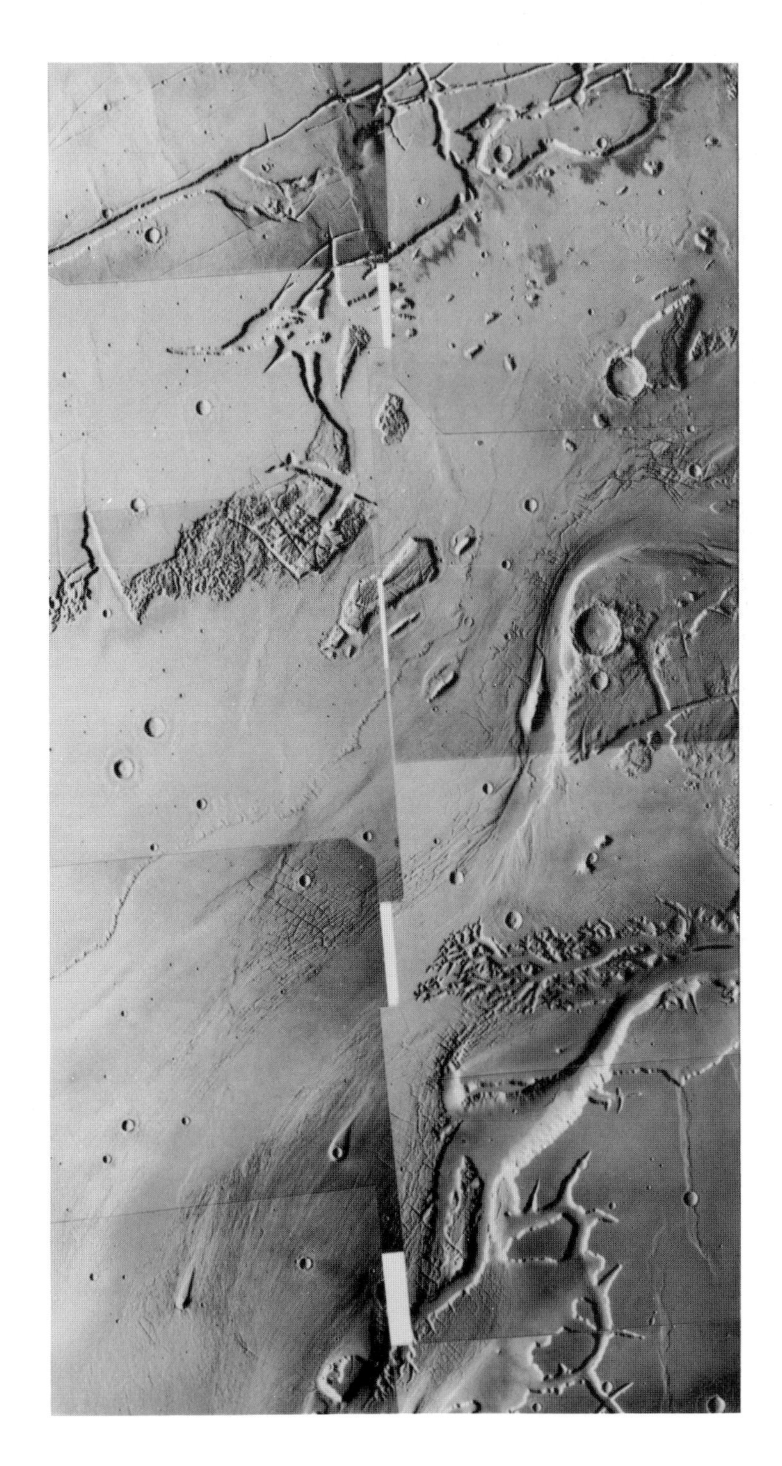

Figure 10.14. Detail from the previous figure, showing the region at 20°N, 73°W, where the flow in Kasei Vallis changes from north-south to east-west. Flow from the lower left appears to have overshot the deeper channel and scoured the region beyond the channel to the north. The narrow cracklike features appear to have formed by preferential plucking along lines of structural weakness, as in the channeled scablands of eastern Washington (see text). The deep channel is 15 km across. (519A10)

Figure 10.15. The source of Mangala Vallis, at 16°S, 149°W. The channel originates at a graben, then continues northward for several hundred kilometers, its path marked by scour and streamlined islands, as in the upper part of the figure. The scene is 250 km across. (639A11)

irregular traces, with linear segments and angular junctions. In many places layering is distinctly visible in the escarpment, and an upper layer has commonly been stripped back. Talus, landslides, and other mass-wasting features are common. Rarely are the escarpments streamlined to conform with the channel flow patterns. Mass wasting, therefore, may have modified many of the steep slopes originally cut by the channels.

MANGALA VALLIS. Mangala Vallis has almost identical characteristics to the circum-Chryse channels, except that it arises not in areas of chaos but at a notch in a graben wall, at 18°S, 149°W (figure 10.15). Flow was to the north, where streamlined features and incised north-south channels cover a swath up to 80 km wide that can be traced northward for almost 1,000 km. Where flow was over plains, the scoured area is relatively shallow, and numerous streamlined islands and barlike forms are present. Where flow was over cratered terrain, the main channel breaks up into numerous narrow, deeply incised channels, with a crude anastomosing pattern. The differences suggest that the cratered terrain is more erodible than the plains. A similar contrast was seen above in the case of Maja Vallis. Approximately 500 km north of the Mangala source the channel breaks in two, with one section branching off to the northwest. The various branches of the channel cease abruptly at the plains/uplands boundary at around 4°S, possibly because of burial by later deposits.

AMAZONIS CHANNEL. In the knobby terrain just to the west of Amazonis Planitia is a broad, shallow, north-south–trending channel. It first becomes visible at 5°N, 183°W, just south of Orcus Patera, where it is about 70 km across. From here it can be traced for about 1,000 km, first northwestward to 176°W longitude, then northward to approximately 21°N, where it turns eastward into Amazonis Planitia and fades away. The source is unknown and may be buried by younger eolian deposits, which are common in this general region. The main characteristics of the channel are shallow depth and numerous low, but well formed, streamlined islands.

Confined outflow channels

ELYSIUM CHANNELS. Several channels start 200–300 km northwest of Elysium Mons and Hecates Tholus and extend about 1,000 km to the northwest, where they disappear in some complex terrain south of the Viking 2 landing site. These channels differ significantly in their general morphology from the circum-Chryse and Mangala channels just described. They start in several west-northwest–east-southeast–trending, grabenlike depressions up to 15 km across (see figure 7.24). As the graben are traced downslope, they assume characteristics that are more fluvial-like than tectonic, commonly breaking up into several sinuous steep-walled channels, which resemble lunar rilles in their uniform sinuosity, width, and depth of incision over long distances (figure 10.16). The overall pattern of channeling, however, is quite different from that on the Moon. The Elysium channels frequently branch and rejoin. In places the surface between closely spaced channels appears to be intensely fractured in a manner not seen on the Moon. Although most of the channels are less than 3 km across, little detail is visible on their floors; some sections are as much as 15 km wide, and in these, teardrop-shaped islands, much like those at the mouths of the Chryse channels, are present. Apart from the islands, however, the Elysium channels have little in common with the channels around Chryse. Flow appears to have been totally confined to narrow discrete channels, so that the interchannel areas are completely unscoured. Longitudinal striations, inner-channel headcuts, plucked fea-

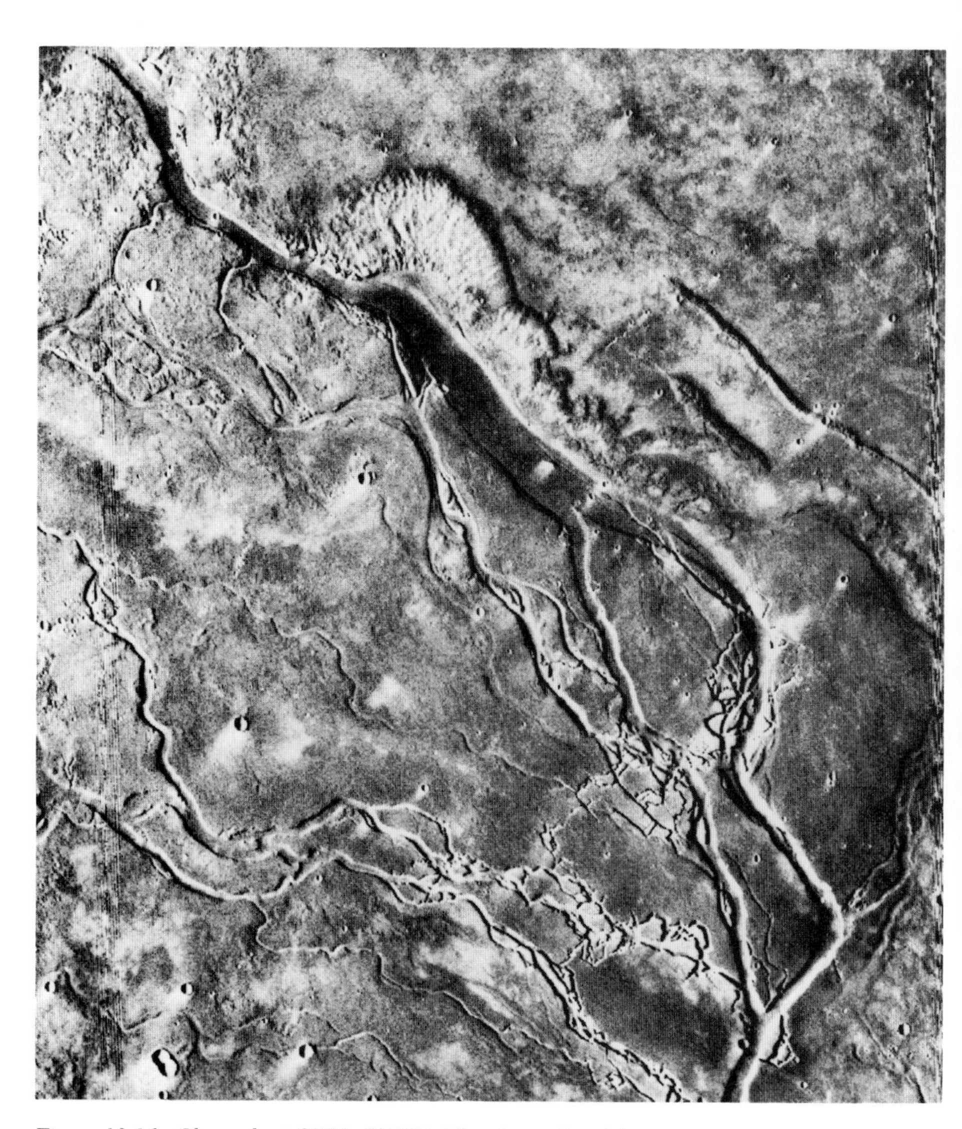

Figure 10.16. Channels at 29°N, 224°W. The channels originate at some grabenlike troughs to the northwest of the Elysium Mons volcano (see figure 7.24). Flow appears to have been confined to the channels leaving the ground between unscoured. The frame is 170 km across. (541A20)

tures, and expansion bar complexes are all absent. Moreover, in contrast to the channels around Chryse, the Elysium channels have a high sinuosity and high depth-to-width ratios. Thus, most of the features that strongly support a catastrophic flood origin for the Chryse channels are absent. The Elysium channels are more explicable in terms of slow erosion, although what fluid was involved—water, lava, or something else—is unclear.

HELLAS CHANNELS. Dao Vallis, which starts on the northeast rim of Hellas, at 30°S, 264°W, is also associated with a large volcano, in this case, Hadriaca Patera (see figure 7.25). The channel starts in two steep-walled spatulate depressions near the southeast edge of the volcano and extends southwest for almost 1,000 km into the Hellas basin. It is 20 km wide at its start but narrows downstream. Within the channel at its upstream end are some islands, but these are jagged in outline and show little, if any, evidence of streamlining by fluid flow. Roughly parallel to Dao Vallis and 200 km to the southwest is another, similarly sized channel, Harmakhis Vallis, which also starts in an irregularly outlined, flat-floored depression. Curiously, Harmakhis Vallis starts close to the termination of yet another channel, Reull Vallis, which is one of the larger channels in the southern hemisphere. Reull Vallis has characteristics of all three channel types: runoff, outflow, and fretted. It has a small number of tributaries like a runoff channel, but some start full-born like outflow channels, and the main channel has a flat floor and steep walls like a fretted channel. The relation between Reull Vallis and Harmakhis Vallis is unclear, but their close juxtaposition seems hardly coincidental, and a subsurface connection appears likely.

Ages of outflow channels

It is clear that the outflow channels postdate the old cratered terrain and many of the plains, since they dissect both units. They therefore differ from the runoff channels, which dissect only the cratered uplands. Crater counts by Masursky and co-workers (1977) show that most of the large Chryse channels are close in age to the plains that form Chryse Planitia. Counts on the undissected and dissected parts of Chryse Planitia are identical in all cases within the counting errors. This observation led Schonfeld (1978) to conclude that they were penecontemporaneous and were cut by the lava that forms the plains. Gregory (1979) gave a value of 2,100–2,400 craters larger than 1 km per million km^2 for Chryse Planitia. The counts by Masursky and colleagues (1977) on the Kasei, Maja, and Bahram Valles are almost identical; only the Vedra and Tiu Valles and possibly the inner channel of Kasei give lower counts, indicating a significantly younger age. Even these counts are suspect as far as indicating the actual age of the channel is concerned. No low counts are found on the Chryse plains. It is only in steep, narrow gorges that the lower counts occur, yet the gorges and the dissected plains of Chryse should be contemporaneous, in that they were dissected during the same events. The low counts in the steep valleys may therefore be the result of mass-wasting processes or eolian infilling after the channel was cut. As we saw in chapter 5, a crater number of 2,100–2,400 is roughly equivalent to an age of 3 to 3.5 billion years. Thus most of the Chryse outflow channels are relatively old, although the lower counts, if valid, indicate that flooding could have continued episodically to much later dates. Preliminary, unpublished data on Mangala Vallis and the Amazonis channel suggest that these are both significantly younger than the Chryse channels.

Formation of outflow channels

CATASTROPHIC FLOODS. The circum-Chryse channels have been compared with the largest known terrestrial flood features on Earth, the channeled scablands of eastern Washington (Baker and Milton 1974; Baker and Kochel 1978a,b). The scabland is an area roughly 300 km square that was extensively scoured during the Pleistocene period by several catastrophic floods of enormous magnitude (Bretz 1923, 1932, 1969; Baker 1973, 1974). The last such flood occurred around 18,000 years ago as a consequence of the collapse of an ice dam in western Montana, behind which had ponded Lake Missoula, which had a volume of 2×10^{12} m^3 and was 600 meters deep in places (Pardee 1942). When the dam burst, water up to 120 meters deep flowed from the lake down the regional slope of the Columbia plateau, almost completely burying the preflood valleys. From depth estimates based on high water marks and the general channel geometry, Baker (1973) calculated that peak discharges were close to 10^7 m^3/sec. By comparison the mean annual discharge of the Amazon is 10^5 m^3/sec. At these high discharge rates the lake must have drained within several days, during which enormous amounts of erosion were achieved, with channels as deep as 200 m being cut into the basaltic bedrock. Because the flood dwarfed the preflood drainage system, it was largely unconfined, splitting into numerous branches and rejoining as it spilled over gaps between adjacent valleys. The result is an anastomosing pattern of enormous scale, in which partly streamlined islands separate the various branches.

The landforms produced by such floods are markedly different from those produced by slow, steady erosion in normal river systems, where flow is highly channelized, where a regular hierarchy of tributaries forms, and where the slopes of both the river and the surrounding terrain are in quasi-equilibrium throughout the drainage basin. During massive floods flow is dispersed and is of such short duration that equilibrium is not approached, although massive erosion occurs as the flood stream attempts to establish a new profile and a channel more commensurate with the high discharge rates. The flood features have low sinuosities and high width-to-depth ratios. Their characteristic erosional style produces anastomosing patterns of channels, with pronounced expansions and contractions and residual islands between. Features that are normally seen only on riverbeds achieve such large dimensions that they produce distinct landforms, including ripples, longitudinal grooves, inner channels, cataracts, and scour patterns around obstacles (Baker 1978).

Perhaps the most striking similarity between the channeled scablands and the martian outflow channels is the presence of teardrop-shaped islands, which are especially common in the eastern part of the scablands, where easily erodible loess deposits were cut into. An object within a flow stream creates two types of resistance: pressure drag, which increases directly with the cross-sectional area of the object perpendicular to the flow; and shear drag, which increases with the total surface area exposed to the flow. A lemniscate shape tends to minimize the sum of these two forces. For the scabland and martian lemniscate islands, Baker and Kochel (1978b) plotted various parameters, such as length, width, and area, against one another and showed that the relations are similar for both kinds of island. The significance of this in terms of the processes involved is unclear, however, since similar shapes are formed by glaciation (drumlins) and wind (yardangs). Many of the islands in the scablands are partly eroded bedrock and partly pendant bars. The streamlined forms, therefore, result from both deposition and erosion, but viewed vertically it is difficult to distinguish between the two parts. The same may be true on Mars, where many of the "tails" of the islands may be partly depositional. Another obvious similarity between the scablands and the martian channels is the presence of numerous, closely spaced grooves parallel to the length of the channel. In the scablands these are largely restricted to areas where flow is over basalt; they are probably formed as a result of scour by vortices with filaments parallel to the direction of flow.

Two processes that appear to be considerably more effective during large floods, as compared with normal river flow, are cavitation and hydraulic plucking, which together may account for much of the extraordinary erosive power of the scabland flood in basalt areas. The enormous flow velocities within a flood can result in such large pressure reductions within the flow that exsolution of dissolved gases or vaporization of the water occurs. Collapse of the resulting gas bubbles can produce enormous pressures, possibly as high as 30 kb (Barnes 1956). Such intense hammering can shatter almost any solid material and significantly enhances the erosive power of a flood. Hydraulic plucking is caused by macroturbulence and the development of vertical vortices near obstacles. Low pressures at the bases of the vortex filaments provide lift, which plucks bedrock out of the channel floor. The typical "scabland" of hollows and mesas probably forms this way. Possible analogues in the martian channels are the gridlike arrays of crevasses in Kasei Vallis (see figure 10.14) and the jagged elongate depressions at the mouths of the Chryse channels. These examples illustrate only a few of the large number of similarities between the channeled scablands and many of the martian outflow features (for a more detailed discussion see Baker and Milton 1974 and Baker 1978). Baker and Milton (1974) argued that the similarities between the large martian outflow channels and the scablands are so great that a common origin—erosion by massive catastrophic floods—is almost inescapable.

If the large outflow channels were cut by running water, then a mechanism is required for massive release of the water. Several possibilities have been suggested. Nummedal (1978) proposed that the water, possibly in the form of sediment-laden slurries, was derived by liquefaction of the rocks in the source regions. He pointed out that sediments can behave as fluids if the pore pressure of the interstitial water is close to the overburden pressure. Under these conditions the sediment has no shear strength and can flow even on gentle gradients. Two types of liquefaction phenomena are recognized (Andresen and Bjerrum 1968). Retrogressive flow slides start as a result of seepage at the base of a slope. Formation of one slide triggers formation of others, and the scar moves progressively into the high ground, incorporating larger and larger areas of the upland. Such slides are common in Pleistocene marine sediments in Norway and may involve millions of cubic meters of material. According to Nummedal (1978) the slides can propagate at rates as high as 1 km/hr. The scars of individual slides are alcove-shaped and contain large flakes of the original dry surface. The downvalley flow typically has a density of 1.9 gm/cm^3 and can flow at high velocities, causing severe erosion. The second type of liquefaction phenomenon is a spontaneous liquefaction, in which metastable sediments may almost instantaneously liquefy over large areas as a result of vibration or shock. While this process may occur on the ocean floor, massive subaerial spontaneous liquefaction is rare on land, and the conditions under which it might occur are not known.

Liquefaction may have played an important role in the disintegration of the upland terrain in the headwater regions of the Chryse channels. As pointed out by Nummedal (1978), liquefaction scars, with their blocky remnants of nonliquefied debris, closely resemble chaos regions. More difficult to reconcile with a purely liquefaction origin are the scoured features downstream. Can the high-density slurries produced by liquefaction travel the hundreds of kilometers required to erode the large Chryse channels, and can they produce the wide variety of erosional features within the channels? Could a slurry, with a density close to 1.9 gm/cm^3, travel a thousand kilometers downstream, then cut a deep gorge such as Maja Vallis, as apparently happened with the large flood that started at Juventae Chasma? The answers to these questions are not known, and judgment of the liquefaction hypothesis must be deferred until it has been explored in greater detail.

Some authors have suggested that the water which caused the floods was released by geothermal melting of ice. Glacial bursts, or jökulhlaups, form by this means in Iceland and have discharges as high as 10^5 m^3/sec (Thorarinsson 1957). The Icelandic glacial bursts form as the base of the Vatnajökul ice sheet melts owing to the steep geothermal gradient in this highly volcanic area. The meltwater slowly accumulates as a subglacial lake, then periodically breaks out to form floods with very high discharges.

McCauley and Masursky and their colleagues (McCauley et al. 1972; Masursky et al. 1977) proposed a similar mechanism for the martian flood features. A clear distinction should be made, however, between ground-ice and continuous ice sheets. For high discharges to be achieved, outflow must be unimpeded. If the ice is dispersed as ice lenses or as interstitial ice throughout a host rock, as is normally the case in permafrost terrain, then rapid release will be prevented by the permeability of the host rock, and the necessary discharge rates cannot be achieved (Carr 1979). Only if the ice exists as continuous sheets, or glaciers, can the glacial bursts form. This somewhat undermines the credibility of the mechanism as applied to Mars, for although there is abundant indirect evidence of ground-ice, the evidence for ice sheets is less compelling. Another argument against the glacial burst hypothesis is the lack of volcanic features in the source regions of the large outflow channels. Floods might be expected around Tharsis, where there has been abundant volcanic activity, but they are totally lacking. A more convincing case can be made for the channels that start near Elysium and Hadriaca Patera (discussed below), where a connection with volcanic features is clear.

As an alternative, Clark (1978) proposed that the water is derived not by geothermal melting of ice but by geothermally induced decomposition of hydrated minerals in the regolith. He suggested that a thick regolith of hydrated minerals is heated from below, and that water from the deep-baked zone migrates upward to form a layer rich in water and ice between the dehydrated regolith below and the still hydrated frozen regolith above. The water within the zone could be under considerable confining pressure as a result of capping by ice and so could break out to form floods. As pointed out by Clark, however, the kinetics of the process are poorly understood, and judging by the amount of collapse (1–2 km) in the chaos regions, a regolith consisting mainly of hydrated minerals and extending to depths of kilometers would be required.

I proposed that the large flood features are caused by breakout of water under high pressure from confined aquifers (Carr 1979). Early in the planet's history, when most of the runoff channels formed, climatic conditions must have been more temperate than at present and must have permitted surface flow at modest discharge rates. Much of the water that cut the runoff channels may have been lost not to the atmosphere but to an extensive groundwater system within the highly brecciated and porous old cratered terrain. Subsequently, climatic conditions became more like those that currently prevail. As a result a thick permafrost developed, trapping the groundwater beneath. Relief on the old cratered surface, particularly the Tharsis bulge, caused a slow migration of groundwater to low areas, where large pore pressures developed within the aquifers. The low-frequency relief around the Chryse basin is high enough for pore pressures within the aquifers to exceed the overburden pressure. Such a situation is unstable, and breakout could ensue, either spontaneously or triggered by faulting or impact. Because of the high pore pressures evacuation of the water would be extremely rapid. The aquifer itself would tend to disintegrate and be withdrawn along with the water, causing undermining of the region around the vent and formation of chaos. I calculated discharge rates for various thicknesses of aquifer, depths of burial, and pore pressures (Carr 1979) and concluded that discharge rates of the magnitude required by the channel dimensions (10^8 m^3/sec) could be achieved only with extremely large permeabilities (1,000 darcies) if the aquifer maintained its integrity. The only places on Earth where such permeabilities are found are volcanic terrains, such as the Snake River plains, Idaho, and Hawaii, where lava tubes provide very efficient means of subsurface transport. If the aquifer disintegrates, however, much smaller permeabilities suffice. Flow ceases when the aquifer is depleted or when the local pore pressures—and hence discharges—fall to such low values that the flow freezes. The aquifer can be recharged by migration of water into the low areas from more distant parts of the aquifer system, but all the water involved in the floods is lost from the system and cannot reenter because of the permafrost seal. It may freeze and remain as ice sheets in high polar latitudes, or it may sublime into the atmosphere to be ultimately frozen out at the poles. Floods could thus form under present climatic conditions.

This mechanism is consistent with the relatively old age of the runoff channels and the somewhat younger ages of the outflow channels. It also provides an explanation for the connection between the canyons, which appear to be largely tectonic, and the flood features. The west end of the equatorial canyon system is several kilometers higher than the eastern end. If there has been subsurface drainage downslope to the east, then the faulting that created the canyons to the west may have cut old cratered terrain that was largely drained of water, so that there were no fluvial side effects. In the lower eastern part, however, the faults may have cut water-charged aquifers and caused formation of chaos and large channels; hence the transition from a canyon with largely tectonic features to the west to a region dominated by chaos and fluvial features to the east. Mangala Vallis, which starts at a graben, could similarly have formed by disruption of an aquifer by faulting.

The different hypotheses just discussed are not necessarily mutually exclusive. It is not uncommon in geology that contending theories each contain an element of the truth, and that the complete story is a complex interweaving of different processes. The liquefaction and aquifer models, for example, are not greatly different. If the aquifers were fine-grained sediments, not basaltic flows with lava tubes as I originally implied (Carr 1979), then the liquefaction and aquifer models could be merged. Pore pressures would build up in the lower parts of the aquifers to create instabilities, as in the original model, but breaking of the permafrost seal would result in liquefaction of the entire aquifer and outflow of a slurry. Indeed the liquefaction hypothesis originally proposed by Nummedal (1978) almost requires a confined aquifer system for pore pressures to build up so that liquefaction can occur. Similarly lenses of ground-ice could segregate with the aquifers and be melted by geothermal heating, further increasing pore pressures and causing breakouts. Finally, if breakout were slow, then ice could have accumulated at the surface and given rise to glaciers. Thus most of the elements of the complete story may already have been perceived.

Although most geologists who have examined the martian outflow channels support the catastrophic flood hypothesis, particularly for the unconfined types, other modes of formation have been proposed. These are discussed briefly below to give the reader a sense of the issues involved. The summary of the glacier hypothesis is especially brief, not because of its lesser merit, but because it was proposed only recently and has not been discussed fully in the literature.

GLACIERS. If the water flowed away from the source region mostly as ice, as suggested by Lucchitta (1980), then much lower discharges are required. Lucchitta pointed out that many of the channel features, such as U-shaped cross profiles, hanging valleys, long parallel scour marks on valley floors and walls, and streamlined islands, are consistent with a glacial origin. She suggested that artesian springs are the most probable source of the water, and that over such springs ice domes would form, from which the glaciers would flow. In cold-based glaciers the base is frozen to the ground, and movement is by shear within the ice. Lucchitta calculated that the ice in this type of glacier must be at least 4 km thick to move under martian conditions. Since flow is mostly within the ice, such glaciers have limited erosive capability. In contrast most of the movement in warm-based glaciers is at the base, where temperatures are close to melting and a layer of liquid may form. Since much of the flow is within the interface layer, such glaciers can erode powerfully. One attractive feature of this hypothesis is that extreme discharges are not required. Erosion rates are low, but presumably they were sustained for long periods of time.

WIND EROSION. Despite a strong resemblance between some martian outflow channels and large terrestrial flood features, Blasius and Cutts and their colleagues (Blasius et al. 1978; Cutts et al. 1978a,b) argued that the martian channels cannot have been cut by water. Two observations led them to reject the water hypothesis. They assumed that the maximum amount of water that can originate from the chaos is equivalent to the volume change caused by chaos collapse. This falls far short of that needed to erode the channels, being considerably less than the volume of material eroded away. Secondly, they pointed out that some channels increase greatly in size downstream. This, they claimed, is incompatible with the water hypothesis, in that it requires that the erosive and load-carrying capabilities of the stream increase downstream, which is contrary to what might be expected on Mars, where water should be lost continually from the stream by freezing and evaporation.

The volume discrepancy between chaos and channels is not necessarily fatal to the water hypothesis. If the aquifer model is valid, then the flood waters are derived by underground drainage of large areas. The volume change in formation of chaos is therefore much smaller than the total volume of water released. Furthermore, if, as Nummedal (1978) proposed, channels form by liquefaction of a near-surface, water-charged layer, then they form more by mobilization than erosion. Thus the volumes of channel and source are irrelevant. The problem of the increase in channel size downstream may have been overstated by Cutts and co-workers (1978), for loss of water by freezing and evaporation, even under present conditions on Mars, may be quite small (Wallace and Sagan 1979).

As an alternative to water Cutts and Blasius and their co-workers (Cutts et al. 1978a,b; Blasius et al. 1978) hypothesized that most large outflow channels were cut by either wind or lava. Those that arise in chaos, they suggested, are wind-eroded. They proposed that the chaos areas are local sources of fine-grained debris. Wind blowing across the chaos picks up debris, enhancing its erosive capability and causing channel-like features to be cut downwind. As erosion proceeds, debris continues to be entrained in the near-surface saltating layer, so that long channels can be cut over long periods of time.

Several arguments can be raised against the wind hypothesis. In general wind tends to affect broad areas rather than being highly channelized, and features produced by wind erosion tend to maintain a constancy of direction irrespective of the local topography, because of strong coupling between the movement of air near the surface and the regional flow of the atmosphere above the boundary layer. Thus, eolian scour tends to cut across the topography, in marked contrast to water channels, which always follow the local intricacies. The martian channels all run downslope. In the case of the Chryse channels, this leads to convergence of flow from several directions. This is not the pattern shown by wind indicators such as crater streaks, nor is it the pattern expected from the interaction of the global circulation and the regional topography (Webster 1977), which is more likely to produce cross-slope winds than downslope winds. It could be argued that near-surface eolian debris flows form in a somewhat analogous manner to turbidity currents under the ocean, but turbidity currents have never been observed in Earth's atmosphere except around volcanoes. Moreover, some broad scoured features that originate in chaos, and that are wind-formed according to the hypothesis of Cutts and co-workers (1978a), become highly channelized downstream. The channel originating in Juventae Chasma, for example, converges on Maja Vallis as it cuts through the old cratered terrain between Lunae Planum and Chryse Planitia. Such convergence into a narrow channel by wind action is unlikely. Cutts and colleagues recognized this problem and invoked a dual origin for this channel, suggesting that both wind and lava were involved. Lastly, the wind hypothesis fails totally to explain Mangala Vallis, which resembles the chaos-derived channels in every way, except that it arises in a graben, not chaos. Thus, while many of the constituent features of channels can be plausibly explained by wind erosion, the regional patterns of channels argue convincingly against it.

LAVA EROSION. The presence of unambiguous small lava channels on the large volcanoes (see chapter 7) and the association of some outflow channels with volcanoes lead naturally to the question of whether lava could form some of the large outflow channels. While most of the evidence suggests that lava is unlikely to have formed the large circum-Chryse channels, a more plausible case can be made for the Elysium channels. As McGetchin and Smyth (1978) pointed out, martian lavas could have very low viscosities because of their high iron contents. During terrestrial eruptions of low-viscosity lava molten lava is commonly transported away from the eruption site in channels. These may crust over and form tubes, and it is through such tubes and channels that lava is transported for large distances from eruption sites to active fronts. In most cases erosion plays a relatively minor role in their formation. The channels form initially as movement within the flow becomes restricted to a narrow zone, as much of the flow solidifies. The channel thus forms contemporaneously with the terrain and is not cut into it. Once a channel is formed it may overflow repeatedly to form levees or side branches, or it may become choked with lava or drain as the magma supply is cut off by cessation of eruption or diversion of the flow. Channels of this type are common on the martian volcanoes and lava flow fields and have been discussed previously. However, many sections of large martian channels, such as those incised into the old cratered terrain around Chryse, are clearly younger than the terrain in which they occur. If caused by lava, they must have formed by erosion and not by the mechanism responsible for most terrestrial lava channels.

Evidence of lava erosion in terrestrial eruptions is sparse. Swanson (1973) reported that during the 1969–1971 eruptions of Mauna Ulu, the level of lava within a major tube fell approximately 1.6 m per month, which, he suggested, could have been caused by lava cutting into its bed. Greeley and Hyde (1972) invoked lava erosion to explain asymmetric cross sections within lava tubes at meander bends. The best evidence for lava erosion, however, comes from the Moon, where several sinuous rilles are deeply incised into nonvolcanic highland surfaces (Carr 1974b). It seems inescapable that these were eroded by lava. Moreover, large sinuous rilles, such as the Hadley Rille, are much more plausibly explained as resulting from erosion by lava only a few meters thick than by having the whole rille fill with lava to depths of several hundred meters, which is what is required without erosion.

I explored possible mechanisms whereby lava within a channel could erode into its bed and concluded that thermal incision is more probable than mechanical plucking (Carr 1974b). A normal river either erodes its bed or deposits sediments according to subtle changes in gradient and discharge, and it may be forced repeatedly to change its position as sediments accumulate within the channel. In contrast, even if a lava stream had a bed load, it would probably have a density comparable to the lava itself, so it would not saltate or settle out. Bars, therefore, cannot form, and lateral shifts in channel position cannot occur. Furthermore, erosion by mechanical plucking is unlikely, because flow of the wall materials tends to smooth the walls and floors, as is well demonstrated in drained lava tubes. If erosion occurs, it is probably by thermal incision, in which the wall and floor of the channel are heated to such temperatures that the materials flow and become incorporated into the lava stream. The process is slow, requiring uninterrupted flow for long periods, which may explain why lava erosion on Earth is rare, most terrestrial eruptions being episodic and of short duration. If this is the dominant process of lava erosion, then once a lava channel becomes established, it remains essentially fixed in position, merely becoming more incised. This leads to the relatively simple shape of lunar rilles, which are mostly narrow sinuous channels with few branches. The sinuosity is presumably inherited from the initial flow of lava over a relatively flat surface, and incision followed as flow was sustained. The contrast in the processes whereby fluvial and lava channels form leads to differences in morphology. Water channels show abundant evidence of deposition and lateral shifts in channel position, while lava channels are relatively simple in outline, with no depositional features.

The channels northwest of Elysium have many characteristics suggestive of lava erosion. They start in spatulate depressions and grabenlike features on the flanks of the Elysium volcanoes. Many are highly sinuous and deeply incised, strongly resembling lunar rilles. Major differences are a complex branching pattern, which is never seen in lunar rilles, and the presence of streamlined islands in some of the wider channels. The branching could be explained on the basis of eruptive surges or shallow longitudinal profiles, either one of which could facilitate overflow and establishment of new branches. The islands presumably form in the same manner as with water. The large channels on the northeast rim of Hellas near Hadriaca Patera also have simple rillelike forms and originate near a volcano, so could similarly be interpreted as lava channels. Water is still a strong possibility, however, in all these cases. Eruptions into ground-ice around the volcanoes could have released water that cut channels. The evidence is thus inconclusive, and both lava and water must be retained as possibilities.

Formation of the large circum-Chryse channels by lava erosion, as suggested by Schonfeld (1978), is much less likely. In these channels features that could be interpreted as bars are common, and there is evidence of lateral shifts in channel position, particularly in the inner channel of Kasei Vallis. The evidence of deposition in the lower reaches of Kasei Vallis and at the mouth of Maja Vallis is especially strong. In both these areas the terrain appears to have formed by cut and fill, with erosion localized in specific channels with depositional features between. Another argument against a volcanic origin is the lack of evidence of volcanic activity in the source regions, although such features should be highly

Figure 10.17. Fretted channel at 41°S, 257°W, to the east of Hellas. Many of the nearby hills are surrounded by debris flows, and mass wasting may have played a major role in enlargement of the channel and formation of its flat floor. Compare with figure 6.8. The frame is 280 km wide. (97A62)

visible in the chaotic terrain. We can conclude, therefore, that some of the outflow channels of the confined type may have formed by lava erosion, but that the unconfined channels, typified by those around Chryse, are unlikely to have formed in this way.

FRETTED CHANNELS

Fretted channels were described in chapter 7. The term applies to wide, flat-floored channels with steep walls, which differ from runoff channels mainly in size but also in that they have clearly delineated flat floors. They differ from outflow channels in that bedforms, such as longitudinal scour, jagged depressions, and cataracts, are completely lacking, and sculpted features, such as teardrop-shaped islands, are rare. They also have tributaries and increase in size downstream. Most occur in two bands of fretted terrain roughly 25° wide and centered at 40°N and 45°S (Squyres 1979), in which mass wasting features are common. Major exceptions are Ma'adim and Al Qahira Valles, at 20°S and 200°W and 182°W, respectively. Formation of the fretted channels appears intimately connected with the formation of the debris flows that are observed at the bases of most steep slopes in these areas (figure 10.17). A plausible explanation is that the channels are runoff channels that have been subsequently enlarged by mass wasting. The enlargement process is probably a slow one, in which material is removed from the bounding escarpment and carried downstream by creep (Squyres 1978). The channel floors have low crater counts (Masursky et al. 1977), which almost certainly date the mass-wasting features. The counts suggest that the mass wasting is geologically recent; it could still be continuing. Most of the fretted channels are restricted to areas of the upland close to the plains/upland boundary. It is possible that only in these areas is there a sufficient downchannel gradient to transport the mass-wasted material downstream.

Summary

Runoff channels provide the most convincing evidence that water once flowed across the martian surface. Many of the channels appear to have formed by sapping as a result of release of water or melting of ground-ice, but rainfall cannot be ruled out as a means of directly providing runoff or charging the groundwater system. Runoff channels are restricted to the old cratered terrain, which suggests that they too are old. Alternatively, the older terrain may be more susceptible to being channeled by virtue of its erodibility or the presence of groundwater. The relatively small discharges suggested by the morphology of the runoff channels imply that surface temperatures and pressures had to be higher when the channels formed than they are at present. The runoff channels, therefore, suggest more temperate climatic conditions early in the planet's history, almost certainly before 3 billion years ago.

The outflow channels may be of diverse origins. Mangala Vallis and the circum-Chryse channels appear to have formed either by catastrophic floods of enormous magnitude or by artesian-fed glaciers. Instantaneous release of large volumes of water is difficult to accomplish by known geologic processes. Plausible possibilities are release of groundwater confined under high pressures, liquefaction of water-rich sediments, and geothermal melting of surface or buried ice sheets. Formation of the outflow channels, if by catastrophic flooding, could have taken place under present climatic conditions. Conditions required for a glacial origin are unclear. The outflow channels dissect the plains, so they are mostly younger than the runoff channels, which mostly do not. Even so, crater counts on most outflow channels indicate a relatively old age (>2.5 billion years). Possible exceptions are Mangala Vallis and a large channel in Amazonis Planitia. The channels near Elysium and Hadriaca Patera are less compelling as large waterworn features, although water may still be the most plausible agent. Their deeply incised, highly channelized forms and their origin near large volcanoes suggest lava erosion as an additional possibility. Fretted channels are formed predominantly by enlargement of preexisting channels by mass wasting. Their formation is mainly restricted to two latitude belts, centered on 40°N and 45°S. The process of enlargement may still be continuing.

11 WIND

That wind erodes the martian surface, producing debris that is continually being recirculated, cannot be doubted. We have observed storms that stir up so much dust that most of the surface of the planet becomes hidden from view. We have monitored changes in surface markings that result from the storms and have measured wind speeds at the landing sites. The uncertainty regarding wind is not whether it has modified the surface but where and to what extent. Given the violence and frequent occurrence of the dust storms, it is somewhat surprising that eolian effects are not more pervasive. An ancient cratered topography, probably over 3 billion years old, survives over much of the planet. Yet it has been exposed to wind action for billions of years, and perhaps to hundreds of millions of dust storms comparable to the ones we observed in 1971 and 1977. Many of the plains, such as Chryse Planitia, may also be billions of years old, but at a scale of 100 m they do not look very different from the lunar maria, having crisp wrinkle ridges and well-defined craters. Wind erosion appears to have been highly selective, being very obvious in some areas and negligible in others.

The main effect of the wind may be simply the continual redistribution of already eroded debris. Addition of new material to the generally circulating mass must be slow, otherwise after billions of years of activity erosional effects would be widespread, and more of the surface would be covered with thick eolian blankets. The effects of wind become more obvious at finer scales. At resolutions of 10–50 m many parts of the surface have an etched and pitted appearance, which is most likely caused by wind; at a scale of 100 m, however, the erosive effects of wind are relatively minor, except in a few local areas. Thus, at coarse scales the topography of most of the surface is dominated by processes such as impact, volcanism, and tectonism; only at scales of a few meters do the effects of wind begin to take over. In the following discussion the processes of wind transport and erosion are discussed briefly with reference to some terrestrial examples. The relatively straightforward evidence for wind transport is then examined, particularly with reference to wind streaks and dunes. Finally, the more controversial aspects of wind, specifically erosional landforms and erosion rates, are addressed.

WIND TRANSPORT

Material is transported by the wind, either as a suspended load within the atmosphere or by saltation as the entrained particles bounce and jump across the surface, or by creep—the slow, almost imperceptible movement of surface debris as it is impinged upon by other particles. The three types of movement can result in efficient separation of particles of different sizes. Most terrestrial desert sands, which move mainly by saltation, are in the 0.1 mm to 0.5 mm size range (Wentworth 1931; Bagnold 1941). Finer materials are blown away in suspension; coarser debris is left behind as a lag.

On Earth particles around 0.08 mm in diameter are those most easily moved by the wind (Bagnold 1941). Above this diameter larger wind speeds are required to overcome the inertial resistance due to the particle mass. The increased resistance to movement for smaller diameters is less well understood. Contributing factors may be increased intergranular cohesion, drag inefficiencies caused by the small cross-sectional area of the particles, and immersion of the particles within the laminar flow regime very near to the surface. If intergranular cohesion is a strong factor, as appears likely, then the threshold velocity for particle movement will be sensitive to the moisture content of the materials being moved.

On Mars the threshold velocities are larger than those on Earth because of the thinner atmosphere. They also show a wider range of values, because variations in surface temperatures and pressures cause significant changes in atmospheric density. Wind-tunnel studies by Greeley and coworkers (1977) showed that the optimum size for particle movement on Mars is near 0.1 mm (figure 11.1), close to the size for Earth. Threshold velocities required to move the 0.1-mm particles range from 2.4 m/sec for a 10-mb atmosphere to 4 m/sec for a 5-mb atmosphere. On Earth threshold velocities at the optimum size are close to 0.2 m/sec.

The velocities just given are those right at the surface. Equivalent wind velocities above the surface can be determined from the Von Karman relation (Ryan 1964):

$$V_z = 5.75\ V_t \log Z/K$$

where V_z is the wind velocity at height Z, and K is the surface roughness. The 2.4 m/sec threshold velocity is equivalent to free-stream (above the boundary layer) winds of 45–125 m/sec, depending on the surface roughness (Greeley et al. 1976). Wind speeds at the 1.6-m height of the meteorology boom of the Viking landers should be close to half these values. At the Viking landing sites during the dust storm period, occasional gusts as

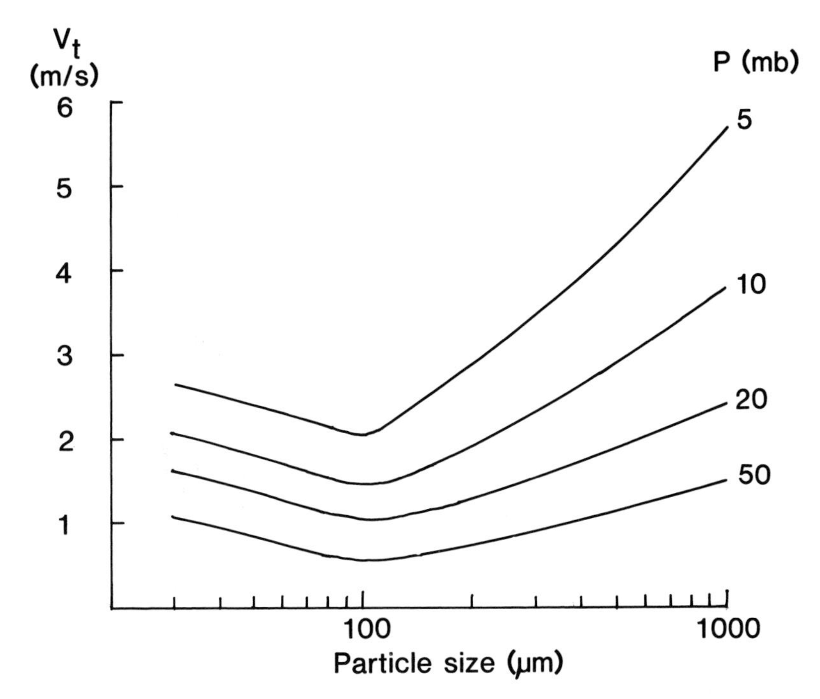

Figure 11.1. Threshold velocities for initiation of particle movement as a function of particle diameter. The data are from wind-tunnel studies at four pressures representative of conditions at the martian surface. (Adapted from Greeley et al. 1976.)

high as 30 m/sec were recorded. This is only just within the threshold velocity limit, so the lack of detection of movement of local materials at the site is not surprising. Clearly the threshold velocities are regularly exceeded closer to the areas of storm initiation in the southern hemisphere.

The size threshold for saltation, as opposed to suspension, depends on the settling velocity of the particles and local eddying effects at the surface. Arvidson (1972) showed that settling velocities on Mars are similar to those on Earth, so that the size difference between saltating and suspended loads should be similar on both planets for the same load characteristics and wind regimes. Once particle movement starts, it is to a certain extent self-sustaining, in that particles which fall back to the surface dislodge other particles. Generally particles are ejected from the surface at high angles, then are swept downward to impinge the surface again at low angles. Computer simulations of saltation on Mars show that mean saltation heights for particles in the 0.05 mm to 1.75 mm size range are 0.4 m to 1 m for surface wind velocities of 5 m/sec (White et al. 1979), which are comparable to terrestrial saltation heights. Saltation path lengths are 3 m to 10 m. Because of the limited height for saltation and the effective size separation by the wind, sheets of saltating sand are occasionally observed in terrestrial deserts, even when the atmosphere a few meters above the surface is clear. Moving sand sheets could also occur on Mars, provided a supply of appropriate size particles is available.

The ability of terrestrial winds to move sand varies approximately with the cube of the excess of the surface velocity over the threshold velocity. It is inhibited, however, by any moisture content of the sand (Woodruff and Siddoway 1965). Whether the latter effect is significant on Mars is unclear. Sagan and Bagnold (1975) claimed that absorbed moisture plays a negligible role, but Breed and colleagues (1979) suggested that frosting by either carbon dioxide or water-ice could be significant in suppressing sand movement, at least in some seasons.

Sand dunes

The global distribution of "sand" differs greatly on Earth and Mars. The most extensive terrestrial sand seas, or ergs, are in the mid- to low-latitude deserts. On Mars most dunes are in the high latitudes. In the north an almost continuous expanse of dunes forms a collar, in places 500 km across, around the layered terrain (Cutts et al. 1976); in the south dune fields mostly form discrete deposits within craters. In the equatorial latitudes (± 30°) dunes are much less common but are still present, particularly in sheltered areas such as parts of Valles Marineris and in some depressions along the plains/upland boundary. Dunes form where sand, moving largely by saltation, accumulates (Bagnold 1941). Sand grains saltate up the low-angle windward face then fall over into the wind shadow on the lee slope. As sand accumulates on the lee slope, it becomes unstable and slips, keeping the lee slope close to the angle of repose. Where sand is abundant, dunes tend to form regular patterns of sinuous ridges. These ridges are commonly aligned transverse to the prevailing winds, but relations between dune patterns and seasonally and diurnally shifting winds are often complex, so that the patterns cannot be correlated simply with the observed winds. Where sand supply is more restricted, solitary dunes form. The most common type is the crescent-shaped barchan (figure 11.2), but there are other types, which have been variously termed star, whaleback, or sigmoidal, according to their shape. Longitudinal, or seif, dunes extend in parallel arrays, often for hundreds of kilometers. Their mode of formation is poorly understood. They appear to be the principal means by which sand is transported from a source region to a region of accumulation, but the topography and the wind and sand supply conditions necessary for their growth are not clear.

The north polar dunes between 75°N and 85°N cover an area of about 10^6 km^2, about the same as the total area of active basin ergs in North Africa. The north pole erg is remarkable for the consistency in spacing and orientation of the mostly crescentic dunes over large areas (figure 11.3). Breed and co-workers (1979) showed that the width, length, and spacing of the martian dunes fall toward the upper part of the range for these parameters in terrestrial deserts. The mean length, width, and crest-to-crest spacing for the martian circumpolar dunes are 0.34 km, 0.54 km, and 0.55 km, respectively; the corresponding numbers for an individual dune field in the southern hemisphere are 0.61 km, 1.06 km, and 0.92 km. Gaps

Figure 11.2. Barchanoid dunes in the Sechura Desert, Peru. Where sand supply is abundant, transverse dunes form; where supply is limited, barchans form. Here we see a transitional region, in which barchanoid types are merging. (From Grolier et al. 1974.)

within the northern dune belt are mostly associated with craters and other topographic features, which are commonly surrounded by a dune-free halo. The relations between dunes and craters are complicated, however, for in some cases the dunes fill craters, in other cases dunes extend up to the crater rim—possibly on just one side—and elsewhere vague circular features suggest burial of craters by the dunes. Along the edge of the dune field and around gaps within the field the continuous sand sheet breaks up into strings of barchans and other equant types (Cutts et al. 1976). The relation of the northern dunes to the layered deposits is unclear. Some, such as those within Chasma Borealis, appear to overlie the layered deposits, but the reverse relations may also occur. The dunes are completely devoid of superposed impact craters. Their absence indicates a young age, either as a result of recent deposition or, more likely, as a result of continual reworking of the sand by the wind. The dune fields in the southern polar regions are mostly restricted to crater floors but are otherwise similar to the northern dune fields, being mainly transverse but with barchanoid types on the periphery (figure 11.4).

The source of the materials that formed the dunes is unclear. Most researchers agree with Breed and colleagues (1979) that the similarity in size and form between terrestrial and martian dunes indicates that the surface materials have responded to wind action in the same way on both planets, despite differences in particle size and composition, atmospheric density, and wind speeds. Saltating sand-size particles are almost certainly required. Sagan and co-workers (1977) pointed to the difficulty of retaining sand-size particles on Mars, because they are likely to self-destruct at a rate that is high compared with their rates of renewal by erosion. On Earth most sand-size particles form initially by running water. With the limited amount of water erosion on Mars the rate of destruction should far exceed any formation by fluvial processes. An additional difficulty is that most terrestrial sand is quartz, which is largely derived from highly differentiated silicic rocks such as granites and granodiorites, which are unlikely to be abundant on Mars (chapter 13). Other rock-forming minerals tend to weather or cleave easily and so do not normally form sands on Earth. Garnet is a possible mineral candidate for martian soil, since it is more consistent than quartz with probable martian petrologies (McGetchin and Smyth 1978). Krinsley and colleagues (1979) proposed that sand-size particles could be produced by electrostatic aggregation of smaller particles. Frost may also play a role in cementing smaller particles, particularly in the polar regions (Pollack et al. 1978). Yet another possibility is that the sand-size particles originate within the soil by the same cementation processes that form the duricrust observed at both Viking landing sites (Mutch et al. 1976). Breed and colleagues (1979) agreed with Sagan and colleagues (1977), however, and argued that the sand particles are unlikely to be aggregates, because they would be destroyed by continual reworking over the hundreds of thousands—even millions—of years that are probably required to form the ergs.

Figure 11.3. Field of transverse dunes at 81°N, 144°W, peripheral to the north polar ice cap. The dunes have a consistent trend, approximately north-south, with only minor sinuosity branching and merging. White dots are frost. Such dunes almost completely encircle the north pole just south of the layered terrain. The frame is 62 km across. (59B32)

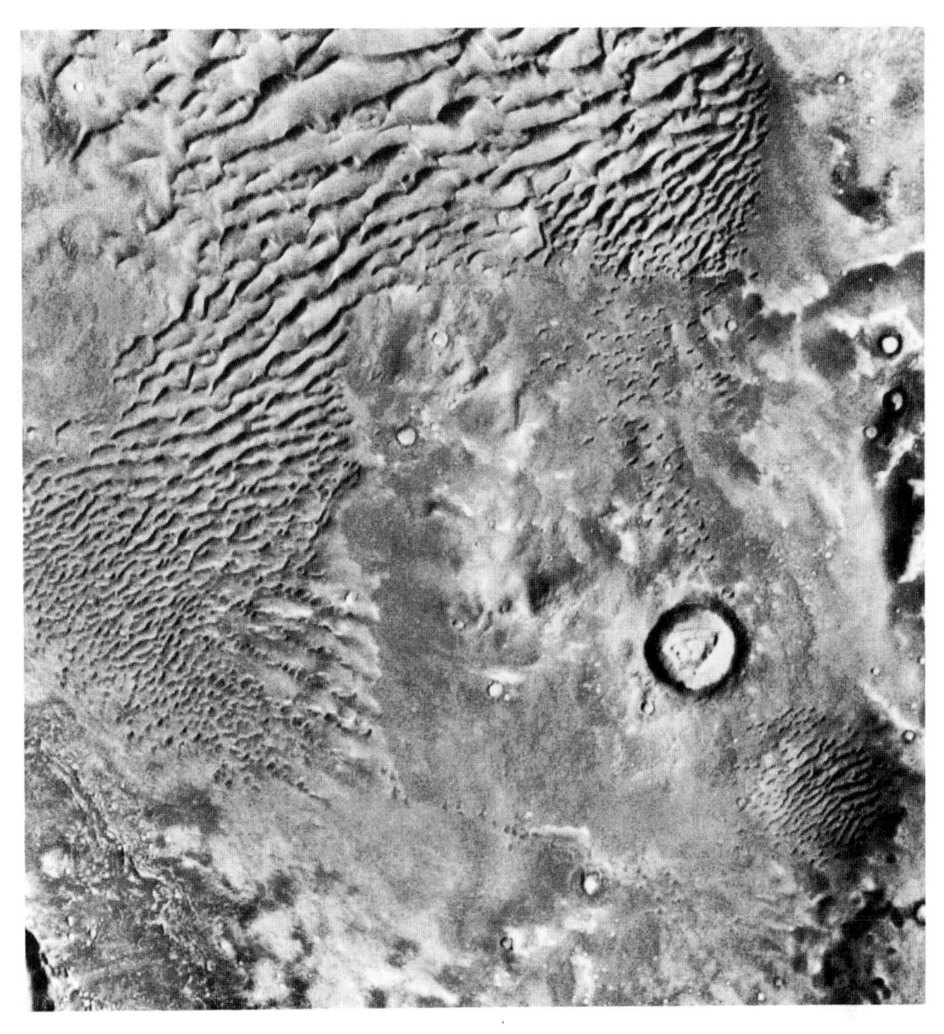

Figure 11.4. Dunes within the old cratered terrain at 47°S, 340°W. An array of transverse dunes in the upper left breaks up into isolated barchanoid types at right center. The frame is 57 km across. (575B60)

The major difference between the patterns of dunes on Earth and Mars is the almost complete absence of longitudinal dunes on Mars (Breed et al. 1979). As indicated above, longitudinal dunes on Earth tend to form in regions through which sand is being transported, in contrast to transverse dunes, which form in areas of accumulation. Possibly Mars has a more mature system of dunes, in which most of the available sand is already in depositional sinks. The question of maturity has implications for the age of the dunes. The large ergs in the Sahara are estimated to have accumulated over periods of 10^6 years or more (Wilson 1973), and comparable or longer periods are possible for the martian ergs. A distinction must be made between the time required to accumulate the sand and the time required to develop a particular dune pattern; the latter is likely to be considerably shorter. The times for imposing a new dune pattern on the sand may still be long, however. Reconstitution times (time required for a dune to migrate a dune-length downwind) for medium-size terrestrial dunes may be as long as 10,000 years (Wilson 1971), and comparable times would be required to superimpose a new pattern over an older pattern that formed in response to a former wind regime.

Another unresolved issue is whether the present prevalence of dunes in the north is a permanent feature of the planet, or whether it is the result of the planet's current position in the precessional cycle. We will see in the next chapter that dust is being preferentially deposited in the north at present. If the sand is composed of bonded aggregates, and if the aggregates can readily disintegrate and re-form, then interchange of dust, and hence sand, between the two poles could readily occur on the precessional cycle, every 25,000 years, by means of dust storms. If, however, the sand grains are permanently bonded or are individual mineral grains, then interchange could not be achieved by suspension in the atmosphere. The material would have to saltate from pole to pole, which seems unlikely. The prevalence of sand in the north may therefore be a permanent feature of the planet. This conclusion is supported by the long times that are probably required to accumulate the sand in the ergs and form the dunes. Sand may be more available in the north because of fluvial processes. As we saw in chapter 10, most large flood features drain northward. Those that funnel into the Chryse basin fade away at around 40°N and could have supplied copious amounts of fragmental debris for dune formation. There is no comparable drainage of large flood features to the south.

Streaks and splotches

In many parts of Mars light or dark streaks extend downwind from craters or other features that present obstacles to the uniform flow of air across the surface (figure 11.5). In some areas, particularly in high southern latitudes, irregular dark splotches are common. Most are within craters or are adjacent to low cliffs and escarpments. Usually the splotches in any one region are uniformly displaced to one side of the crater floor. In still other areas sharp albedo contrasts, commonly with jagged outlines, extend large distances across the surface and are seemingly unrelated to local relief such as craters. All these features appear to be caused by differential accumulation and erosion of wind-blown debris, and changes in their pattern have been observed, particularly before and after the incidence of major dust storms. The different types of markings vary in their stability, the light markings being generally more stable. Some markings represent areas of deflation; others may be accumulations of debris that differ either in grain size or thickness from the surrounding terrain. The orientations of the streaks and splotches are excellent indicators of the direction of the wind at the time they formed. They therefore provide a basis for comparing actual winds with those predicted by theoretical models.

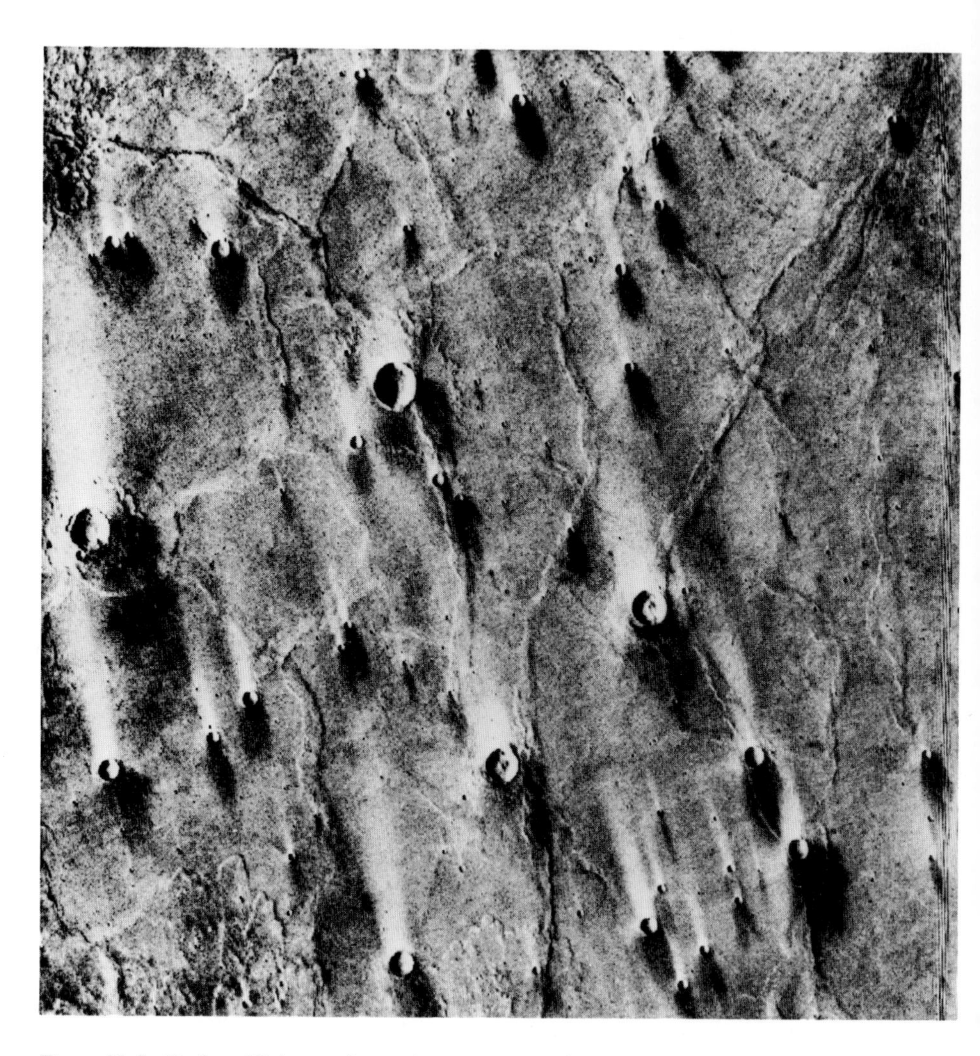

Figure 11.5. Dark and light streaks trending in opposite directions at 28°S, 245°W. Typically, dark streaks are broader and shorter than light streaks. The picture is 220 km across. (553A54)

Attempts have been made to model crater-related streaks in wind-tunnel experiments (Greeley et al. 1974a,b). As air flows over and around a crater, two vortices separate, wrap tightly around the upwind half of the crater rim, then trail off in a slightly convergent pattern downwind (figure 11.6). Some reverse flow occurs at the upwind point of vortex separation, and deposition may occur locally near the null-point. The highest wind velocities are within the vortices, so that erosion should be preferentially enhanced along their path. Within the vortices motion is outward at ground level, so that scour by the vortex displaces material outward. Between the two trailing vortices is a "shadow zone" of low wind velocity, in which deposition preferentially occurs. Within the crater itself there may

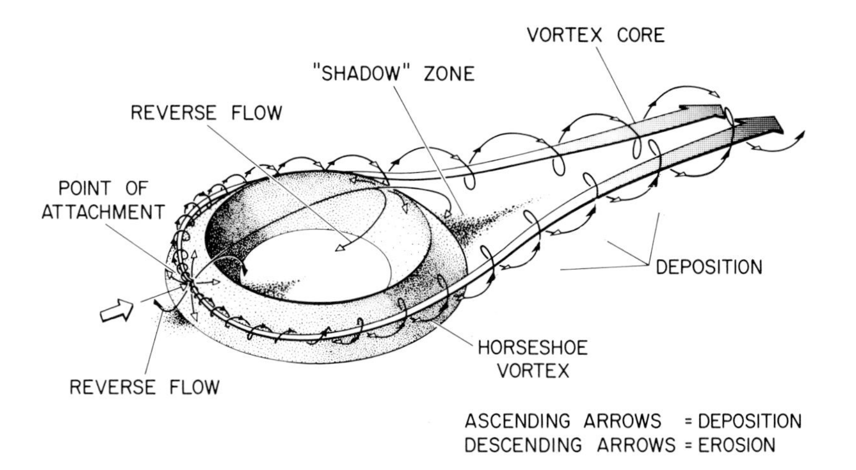

Figure 11.6. Flow over a raised rim crater, showing the vortex pattern. Deposition tends to occur on the windward side within the crater and in the shadow zone in the lee of the crater downwind. Erosion occurs preferentially along the vortices. (Reproduced with permission from Greeley et al. 1974a, copyright 1974, American Association for the Advancement of Science.)

be reverse flow, with null velocities biased toward the upwind or downwind part of the crater, according to the wind velocity.

With relatively low winds the trailing vortices converge rapidly downwind, and the "shadow" depositional zone between them is suppressed. With higher winds the trailing vortices become more nearly parallel, and the triangular-shaped depositional zone between becomes enlarged. Thus the modeling suggests that the streaks may be either depositional or erosional, depending on wind conditions. The models also indicate that streaks are more likely to form if the crater is bowl-shaped and has a distinct rim, which is what is observed (Mutch et al. 1976). Formation of splotches, on the other hand, is not dependent on the separation of vortices around the rim, but on reverse flow within the crater. Splotches can therefore form in relatively subdued craters without rims. This again is consistent with what is observed: most splotches occur in large flat-floored craters with low rims. Many are resolvable into fields of dunes.

Thomas and Veverka (1979), following Sagan, Veverka, Greeley, and colleagues (Sagan et al. 1973a; Veverka et al. 1977; Greeley et al. 1978), recognized three main types of streaks (figure 11.7):

1. "Bright streaks," which are tapered bright markings associated with craters and hills and are interpreted as deposits of dust in the lee of obstacles. The albedo contrast is caused by the lack of fragmental debris in adjacent areas or by the preferential enhancement of the fine fraction (< 40 μm) in the streak.

2. "Dark erosional streaks," which are interpreted as zones in which a thin veneer of bright dust has been removed by vortical or turbulent motion downwind of topographic obstacles.

3. "Splotch-related dark streaks," which originate at low-albedo features inside craters. Many have bright margins and have also been called mixed-tone streaks (Arvidson 1974a; Veverka et al. 1978). Thomas and Veverka (1979) interpreted most of these as deposits of material deflated from the dunes that constitute the splotches within the crater. Soderblom and co-workers (1978), however, interpreted these dark streaks as erosional also, largely on the basis of color differences.

In addition two different types of splotches may be present: erosional splotches, analogous to erosional dark streaks, in which a fine-grained veneer of dust has been locally removed; and depositional splotches, which can commonly be resolved into fields of dunes. Erosional splotches tend to

Figure 11.7. Three possible types of streaks: (a) bright streaks, composed of fine-grained, high-albedo material preferentially deposited in the "shadow zone" downwind of craters and other obstacles, as observed at the Viking 1 landing site; (b) dark streaks, which form where the dust is removed downstream from a crater to expose a dark substrate; (c) splotch-related dark streaks, in which materials within a crater feed the streak downwind. (Adapted from Greeley et al. 1978.)

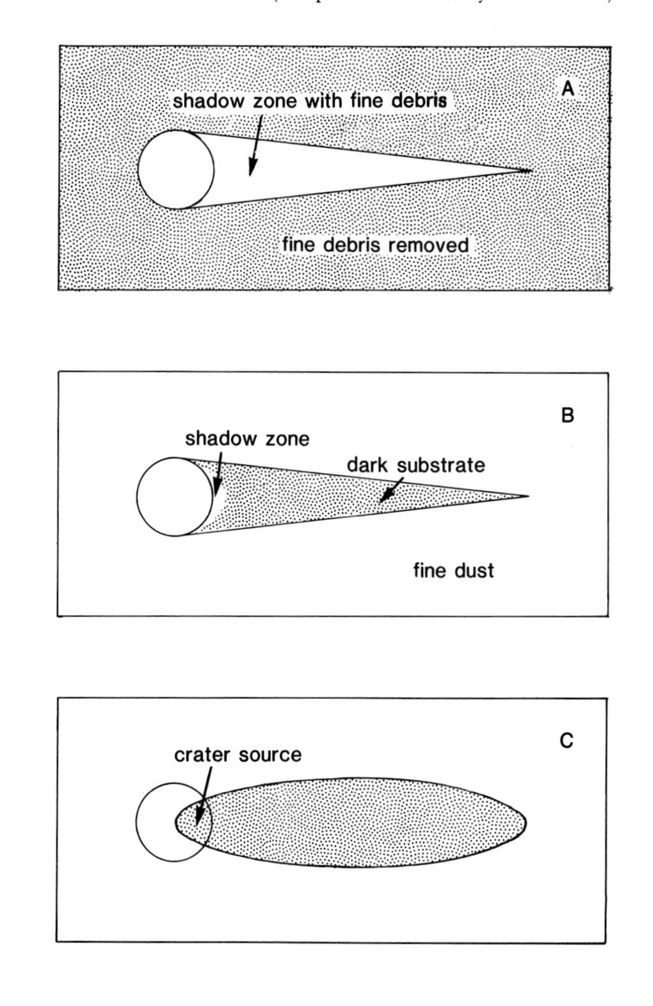

Figure 11.8. "Big Joe," a 2-m–wide rock near the Viking 1 lander. Fine-grained eolian debris is draped over the rock, and drifts of fine-grained debris may be in local wind shadows. (Viking lander 1 event no. 11B097)

form within craters on the upwind side; depositional splotches tend to form on the downwind side, although this may depend on wind conditions.

Some indication of what light streaks might look like on the ground is provided by the view from the first Viking lander (figure 11.8; see also figure 2.7). Dunelike features are seen in several locations. These are thought to be not dunes, but drifts of fine-grained material that have accumulated locally in the lee of obstacles by deposition of material from suspension, rather than by saltation as in the case of a true dune (Sagan et al. 1977). The debris may be similar in size to that sampled at both sites and may be in the micron, rather than the millimeter, size range. Presumably the interdrift areas are regularly swept clean of fine debris as the drift slowly accumulates. A small slump on the side of one accumulation near the Viking 1 site reveals a thin crust, which suggests that the surface of the drifts may stabilize when a duricrust forms as a result of migration of volatiles within the deposit (Jones et al. 1979).

Thomas and Veverka (1979) examined the global pattern of streaks to determine how it changed between 1976 and 1978, while the Viking mission was in progress, and how it had changed since 1972, when it was observed by Mariner 9 (figure 11.9). As demonstrated by Sagan and colleagues (1973a), the bright streaks are quite stable. Prominent light streaks in Amazonis and Elysium remained essentially unchanged between 1972 and 1978. Others, in Hesperia, Lunae Planum, Syrtis Major, and Mar-

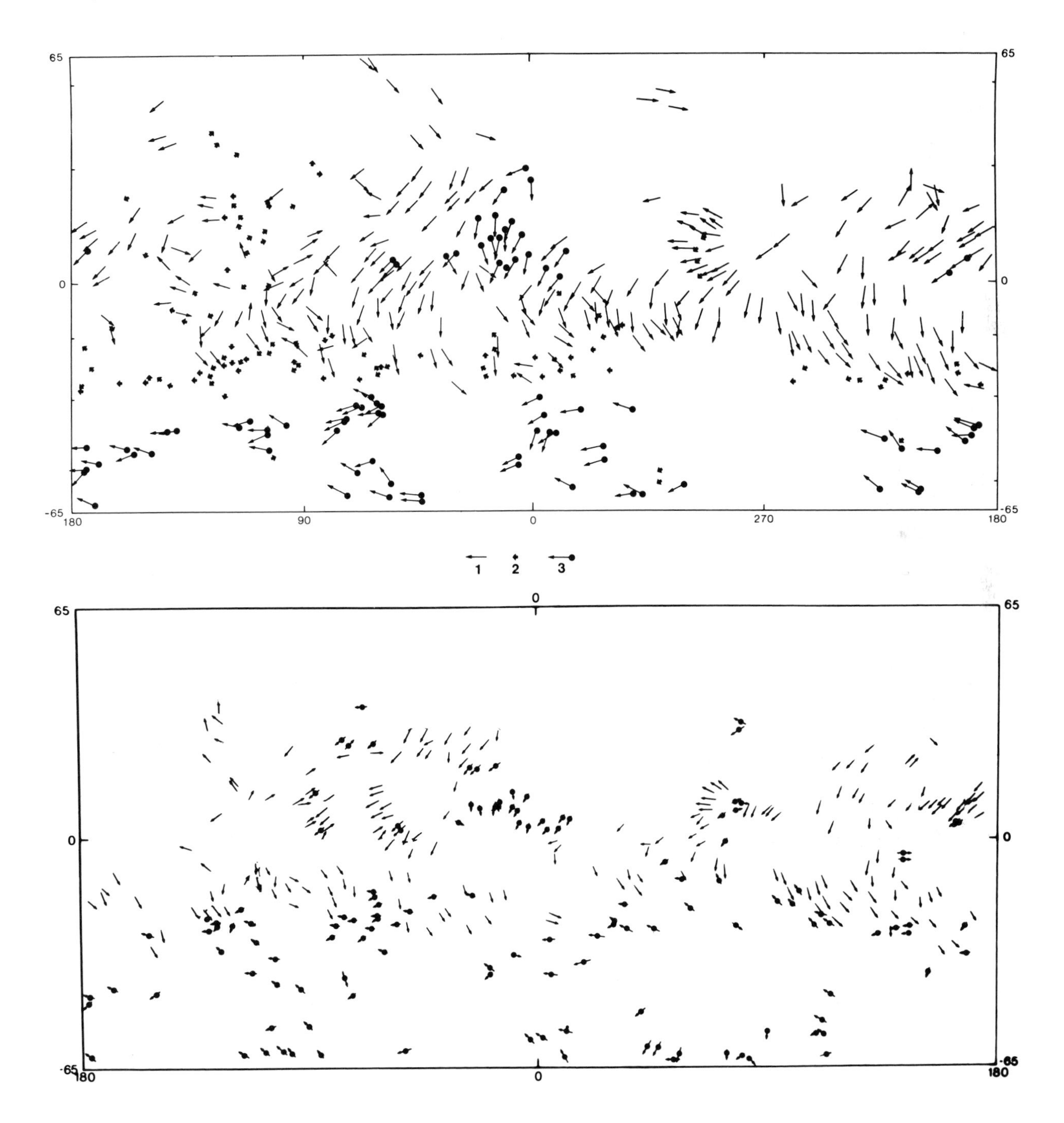

Figure 11.9. Global pattern of wind streaks. Average local trends are plotted. The upper map shows the pattern established after the 1977 dust storm, the lower the pattern after the 1971 dust storm. 1—bright streaks; 2—dark erosional streaks; 3—splotch-related streaks. (Reproduced with permission from Thomas and Veverka 1979, copyrighted by the American Geophysical Union.)

garitifer Sinus, showed minor variations in form but no detectable change in direction. One area where large changes in bright streaks were observed is between Syria Planum and Arsia Mons. This area is characterized by well-defined, rectilinear albedo boundaries, which extend for hundreds of kilometers. Most of these boundaries are not associated with craters. In this region the pattern changed constantly, before, during, and after the 1977 dust storm, with many of the streaks disappearing entirely during the storm. The patterns were also distinctively different from those seen by Mariner 9 in 1972.

Dark streaks show the greatest variability. Thomas and Veverka (1979) described a consistent pattern of changes in the Noachis region, centered at 27°S, 350°W, and the Daedalia region, centered at 30°S, 120°W. In both these areas the dark streaks disappeared during southern spring and summer while the 1977 dust storm was in progress but had re-formed by early fall. The pattern after re-forming was essentially the same as that which had been observed prior to the dust storm and during the Mariner 9 mission. The dark splotch-related streaks are more stable. Many remained visible during the 1977 dust storm, although changes in detail were observed, as in the 1971–1972 dust storm (Sagan et al. 1973a). Small changes in the length of some streaks and in their outline may be due to long-term secular changes.

The global pattern of streaks and splotches appears to be controlled by the winds that prevail during, and immediately following, the global dust storms. Not only are most changes seen during this period, but the pattern is a clear reflection of the global circulation pattern expected during southern summer and early southern fall, when the dust storm is in its final stage. The pattern is fairly stable. As we have just seen, changes observed between 1971 and 1977 are in detail and not in overall trends. The general global pattern is as follows (Thomas and Veverka 1979):

1. Between the equator and 30°N the streaks indicate winds mostly from the northeast. South of the equator the winds become more northerly, then turn northwesterly between 20°S and 30°S. The flow is that expected during southern summer (Leovy and Mintz 1969).

2. At latitudes south of 30°S the streaks indicate flow mostly from east to west but may include a southerly component, particularly at higher latitudes.

3. At high northern latitudes (> 40°N) the streaks indicate flow mainly from the west, which is consistent with the wind pattern for northern winter (southern summer), as indicated by frost streaks (Thomas et al. 1979) and lee wave clouds (Leovy et al. 1973).

4. The general pattern just outlined is perturbed locally by topographic features such as the Tharsis volcanoes, Syrtis Major, and Elysium. While the overall pattern does not change greatly from year to year, changes in albedo patterns on a time scale of decades have been seen from Earth (chapter 1). These may in part be the integrated effect of large changes in the distribution and density of streaks and splotches in different regions. The pattern of streaks and splotches almost certainly changes on a time scale comparable to the precessional cycle, as the center of dust storm activity shifts from hemisphere to hemisphere.

The amount of material involved in the changes may be quite small. Little dust is needed to change the albedo of the surface. Wells and Veverka (1978) estimated that a depth of only 1 μm is required on a smooth surface. For comparison the average annual fallout from the dust storms is estimated as about 10 μm, although much of this may fall on the poles (Pollack et al. 1979). The dark erosion streaks are apparently dusted over during the heights of the storms and are then swept clean in the waning phases. Conversely the long survival of bright streaks through several dust storms indicates that the intervening areas are being continually or episodically cleared of dust, so that the albedo contrast between the streaks and the surroundings is maintained. No discernible relief has been detected on any of the streaks. Many splotches are resolvable into fields of dunes, however, and these and other dunes may constitute a much larger fraction of the fragmental debris present on the surface than the dust that causes most of the observed changes.

WIND EROSION

The conventional view of wind erosion on Earth is that it is extremely slow, at least in the case of competent rocks, and that even in the most arid of deserts most of the erosion is accomplished by running water during the rare fluvial episodes. A corollary view is that the limited amount of wind erosion that does occur is due mainly to abrasion by saltating sand grains, whereas above the saltating layer there is essentially no erosion. This view has been recently questioned by Whitney (1978), who showed through experimental studies that the effects of the wind can be greatly enhanced by local lift and vorticity; by Whitney and Dietrich (1973), who demonstrated the effectiveness of entrained dust in producing ventifacts; and by McCauley and colleagues (1979), who showed similarities between pitted and fluted rocks in hyperarid terrestrial deserts and those at the Viking lander sites. The data on Mars appear to support the conventional view. Over most of the surface large-scale wind erosion features are rare, despite billions of years of exposure. Wind erosion may be very selective, however, for intense scour and pitting do occur locally. Moreover, removal of loosely consolidated debris by the wind may be quite efficient, despite the slow erosion rates for competent rocks.

Arvidson and colleagues (Arvidson 1979; Arvidson et al. 1979) pointed out that the preservation of craters and wrinkle ridges in Chryse Planitia limits to a few meters at most the amount of erosion that can have occurred since the plains formed (chapter 5). Assuming an age within a factor of two or three of 3.5 billion years, Arvidson estimated that the rate of rock breakdown is no more than 10^{-3} μm/yr. Even if this estimate is off by a factor of a hundred, it is remarkably low in view of the repeated exposure of the surface to violent dust storms. Chryse Planitia does not appear to be unique. Other plains in the equatorial region also have abundant crisply preserved small craters, and the floors of many craters in the uplands resemble those in the lunar highlands in preserving myriads of secondary craters, ranging down to sizes as small as 10 m across (see figures 6.5 and 6.6). For any reasonable flux rate these must be at least hundreds of millions of years old, again indicating erosion rates of 10^{-2}–10^{-3} μm/yr.

In some areas wind erosion may have been more efficient. As we saw in chapters 4 and 6, many craters in high latitudes are surrounded by low platforms or pedestals. These are interpreted as having formed by removal of material by the wind in the intercrater areas and by preferential retention around craters, where the surface is armored by ejecta (McCauley 1973; Arvidson et al. 1976). A difference in erosion rates between pedestal and intercrater areas is required. The intercrater materials may be poorly consolidated or may possibly be temporarily cemented by volatiles, which dissipate when climatic conditions change. In contrast the ejecta around the craters is probably blocky, resistant debris, such as is observed around the Viking 2 landing site. As we saw earlier, the regions in which the pedestal craters occur may have been subject to repeated deposition and removal of eolian debris blankets in response to planetwide climate changes (Soderblom et al. 1973). Accurate measurements of pedestal heights are lacking, but estimates based on shadows suggest that many are in the range of 100 m. The ages of the cratered surfaces are almost certainly in excess of 1 billion years, which gives erosion rates for the intercrater areas of around 10^{-1} μm/yr, possibly slightly higher. Even these relatively high rates for what are likely to be poorly consolidated materials fall far short of early estimates for martian erosion rates, which were in the range of 3×10^{3} μm/yr (Sagan 1973).

While erosion rates are low, and wind in most areas has had a relatively minor effect on landscape evolution at scales of hundreds of meters or more, at smaller scales its effect may be pervasive. High-resolution (< 25 m/pixel) photographs of the old cratered terrain north of Syrtis Major show that the surface has a scabby appearance over large areas. Surface layers appear to have been partly stripped away, leaving numerous, low, flat-topped mesas with highly irregular outlines (figure 11.10). Many of the remnants lie within craters, suggesting that what has been removed is not the rock surface in which the craters formed but something deposited from above. A probable explanation is that an old eolian deposit, possibly also including volatiles, has been partly stripped away, much in the manner just described for pedestal craters. Similar patterns are seen at a larger scale in the etch-pitted terrains at the south pole (Sharp 1973c) and locally within the old cratered terrain, as, for example, "white rock" (figure 11.11). Such etch patterns are rare at a scale of 100 m. If the patterns are produced by the wind, as appears likely, then the 100 m to 10 m scale range may be a transitional one, in which primary processes such as impact, volcanism, and tectonism dominate at the large scale, whereas wind and possibly other erosional processes are beginning to take over at the lower end of the scale. The effect is particularly evident where easily erodible surficial deposits are at the surface.

Figure 11.10. High-resolution view of cratered terrain at 25°N, 290°W. The surface has an "etched" appearance, as though near-surface layers had been partly removed, presumably by the wind. The frame is 42 km across. (184S11)

The pattern of erosion in the etched terrain and also that in the fretted terrain (chapter 6) suggest that erosion rates are faster on vertical faces than on horizontal surfaces. Once the surface is breached, erosion proceeds relatively rapidly by escarpment retreat, to produce erosional remnants or pits with irregular outlines (Sharp 1973b,c). In the fretted terrain channels appear to have been widened tens of kilometers by escarpment retreat. The extent to which wind participates in the process is unclear. We have seen that mass wasting plays an important role, but wind may also be involved. Wind is almost certainly responsible for the etch-pitted patterns near the south pole and in the "scabby" terrain north of Syrtis Major just described. The only plausible alternative is sublimation of volatiles. Escarpment re-

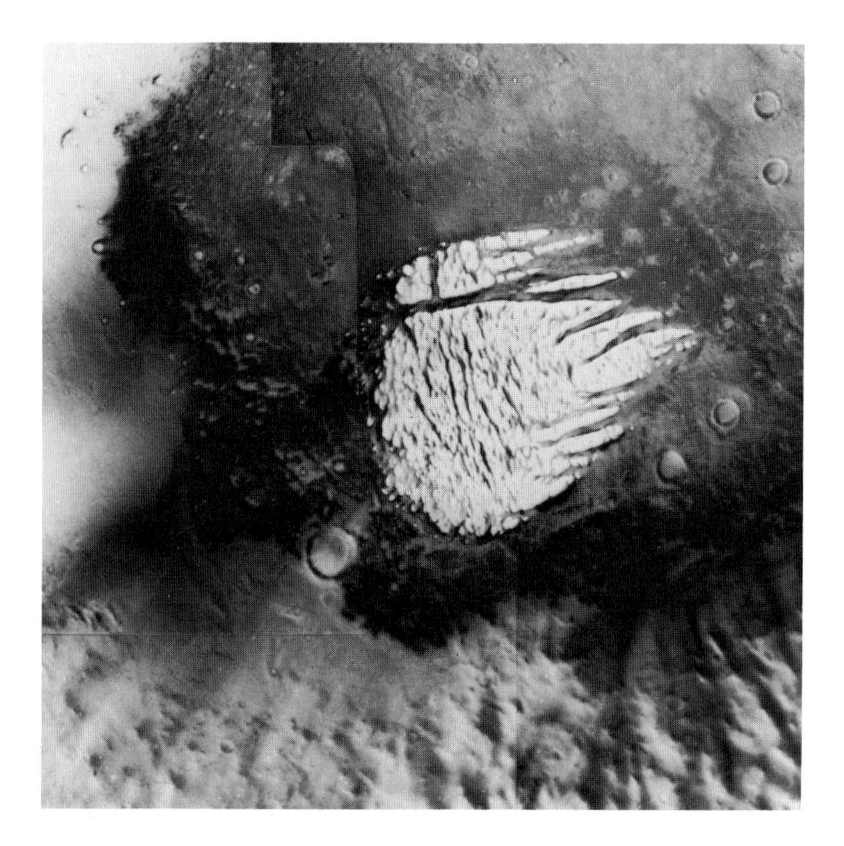

Figure 11.11. "White Rock," at 8°S, 355°W. The origin of this 18-km-wide feature is not known, but the rock appears to be the wind-eroded remnant of a layered deposit within a 93-km-diameter crater. (P-20696)

treat of tens of kilometers over hundreds of millions of years is plausible in some of these areas, which implies erosion rates of 100 μm/yr, at least a factor of a thousand times that on horizontal surfaces.

Yardangs

Yardangs—elongate, streamlined, wind-eroded ridges, aligned parallel to the dominant wind direction—provide unambiguous evidence for wind erosion (Hedin 1903; Blackwelder 1934). They result from a combination of abrasion and deflation and occur in all the major deserts of the world (Blackwelder 1934; McCauley et al. 1977; Mainguet 1972; Ward et al. 1979). They generally form when topographic irregularities on a surface become emphasized by sustained wind action, which widens and deepens hollows. Ultimately discrete ridges become isolated and then streamlined into the characteristic inverted hull shapes. The ridges may be tens of kilometers long and up to one hundred meters high and commonly occur in fields.

The largest array of yardangs on Mars is in southern Amazonis, between latitudes 5°S and 10°N and longitudes 135°W and 185°W (Ward et al. 1979). This large area abounds with wind erosion features, including irregular hollows, fluted escarpments, and pedestal craters, in addition to the yardangs (figure 11.12). The yardangs are mostly double-tapered, flat-topped ridges, up to 50 km long and 1 km across, which are generally in closely packed arrays with a slight sinuosity. In the northern part of the area they are aligned approximately east-west and cut across the dominant northwest-southeast structural grain of the terrain. Elsewhere the yardangs

Figure 11.12. Yardangs and pedestal craters in southern Amazonis. The picture is a detailed view of the sparsely cratered deposit in the upper part of figure 6.7. A friable, easily erodible deposit has been deeply eroded by the wind, to leave isolated pedestal craters and fields of yardangs. The picture is 320 km across. (635A84)

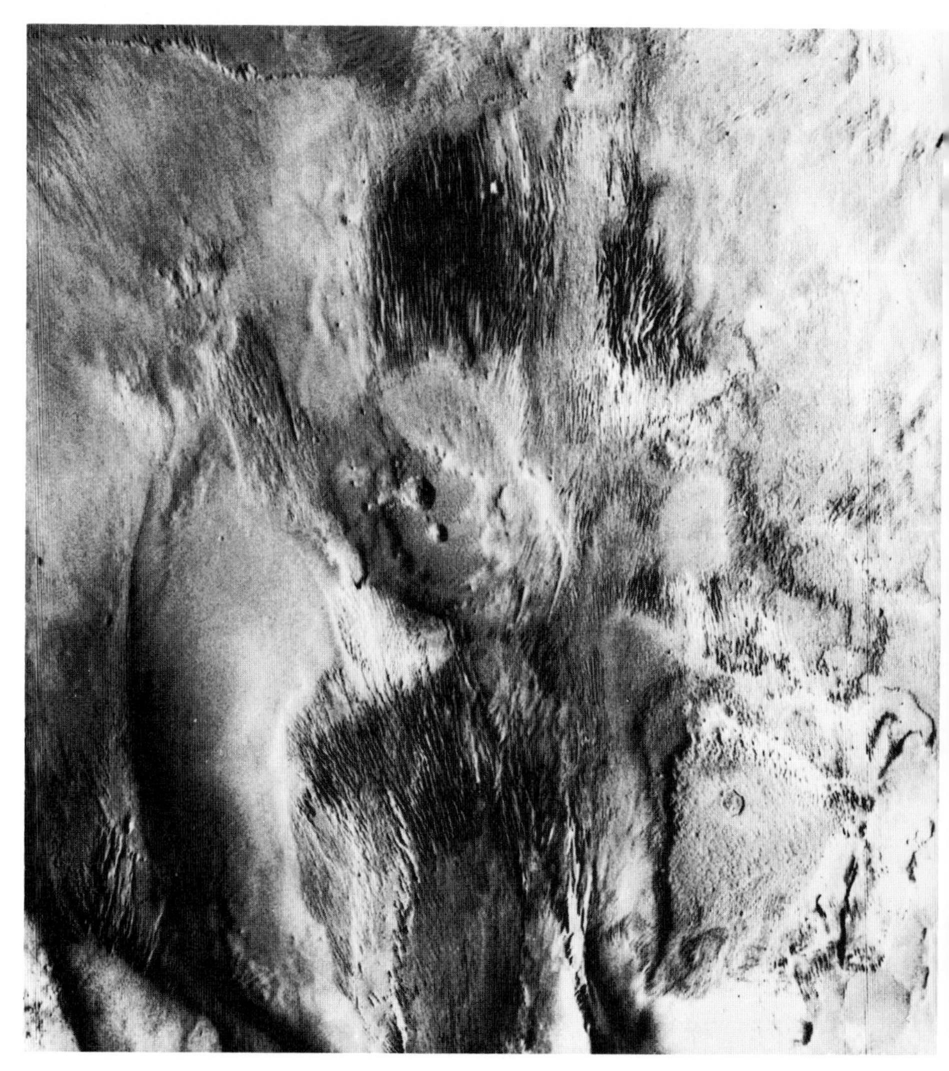

Figure 11.13. Yardangs south of Olympus Mons, at 5°N, 138°W. The southern edge of the Olympus Mons aureole is visible at the top of the frame. The picture is 200 km across. (44B37)

and other erosion features are oriented in other directions, mostly north-south and northwest-southeast (figures 11.13 and 11.14). Most, but not all, are in a sparsely cratered, layered deposit that appears to overlie both the old cratered terrain and the local lava plains. The surface of this unit is in places etched out to form low mesas surrounded by escarpments, with a fluting parallel to nearby yardangs. Ward and colleagues (1979) suggested that preferential development of yardangs in the younger, cross-cutting deposit indicates that it is friable and easily eroded. He proposed ignimbrites, porous lava, and palagonite-rich mudflows as possible materials. The deposit could also be eolian in origin, although why such a thick accumulation would form preferentially in this area is unclear.

Yardangs are relatively uncommon in other areas. Ares Vallis contains some fluted escarpments suggestive of incipient yardang development (Ward et al. 1979), as does Valles Marineris. Ward and co-workers also pointed to some elongate hills in the Iapygia region. McCauley (1973) suggested that some of the inverted hull forms within the aureole of Olympus Mons could have been shaped by the wind, and Cutts and colleagues (1978a,b) argued that wind has formed many of the large outflow channels

Figure 11.14. Terrain etched and fluted by the wind, at 11°N, 147°W, southwest of Olympus Mons. The frame is 24 km across. (693A38)

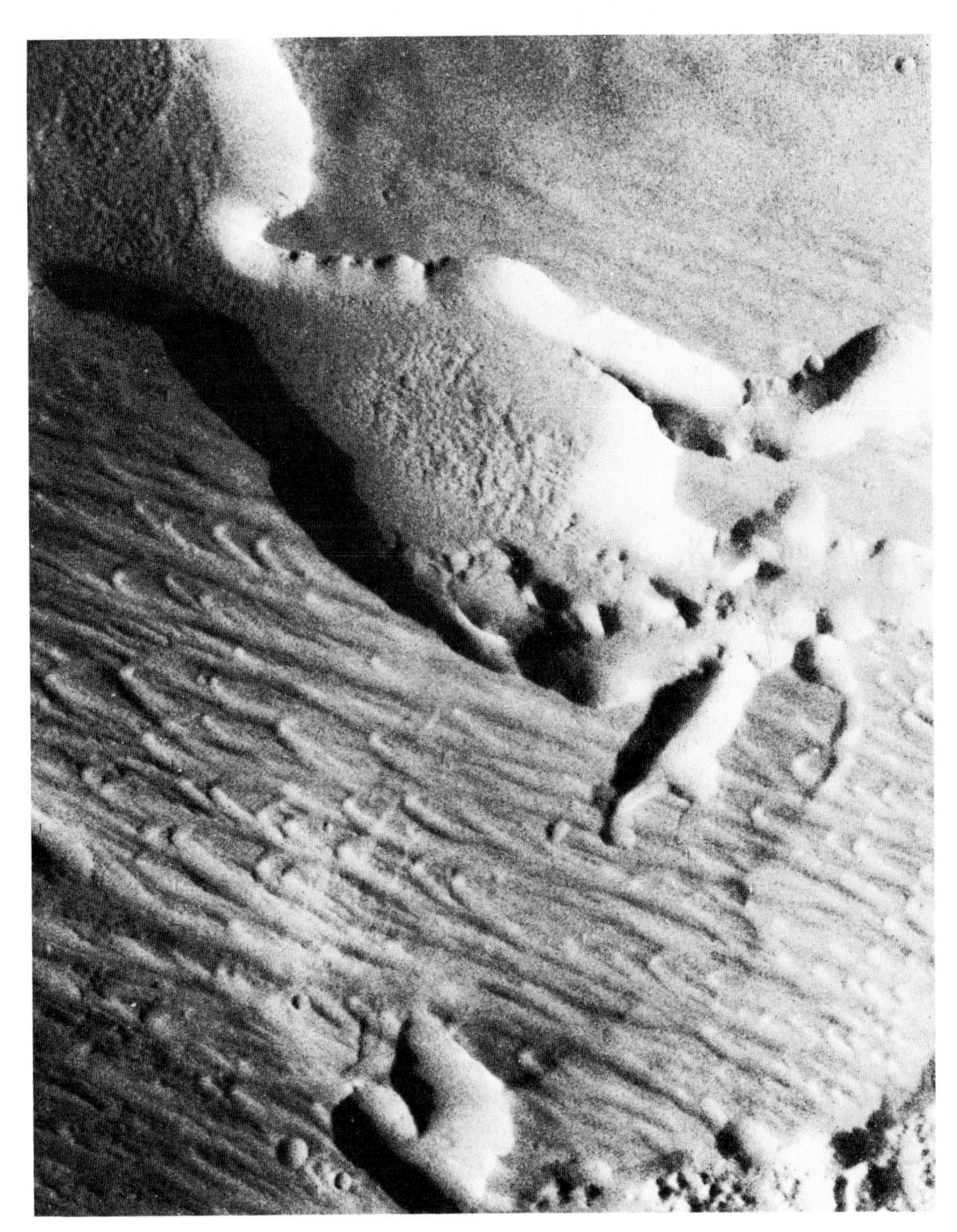

around the Chryse basin and, by implication, that their associated streamlined islands are yardangs. In general, however, yardangs are rare outside southern Amazonis, which makes their profuse development in that region puzzling. Possibly some regional wind-trends are emphasized by the topographic effects of Olympus Mons just to the northeast, and this wind enhancement, combined with the friable nature of the surface materials, has led to unusually intense erosion.

In conclusion the morphology of the surface suggests that wind erosion of consolidated rocks is extremely slow ($\sim 10^{-3}\ \mu$m/yr), despite repeated exposure to violent dust storms. This conclusion is corroborated by the views from the Viking landers, which show numerous fresh-appearing rocks exposed on surfaces that are probably billions of years old. Wind erosion is also highly selective. Many rocks, particularly at the Viking 2 site, have a pitted surface, which could be caused by wind etching (Sagan et al. 1977; McCauley et al. 1979), and such pitting may be one mechanism by which fine-grained debris is added to the generally circulating mass of eolian debris. Erosion rates in poorly consolidated materials are higher, perhaps by a factor of a hundred or more. Erosion is also faster on vertical faces exposed to mass wasting than on horizontal surfaces. The net result of all these factors is that in most areas eolian erosion is relatively minor at scales of 100 m or more, but that it becomes more evident at finer scales. Large-scale wind erosion features are pervasive in southern Amazonis but are relatively rare elsewhere. In contrast large-scale depositional wind forms are common, although they too occur in well-defined zones. Dune fields in the high latitudes indicate an abundant supply of sand-sized debris from unknown sources. More comprehensible are the streaks and other markings, which result largely from continual redistribution of dust by global storms.

12 THE POLES

The polar caps were among the first features of the martian surface to be recognized from Earth. The caps advance and recede with the seasons, and at both poles a small remnant cap is left at midsummer. By analogy with Earth, the caps were long assumed to be composed of water-ice. In the mid-1960s, however, as CO_2 came to be recognized as the main component of the atmosphere, and as more abundant data on surface temperatures became available, Leighton and Murray (1966) and Leovy (1966) proposed that the seasonal caps are mostly CO_2. The presence of CO_2-ice was subsequently confirmed by measurements of temperatures as low as 150°K by Mariner 7 (Neugebauer et al. 1971). In addition Leighton and Murray correctly predicted that the seasonal variations in the cap dimensions cause variations in atmospheric pressure, because such a large fraction of the CO_2 present in the atmosphere freezes onto the winter cap.

During the Mariner 9 mission in late 1971 and early 1972 the southern cap was at its minimum, and the spacecraft was able to get a good view of the surface around the south pole. The investigators were startled to observe a thick sequence of layered sediments resting unconformably on the old cratered terrain. The sediments have since been interpreted as mixtures of volatiles and wind-blown dust, with the differences between adjacent beds being caused by differences in the proportions of the two components. The layering implies cyclic variations in the sedimentation rate. The most likely causes of the variations are climate changes, which influence both surface temperatures and winds, affecting both the stability of the volatiles and the mobility of the entrained debris. Discovery of the layered deposits therefore led directly to speculation on possible causes of climate change. Among the more likely are variations in the rotational and orbital motions of the planet, which cause differences in the seasonal fluctuations of solar energy and how it is distributed by latitude.

The Viking mission has since shown that layered terrains also exist at the north pole. Close-up pictures show that they are very sparsely cratered, if cratered at all, almost certainly indicating a young age ($< 10^8$ years) and reinforcing the impression that the terrains at both poles preserve a record of recent events. Water-vapor abundance and temperature measurements made by the Viking orbiters show that the residual north polar cap is largely water-ice, although the seasonal cap—that which forms in the fall and dissipates in the spring—is confirmed as carbon dioxide. The composition of the remnant southern cap, whether water, carbon dioxide, or both, is uncertain. In addition Viking lander measurements of the amount of dust in the atmosphere during dust storms allow better estimates to be made of the rate of sedimentation of eolian debris at the poles under present climatic conditions. In this chapter we review the characteristics of the caps and the layered terrains and examine some of the mechanisms proposed to explain their peculiar characteristics. The puzzling question of why all the layered terrain appears to be so young is addressed, and climate changes, and the planetary motions that could cause them, are briefly examined.

SEASONAL CAPS

The most obvious features of the poles are the seasonal caps. Mars is close to aphelion during southern winter, moving relatively slowly in its orbit, and as a consequence winters in the south are colder and longer than those in the north (see chapter 1). The south polar cap is therefore more extensive at its maximum. In contrast southern summers occur close to perihelion, so are relatively hot, causing the southern cap to contract further. At its smallest the southern cap is about 350 km across, compared with 1,000 km for the remnant northern cap. The recession of the southern cap has been observed through the telescope for many years, and the fact that temporary outliers are left as the cap retreats has long been noted. Telescopic viewing is better for the retreat of the southern cap than it is for the northern cap, because the closest oppositions occur when Mars is close to perihelion, during southern spring, and at this time Mars's southern hemisphere is pointing toward Earth. Formation of the caps is not seen, because during the fall clouds of CO_2—the polar hood—form over the polar regions as the CO_2 condenses onto the caps. The north polar clouds were thick in northern falls during the Viking mission, but in the south the hood was much thinner in the equivalent season, leading to the suspicion that thick polar hoods do not form in the south (see chapter 1). One of the Viking orbiters was placed in a high-inclination orbit specifically to observe the polar regions better. Because its periapsis was in the north, we have good high-resolution coverage of the northern cap, but poor synoptic coverage. In the south we have the reverse, good regional coverage but poor high-resolution coverage.

The retreat of the south polar cap, as observed from Earth, is nonuniform in longitude but similar from year to year. At its maximum extent the cap is roughly circular around the pole. Its edge follows the 60°S latitude from 120°W longitude westward to 300°W but extends a few degrees further north around the rest of its circumference (Fischbacher et al. 1969). Within the Argyre basin the edge extends as far north as 50°S (James et al. 1979). These latitudes are only approximate; discontinuous frost almost certainly occurs further north. Along the cap edge are numerous frost outliers, mostly within craters. Entire crater floors may be covered or just those parts adjacent to the northern rims. Not all outliers are associated with craters. The largest and most well-known from telescopic observations are the so-called "Mountains of Mitchel," centered at 75°S, 320°W. As the cap edge retreats southward past this area, it leaves behind a large outlier, which remains frost-covered for another 20 to 30 days. The frost probably remains because of a southward regional slope; no obvious topographic features coincide with the outlier except at its southernmost part, where it runs along the southward-facing scarp of a large impact basin centered at approximately 84°S, 270°W. Even before the cap has retreated and while the "mountains" lie totally within the cap, they are brighter than the surroundings, suggesting a more continuous frost cover at fine scale. The location of outliers in general appears to be slope-controlled, as indicated by preferential frost retention just within the north walls of craters. But wind is also a factor. The frost appears to be readily moved by the wind, as indicated by numerous frost streaks around the cap and also by the occasional preferential retention of frost inside the east rims of craters (Thomas et al. 1979).

As the southern cap approaches its minimum, it becomes more asymmetrically positioned with respect to the pole. At its smallest it is centered at 30°W, 86°S, with the pole just touching its edge. The seasonal cap has a slightly reddish color, almost certainly because of admixed dust (James et al. 1979). The perennial cap is less red and slightly brighter. The permanent cap thus becomes visible as a brighter area before the seasonal component has dissipated. As the seasonal cap contracts, the characteristic "swirl" texture of the permanent cap emerges (figure 12.1). The pattern is caused by preferential removal of frost on northward-facing slopes in valleys and on escarpments within the layered terrain. The slopes have a roughly spiral pattern, curving in a clockwise direction around a center on the 80°W longitude that coincides neither with the center of the cap nor the geographic pole.

The retreat of the northern cap has been less well observed (Briggs 1974). At its maximum, when it emerges from the polar hood at the start of spring, it extends to approximately 65°S. Discontinuous frost, however, extends much further south. Pictures of the Viking 2 landing site at 48°N during the late winter show local frost patches on the north side of boulders where the ground is in shadow for much of the day (Jones et al. 1979). The patches, believed to be H_2O-ice (chapter 2), first appeared in northern fall around $L_s = 240°$ and remained until the end of winter, $L_s = 360°$.

Figure 12.1. Mosaic of medium-resolution photographs of the south pole, showing the remnant southern cap at its minimum. The black area around the cap is an artifact of the picture processing. Layered deposits, with smooth upper surfaces and outward-facing escarpments, cover most of the area and partly fill an old impact basin in the upper right. Chasma Australe, a large valley within the layered terrain, is at the center. Compare with figure 12.4b. The scene is 1,200 km across. (211-5541)

As the northern cap retreats in spring, it leaves local outliers, as in the south. In the north, however, because of the less rugged terrain across which the cap retreats, the outliers tend to be more local, and large outliers like the "Mountains of Mitchel" do not form. Most of the outliers are associated with craters, but bright frost remnants on the northern slopes of dunes within the dune fields that surround the north pole can create some deceptive effects by simulating highlighting of the topography. The retreat of the northern cap is also more symmetric than that of the southern cap; even so the northern residual cap is not centered exactly over the pole. It extends to approximately the 80°N latitude from 270°W westward to 30°W and to approximately 85°N around the rest of its circumference. The extent of the northern residual cap approximates that of the layered terrain, so that the northern layered deposits are never seen without a discontinuous frost cover. As in the south a swirl texture emerges during the final stages of retreat, as valley walls and escarpments within the layered terrain become preferentially defrosted (figure 12.2). Hess and colleagues (1979) calculated the average thickness of the seasonal CO_2 cap, based on the seasonal variations of atmospheric pressure. They estimated a minimum value of 23 cm but suggested that the actual value may be as high as 50 cm.

RESIDUAL CAPS

The two residual caps differ from one another, probably as a result of hemispheric asymmetries caused by global dust storms. The northern remnant is almost certainly water-ice. Brightness temperatures over most of the cap during late 1975, when it was summer in the north, were near 205°K, close to the frost point of water in a well-mixed atmosphere containing a few tens of precipitable microns of water (Kieffer et al. 1976). Darker areas were warmer, with temperatures as high as 235°K. The sublimation temperature of CO_2 at 6.1 mb is 148°K, so that solid CO_2 is unlikely to have been present in large amounts. The temperatures are consistent with the observation of large amounts of water vapor (80–100 pr μm) in the atmosphere over the poles at the same time (Farmer et al. 1976).

Half a martian year later, during southern summer, similar measurements were made over the southern cap. The temperatures there were around 160°K, substantially colder than the northern cap at the same season, despite the closer proximity of Mars to the Sun as a result of coincidence of southern summer with perihelion. In addition no significant increase in water vapor was detected over the south pole; the level was the same as elsewhere on the planet at this season ($\approx$6 pr μm) (Farmer and Doms 1979). At the 160°K temperatures measured at the pole, the vapor pressure of water is less than 1 μm of water. The 6 pr μm measured must therefore be well above the surface, where the temperatures are higher. Any contribution from the cap at the measured temperatures would be small, so that the lack of detection does not necessarily mean that

Figure 12.2. Oblique view of the residual northern cap, showing the swirl texture caused by preferential removal of frost on south-facing slopes. (710A74)

water-ice is not present. Kieffer and Palluconi (1979) interpreted the 160°K temperatures as caused by both CO_2-ice at its frost point (148°K) and warmer defrosted areas within the field of view of the detectors and concluded that the southern cap is predominantly CO_2.

Kieffer and Palluconi ascribed the differences between the caps to the effects of global dust storms. The dust storms start in the southern hemisphere close to perihelion, raise large amounts of dust into the atmosphere, and cause its optical depth to increase from values close to 1 to values between 5 and 10 (Pollack et al. 1979). The dust partly shields the remnant southern cap from the sun's radiation, thus affecting the energy balance at the surface. At the same time CO_2 is condensing out at the north pole in a dusty atmosphere. The CO_2 probably nucleates around the dust grains, and Pollack and co-workers suggested that condensation of the northern seasonal cap plays a significant role in removing dust from the atmosphere.

Figure 12.3. Detail of the partly frosted layered terrain in the north. The frost creates strange patterns that mask the topography. A fine layering is visible on partly defrosted slopes, and a small unconformity can be seen at left center. The frame is 65 km across and is centered at 80°N, 347°W. (56B84)

In contrast, in the opposite season when the south polar cap is forming, the atmosphere is clear, so little dust is incorporated into the seasonal cap. The south pole, therefore, has clean, high-albedo ice that is partly shielded from the summer sun, whereas the north has dirty ice over which the atmosphere is clear in summer. The results are warmer temperatures on the remnant northern summer cap than on the remnant southern cap, complete volatilization of CO_2 at the north pole during summer, and retention of a small remnant CO_2 cap in the south.

Whether water is present in the remnant southern cap or not is still uncertain, for the cold southern summer temperatures prevent detection of water. It appears probable, however, that the southern cap acts as a trap for both CO_2 and H_2O. Once water is in the remnant southern cap, it cannot get out, because the temperatures never reach the frost point, at least under present climatic conditions. Furthermore, because of the low sublimation rate of CO_2 during summer, CO_2 may be accumulating in the southern cap at the expense of any CO_2 in the north. In other words, the remnant southern cap may be slowly growing in size.

The behavior of the caps presumably alternates with the precessional cycle. At present the northern residual cap consists largely of dirty water-ice and has dust added to it in the aftermath of the global dust storms, whereas the southern cap is cleaner and colder and is accumulating CO_2. In 25,000 years the conditions will be reversed. Dust storms will occur during northern summers, and the debris will be preferentially deposited onto the south pole, which will become relatively warm and dirty. The dust-storm cycle may also be responsible for the asymmetric distribution of eolian debris around each pole. As we saw in the last chapter, the north pole is surrounded by vast fields of dunes, whereas in the south dunes are localized, being mostly within craters. Kieffer and Palluconi (1979) suggested that this could result from net transport of atmospheric dust into the region of the north pole, as a consequence of the dust being scavenged from the atmosphere by CO_2 condensing on the pole. Most of the dust would then be remobilized in spring as the transient CO_2 cap dissipates. If this explanation is valid, then the circumpolar dune fields alternate between the poles on the precessional cycle, the dunes presently observed at the north pole are no more than 25,000 years old, and 25,000 years hence a comparable array of dunes will surround the south pole. As we saw previously, however, such a model appears unlikely because of the long times that are probably required to form the dune fields. Moreover, as we noted in the last chapter, sand-size particles are needed to form dunes. If the dunes are formed from dust scavenged from the atmosphere, then the particles must aggregate in some way, and as we saw, such aggregates are unlikely to survive the continued reworking that is involved in dune formation.

LAYERED TERRAIN

Layered deposits occur at both poles and cover most of the area within the 80° latitude circles. They lie on the old cratered terrain in the south (see figure 12.1) and on plains in the north. During summer the southern deposits are largely uncovered because of the small southern cap, but in the north, where the remnant cap is more extensive, the deposits have not been seen without at least a partial frost cover (figure 12.3). They mostly form a smooth, crater-free surface, with gentle undulations that contrast markedly with the generally crisp, cratered topography of the surrounding areas. Incised into the smooth upper surface of the deposits are numerous valleys and low escarpments. These curl out from the pole in a consistently counterclockwise (eastward) direction in the north (figure 12.4, *overleaf*) and in a predominantly clockwise (westward) direction in the south. Between the linear valleys and escarpments the terrain is mostly flat and featureless, although some low undulations have been detected that parallel the larger topographic features (Cutts et al. 1979). The valleys and escarpments are approximately equally spaced and are roughly 50 km apart. The minor undulations are close to 10 km apart. Preferential removal of frost on the equator-facing slopes gives the poles their characteristic summer swirl texture. The valley walls and escarpments range in height from 100 m to 1,000 m and have slopes that range as high as 6° (Dzurisin and Blasius 1975). An especially deep, wide valley in the south is named Chasma Australe; a similar broad valley in the north is called Chasma Boreale. The minimum thickness of the deposits has been estimated at 1–2 km in the south and 4–6 km in the north, with the northern deposits at an absolute elevation that is approximately 4 km lower than that of deposits in the south (Dzurisin and Blasius 1975). The outer boundary of the deposits is generally very sharp, consisting of an outward-facing, slightly convex-upward escarpment (figure 12.5).

The layering is best seen as a fine horizontal banding on defrosted slopes. Individual layers have a remarkable continuity, being traceable for hundreds of kilometers (Cutts 1973a; Murray et al. 1972; Soderblom et al. 1973). Thicknesses of 10 m to 50 m have been measured (Dzurisin and Blasius 1975), but the layering almost certainly continues below the limiting resolution of the available photography. The layering is visible mainly because of albedo contrasts between individual layers; a fine terracing is also visible in places, particularly in the south where larger areas are defrosted. Unconformities are present (see figure 12.3) but rare. More generally the banding maintains a continuity in relative thickness of adjacent layers over long distances. Disruption of the layering, such as might be expected from impacts, is also rare, suggesting that the small number of superimposed craters is truly indicative of a young age and is not due to some self-annealing process or to infilling.

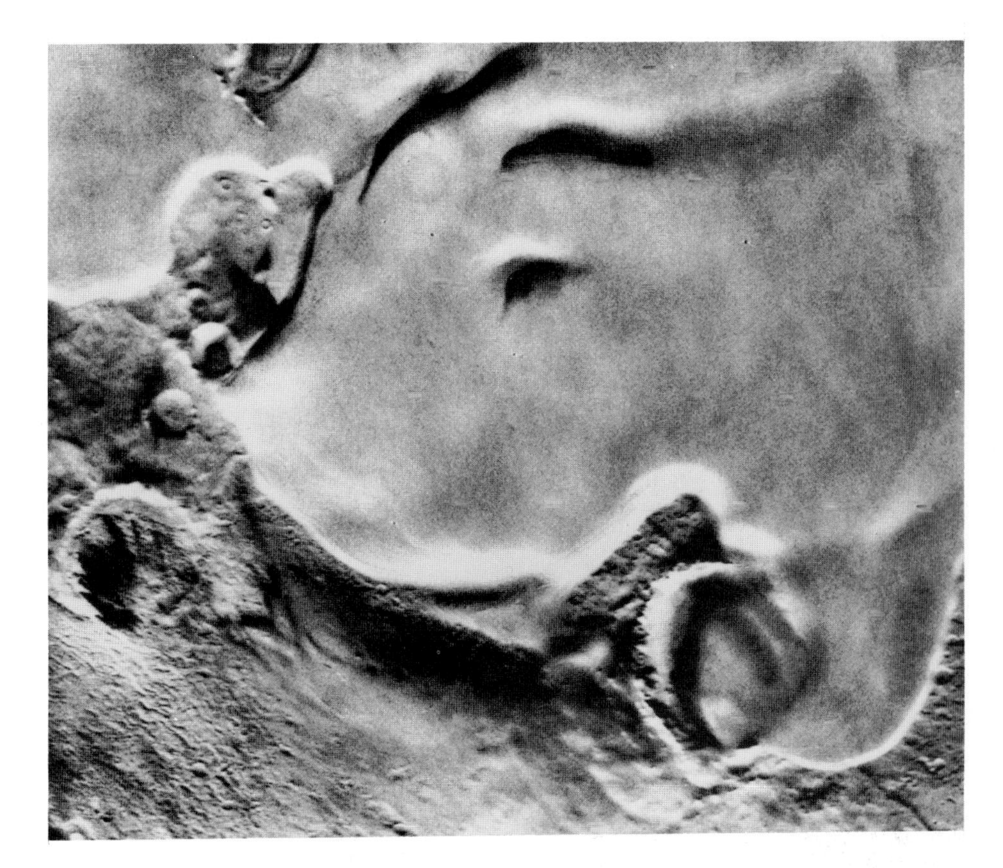

Figure 12.5. The edge of the layered deposits in the south. The deposits rest unconformably on an old cratered surface but are themselves sparsely cratered. The frame is 200 km across and is centered on 80°S, 255°W. (383B49)

Climatic variations as a cause of the layering

The layered deposits are believed to be accumulations of volatiles and dust, with both the relative and the absolute accumulation rates of each component being somehow modulated by climatic variations (Murray et al. 1972; Cutts 1973a; Cutts et al. 1976; Sharp 1974). The most obvious causes of such climatic variations are variations in the orbital and rotational motions of the planet. We have seen that because of precession of the orbital and rotational axes, the hot and cold poles alternate every 25,500 years—that is, on a 51,000-year cycle—and that at present dust is probably accumulating at the north pole faster than it is at the south pole. But orbital motions other than the precessional cycle may affect accumulation. The eccentricity of the martian orbit (figure 12.6) is currently 0.093, but it

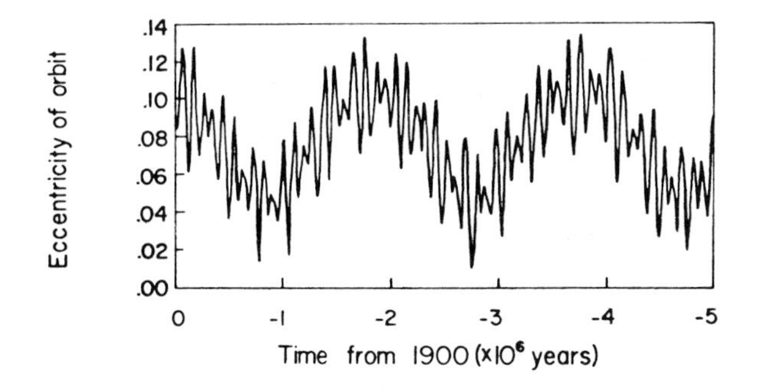

Figure 12.6. Variation in eccentricity of Mars's orbit over the last 5 million years. Eccentricity variations modulate the amplitude of the hemispheric differences in climate caused by precession. (Reproduced with permission from Murray et al. 1973, copyright 1973, American Association for the Advancement of Science.)

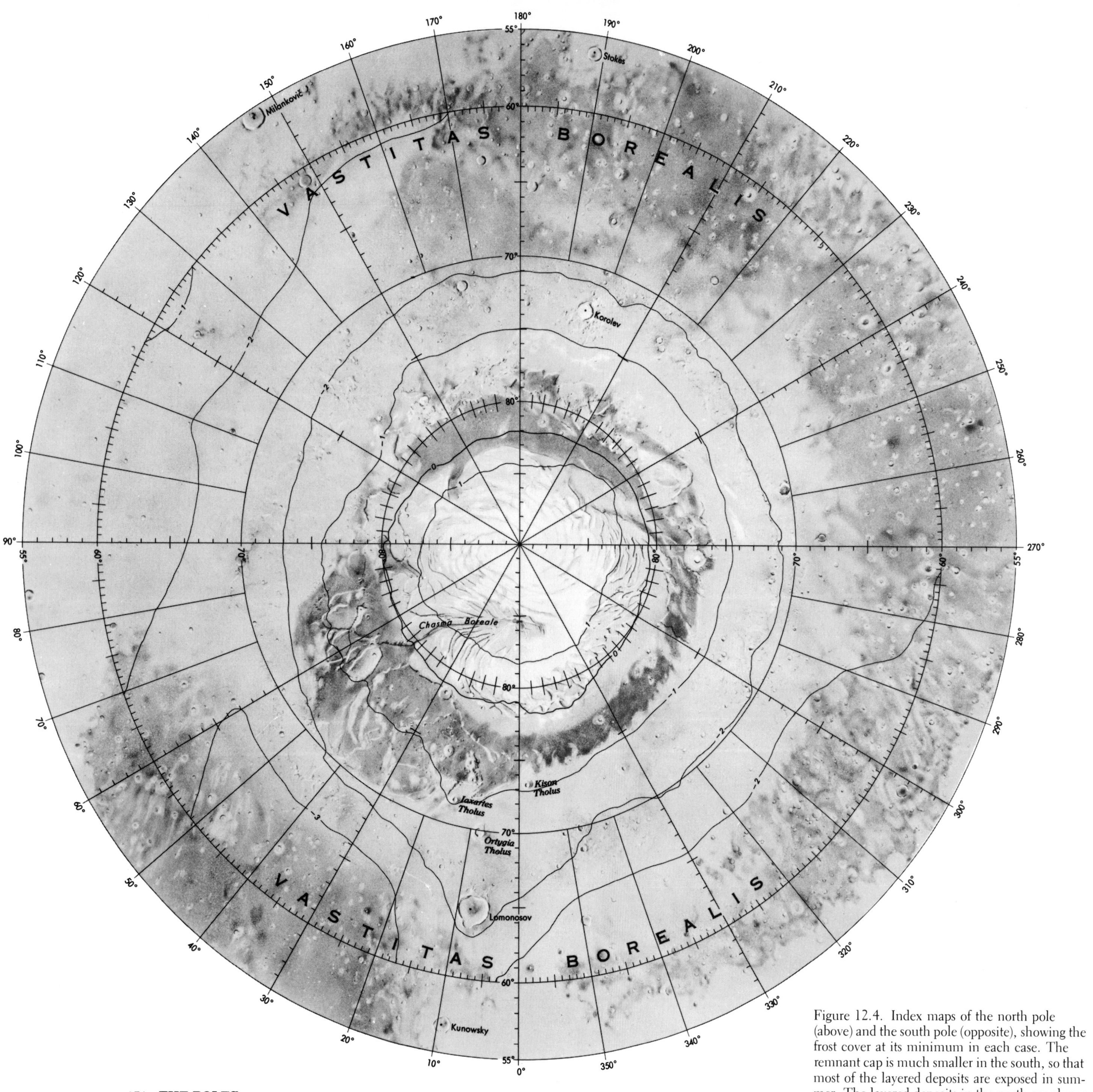

Figure 12.4. Index maps of the north pole (above) and the south pole (opposite), showing the frost cover at its minimum in each case. The remnant cap is much smaller in the south, so that most of the layered deposits are exposed in summer. The layered deposits in the north are always frost-covered. The dark collar around the northern cap is composed mostly of dunes. (Courtesy U.S. Geological Survey.)

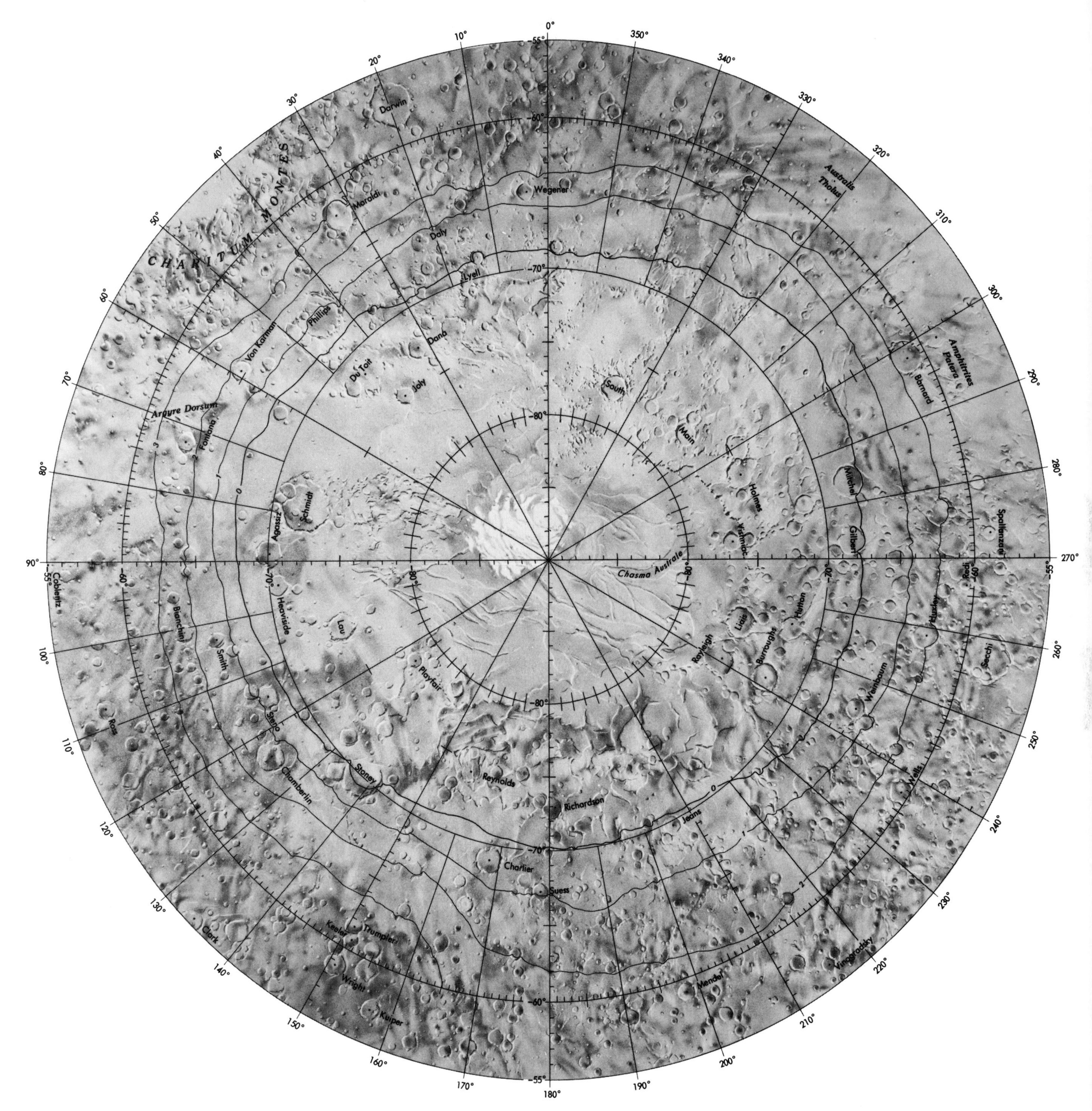

CHARITUM MONTES
Argyre Dorsum
Australis Tholus
Amphitrites Patera
Chasma Australe
Darwin
Maraldi
Wegener
Daly
Lyell
Phillips
Dana
Von Karman
Du Toit
Joly
South
Barnard
Main
Fontana
Mitchel
Holmes
Schmidt
Agassiz
Spallanzani
Gilbert
Coblentz
Heaviside
Hutton
Bianchini
Lau
Liais
Huxley
Smith
Rayleigh
Burroughs
Secchi
Playfair
Weinbaum
Ross
Stoney
Chamberlin
Reynolds
Wells
Richardson
Jeans
Charlier
Suess
Clark
Keeler
Trumpler
Wright
Kuiper
Mendel
Vinogradsky

varies, with two characteristic periods. Over a period of 2×10^6 years eccentricity can change from 0.14 to 0.01 (Murray et al. 1973). Superimposed on this long-term change is a smaller oscillation—less than 0.06—with a period of 95,000 years. Clearly when the orbit is nearly circular (eccentricity of 0.01), the effects of the precessional cycle are minimized, and the behavior of the poles is relatively symmetrical. At the maximum eccentricity of 0.14 the effects of the precessional cycle are amplified, and the two poles become very different. As a consequence, in periods of high eccentricity deposition rates at the poles should change on a 51,000-year cycle, whereas during periods of low eccentricity, deposition should be relatively uniform. Furthermore, during periods of low eccentricity the maximum insolation received by the summer hemisphere at perihelion may fall below the threshold required to generate global dust storms (Briggs 1974), so that dust deposition at the poles could decrease to a low value. Thus the combination of eccentricity variations and precession cause periodic variations in deposition on time scales ranging up to 2 million years.

The energy balance at the poles is also affected by the planet's obliquity—the angle between the equatorial plane and the plane of the orbit. Variations in obliquity affect the latitudinal distribution of the Sun's insolation. The obliquity of Mars ranges from 15° to 35° (Ward 1974). It oscillates with a period of 1.2×10^5 years (figure 12.7) and with an amplitude that varies on a time scale of 1.2×10^6 years. The present obliquity is 25°, close to the midpoint of the full range. At the maximum obliquity of 35° approximately twice as much insolation falls on the poles as at the minimum obliquity. Ward and co-workers (1974) estimated that by controlling the amount of CO_2 frozen out on the poles, obliquity variations could cause changes in the atmospheric pressure that range from 0.3 mb when obliquity is at its minimum to 30 mb when it is at its maximum. These calculations were based on the assumption that the

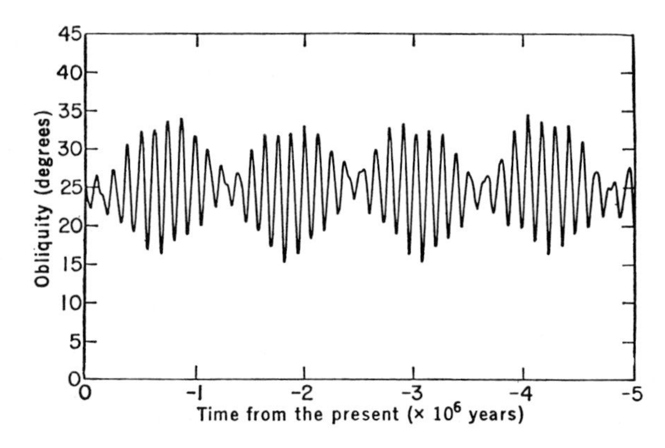

Figure 12.7. Variations in the obliquity of Mars over the last 5 million years. Obliquity variations cause differences in the latitudinal distribution of the Sun's radiation. (Reproduced with permission from Ward 1973, copyright 1973, American Association for the Advancement of Science.)

present atmosphere is buffered by a remnant CO_2 cap in the north, which, as we now know, is not true. They do give some indication of the magnitude of the obliquity effects, however.

An additional complication is the effect of obliquity on the adsorption and desorption of volatiles in the regolith (see chapter 13). High obliquities will raise the temperature of the regolith in the polar regions, causing desorption. At the same time the equatorial temperatures will be lowered, promoting adsorption. Interchange of volatiles between the polar and equatorial regions will thus take place on a time scale corresponding to the obliquity variations. How this would affect the total amounts of volatiles within the atmosphere is not clear, but such interchanges may buffer the atmosphere and reduce the variations in atmospheric pressure that would be caused by changes in the size of the polar caps.

In summary, the eccentricity of Mars's orbit causes an asymmetry in the behavior of the two poles. Their characteristics change on a 51,000-year cycle because of precession of the orbital and rotational axes. The differences between the two poles are accentuated or diminished by variations in eccentricity, which have periods of 95,000 years and 2×10^6 years. An additional periodic perturbation is caused by variations in obliquity, which affect the magnitude of the differences between the poles and the equator. A final complication is how all these variations affect the threshold for initiating global dust storms, which may control the rate of deposition of dust at the poles.

Depositional rates

Pollack and colleagues (1979) calculated the present rate of dust deposition in the polar regions on the basis of the amount of dust observed in the atmosphere during the 1977 dust storm. Observations made that year suggested that the amount of dust added to the atmosphere was equivalent to increasing the optical depth by a factor of 7.5, which is equivalent to 4×10^{-3} gm/cm^2. Since dust storms occur only every other Earth-year, the average sedimentation rate over the whole planet is 2×10^{-3} gm/cm^2/yr. Assuming that half of this falls in the 6 percent of the planet that is north of 60°N, and that the final deposit is half water-ice with a 50 percent porosity, Pollack and co-workers estimated that the sedimentation rate is 4×10^{-2} cm/yr, at which rate it would take one hundred thousand years to deposit a 30-m-thick layer. To deposit 2–3 km would take 6–10 million years.

Cutts (1980) pointed out that the actual sequence of events may be far more complicated than that just outlined, in that long gaps may occur between periods of sedimentation. The present rates depend largely on dust raised into the atmosphere by dust storms. The initiation of global storms depends on eccentricity and obliquity, which interrelate in a complicated way, in that they vary with different periods. Episodic gaps may occur in the sedimentary record, which represent periods when conditions were such that dust storms did not occur. Similarly the condensation and stability of volatiles depend on obliquity, eccentricity, and precession in a complex manner not fully understood. We can conclude, however, that at the present rate the observed deposits could have accumulated in a relatively short time, a few tens of millions of years at the most. This is consistent with the young ages implied by the lack of superimposed impact craters.

The relatively young age is one of the more puzzling aspects of the layered terrain. As indicated previously, the scarcity of craters appears in large part to be due to this young age and not to self-annealing or infilling, because impact scars do not interrupt the exposed layers. Cutts and colleagues (1976) noted that in an area of 800,000 km² of layered terrain in the north, not one crater as large as 300 m was observed. If this lack of craters was due solely to age it would imply a surface less than 1 million years old. More probably some annealing and infilling of craters does occur, but scars from craters as small as 300 m are not detectable with the present photography. Even with this concession ages much in excess of a few tens of millions of years are unlikely. What then was occurring at the poles prior to the last few million years? Either the conditions were much like they are at present, and the old layered terrain has been removed, or the present era is unique. The latter possibility is unlikely, for it would require us to be coincidentally observing the planet just when the layered deposits started to accumulate. Some clues as to how the older layered sequences might have been erased may be provided by the present topography of the layered terrain.

Origin of the topography of the layered deposits

While there is general agreement that the layered sequences at the poles originate by deposition of volatiles and wind-blown dust, modulated by climatic variations, no such consensus has been achieved on the origin of the valleys and scarps that form the distinctive swirl texture of the deposits. We have seen that dark bands spiral out from the north pole in a generally counterclockwise direction (anti-Coriolis), although the pattern is interrupted by Chasma Borealis, which runs diagonal to the general pattern and around which the bands curl. In the south the pattern is less consistent, but a crude circularity centered at 87°S, 0°W, can be distinguished. The bands are caused by the absence of frost on linear, sunward-facing valley walls and escarpments.

A question of some interest is whether the relief is forming contemporaneously with the deposits, or whether periods of deposition alternate with periods of erosion, during which the walls and escarpments are formed. In the first case climatic variations comparable to those that cause variations in sedimentation rates are implied; in the second case larger climatic variations are needed, with depositional and erosional periods alternating on a time scale that is long compared with that represented by the layers in the sediments.

Murray and Malin (1973) suggested that the swirl pattern was caused by a shift in the position of the spin axis. They postulated that the layered sequence was deposited as roughly circular plates centered on the geographic pole, and that as the pole shifted, so did the plate edges. The result is a roughly spiral pattern of escarpments. This hypothesis now seems quite improbable. The swirl pattern is not hemispherically symmetric; the pattern in the north curls around the present pole, while that in the south is offset. The pole shifts should result in numerous unconformities, which are not seen. Moreover, the layered deposits appear to be young, and pole offsets of several degrees within a few million years are unlikely. Clark and Mullin (1976) suggested that the laminated deposits are largely CO_2-ice, and that the stratification and swirl pattern were largely the result of outward flow from the center of the ice sheet. However, we now know that the northern cap is water-ice, and that CO_2-ice is unstable at the surface. Although CO_2-ice persists in the southern residual cap, it does so partly because it is relatively clean and has a low albedo. Any CO_2-ice near the surface in the surrounding unfrosted layered terrain, with temperatures around 230°K, would rapidly sublime (Ingersoll 1974; Kieffer et al. 1976). The layered deposits could still contain large fractions of water-ice, but whether water-ice could flow appreciably at polar temperatures (140–235°K) is doubtful (Sharp 1974). Polar wandering and carbon dioxide glaciations, therefore, appear to be ruled out as explanations of the swirl texture. More probable mechanisms for creating the relief are wind erosion and differential sublimation and condensation.

Evidence for wind erosion at the poles is widespread. Cutts (1973b) gave several examples of a fine linear fluting on the surface of the layered deposits, which he plausibly interpreted as caused by wind. In addition adjacent to the main body of layered deposits at the south pole are some etched and pitted terrains (Sharp 1973c). In these areas the surface is generally smooth, except for numerous irregularly shaped, steep-sided pits (figure 12.8). Many of the pits are windows through which the underlying cratered terrain is visible. Both Cutts (1973b) and Sharp (1973c) considered wind to be the most likely agent for removal of material to form the pits, although sublimation of volatiles could also have been involved. No comparable pitted terrain occurs in the north. In other areas complex patterns of intersecting ridges suggest etching out of resistant dikelike bodies by the wind. The curvilinear form of the large-scale topography and the fine-scale terracing caused by individual laminae on slopes are both consistent with wind erosion. Finally depositional wind features are common. At the north pole the layered terrain is almost everywhere surrounded by dune fields, and the large reentrant Chasma Borealis is filled with dunes. Dunes are less common in the south but are still present locally within craters around the periphery of the polar regions. Thus wind appears to have played a significant role in sculpting the polar landscape.

Differential sublimation and condensation may also have played a role in shaping the layered deposits. Howard (1978) suggested that the layered terrain is composed of ice and dust and is in a state of dynamic equilibrium with present climatic conditions. According to his hypothesis erosion and deposition are occurring simultaneously. Ice sublimes on sunward-facing

Figure 12.8. Etch-pitted terrain at 76°S, 74°W. A former horizontal surface has been extensively eroded, resulting in the formation of numerous irregular pits. The deposit appears to be layered and may be the remnant of a formerly more extensive cover of layered deposits. (390B90)

slopes, causing the layered materials to disaggregate. The dust released is then readily mobilized by the wind and becomes cemented onto the adjacent flats where volatiles driven off the dark slopes refreeze. The sequence exposed on the escarpment slopes is thus eroded away, and new deposits are built up on the adjacent flats. Howard demonstrated by computer modeling that this process will cause minor variations in slopes to become accentuated and ultimately to become consolidated into discrete linear escarpments separated by broad smooth areas. In effect a growing escarpment eats into the higher ground, sweeping up the local slope variations and leaving behind a smoothed-off surface. The curling spiral pattern, he suggested, is caused by a slight preferential erosion on westward- as opposed to eastward-facing scarps, because afternoon temperatures are higher than morning temperatures as a result of a slight atmospheric warming during the day. Howard thus envisaged a terrain that is constantly being renewed as the escarpments slowly march across the landscape. The young age of the terrain is merely a measure of the rate at which erosion is taking place and at which the whole sequence is being re-formed.

Cutts and co-workers (1979) also invoked preferential ablation to explain the topography of the layered deposits. They noted first that most of the linear features that are defrosted in summer are valleys, and not escarpments as formerly thought, and second that the intervening flats have slight undulations that follow the general trend of the valleys. These observations were based on photographs of the layered terrain taken in winter when the surface was completely frost-covered and could be viewed without the confusing albedo effects caused by frosted and unfrosted ground. Cutts and colleagues suggested that the minor undulations mark former positions of the edge of the layered terrain. The undulations are muted, because later layers drape over the former edges as the region of deposition increases in size. On the poleward side of an undulation the sequence is slightly thicker and has accumulated for a slightly longer time than on the equator side. The differing extent of the layered terrain, as indicated by the undulations, he ascribed to climatic changes. He further postulated that the valleys form in a manner analogous to the escarpments in Howard's hypothesis. On some of the undulations the slopes exceed those required to retain frost in summer. Any material deposited on these slopes in winter is therefore removed when the ice sublimes in summer, whereas dust and ice are retained on the adjacent frosted ground. As a result, linear zones of nondeposition cut across the layered terrain, following former positions of the outer margin of the deposits. Over an extended period of time these linear zones will develop into valleys as material accumulates on the adjacent flat areas. Only every third or fourth undulation is steep enough to be defrosted, and Cutts speculated that some periodic modulation in the amplitude of the undulations could have been caused by periodic climatic variations of astronomical origin. In the Cutts hypothesis, therefore, the spiral pattern is static and follows former positions of the edge of the layered terrain; in the Howard hypothesis, the pattern is dynamic, and the terrain is constantly being renewed.

The Howard and Cutts hypotheses do not exhaust the possibilities. There is good evidence in the south for erosion around the periphery of the layered deposits. At 85°S, 353°W, a 25-km–diameter crater and half its secondaries—those that formed in the layered terrain— survive (figure 12.9). The relations suggest that the layered deposits have been stripped back, and that where secondaries are missing the ground was formerly covered. Extensive erosion of the layered deposits is also implied by the etched and pitted terrains (Sharp 1973c), which appear to be eroded remnants of the layered terrain (see figure 12.8). In addition the huge valleys (Chasmas) incised into the sequences at both poles suggest amounts of erosion that are difficult to explain by either the Howard or the Cutts hypothesis. Other erosional mechanisms are therefore implied.

Erosion and deposition may be occurring penecontemporaneously, as periods of erosion alternate with periods of deposition. The alternation could be on both seasonal and precessional cycles. This almost certainly happens. Fine layers are visible on defrosted ground as albedo differences, even in the north where active deposition is taking place. Their visibility indicates that the dark bands are being kept clean of dust, probably by the wind. The question is not whether erosion is taking place but the scale and rate at which it is occurring. Can the large Chasmas, the apparent retreat of the edge of the layered terrain in the south, and the lack of old layered terrain all be explained on the basis of present conditions at the pole, perhaps modulated by the precessional cycle, or are larger, longer-term climatic changes required?

Figure 12.9. Secondary craters in the southern layered terrain. The crater at the bottom of the picture has caused secondary craters in the layered terrain to the right, but the secondaries are missing to the left where there is no layered terrain. These relations and the survival of only half the crater suggest that the layered terrain was formerly more extensive and has been eroded back. (421B79)

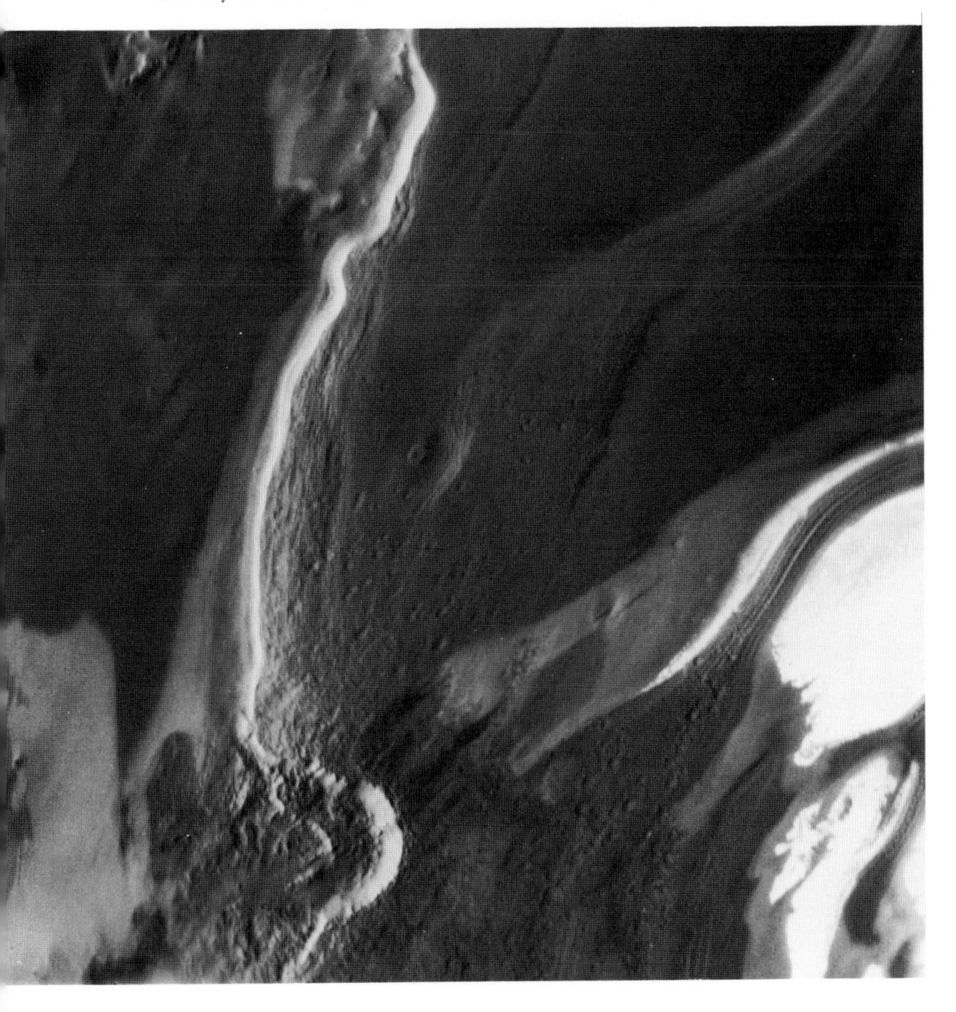

The question cannot be answered with any certainty, but it is interesting to speculate on some possibilities. At present Mars is at the midpoint of its obliquity range. At the maximum value of 35°, the amount of insolation falling on the poles is 35 percent higher than at present. If the layered terrain is cemented by water-ice, then the increased insolation could result in increased disaggregation and erosion. Erosion would occur preferentially around the periphery of the deposits on slopes facing the equator, because here the amount of insolation falling on the deposits would be highest and any "lag" would be readily removed, allowing erosion to proceed. It is unlikely, however, that all the layered terrain could be removed every 600,000 years, following the 1.2×10^6-year amplitude variations in the obliquity. To have accumulated the deposits in the short period of time since the last erasure would require depositional rates close to 0.5 cm/yr, which are excessively high. It should be noted in passing that the very high obliquities calculated by Ward and co-workers (1979) for the time before the formation of the Tharsis bulge are not relevant to this problem, because they would have occurred early in the planet's history, not just a few million years ago. If the layered deposits are periodically removed, as appears likely from the young ages and the presence of eroded remnants, then it is probably on a time scale of at least several million years. Perhaps a combination of high obliquity and high eccentricity is required, and the combination occurs only once every several million years. This is admittedly speculative, but the observations appear to demand erasure of the layered terrain on a time scale that is long compared with both the time of deposition of the observed layers and the perturbations in the rotational and orbital motions.

Summary

Each pole has a seasonal cap of CO_2, which in winter extends to the 50–65° latitudes, and a summer remnant cap a few degrees across. The coincidence of global dust storms with the accumulation of the northern cap results in differences between the two remnant caps. That in the north consists of water-ice and probably contains large amounts of dust; that in the south is mainly CO_2 and dust-free. At the poles are layered deposits up to 5 km thick, which extend out to about the 80° latitude. They probably consist of dust cemented by ice, and they are among the youngest features of the planet. They appear to form as dust from the global dust storms is scavenged from the atmosphere by CO_2 condensing on the cap. The northern cap is currently accumulating debris and adding to its layered deposits, but the pole of accumulation may alternate with the 51,000-year precessional cycle. Obliquity and eccentricity variations could modulate depositional rates in a manner not understood. The layered deposits appear to be partly eroded, being cut by numerous curvilinear valleys and escarpments, although these could be primarily depositional features. Erosion is probably also modulated by precession, obliquity, and eccentricity and may periodically lead to complete removal of the deposits.

Figure 13.1. Rock-pushing at the Viking 2 landing site. After anlayzing several samples of the fines at both landing sites, attempts were made to obtain samples from under rocks, in an effort to find organic compounds. These pictures show the sample arm pushing a 20-cm–diameter pitted rock, so as to gain access to soil protected from the Sun's radiation. In the picture to the right magnetic minerals attached to the magnets on the underside of the sample scoop can be seen. (Viking lander 2 event nos. 22B038 and 22B030.)

13 SURFACE CHEMISTRY AND THE DISTRIBUTION OF VOLATILES

Evidence for the chemical and mineralogic composition of the surface comes from four main sources: (1) direct chemical analyses by the Viking landers; (2) spectral reflectance measurements; (3) surface morphology; and (4) theoretical studies. This chapter is divided into two parts. The first deals primarily with rocks and fragmental debris at the surface. The chemical analyses are examined, and different possible interpretations of the data are discussed. Emphasis is mainly on the fine-size fraction, since only this fraction was directly analyzed, but some implications for coherent rocks are alluded to, and, by extension, implications for the composition of the martian mantle. The second part of the chapter concerns volatiles, mainly CO_2 and H_2O. In previous chapters ground-ice and groundwater were frequently invoked to explain surface features such as channels, ejecta, collapse pits, and so forth. Here we examine the physical constraints on the distribution of ice and water. The adsorption and desorption of CO_2 in the regolith are also discussed, together with the possible effects of long-term climatic changes on the interchange of CO_2 between the regolith and the atmosphere.

SURFACE CHEMISTRY

Each Viking lander carried an x-ray fluorescence spectrometer to analyze the surface materials (Clark et al. 1977). Debris was scooped from the surface and was introduced into a sample chamber on the lander, where it was subjected to irradiation from two sources (^{55}Fe and ^{109}Cd). The x-rays emitted were detected by proportional counters, and the resulting x-ray spectra were then computer-matched to spectra of known mixtures to obtain a chemical analysis. Only elements of atomic number greater than 11 were analyzed, leaving several common elements, such as H, C, N, O, and Na, undetermined.

The materials at the two sites had remarkably similar compositions, considering their wide separation (table 13.1). What was analyzed was probably a relatively thin, easily mobilized veneer of particulate debris (figure 13.1) that has been homogenized over most of the planet by repeated involvement in planetwide dust storms. The main cause of variations between samples is different amounts of "duricrust," the calichelike deposits up to a few centimeters thick, close to the surface at both landing sites; the duricrust contained significantly more sulfur (and possibly chlorine) than the loose debris. It is almost certain that no coherent rocks were analyzed, despite the repeated introduction of pebble-size fragments into the sample chamber. These all had approximately the same disturbed density (1.1 ± 0.15 gm/cm^3) as the fines as well as a higher sulfur content, suggesting that they were clods rather than rock fragments (Baird et al. 1977; Toulmin et al. 1977).

The analyses showed a deficit of several percent, which is real. It probably results mostly from the presence of H_2O and CO_2 and possibly from small amounts of Na_2O and nitrates. The amount of water driven off from soil samples by the Viking gas chromatograph–mass spectrometer (GCMS) suggests H_2O contents of several tenths to several percent (Biemann et al. 1977), so water is present. Huguenin (1974) and Booth and Kieffer (1978) have shown that under the stimulus of ultraviolet radiation, carbonate minerals can be produced under martian conditions at geologically significant rates, so carbonates may be present among the weathered products and may constitute part of the deficit. Their presence is supported by the detection of CO_2 by the GCMS. While this may have been partly adsorbed

TABLE 13.1. COMPOSITION OF MARTIAN SAMPLES

	Chryse Fines	*Chryse Duricrust (1)*	*Chryse Duricrust (2)*	*Utopia Fines*	*Estimated Absolute Error*
SiO_2, wt %	44.7	44.5	43.9	42.8	5.3
Al_2O_3, wt %	5.7	n.y.a.	5.5	n.y.a.	1.7
Fe_2O_3, wt %	18.2	18.0	18.7	20.3	2.9
MgO, wt %	8.3	n.y.a.	8.6	n.y.a.	4.1
CaO, wt %	5.6	5.3	5.6	5.0	1.1
K_2O, wt %	<0.3	<0.3	<0.3	<0.3	—
TiO_2, wt %	0.9	0.9	0.9	1.0	0.3
SO_3, wt %	7.7	9.5	9.5	6.5	1.2
Cl, wt %	0.7	0.8	0.9	0.6	0.3
Sum	91.8	n.y.a.	93.6	n.y.a.	—
Rb, ppm	≤ 30			≤ 30	
Sr, ppm	60 ± 30			100 ± 40	
Y, ppm	70 ± 30			50 ± 30	
Zr, ppm	≤ 30			30 ± 20	

SOURCE: Toulmin et al. 1977.
NOTE: The abbreviation n.y.a. means not yet available.

CO_2, some may also have come from decomposition of carbonates, such as dolomite and magnesite. Na is unlikely to constitute a significant part of the deficit, since it should closely follow K in abundance, and K, which was analyzed, is low. Nitrates may be present in small amounts, as predicted from models based on the isotopic composition of nitrogen in the atmosphere (chapter 3).

Sulfates appear to be a major component of the duricrust. In all cases indurated samples had higher sulfur contents than the loose fines. While the oxidation state of the sulfur could not be determined directly, a number of lines of evidence suggested that the surface soils are highly oxidized. The red color is generally regarded as resulting from ferric oxides or hydroxides (Salisbury and Hunt 1969; Adams and McCord 1969). The properties of the magnetic minerals in the soil are more consistent with maghemite (γ-Fe_2O_3) than the less-oxidized magnetite and pyrrhotite (Hargraves et al. 1976), and the results of the Viking biology experiment (see chapter 14) indicate that elemental oxygen is released rapidly from the soil by water. Finally Gooding (1978) showed that weathering of sulfides on Mars should produce sulfates. Moreover, if the sulfur exists as a sulfide, some sulfur should have been detected by the GCMS, and it was not. Thus the sulfur almost certainly exists as a sulfate. The duricrust may form as a result of leaching of soluble sulfate ions from solid particles in the regolith by thin intergranular films of water. The moisture migrates to the surface, where it evaporates into the atmosphere, leaving the soluble salts as a cement (Toulmin et al. 1977). Small amounts of NaCl and some nitrates may also be involved, but this has not been confirmed.

Interpretation of the chemical analyses

The main characteristics of the analyses compared with most terrestrial igneous rocks are high Fe, Mg, and S; low Al; and extremely low K. Because the landers carried no means of determining mineralogic composition, the results are not susceptible to a unique interpretation. The one preferred by the Viking experimenters is that the fines consist largely of Fe-rich smectite clays, with minor amounts of sulfates, carbonates, and oxides, but before exploring this model in detail, some alternatives will be discussed briefly.

The material could consist of igneous minerals in roughly the same proportion as in some primary igneous rock. This, however, is implausible on several grounds. Assuming unaltered rock-forming minerals are present, their proportions can be computed from the chemical analyses. The resulting assemblages do not correspond to any known igneous rocks on Earth or the Moon, nor do they fall near any known eutectics (Toulmin et al. 1977). Al_2O_3 is particularly anomalous. On Earth only very primitive melts, those derived by almost complete fusion of mantle rocks, are as low in Al_2O_3, and these are generally low in SiO_2 and are not quartz-normative like the martian analyses. In addition detection of water by the Viking GCMS, when samples are heated to 200–500°C, indicates that the samples contain significant amounts of chemically bound water, so are unlikely to be entirely primary igneous minerals.

A second model is that the fines are a mixture of oxides and contain few silicates. The main argument for this model is thermodynamic. Most silicates are unstable under present conditions at the martian surface and should break down into oxides and carbonates (Gooding 1978) in the absence of liquid water. It is not known, however, whether the reaction rates are such that significant amounts of the oxides could accumulate in the time available. The homogeneity of the samples argues somewhat against an oxide model. Discrete oxide grains would be susceptible to mechanical fractionation by the wind, yet the analyses at both sites are essentially identical, and the bright areas of the planet and the dust in the atmosphere have similar reflectance spectra (Toon et al. 1977; McCord et al. 1977; Houck et al. 1973; Pollack et al. 1977), indicating that the composition of the wind-blown dust is everywhere similar. The wind has had a homogenizing, rather than a fractionating, effect. Probably a stronger argument against oxides is the detection of water, since this affects the stability relations.

The analyses can also be approximated by mixtures of mafic rocks and some meteorites. Equal parts of carbonaceous chondrites and tholeitic basalt, for example, provide a good match. The proportions of meteorite required, however, are improbably large (Toulmin et al. 1977). For comparison, the materials on the surface of the Moon have less than 1 percent meteoritic material, despite exposure to billions of years of infall.

The presence of montmorillonitic clays was initially suggested by the reflectance spectra. Hunt and co-workers (1973) and Logan and co-workers (1975), on the basis of infrared spectra of dust in the martian atmosphere during the Mariner 9 mission, suggested that montmorillonitic clays were a major component of the dust. They suggested that sulfate could also be present in significant amounts. Aronson and Emslie (1975) showed that feldspars also gave a good fit to the spectral data but acknowledged that mica and montmorillonite were also plausible. Toon and colleagues (1977) found that the dust spectra were best explained by silicates with SiO_2 contents in excess of 60 percent or by weathering products such as clays. They also indicated that some carbonate could be present, but no more than a few percent. Finally, Houck and colleagues (1973) pointed out the presence of an adsorption feature at 2.85 μm, which could be interpreted as caused by bound water, and suggested that the regolith could contain 0.3 percent to 3 percent water by weight.

These considerations and the detection of water in the soil at the two Viking landing sites by the GCMS (Biemann et al. 1977) led the investigators on the Viking inorganic analysis team to consider a significant clay content for the soils analyzed. The interpretation of the chemical analyses favored now (Toulmin et al. 1977) is that the soils are a mixture of smectite clays, with lesser amounts of sulfates, carbonates, and oxides, such as kieserite, calcite, rutile, and maghemite (table 13.2). Such mixtures give

TABLE 13.2. CHEMICAL COMPOSITIONS OF COMPUTER-MODELED MIXTURES COMPARED TO CHRYSE FINES

Item	Composition, wt %			
	Mixture 1	Mixture 2	Mixture 3	S1
Oxide				
SiO_2	55.1	46.0	43.6	44.7
Al_2O_3	8.3	8.0	6.9	5.7
Fe_2O_3	19.5	19.0	18.4	18.2
MgO	10.1	9.6	9.0	8.3
CaO	2.4	2.0	5.6	5.6
K_2O	0.0	0.0	0.0	0.1
TiO_2	0.0	0.0	0.9	0.8
SO_3	0.0	9.4	7.3	7.7
Mineral				
Nontronite ($Fe_2\ Al_{0.5}\ Si_{3.5}\ O_{10}\ (OH)_2\ Ca_{0.25}$)	51	52	47	
Montmorillonite ($Mg_{0.3}\ Al_{1.7}\ Si_4\ O_{10}\ (OH)_2\ Ca_{0.15}$)	19	21	17	
Saponite ($Mg_3\ Al_{0.5}\ Si_{3.5}\ O_{10}\ (OH)_2\ Ca_{0.25}$)	30	13	15	
Kieserite ($MgSO_4.H_2O$)		16	13	
Calcite ($CaCO_3$)			7	
Rutile (TiO_2)			1	

SOURCE: Toulmin et al. 1977.

an excellent fit to the chemical analyses and also explain why the soil gives off water when heated. The dominant mineral in the model is the Fe-rich clay nontronite. Differences between crustified and loose samples are due largely to differences in the amounts of sulfates and halide salts. Madarazzo and Huguenin (1977) alternatively showed that the analyses can be matched with an assemblage consisting of 3–5 percent magnetite, 5–10 percent unweathered basalt, ~13 percent $MgSO_4$, with the rest being weathered picritic (olivine-rich) basalt. However, the precise mineralogy of the weathered products was not determined, and this model may not be greatly different from the smectite clay model, in that the dominant component is an Fe-rich weathering product.

Formation of clay minerals

Several plausible mechanisms exist for the formation of smectite clays on Mars. On Earth smectite clays form mostly by hydrothermal alteration and chemical weathering, in which liquid water plays a crucial role. As we have seen, there is abundant evidence for the availability of liquid water on Mars. The presence of channels, the morphology of impact ejecta blankets, and the occurrence of numerous possible permafrost features all suggest that liquid water could exist below a permafrost layer. In addition the presence of a duricrust at both landing sites implies the presence of intergranular or adsorbed water, in which ionic reactions and transport could take place.

Toulmin and co-workers (1977) pointed out that the clays could have formed by interaction of magma and ground-ice. On Earth lava-ice interaction is commonly accompanied by the formation of a highly oxidized and altered clay-rich volcanic glass, called palagonite. Toulmin and colleagues suggested that on Mars, violent interaction of iron-rich basaltic magma and subterranean ice could give rise to large quantities of palagonite. Clays could form both simultaneously with the glass and later as the glass devitrifies. As Soderblom and Wenner (1978) pointed out, the infrared reflectance spectra of many parts of the martian surface are similar to the spectrum of palagonite, although they are not uniquely diagnostic. Further possible evidence of ice-magma interaction is the existence of some pedestal craters that resemble terrestrial table mountains, which form when lava is erupted into and onto ice (Allen 1979b; Hodges and Moore 1978).

A second potential source of clay minerals is the alteration of rocks by groundwater below the permafrost zone. Such water is likely to be acidic, being charged with CO_2 and possibly SO_2, and so could be an effective weathering agent. Even if the process is slow, billions of years are available for the reactions to occur. The groundwater may be primitive, being introduced into the circulating system from below, or it may have percolated down from the surface before the permafrost formed (Carr 1979). Hydrothermal alteration may also occur locally around discrete volcanic centers. The altered products must be brought to the surface by some mechanism, however, if they are to be moved around the surface by dust storms. Inclusion of clays in the water that is released to form the vast flood features is an obvious way. Indeed the liquefaction model for formation of large channels (Nummedal 1978) requires that large amounts of clay minerals be carried along with the floods (see chapter 10). Impact can also excavate to below the base of the permafrost and bring weathered products to the surface.

The impact process itself provides another way of producing clay minerals. Direct chemical interaction of materials during the impact event is unlikely to produce significant quantities of clay minerals, even if water or ice is present, because the time during which mineralogic transformations can take place is so short (Kieffer and Simonds 1979). Hydrothermal alteration of impact melt sheets can produce clays, however. Smectite clays within brecciated rocks at Clearwater Lake, Quebec, for example, have been ascribed to hydrothermal alteration of glass and mafic minerals (Phinney et al. 1978). Newsome (1980) postulated that brecciation caused by impact would allow easy access of groundwater to the hot melt sheets, which would rapidly devitrify to hydrated minerals. The process would be particularly effective early in the history of the planet when impact rates were high.

A fourth possible source of clay minerals are ultraviolet-induced reactions at the surface. Huguenin (1974) postulated that a monolayer of water could form on materials exposed at the martian surface, and that ultraviolet radiation would stimulate water-silicate reactions to produce clay minerals. These would be eroded away by the wind to expose fresh surfaces for further alteration. Other workers (Morris and Lauer 1979) have failed to reproduce Huguenin's experimental results, however, and the efficacy of this mechanism is in doubt. Nevertheless it is clear that several mechanisms could account for the presence of clays on the surface, despite their being in disequilibrium with present surface conditions.

Chemistry of primary igneous rocks

The overall composition of the analyzed samples suggests derivation by weathering of mafic (magnesium- and iron-rich) rocks such as basalts, rather than more highly differentiated sialic (silicon- and aluminum-rich) rocks such as granites. The low K_2O and Al_2O_3 contents are particularly suggestive. It is unlikely that chemical or mechanical processes during weathering and erosion would selectively fractionate K and Al together. K tends to move with the soluble fraction, whereas Al forms resistates (Toulmin et al. 1977). More probably K and Al are low because the surface rocks from which the weathering products were derived are low in K and Al. Toulmin and co-workers (1977) estimated that even 10 percent-granitic rocks at the surface would yield K_2O values several times the limits set by their data. They therefore concluded that highly differentiated alkali-rich, aluminum-rich rocks such as granite are rare.

Mafic or ultramafic rocks are also suggested by the reflectance data, although most of these data are frustratingly unspecific. The spectra of the light areas resemble those of the dust in the atmosphere and are dominated by strong Fe^{3+} absorption bands between the ultraviolet and 0.75 μm. It is this absorption that gives Mars its characteristic red color. Spectra from the darker areas are less dominated by Fe^{3+} absorption below 0.75 μm and may be more indicative of unaltered rocks. Soderblom and colleagues (1978) showed that significant spectral variability exists within areas of low albedo. They suggested that the variations are due to the different rocks exposed at the surface and not to different amounts of eolian debris, because the differences correlate well with topography. In the dark parts of the old cratered terrain, for example, crater rims tend to be more red than the intervening volcanic plains. Some of the reddest and darkest features of the planet occur on the large volcanoes in Tharsis. A weathered component, however, is probably still present in most areas. Singer and co-workers (1979) attempted to correct the observed spectra of dark areas for the effects of light weathered debris by stripping out various fractions of the typical bright spectra. This process brings out details in the dark spectra, particularly above 0.8 μm. Of special importance is an Fe^{2+} absorption band near 1 μm, which is diagnostic of mafic materials such as pyroxene and possibly olivine. Singer and colleagues suggested that a 20–30 percent contamination of dark areas by bright material may be typical.

Huguenin and colleagues (1978), in a similar study, showed that there was significant spectral variability within the dark areas (mainly in the 0.5–2.0 μm range). Some of the variability could be attributed to different proportions of bright materials, but other differences, they suggested, were intrinsic to the dark spectra. They attributed these characteristics to mixtures of orthopyroxene (or pigeonite), clinopyroxene, and olivine (or glass) and suggested that the differences from area to area were due to differences in the proportions of these components. The assemblages proposed are those typical of mafic and ultramafic rocks.

Iron-rich basic (45–52 percent SiO_2) and ultrabasic (< 45 percent SiO_2) rocks are expected from petrologic considerations. McGetchin and Smyth (1978) suggested that the mantle of Mars is more Fe-rich than that of Earth and that it should result in iron-rich basic and ultrabasic rocks at the surface. Their argument hinged largely on an inferred density of 3.55 gm/cm^3 for the martian mantle, which is derived from the moment of inertia and modeling of the interior density distribution (Johnston and Toksoz 1977). The inferred density is significantly higher than that for Earth's mantle (3.3–3.4 gm/cm^3). Most of the iron in Earth is within the core. Mars appears to have a much smaller core proportionately (Johnston and Toksoz 1977), so that a larger fraction of its iron is distributed throughout the rest of the planet. Such reasoning led McGetchin and Smyth to suggest that the composition of Mars's mantle can be computed by adding sufficient FeO to the inferred pyrolitic composition of Earth's mantle (Green and Ringwood 1963) to bring the density up to 3.55 gm/cm^3. They then examined the mineral assemblage and phase relations expected of such a composition. Their conclusion was that the mantle is likely to produce ultrabasic melts with low Si contents (41–44 percent), high Fe contents, and extremely low viscosities. They also suggested that if Mars is volatile-rich, then abundant ultrabasic pyroclastic eruptions could occur through diatremes, much in the style of kimberlite eruptions on Earth.

Mafic melts are also consistent with the observed volcanic features. As we saw in chapter 6, most of the plains regions of the planet, such as Lunae Planum and Syria Planum, appear to be volcanic and the result of emplacement of fluid lava. In addition most of the large volcanoes appear to have formed from lavas with fluidities comparable to those that form the large shield volcanoes on Earth. Lavas as fluid or more fluid than tholeiitic basalts are implied in most cases. Such low viscosities are most easily explained by iron-rich basaltic magmas. There are exceptions, and more silica-rich melts may have erupted locally, but the vast majority of the surfaces that are demonstrably volcanic give indications of having formed from fluid lavas. There is thus a convergence of evidence from several sources—from direct analyses of the weathered products, from interpretation of reflection spectra, and from petrologic and geomorphic considerations—that the predominant primary igneous rocks on the martian surface are Fe-rich and Si-poor. The most probable candidates are olivine-rich basalts, with olivine being rich in the Fe-rich end member fayalite, although the reflection spectra suggest that there may be considerable variation in the proportions of the constituent mafic minerals. Although intermediate and acidic rocks cannot be excluded, all the evidence suggests that they constitute only a small fraction of the total rocks exposed.

NEAR-SURFACE STORAGE OF CARBON DIOXIDE AND WATER

Several authors, most notably Fanale and Cannon (1974, 1979) and Fanale (1976), have pointed to the importance of the regolith as a sink for volatiles, especially CO_2 and H_2O. In chapter 3 we saw from several lines of evidence that the present atmosphere represents only a fraction of the volatiles that have outgassed from the planet. Some of the volatiles have been lost by exospheric processes at the top of the atmosphere; others must exist either within interstices, adsorbed on particles, or chemically bound within the surface rocks or regolith. Particularly significant is the lack of any observable fractionation of the oxygen isotopes despite enormous enrichment of heavy nitrogen (^{15}N). This implies that the oxygen currently in the atmosphere constitutes only part of the total, which is subject to isotopic modification by exospheric processes. The rest must exist close to the surface, where it can interact dynamically with the atmosphere and so dilute the fractionation effects. The most plausible forms are CO_2 or H_2O, which could be added to, or removed from, the atmosphere as climatic conditions change. The exchange cannot be restricted to some more clement episode in the planet's distant past; it must be either continuous or recent, otherwise fractionation would be detected.

The storage capacity of the regolith depends on its thickness and structure and may be quite different for H_2O and CO_2, in that most of the CO_2 may be adsorbed, while most of the H_2O may be interstitial or chemically bound. The term regolith refers rather loosely to a layer of fragmental debris close to the surface. The thickest regolith was produced early in the planet's history, when the surface was deeply "gardened" by intense meteorite bombardment to produce a "megaregolith" several kilometers thick. On terrain that survives from this time there is probably a continuous transition from a fine-grained soil at the surface through coarse fragmental debris at slightly greater depths to fractured rocks roughly in place to largely undeformed rocks below. The brecciated zone is probably porous, with its thickness limited by annealing and closure of pore spaces at depth, as a result of self-compression and volcanic processes. On the Moon seismic evidence (Latham et al. 1971) indicates that the brecciated zone extends to depths of about 25 km. The depth for Mars is almost certainly shallower (~10 km) because of higher gravity and possibly higher heat flow (Carr 1979). The thickness of unconsolidated materials near the surface is probably considerably less than that of the brecciated zone, but thicknesses of 1–1.5 km are plausible (Fanale and Cannon 1974). The simple picture just outlined is locally complicated by deposits which postdate the early high impact flux. Such deposits develop their own regolith, but this is likely to be only a few meters thick, like the regolith on the lunar maria. The old thick megaregolith, however, must exist at depths below the younger deposits, except in those, probably rare, cases, such as beneath large volcanoes, where the regolith may have been destroyed by metamorphism. Another complication is that parts of equatorial regions may be largely swept free of fines, whereas the polar regions may be sites of preferential accumulation.

The capacity of the regolith to hold volatiles must vary greatly with depth. Below one or two kilometers the regolith probably consists largely of unweathered coarse-grained breccias with a limited adsorption capacity. It may still be porous, however, and permit the presence of interstitial ice or groundwater. The near-surface layers, on the other hand, may contain large fractions of weathered debris capable of efficiently adsorbing atmospheric gases but be largely devoid of interstitial water because of disequilibrium with the atmosphere.

CO_2 adsorption in the regolith

Fanale and Cannon (1974, 1979) showed that the capacity of nontronite and finely ground basalt to adsorb CO_2 depends on temperature and the partial pressure of the CO_2 it is in contact with. Adsorption increases with both higher CO_2 pressures and lower temperatures (figure 13.2). Nontronite at 158°K under 6.1 mb CO_2 pressure, for example, can adsorb 11 cm^3 CO_2/gm, measured at standard temperature and pressure (STP), whereas it can adsorb only 2.5 cm^3 at 196°K. The equivalent figures at 10 mb are 15 cm^3 and 3.5 cm^3. Ground basalt has approximately one tenth the adsorption capacity of nontronite. Fanale and Cannon (1979) ascribed the high adsorptive capacity of nontronite and the strong dependence on temperature to: (1) the high free surface area characteristic of clays, (2)

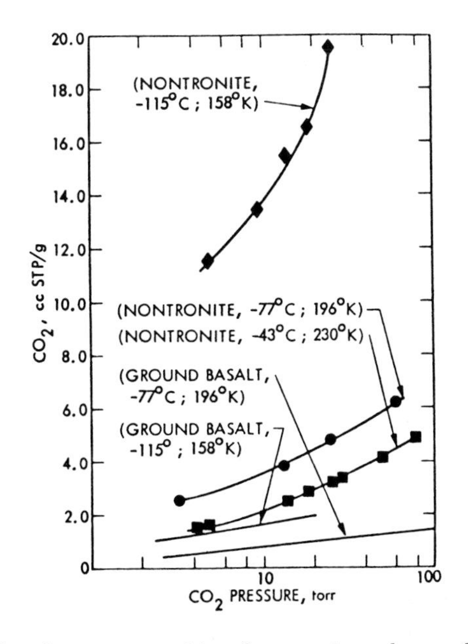

Figure 13.2. The adsorptive capacities of nontronite and ground basalt for CO_2 at different temperatures and pressures. The adsorptive capacities increase with CO_2 pressures and lower temperatures. (Reproduced with permission from Fanale and Cannon 1979, copyrighted by the American Geophysical Union.)

partial penetration of CO_2, particularly at low temperatures, into spaces between sheets in the atomic structure that are normally occupied by water molecules, and (3) "capillary" condensation of CO_2 in micropores within the clays. A small amount of CO_2 may also exist as carbonates within the soil.

The dependence of adsorption on temperature must result in exchange of CO_2 between the atmosphere and the regolith. As surface temperatures fall, atmospheric CO_2 will tend to be adsorbed into the regolith; as temperatures rise, the CO_2 will tend to be driven off. The depth of the regolith affected will depend on the time scale of the temperature changes. Diurnal temperature fluctuations affect only the top few centimeters, annual changes approximately the top meter. The exact values depend on the thermal inertias and thermal conductivities of the materials involved (Fanale and Cannon 1974). Long-term climatic changes can penetrate to much greater depths. Fanale and Cannon estimated that changes in the mean annual temperature due to precession variations and to changes in obliquity and eccentricity, which have time scales of 5×10^4 years to 10^6 years, should penetrate to depths of a few hundred meters to one kilometer. The effects on adsorption, however, should be greatest close to the surface, where temperature fluctuations are largest, and the regolith has the smallest grain size and the largest proportion of weathered products.

Fanale and Cannon (1979) modeled the effects of long-term climate changes on the adsorption and desorption of CO_2 on the basis of a mean regolith thickness of 150 m. They suggested that the regolith could be thought of in terms of three zones: a polar zone with a mean annual temperature of 158°K, an intermediate latitude zone at 196°K, and an equatorial zone at 233°K. They assumed on geologic grounds that the regolith is thicker at the poles than in the equatorial regions. A regolith of pure nontronite, they calculated, would contain 390 gm CO_2 per cm^2 of the martian surface, or the equivalent of an atmospheric pressure of 144 mb. This is probably an upper limit, for a 150-m regolith composed entirely of montmorillonitic clay is unlikely. More probably there is a transition from a highly adsorbing, dominantly clay regolith at the surface to a less adsorbing regolith of less altered rock fragments at greater depths. Nevertheless it is clear that the regolith can potentially adsorb many times the amount of CO_2 that is currently in the atmosphere. Some CO_2 is also condensed in the remnant south polar cap, but the cap covers only 0.01 percent of the planet's surface and, even if very thick, would constitute only a fraction of the adsorbed inventory.

Interchange of CO_2 between atmosphere and regolith

The CO_2 near the surface is distributed three ways, between the atmosphere, the cap, and the regolith, with most of it being in the regolith. The balance between the three components depends on the global temperature distribution, and any changes will result in interchange of CO_2 between the three sinks. The most significant long-term changes are those that result from changes in obliquity, because such changes strongly affect temperatures in the polar regions, where most of the CO_2 is adsorbed. Seasonal temperature changes affect only a thin skin; changes due to precession are mostly between opposing hemispheres, not between poles and equator; and eccentricity changes mainly affect the magnitude of the hemispheric differences caused by precession. Mars is currently close to the middle of its obliquity range. At maximum obliquity, temperatures at the poles would be approximately 15°C higher than at present; equatorial temperatures would be a few degrees lower. At minimum obliquity polar temperatures would be 15°C lower than at present, and equatorial temperatures would be slightly higher. The changes take place with a period of 1.2×10^5 years (Ward et al. 1974). In response CO_2 is exchanged between polar and equatorial regoliths, and a new equilibrium is established with the atmosphere.

Fanale and Cannon (1979) showed that for a nontronite regolith, an increase of temperature of 15°C at the pole would result in an increase in atmospheric pressure to 21 mb and a 10 percent decrease in the total CO_2 in the regolith. The depletion is all at high latitudes; at the equator there is a slight increase in the adsorbed CO_2. Conversely, if the obliquity decreases, and the polar temperatures fall 15°C, then CO_2 is desorbed near the equator and resorbed at the poles. In addition the permanent CO_2 caps enlarge, taking up significant amounts of CO_2 and reducing the atmospheric pressure to 0.5 mb or less. So despite buffering of the atmosphere by both regolith and caps, the atmospheric pressure may swing from 0.5 mb to 22 mb as a consequence of obliquity changes. The high-pressure values are maxima, being based on nontronite models. A regolith of ground basalt results in smaller variations. The top 150 m of the regolith is thus subject to a slow flushing action by CO_2, on a cycle of approximately 10^5 years. The slow interchange may in part explain why the oxygen in the atmosphere shows little enrichment in O^{18}, as anticipated from considerations of rates of exospheric escape. The enrichment has been diluted by the large regolith reservoir. Although interchange of oxygen with the atmosphere is largely effected by the movement of CO_2, a large fraction of the oxygen in the regolith may be combined as water, which is not as mobile. Fanale and Cannon therefore suggested that the CO_2 may act as a carrier for oxygen and ensure that all the oxygen, combined as volatiles within the regolith, maintains isotopic identity with the atmosphere. The flushing action may also affect the stability of water near the surface, facilitating migration of water and establishment of thermodynamic equilibrium with the atmosphere.

Near-surface water

Although little water is present in the martian atmosphere, and stability relations forbid the presence of liquid water or ice near the surface at most latitudes, there is persuasive geologic evidence that abundant water (or water-ice) exists near the surface. As we saw in chapter 4, the peculiar morphology of martian impact craters is most plausibly explained by incorporation of water in the ejecta. Channel networks suggest slow erosion by running water, and large channels suggest episodic eruption of artesian water or fluidization of near-surface materials. Formation of collapsed ground, large landslides, closed depressions, debris flows, and fretted terrain are all more comprehensible if liquid water or ice is involved.

Geochemical arguments as to the amount of water outgassed from the planet are based on two main lines of evidence: first, the isotopic abundances in the martian atmosphere, and second, reconstruction of the total volatile inventory from a comparison of the composition of the martian atmosphere with the compositions of the atmospheres of other planets and meteorites. The subject was discussed fully in chapter 3, but the conclusions will be summarized briefly here. Mars's atmosphere is strongly enriched in ^{15}N. To achieve the observed enrichment, McElroy and coworkers (1977) estimated that the initial pressure of nitrogen 4.5 billion years ago must have been 10 to 150 times its present value. Assuming H/N ratios similar to those on Earth, they concluded that the amount of water evolved can range as high as the equivalent of 200 m distributed over the entire surface of the planet. In comparison the equivalent value for Earth is around 3 km (Turekian and Clark 1975). The oxygen in the martian atmosphere shows no detectable enrichment (< 5 percent) in the heavy isotope ^{18}O, despite preferential exospheric escape of ^{16}O. McElroy and colleagues attributed the lack of detection to dilution of the anomaly by interchange of atmospheric oxygen with a surface reservoir, probably mostly water, and concluded that a minimum of 13.5 m of water spread over the entire planet is needed to hide the light isotope loss. From cosmologic considerations Pollack and Black (1979) estimated that the amount of water outgassed from Mars is equivalent to a layer 80–160 m deep. Thus, the isotopic and the recent geochemical data are somewhat convergent in suggesting surface water inventories of the order of 100 m/unit area. The estimates, however, are still tentative. The isotopic data give only lower limits, and the cosmologic arguments are rather speculative. It is interesting to note that if Mars had outgassed 100 m/unit area of water and had a CO_2/H_2O outgassing ratio similar to Earth's, then it would have a CO_2 inventory of approximately 500 gm/cm^2, which is similar to the inventory of adsorbed CO_2 estimated by Fanale and Cannon (1979).

Before discussing where the water might be, we will examine briefly the stability relations of water under conditions that prevail at the martian surface. The amount of water vapor in the atmosphere in the equatorial regions ranges from 5–15 pr μm (Farmer and Doms 1979), with an average of 12 pr μm. For this amount in a perfectly mixed atmosphere, the frost-point temperature—the temperature at which ice will start to precipitate out—is 198°K. At temperatures higher than this ice is unstable with respect to the atmosphere. A block of ice placed on the surface at a temperature of, say, 230°K (saturation vapor pressure equivalent to 2,300 μm H_2O) will start to sublime and will continue to do so until the vapor pressure in contact with the ice reaches the saturation vapor pressure for that temperature. Since diffusion and convection will carry water vapor away from the immediate vicinity of the block, it will ultimately evaporate. Similarly ice buried within the soil will tend to sublime if the mean annual temperature is above the frost point and if water vapor can diffuse through the soil to the atmosphere. Farmer and Doms (1979) examined the stability of water in martian soil down to depths of 10 m (figure 13.3). They showed that at latitudes within 20° of the equator, near-surface temperatures are always above the frost point, so that ice can never be present in equilibrium with the atmosphere. The soil in these latitudes is therefore devoid of free ice (and water), at least to depths that maintain diffusive contact with the atmosphere. Water can still exist if chemically bound or strongly adsorbed, as demonstrated by the GCMS results (Biemann et al. 1977). At latitudes higher than 40° parts of the soil never reach the frost point at any time during the year, so can contain permanent ice.

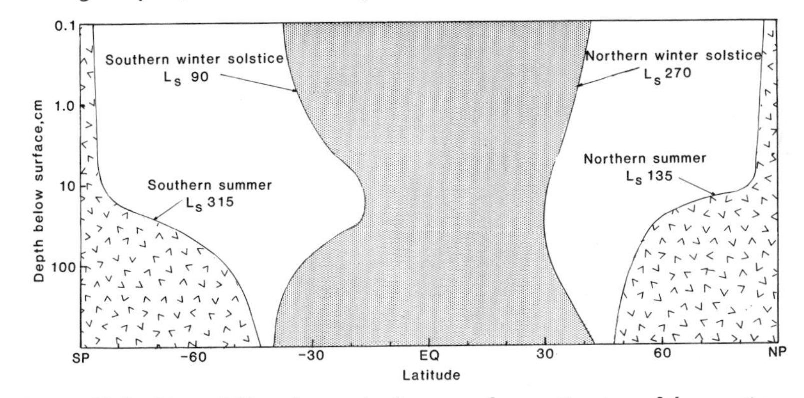

Figure 13.3. The stability of water in the upper few centimeters of the martian surface as a function of latitude. In the equatorial regions that are shaded in the diagram, water is never stable with respect to a well-mixed atmosphere containing 12 pr μm of water. In the polar regions that are shaded, water is always stable as ice. In the intervening latitudes, water-ice is stable for parts of the year. (Adapted from Farmer and Doms 1979.)

The upper boundary of the zone containing permanent ice intersects the surface at the edge of the permanent ice cap. Between the region of permanently frozen soils at the poles and the zone of permanently dehydrated soils (or soils in permanent disequilibrium) at the equator are regions where ice can exist in the soil for parts of the year in equilibrium with an atmosphere containing 12 pr μm of water.

At depths greater than 10 m, the temperature profile is controlled largely by internal heat, although it is cyclically perturbed a few degrees to depths of a kilometer or so by long-term climatic changes. Assuming a thermal gradient of 40°C/km, based on a chondritic Mars and a thermal conductivity of 8×10^4 erg.cm/sec/°K and perfect diffusive contact with the atmosphere, Fanale (1976) showed that permanent ice could exist in equilibrium at the poles to depths of 1 km, and that the base of the equilibrated zone gets shallower toward the equator, such that the zone narrows to zero at around the 40° latitude (figure 13.4). At greater depths the 273° isotherm is encountered, and liquid water can exist. Fanale estimated that the 273° isotherm should be encountered at a depth of about 1 km at the equator, at 2 km at the 40° latitude, and at considerably greater depths at higher latitudes. Thus a 1-km–thick permafrost zone exists at the equator, which expands to over several kilometers thick at the pole. It should be noted in passing that "permafrost" means simply that temperatures are below freezing; the permafrost zone could be completely devoid of ice, although this is unlikely. If water exists below the permafrost as a brine charged with carbonates, sulfates, chlorides, and other salts, then the freezing temperature would be less than 273°, and the depth to the base of the permafrost would be correspondingly shallower.

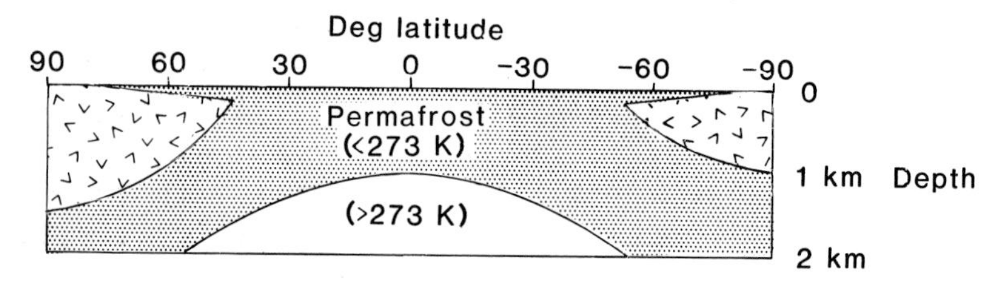

Figure 13.4. The stability of water at different depths (given in km) as a function of latitude. Water-ice is stable with respect to the atmosphere in lenslike regions under the poles. The 273° isotherm indicates the depth below which liquid water might be found. In the shaded permafrost areas, any water that might be present is frozen. (Adapted from Fanale 1976.)

The stability relations just described are somewhat idealized. The actual situation is likely to be complicated. First the water vapor content of the atmosphere is not a constant 12 pr μm; it varies, and as it varies, so does the frost-point temperature and the stability of water in the soil. This, however, is a relatively minor effect, unless the atmosphere was significantly warmer in the past and contained substantially more water. Second and more important, the soil is not necessarily in instantaneous diffusive contact with the atmosphere at all depths. Ice could exist near the surface in the equatorial latitudes at temperatures above the frost point, if water vapor were prevented from diffusing to the surface. Water-vapor pressure could then build up locally within the soil until it was in equilibrium with the ice. Smoluchowski (1968) calculated that, because of slow diffusion rates of water vapor through fine-grained compacted soils, ice could exist for billions of years at relatively shallow depths on Mars. For example, he estimated that for a grain size of 0.5 μm, a porosity of 0.01, and a surface temperature of 220°K, a soil layer 100 cm thick could provide such an effective diffusion barrier that ice could exist below it for 10^9 years. These calculations were based on a simple diffusion model, however, and do not take into account the flushing action of CO_2 (and H_2O) as it is adsorbed and desorbed from the regolith. As we saw above, such an interchange can take place up to a depth of a kilometer in response to long-term climate changes and is more efficient than if water alone were involved, in that diffusion would be driven by much larger pressure gradients. Even so achievement of equilibrium may be sluggish enough to preserve ice in the equatorial permafrost zone for geologically significant periods of time.

Present location of water

We have just seen that about 100 m of water may have outgassed per unit area of the planet. If so, where is the water now? There are several possible sinks in which it could reside:

1. *Permanent caps.* The amount of water stored in the permanent north polar cap is probably small. Assuming an average thickness of 100 m out to the 80° latitude, this constitutes less than 1 m of water distributed over the whole planet. Even if the ice were 1 km thick over this area, which it clearly is not, the permanent cap would be equivalent to only 8 m.

2. *Layered terrain.* The terrain extends from the poles down to 80° latitude at both poles. Estimates of the thickness range up to 4–5 km (Dzurisin and Blasius 1975). The uncertainty regarding the layered terrain is in the fraction of volatiles present. Most researchers (Howard 1978; Cutts 1973a; Cutts et al. 1979; Murray et al. 1972; Sharp 1974) have implied that the layers are comprised mostly of eolian rock debris, with only small amounts of volatiles as cement. Assuming a 2-km thickness, the total volume of the terrain is 4.4×10^6 km^3. If the deposits contain 10 percent water, then this is equivalent to 3 m per unit area of the planet's surface.

3. *Polar ice lenses.* The ground beneath the poles, in which the ice is in equilibrium with the atmosphere, could contain permanent ice. Assuming that they extend to a depth of 1 km below the poles and thin to zero at 40° latitude (Fanale 1976), the lenses each comprise a volume of 1.7×10^7 km^3. We have no way of knowing how much ice, if any, is present within this volume, but to give some idea of the storage capacity, if the lenses contained 10 percent ice, then the amount of water they contained would be equivalent to 23 m over the whole planet. Although permafrost on Earth can locally contain considerably more than 10 percent ice, 10 percent to a depth of 1 km probably represents a reasonable upper bound.

4. *Clay minerals.* The regolith analyzed at both landing sites contained 0.3–3 percent of water. The total amount of water stored in the clays depends on the depth of the clay-rich regolith, which is not known. Outcrops of bedrock at the Viking 1 landing site suggest that the regolith there is very thin. Nevertheless, as we saw earlier in this chapter, a thicker, older regolith may exist at depth beneath the veneer of young lava flows. Whether this consists predominantly of clays, however, is not known. Furthermore, considerably thicker regolith (debris mantles) may be present at high latitudes. If we arbitrarily assume a 100-m–thick, clay-rich regolith everywhere, with 3 percent water, then this would account for another 3 m of the outgassed water.

5. *Megaregolith.* Although clay minerals may be restricted to near the surface, a deeper megaregolith of pulverized rock is likely to extend to

greater depths. Fanale (1976) estimated that ground basalt could adsorb approximately 1 percent of water under typical martian conditions; this would decrease with depth as the temperature increased. However, a 1-km–thick megaregolith containing 1 percent of water would account for another 10 m of water.

6. *Ground-ice.* As indicated above, permanently frozen ground extends from the surface down to depths of one to several kilometers, depending on latitude. Ground-ice, however, is stable with respect to the present atmosphere only in parts of this permafrost layer at latitudes higher than 40°. Ground-ice can exist in the rest of the permafrost zone only if isolated from the atmosphere. The upper 1 km may be purged of ice as a result of adsorption and desorption in response to climate changes, although the efficacy of this process is not known. Ground-ice out of equilibrium with the atmosphere is more likely at greater depths, where the barrier between it and the atmosphere is thicker. Ice cannot exist, however, below the 273° isotherm. Except for the stable ice lenses near the poles ground-ice is thus most likely to be encountered at depths greater than several tens of meters where isolated from the atmosphere and at depths shallower than 1–2 km where it is above the 273° isotherm. No realistic estimates can be made of the water tied up in this way.

7. *Groundwater.* Liquid water can exist below the permafrost layer. The capacity of deeper layers to hold water is large. If the porosity is 1 percent at 1 km, and if the porosity decreases linearly to zero at 10 km, then the entire artesian system could hold 450 m of water. As I recently pointed out (Carr 1979), however, the aquifer system is unlikely to be fully charged. More probably groundwater accumulates in low areas, from which it may have been released episodically to form the large flood features. Again realistic estimates of the amount of water stored below the permafrost layer cannot be made.

From the above discussion, it appears that the equivalent of about 30 m of water per unit area of the planet could be tied up in the poles, either in the permanent cap, the layered terrain, or the ground-ice lenses. Another 15 m could be adsorbed or chemically bound within the near-surface regolith over the rest of the planet. The rest of the 100 m estimated to have outgassed is probably in the deeper (> 500 m) layers, as ground-ice and groundwater. Such a distribution is consistent with most of what we know of the surface geology. All the above numbers, however, should be viewed with considerable skepticism. Estimates of the amount of water outgassed are poorly constrained. The isotopic data give only a lower limit, and the cosmological arguments are speculative. Considerably more water than the nominal 100 m per unit area could have outgassed, leading to underestimation of the amount of groundwater and ground-ice present.

SUMMARY

The fine-grained debris on Mars's surface has about the same composition everywhere. With respect to most terrestrial rocks, the debris is high in Fe, Mg, and S, and low in Al and K. The only differences from site to site and sample to sample are the amounts of S (and possibly Cl). S is especially high in samples of duricrust. The favored interpretation is that the debris consists largely of iron-rich clays with lesser amounts of $MgSO_4$ and possibly $CaCO_3$. The clays could have formed by a variety of processes, such as hydrothermal alteration, interaction of magma with ground-ice or groundwater, slow weathering by groundwater, or ultraviolet-stimulated reactions at the surface. Their composition suggests that they originated primarily from iron-rich basic or ultrabasic rocks.

Large amounts of H_2O and CO_2 may be stored in the near-surface rocks or regolith. Adsorption of CO_2 on rock debris or clay minerals depends on temperature. The effects of long-term climate changes could penetrate to depths on the order of 1 km, causing significant adsorption and desorption of CO_2 and changes in atmospheric pressure. Geochemical arguments suggest that the equivalent of around 100 m of water, averaged over the whole surface, has outgassed from the planet. Because of the thin atmosphere water and water-ice are unstable near the surface everywhere except at the poles. Water-ice, however, can exist at relatively shallow depths (a few tens of centimeters) at latitudes greater than 40°. Any water or water-ice that exists near the surface at lower latitudes must be diffusively isolated from the atmosphere. Only a small fraction of the total water outgassed is represented by that currently in the atmosphere or at the poles. Other possible sinks are chemically bound or adsorbed water in the regolith, ground-ice, and groundwater.

14 THE SEARCH FOR LIFE ON MARS

Harold P. Klein

Modern scientific hypotheses concerning the origin of life generally have as their central theme the idea that living systems arise through "chemical evolution," a process in which simple compounds are generally transformed under the influence of various energy sources into more and more complex molecules, ultimately resulting in a system of replicating molecules (Buvet and Ponnamperuma 1971; Miller and Orgel 1974). Based upon comparisons between the chemical composition of terrestrial living systems and cosmic abundances (Frieden 1972) as well as upon theoretical considerations (Wald 1962), it is further assumed that the key substances in this evolution are all carbon-based compounds. Given a planet with a "hospitable" environment, these processes continue until the stage is set for biological infestation of the planet and further biological evolution follows.

As it became clear that direct experimentation on other planets was technically feasible with the advent of spacecraft technology, attention turned to Mars as the most promising extraterrestrial object upon which to search for evidence of chemical evolution (Pittendrigh et al. 1966). This view was reached despite significant gaps in our knowledge of that planet, most notably about the properties and composition of its surface. Even after the encounters by Mariners 4, 6, and 7 and the more encompassing Mariner 9 mission, virtually nothing was known about the chemistry of the martian surface. Under these circumstances, the status of evolution of carbon compounds and particularly the question of the existence of life on Mars were entirely speculative.

It should not be surprising therefore that, as planning for the Viking mission began, broad approaches to these problems were incorporated into the scientific payload. For elucidating the nature of carbon-containing materials, the combination of gas chromatography, followed by mass spectrometry, offered a survey technique of extreme sensitivity that could, in principle, identify a wide range of compounds (Biemann 1974). The strategy for detecting living systems has been discussed by numerous authors (e.g., Lederberg 1965; Young et al. 1965; Bruch 1966) and the general conclusion from these considerations was that *metabolic* experiments were likely to be the most useful in the initial stages of exploration. But it was also argued that no single technique would be adequate for the search for life on Mars in the face of all of the uncertainty about that planet. Accordingly, a Viking biological payload was selected that contained four separate elements, each based on different assumptions about the nature of martian biology (figure 14.1) (Klein et al. 1972). (Subsequently, during the development of the flight hardware, one of these experiments was eliminated in order to reduce costs.) On the chance

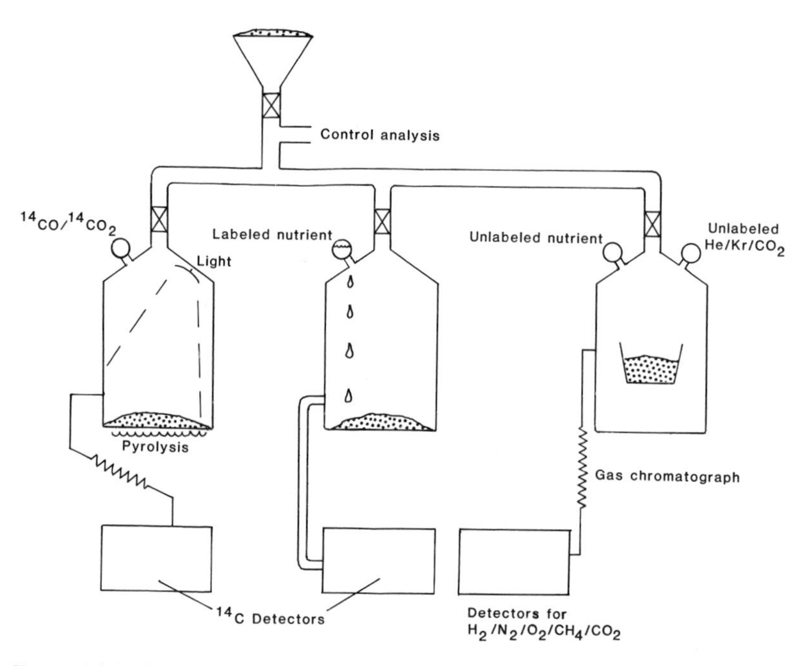

Figure 14.1. Schematic diagram of the three biology experiments on the Viking landers. In the Pyrolytic Release (PR) experiment (*on the left*) soil is incubated under a light in the presence of "labeled" CO and CO_2. After incubation the gases are flushed out, and the soil is heated to see if any of the labeled carbon was metabolized to form complex molecules. In the Labeled Release (LR) experiment labeled nutrient is added to the soil, then the gases are monitored to see if breakdown of the nutrient components occurs. In the Gas Exchange (GEX) experiment changes in the composition of gases in contact with the soil and nutrients are followed.

This chapter was written by Dr. Harold P. Klein at the Directorate of Life Sciences, NASA–Ames Research Center, Moffett Field, California. It is reprinted from *Reviews of Geophysics and Space Physics* 17 (1979): 1655–62, by permission of the American Geophysical Union.

(Sagan and Lederberg 1976) that larger, more complex organisms (than the microorganisms being tested for by the biology experiments) might be present on Mars, the Viking lander cameras with their capability to obtain detailed panoramic images of the local scene over the course of the martian seasons (Mutch et al. 1972), were to provide primary information on this question. In addition to these major investigations which directly addressed the status of organic evolution of Mars, many of the other Viking investigations, notably the mass spectroscopic analyses of the atmosphere (Nier et al. 1972; Anderson et al. 1972) and the analyses of the surface material for elemental composition (Toulmin et al. 1973), were to be important in defining the local environments at the lander sites.

RESULTS FROM DIRECT CHEMICAL ANALYSES

Before the descent of the first Viking lander, one of the open questions regarding the prospects for chemical evolution on Mars was whether—and in what form—nitrogen was present. Earlier Mariner missions had not detected N-containing compounds, using ultraviolet spectrometers with a sensitivity limit in the range of a few percent of nitrogen (Barth 1974), and a total absence of nitrogen would have severely strained optimism about the prospects for organic chemical evolution on Mars, based on current ideas about this process (Miller and Orgel 1974).

Nitrogen was, however, readily detected in the atmosphere by a number of Viking instruments, both in the upper atmosphere (Nier et al. 1976; Nier and McElroy 1976, 1977) and at the surface (Biemann et al. 1976a; Owen and Biemann 1976; Owen et al. 1977; Oyama and Berdahl 1977), using mass spectrometry as well as gas chromatography. All of these techniques converged on surface atmospheric concentrations in the range of 2.5 percent nitrogen, thus easing one important concern about the prospects for complex organic chemistry on Mars. None of the analytical techniques aboard the Viking spacecraft, however, was designed to provide any information about N-containing inorganic constituents and, since no organic compounds were detected in the samples tested (Biemann et al. 1977), there exists no direct information about nitrogen compounds in the surface material of Mars.

The mineralogical models for the martian soil that have been derived from the Viking inorganic analysis experiments (see chapter 13) would in no way be inconsistent with the presence, or even participation, of these materials in the process of chemical organic evolution. Indeed, clay minerals of the type proposed have been found to provide effective catalytic properties for the formation of complex organic compounds in laboratory experiments (Paecht-Horowitz 1971; Lawless et al. 1977; Hubbard 1979), and such materials actively adsorb microorganisms and thus appear to be compatible with large microbial populations on Earth (Müller and Hickisch 1972).

The direct search for organic compounds on Mars failed to reveal the presence of any (Biemann et al. 1976b; Biemann et al. 1977), using instrumentation that was, in principle, capable of detecting virtually all organic compounds, except for highly polymerized kerogenlike matter. Tests using the Viking instrumentation with a large number of known compounds showed the sensitivity limit of the instrument to be at levels of parts per billion or less for compounds containing three or more carbon atoms and in the ppm range for 1 and 2 carbon compounds (Biemann et al. 1977).

RESULTS FROM LANDER IMAGING

The capabilities of the Viking lander camera system in acquiring evidence of biology on Mars have been discussed by Mutch and Levinthal and their colleagues (Mutch et al. 1972; Levinthal et al. 1977). Assuming that at least some organisms in the immediate vicinity of the landers were larger than a few millimeters in size and that their forms (like those of most terrestrial organisms) were discriminable from the surrounding surface objects (like rocks, pebbles, sand dunes, etc.), the Viking cameras should have detected them if they were present. In the many thousands of pictures obtained at the two landing sites, no evidence has yet been obtained revealing any morphological entities that are suggestive of living or fossilized objects (Levinthal et al. 1977); nor have the cameras witnessed any colored areas or color changes in the martian panoramas (as the seasons progressed on Mars) that would suggest biology. Several direct and indirect approaches to the detection of movement also failed to reveal any motion that might be attributable to biological entities (Levinthal et al. 1977).

RESULTS FROM BIOLOGICAL INVESTIGATIONS

The Viking biological investigation was predicated on searching for evidence for metabolism on the basis that metabolic processes are sufficiently "improbable" to distinguish them from ordinary chemical reactions. In table 14.1, the biology experiments are listed along with the different metabolic processes that were to be assayed. Of these, the Pyrolytic Release (PR) experiment (Horowitz et al. 1972; Hubbard 1976) demanded a high level of improbability in that it required the *synthesis* of organic material

TABLE 14.1. VIKING BIOLOGY EXPERIMENTS

Experiment	*Measurement*	*Metabolic Process*
Pyrolytic release (PR)	Incorporation of CO/CO_2 into organic compounds	Photosynthesis and/or chemosynthesis
Gas exchange (GEX)	Uptake or release of various gases	Decomposition of indigenous or added compounds
Labeled release (LR)	Release of radioactive gas from simple organic compounds	Decomposition of added compounds

from Co and/or CO_2 under very benign incubation conditions. Similarly, the Gas Exchange (GEX) experiment (Oyama 1972) required unusual chemistry since the simultaneous or sequential production of *both* oxidized and reduced gases was the primary attribute for a biological "positive" result. Alternatively, the appearance or disappearance of a gas at rates that increased exponentially with time, was another potential "scenario" that could lead to a presumption of biological activity (Oyama et al. 1976). The weakest of the three experimental approaches from this point of view was the Labeled Release (LR) experiment (Levin and Straat 1976a), since this required only the decomposition of one or more simple organic molecules (i.e., containing one to three carbons) with the concomitant release of some volatile C-containing material. While it was recognized that many nonbiological materials could attack and decompose the organic ingredients supplied in the LR nutrient medium, the redeeming feature in this experiment was the intended use of a heated (160°C for 3 hr) "control" in the event that decomposition was observed during any incubation on Mars. On the basis of extensive pre-Viking testing, no *sterile* terrestrial samples had shown any significant ability to decompose the LR nutrient (Levin and Straat 1976a) and it was assumed that if, perchance, the Mars samples were to contain chemically reactive substances (like sulfuric acid or permanganate), these would survive heat sterilization and thus be identified as nonbiological agents.

All of the approaches discussed above were tested and repeated several times at the two Viking lander sites. For these analyses, an instrument of unprecedented complexity, packaged into a volume of 0.027 m^3, was required. Details of the instrumentation can be found in Brown et al. (1978).

The essential findings of the three experiments, which are summarized in table 14.2, lead to the conclusion that only one of the experiments (LR) yielded data that satisfied the *a priori* criteria for metabolism. All of the other experiments yielded information, but these results are best explained as nonbiological processes.

The very first data obtained on Mars came from the GEX experiment and suggested the presence of reactive oxidants in the martian regolith. In its first chromatogram, taken just 2.5 hr after a martian sample had been exposed to a water-saturated atmosphere, a much larger oxygen peak was noted in the test chamber than could be accounted for (Klein et al. 1976). A total of three such determinations was made on the two landers, and it was found that there was significant variation in the capacity of these samples to release oxygen. The range was about 70–770 nmol of O_2 released per cc of sample, and as a first approximation, the amount of O_2 generated appears to be inversely proportional to the available H_2O content (Oyama and Berdahl 1977). Whatever this reactive source of oxygen may be, experiments on Mars showed it to be relatively stable to heating at 145°C for 3 hr. Furthermore, it does not seem to decay with storage for up to approximately 143 days at spacecraft temperatures.

TABLE 14.2. RESULTS OF VIKING BIOLOGY EXPERIMENTS

Experiment	*Major Findings*
Pyrolytic release	Small incorporation of CO/CO_2 into organics
Gas exchange (humid)	Initial rapid release of oxygen; release of CO_2, N_2, Ar/CO
Gas exchange (nutrient)	Slow release of CO_2, N_2, Ar/CO
Labeled release	Initial rapid release of labeled gas; continued slow release

NOTE: In the "humid" mode of this experiment, martian surface material was incubated in the presence of water vapor. In the "nutrient" mode, the samples were in contact with a rich organic nutrient solution.

The origin and nature of this presumed reactive oxidizing material are unknown. Particular attention has been focused—in ground-based work—on metallic peroxides and superoxides. Within a few months after the data from Mars began to be analyzed, two laboratories demonstrated that O_2 was released when superoxides were treated with water under conditions approaching those used in the Viking experiment: Ponnamperuma and colleagues (1977), using KO_2 and ZnO_2, and Ballou and colleagues (1978), using CaO_2.

In other ground-based experiments, Ballou and co-workers (1978) went one step further and showed that a "simulated" martian soil (Amboy no. 6, prepared by the Viking Inorganic Analysis team), subjected to a radiofrequency oxygen plasma for 10 hr and then humidified in a Viking-like gas exchange instrument, rapidly released O_2 with kinetics similar to those obtained in the Viking experiments.

Many investigators studying this phenomenon invoke ultraviolet light as the ultimate generator of this presumed oxidant. Support for this concept has been provided recently by the experiments of Holland and Blackburn (1979). Using a high-intensity UV source (Zill et al. 1979), they irradiated a number of different minerals under a simulated Mars atmosphere and then examined the surface properties of this material by x-ray photoelectron spectroscopy. By this technique, it could be demonstrated that some of these irradiated samples showed changes consistent with the creation of a surface layer of highly oxidizing material. The newly created oxidant(s) appeared to be stable since heating these samples at Viking sterilizing temperatures did not affect their spectra. Finally, preliminary findings by these authors indicated that irradiated pyrolusite produced a strong oxygen release when subsequently humidified in a prototype of the Viking gas exchange experiment (Holland and Blackburn 1979).

The formation of highly reactive (OH^-) radicals by ultraviolet photodissociation of water in the martian atmosphere has been discussed by McElroy and co-workers (1977), Kong and McElroy (1977), and by Hunten (1974, 1979), and the possible operation of these photolytic processes upon water adsorbed on surface particles, by Huguenin and colleagues (Huguenin 1976; Huguenin et al. 1977).

Another suggestion on the origin of the presumed martian oxidant(s) has been offered by Mills (1977), who observed that glow discharges were produced when finely powdered sand was simply rotated under reduced pressure. He suggested that such discharges could facilitate the dissociation of atmospheric CO_2 leading to the production of strongly oxidizing substances of the type under consideration here.

One mechanism for the production of oxidizing substances in the martian regolith that does not require the intervention of ultraviolet or other radiation has been proposed by Huguenin and colleagues (1979a,b). They have argued that in the cold, dry atmosphere of Mars traces of water adsorb on the surface of unweathered particulates as frost and that then the protons migrate *into* the particulates, leaving excess (OH^-) groups on their surfaces. Huguenin has presented laboratory evidence suggesting that surface peroxides are created in this process. His contention is that these peroxides are the source of the molecular O_2 seen in the GEX experiment. He has also reported that freshly ground samples of olivine or pyroxene treated with water at low temperatures and then allowed to warm released molecular O_2 into the atmosphere, with kinetics resembling those seen in the GEX experiment.

In the case of the PR experiments, very small quantities of CO and/or CO_2 appear to have been incorporated into organic matter, but because this process was not affected by prior heating of one of the samples to 90°C for 2 hr, it is extremely unlikely to be the result of biological activity (Horowitz et al. 1977). That this small amount of incorporation is not related to the postulated oxidants discussed above seems clear from one of the PR experiments carried out on Mars, in which a sample of the Chryse material similar to the one that had released most of its available oxygen within 2.5 hr after humidification (in the GEX experiment) was treated with water vapor and then vented after 4 hr. Following this treatment, there was no subsequent reduction of the PR reaction (Horowitz et al. 1977).

One possible, *indirect* effect of oxidant in this reaction has been proposed by Oyama, who suggested that C_3O_2 is formed and then polymerized on Mars (Oyama et al. 1978). Oyama hypothesizes that C_3O_2 polymer was present in all the PR samples tested and that then, during incubation, ^{14}CO was incorporated into this polymer. In ground-based studies, Oyama has, in fact, demonstrated the incorporation of CO into C_3O_2 polymer under conditions approaching the PR experiment and further, he has shown that during the pyrolysis step of this experiment the polymer decomposes, producing some material of low volatility. Oyama's contention is that during this pyrolysis a second peak, which is actually an artifact, may have been produced in the PR experiment on Mars.

Without the necessity of assuming the existence of C_3O_2 or of its polymer on Mars, Horowitz and his colleagues have been studying the possibility that some low-level synthesis of organics may have been catalyzed on Mars by inorganic materials—by reactions similar to those described by Hubbard et al. (1971 and 1973), in which, under a simulated martian atmosphere, CO has been shown to be photochemically fixed to produce low molecular weight organic compounds in the ultraviolet. Since wavelengths below 320 nm were excluded in the Viking experiments, the Horowitz group has been investigating the effects of visible light on this reaction, using iron-rich materials as potential catalytic agents. Several soil samples, as well as minerals (the latter including hematite, magnetite, and maghemite), have been tested under simulated Viking conditions and more recently, iron-charged clays of the montmorillonite class (Hubbard 1979) have been used. In these experiments, the different substrates gave quantitatively different results, but they all yielded significant second peaks under certain conditions, indicating the formation of organics. In addition, some were quite effective catalysts in the dark. While these laboratory simulations thus establish the feasibility of forming small amounts of organic compounds under conditions approaching the Viking experiments, it should be added that none of these, so far, seems to show the heat sensitivity seen in the initial Viking PR results.

Turning now to the only experiment that satisfied the original criteria for a "positive" result—the LR experiment—a review of the essential data from Viking reveals that each time a *fresh* sample was analyzed (this was done four times on Mars) there was a rapid release of labeled gas from the radioactive organic nutrient solution. This was true even for one sample that had been obtained from *under* "Notch" rock. Following the initial period of rapid decomposition, there was a slow continued gas release at a constant rate until the experiment was terminated (the longest duration used was 50 martian days). Heating one sample at 160°C for 3 hr completely destroyed the active agent responsible for the initial, rapid release of gas, while heating at an intermediate temperature, 50°C, reduced this reaction by about 70 percent. Finally, the initial, vigorous LR reaction seemed to be lost upon prolonged storage at spacecraft temperatures—which was not the case with the other two experiments.

Levin and Straat have emphasized that all of these attributes of the LR experiment are consistent with a biological metabolic process (Levin and Straat 1976a,b; 1979a,b). However, since the GCMS experiment has placed an extremely low *upper* limit on the organic content of the Viking samples (Biemann et al. 1976b), it is difficult to rationalize a biological system (especially one with an apparently vigorous ability to decompose organics) that would have evolved into this kind of an environment, where no organics could be detected down to parts per billion levels. It is also difficult to ignore the presumptive evidence for the presence of some oxidant(s) in the martian regolith.

Several investigators have attempted to reproduce the LR Viking data in ground-based simulations, using inorganic model systems. The major efforts in this direction are summarized in table 14.3, from which it can be seen that many of the major characteristics of the LR data have been reproduced. Most of these ground-based studies are based on the assumption that some oxidant, or group of oxidants, was present in the martian samples. (Indeed, a single oxidant cannot account for the main features of the LR experiment (Klein 1978).) Thus, Ponnamperuma and colleagues

Table 14.3. Summary of Results to Simulate LR Initial Reaction

Active Agent	*Substrates Tested*	*Effect of Sterilization**	*Investigators*
H_2O_2, metal peroxides, superoxides	Formate	—	Ponnamperuma et al. (1977)
Pre-irradiation with UV	LR nutrient solution	Inhibited reaction	Ponnamperuma et al. (1977)
Maghemite plus H_2O_2	Formate	—	Oyama et al. (1978)
Maghemite plus H_2O_2	LR nutrient solution	Inhibited reaction	Levin and Straat (1979b)
"Chemisorbed" H_2O_2	Formate	—	Huguenin et al. (1979a, b)
Limonite	Organic acids	Inhibited reaction	Imshenetskii and Murzakov (1977)
Fe-montmorillonite or Fe-nontronite	Formate or LR nutrient solution	Increased reaction	Banin and Rishpon (1979)

*160° for 3 hr.
— not tested.

(1977), Oyama and colleagues (1978), Huguenin and colleagues (1979a, b), and Levin and Straat (1979b) have all described experiments in which peroxides have been shown to facilitate the decomposition of one or more of the LR nutrients. Both Oyama and colleagues (1978) and Levin and Straat (1979b) have found that a combination of γFe_2O_3 and H_2O_2 produces decomposition kinetics that closely parallel those seen in the Viking experiments. The latter workers have even shown this reaction to closely duplicate the heat sensitivity of the original experiments on Mars.

Other investigators have tested models suggestive of a much more benign surface environment, that is, devoid of strongly oxidizing substances. For example, Imshenetskii and Murzakov (1977) have shown that the mineral, limonite, is effective in simulating the Viking LR data in the absence of added peroxides. It is of interest to note that in their studies, ultraviolet treatment inhibited this process, a finding consistent with those of Levin and Straat (1977b), who claimed that UV treatment decreased the activity of their samples in laboratory simulation studies.

Another hypothesis that does not require the intervention of oxidants has been postulated by Banin, who proposed that metallo-clay complexes, operating under the very dry conditions of Mars, may catalyze the decomposition of organic compounds. To support this contention, he has shown (Banin and Rishpon 1979) that the interaction of iron-containing clays with the LR nutrient mixture reproduced the initial active burst of gas seen in the LR experiment. However, the activity of his system, to date, has not been found to be diminished by heating.

IMPLICATIONS AND EXTRAPOLATIONS

The most surprising information to come out of the Viking lander results has been the absence of detectable quantities of organic carbon compounds in any of the samples tested in the organic analysis experiment. Before the Viking landers began their operations on Mars, it was generally assumed that several reasonably well-understood processes would contribute to the presence of organic matter on Mars (Biemann et al. 1977). Meteoritic input from carbonaceous chondrites (Mason 1971) was expected to supply organics at levels that should have been detected by the Viking instrumentation, barring active processes for their destruction (Biemann et al. 1977). Another expected source was from the photochemical fixation of CO into organics. This process, in which low molecular weight organic compounds has been shown to be produced under simulated martian conditions upon powdered surfaces, requires the presence of ultraviolet light (Hubbard et al. 1971, 1973).

The primary objective to be addressed by the organic analysis investigations—from the point of view of organic chemical evolution—was whether indigenous processes leading to the production of organic compounds could be deduced from the data and, furthermore, to assess the complexity of these processes. The difficult problem was to be able to establish what part of the suite of organics was accidental (due to meteorites) and what part represented previous or current active processes of indigenous chemical evolution. Given the nature of the Viking data, this question could not even begin to be answered. Rather, a new set of questions has emerged: Were conditions *ever* conducive to the formation of organics on Mars? If so, are there products of these activities to be found anywhere on the planet? Even if no indigenous synthesis of organics has occurred on Mars, why is there no trace of the organic carbon from prior episodes of meteoritic infall?

That processes destructive to organic compounds may account for at least some of the answers to these questions is clearly suggested in the LR and GEX biology experiments. In addition, ground-based experiments performed under simulated martian conditions have provided ample evidence that organic compounds may not readily survive under these conditions. Chun and co-workers (1978) suggested that the absence of organics might be due to their photodestruction on Mars, catalyzed by TiO_2, and they presented experimental support for this view using samples of the Murchison meteorite (known to contain a variety of amino acids, organic acids, hydrocarbons, and other organics). These were decomposed during ultraviolet irradiation in the presence of TiO_2 (Chun et al. 1979). Similarly, Oro and Holzer (1979) demonstrated that the half-life of a large number of organic compounds irradiated with UV under Mars-like conditions was of the order of hours or days.

Recognizing the striking heterogeneity of martian surface features, it is impossible to say whether the destructive processes suggested by the Viking data can be extrapolated to the entire planet. The samples that were analyzed came from essentially featureless areas of the planet, and the similarity of results between the two Viking landers may attest only to the fact that they were analyzing, in common, fine-grained particulate matter (Moore et al. 1977) that had accumulated in and covered the surface of these regions. If future experiments on Mars, or with returned samples, confirm the absence of organic materials in other more diversified regions of Mars (e.g., in the polar regions or in deeper layers of the surface), this information could significantly affect current ideas about the conditions under which organic chemical evolution may proceed during planetary evolution, and it could well affect probability estimates of the distribution of life in the universe (Drake 1965).

The results of the biological experiments on Viking, taken in isolation, allow the possibility that at least some of the data (i.e., from the LR experiment) might be of biological origin. However, when considered together with all of the other Viking results and when the subsequent ground-based investigations are also considered, it would appear more reasonable to ascribe all of the biology "signals" to nonbiological causation. Given the inherent limitations of the biology instrumentation (Klein 1976, 1978), one can only speculate whether *other* tests or *other* approaches to "life detection" would have provided credible evidence for biology in the samples that were tested. Nor is it possible to rule out that the Viking metabolic tests, applied to samples from other martian sites (perhaps, as suggested by the work by Friedmann (1979), inside of and just beneath the surface of porous rocks) would have given data unequivocally indicative of living systems. Finally, it is not possible, on the basis of the Viking data, to exclude the presence, in the Viking samples, of exotic forms of life not based on carbon chemistry and, therefore, unrelated to current notions about the origins of life.

Only further detailed investigation of martian materials, using the broadest spectrum of experimental techniques, can resolve these questions. However, the available information does not warrant optimism that new approaches will result in different conclusions about biology on Mars.

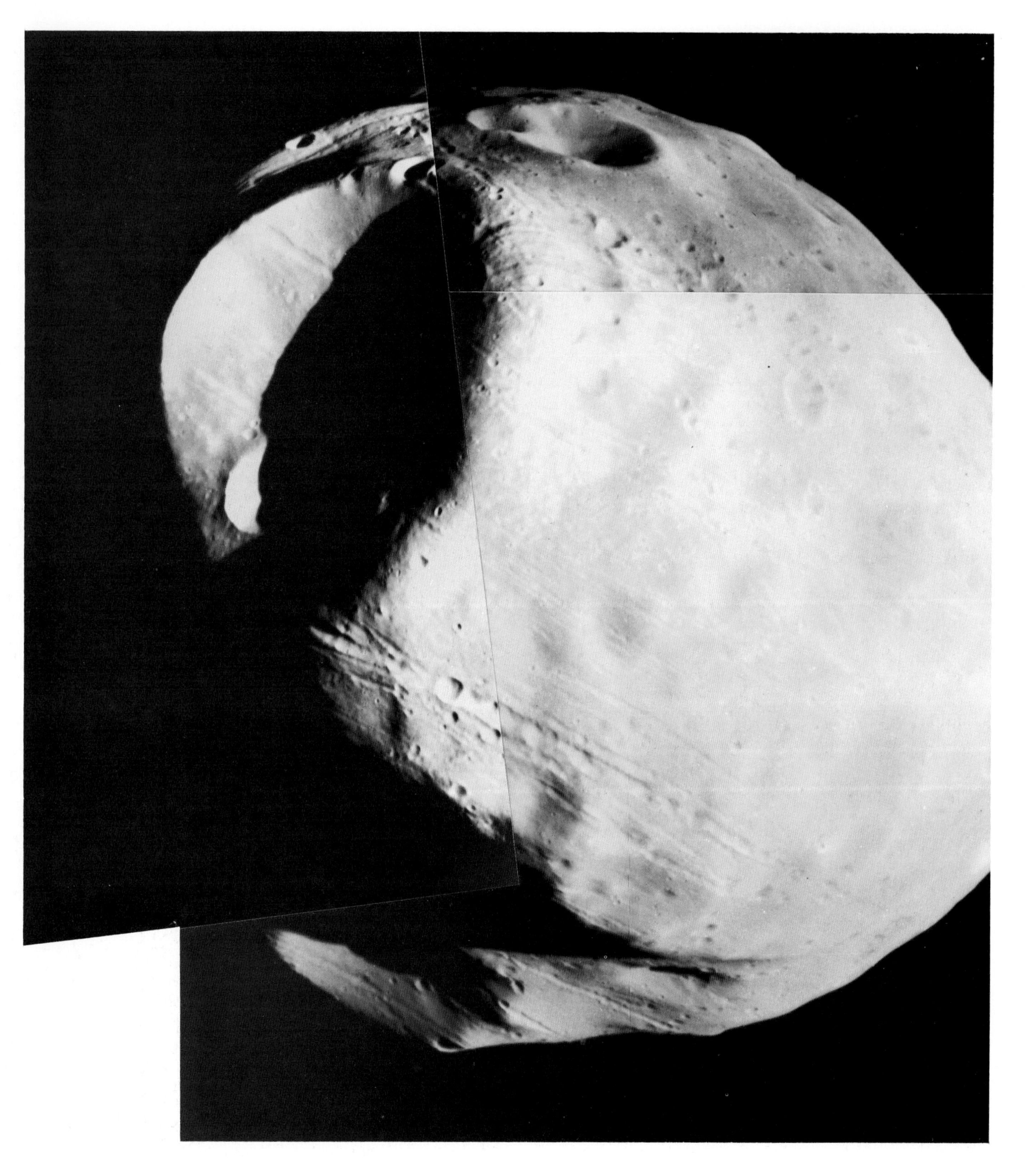

Figure 15.1. View of Phobos from a range of 612 km. The largest crater on Phobos, the 10-km-diameter Stickney, is facing us. Some of the grooves that are radial to Stickney can be seen in the lower half of the image. Below Stickney, and mostly in shadow, is the 5-km-diameter crater Hall. (P20776)

15 PHOBOS AND DEIMOS

The two moons of Mars, Phobos (fear) and Deimos (dread), were discovered by Asaph Hall during the opposition of 1877, the same year that Schiaparelli first reported seeing canals on Mars. Phobos has a mean diameter of 21 km, and Deimos a mean diameter of 13 km. Both are much smaller than Earth's moon and most of the known moons of the outer planets. Curiously the presence of the moons had been predicted 150 years earlier by Jonathan Swift, who wrote in *Gulliver's Travels* that the scientists of Luputa had discovered two moons of Mars. Even more curiously Swift gave a period of 10 hr for the inner satellite, which is not far from the actual period of 7 hr 39 min. The first spacecraft data on the satellites were acquired in 1969 by Mariner 7. One of the pictures of the martian surface contained a silhouette of Phobos, from which Smith (1970) was able to deduce that the satellite was elliptical and twice as large as was formerly thought. In 1971–72 Mariner 9 took numerous pictures of both satellites, providing 200-m resolution for most of Phobos and about half of Deimos. The pictures showed them both to be irregularly shaped, heavily cratered bodies, with Deimos having more subdued relief.

In 1977 and 1978 the Viking orbiters made several close encounters with the satellites. Previous to the encounters the orbital motions were determined precisely from numerous photographs. The encounters were then effected by trimming the orbital periods of the spacecraft so that they were harmonics of the satellite periods (3/1 for Phobos and 5/4 for Deimos) and timing the trim maneuvers so that the satellites and spacecraft would arrive at the intersection points of their orbits around the same time (Tolson et al. 1978; Duxbury and Veverka 1978). The harmonic periods gave repeated passes of the satellites, allowing slight trims to be made to edge near. The passes were so close that care had to be taken not to crash the spacecraft into the satellites. By these means an 88-km flyby of Phobos was achieved in February, 1977, and in May, 1977, the second Viking orbiter flew by Deimos at a distance of 28 km. Some of the pictures taken during the Deimos approach had resolutions of 1 m. The encounters were so close that the masses of the satellites could be determined from their perturbing effects on the spacecraft motions. This information, coupled with volume determinations from the photographic data, allowed the density of each satellite to be computed.

CRATERS

The satellite shapes are very irregular and only approximate triaxial ellipsoids. The silhouette of Phobos (figure 15.1) is dominated by three craters, Stickney (10 km diameter), Hall (5 km), and Roche (5 km). Surface undulations between them are related to smaller craters. In contrast the largest recognizable crater on Deimos is only 2.3 km in diameter. Deimos's shape (figure 15.2) can be thought of as several rounded facets joined by ridges. Veverka and Thomas (1979) suggested that the facets might be spallation scars. The crater size frequencies on both satellites are similar to those on

Figure 15.2. Deimos from a range of 948 km. A ridge between two "facets" can be seen to the left. Bright markings occur along the ridge crest, and bright streaks trail away from the crest line. The surface of Deimos is much more subdued than that of Phobos. (428B22)

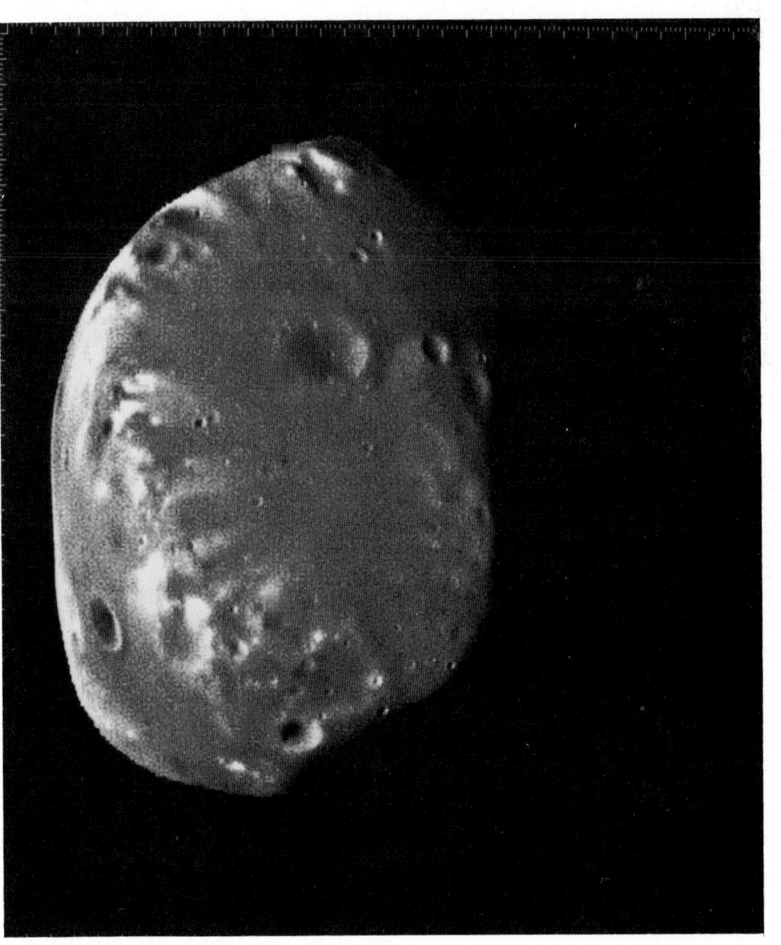

our Moon (Veverka and Thomas 1979), being close to saturation for the full range of diameters present. Deimos looks less cratered than Phobos, but this is deceptive and results from the more subdued shape of the Deimos craters. The fact that the surfaces are saturated suggests ages in excess of 3 billion years for both satellites.

Both satellites are tidally locked, so that for each the rotational and orbital periods are identical, and the long axis of the ellipsoid is always pointed toward Mars. Phobos rotates with a period of 7 hr 39 min at a distance of 9,378 km from the center of Mars; Deimos has a period of 30 hr 18 min and a radial distance of 23,459 km. Burns (1978) suggested that Phobos's orbit is changing and that tidal effects are causing a secular acceleration, bringing Phobos steadily closer to the planet. He estimated that Phobos will either break apart or crash into the planet within the next 10^8 years. In contrast, Deimos may be decelerating and moving away from the planet.

Close up Phobos resembles the lunar highlands (figure 15.3). Not only are the crater densities similar but, as on the Moon, craters are present in various states of preservation, from fresh bowl-shaped types to faint "ghosts." Depth/diameter ratios and rim heights also appear to be close to lunar values (Thomas 1979). Most of the rim height is probably due to upwarping of the surface rather than deposition of ejecta. Ejecta are present in many cases, however, as evidenced by both bright and dark haloes. In addition, Thomas (1979) pointed out that some low hummocks and blocks on crater rims, particularly around the largest crater, Stickney, could be ejecta. The presence of ejecta is somewhat surprising, since the low gravity (accelerations of 0.3 cm/sec^2 to 0.5 cm/sec^2) must allow almost all the ejecta to escape, but it demonstrates that even on bodies as small as Phobos (escape velocity of 15 m/sec) and Deimos (escape velocity of 10 m/sec) ejecta can be retained to build up a regolith. A major difference between Phobos and the Moon is the lack of secondary craters. This may be because material falling back onto the satellite surface does so with such low velocities that craters do not form.

The craters on Deimos (figure 15.4) are more subdued than those on Phobos. Many are flat-floored, with a distinct break in slope at the base of the crater walls. Sediment appears to have ponded in many craters, in some cases almost filling them. High-resolution pictures show numerous blocks 3–20 m across, arrayed in clumps or strings, mostly on crater rims. Both satellites lack complex craters with central peaks and terraces, as would be expected if the transition from simple to complex craters were gravity-dependent (Hartmann 1973a; Pike 1977). On Earth's moon the transition takes place around 20 km. On Phobos and Deimos the transition diameters should be larger than the satellites themselves, so their absence is not surprising.

Figure 15.3. View of Phobos, showing its lumpy shape and a surface saturated with craters, much like that of the lunar highlands. Several crater chains are visible. Although they cross some finer striations, both striations and crater chains are roughly radial to Stickney, as demonstrated by figure 15.5. The large crater near the top of the image is the 5-km–diameter Roche. (39B84)

REGOLITH

The retention of impact ejecta has apparently resulted in the development of a thick regolith on both bodies. Infrared temperatures (Gatley et al. 1974), photometry (Pollack et al. 1973; Noland and Veverka 1977), and polarimetry (Noland et al. 1973) all indicate at least a thin fragmental layer. Thomas (1979) estimated that on Deimos at least 5–10 m of sediment has accumulated in many craters, providing a minimum thickness for its regolith. The actual thickness may be considerably greater. The smooth cross section of grooves on Phobos and the lack of ledges within craters on both bodies suggest no significant change in the strength of materials to depths of 100 m, so that fragmental debris probably continues down to that depth.

Figure 15.4. High-resolution view of Deimos, taken from a range of 45 km and showing an area 2 × 2 km. The surface is heavily cratered, but most of the craters are subdued, some being visible only because of their bright rims. The numerous blocks are presumably ejecta that failed to escape the satellite's small gravity field during impact. (423B63)

Deimos has numerous albedo markings, many of which suggest lateral movement of debris across the surface. Most bright markings are associated with crater rims, but the ridges between the facets also tend to be brighter than other areas. Lines of bright streamers commonly extend from crater rims both into and away from the crater. In any one area streamers external to the crater are uniformally aligned and tend to be oriented at right angles to the ridges (Thomas 1979). Streamers from the ridges are also aligned in the same direction. It appears that material has migrated slowly across the surface, down the regional slopes demarcated by the ridges and facets. The puddles of material in the craters also appear to have been affected by the regional slopes, since many are displaced toward the downslope side. Surface creep is probably caused by thermal effects or shaking of the surface by meteorite impact, but it is surprising that the slight gradients caused by the faceting are sufficient to cause creep in such a weak gravity field ($g = 0.3$ cm/sec^2).

Deimos appears to have a significantly thicker cover of blanketing material than Phobos. Craters on Phobos show the normal spectrum of morphologies expected of a surface in equilibrium, where craters are being destroyed as fast as they form. The fresh craters show little evidence of any fill. In contrast almost all the craters on Deimos are partly filled. The reason for the difference is not known. It has been suggested that Phobos may lose a greater proportion of the material ejected during impact by virtue of its close proximity to Mars. Theoretical studies do not substantiate this view, however. Alternatives are that formation of large craters on Phobos may somehow have cleaned the surface, or that the mechanical properties of the bodies are different, so that they produce different quantities of debris during impact (Veverka and Thomas 1979), but none of these suggestions is well formulated, and the cause of the difference remains uncertain.

GROOVES ON PHOBOS

One of the most striking features of Phobos is the presence of grooves over most of its surface (figure 15.5); none are present on Deimos. Maps of the grooves show that they radiate in all directions from the 10-km–diameter crater Stickney and converge on the opposite side of the satellite at a region close to the Stickney antipode (Thomas et al. 1978). The area around the convergence point, however, is largely groove-free. The grooves are best developed close to Stickney, where some are as large as 700 m across and 90 m deep, although most have widths and depths in the 100–200 m and 10–20 m ranges, respectively. Widths tend to decrease away from Stickney, and most grooves are less than 100 m across near the antipode. The grooves appear to represent the intersection of near-vertical planes with the surface,

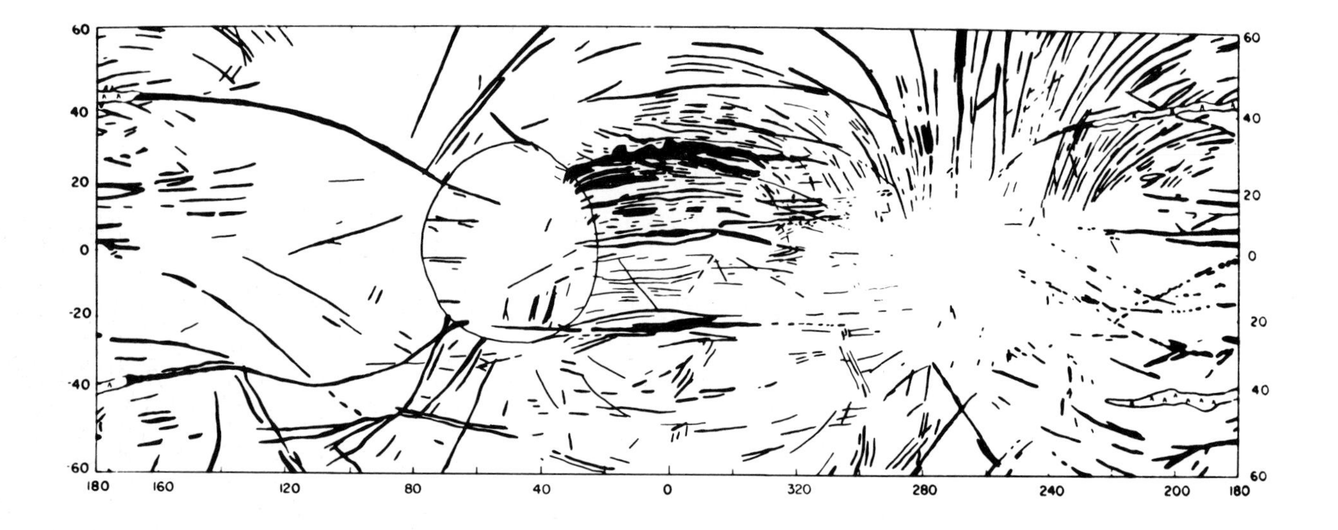

Figure 15.5. Map of the grooves on Phobos. The outline of Stickney is included for reference. (Reprinted with permission from Thomas et al. 1978, copyright 1978, Macmillan Journals, Ltd.)

since they are largely undeflected by undulations in surface topography. They also appear to be old. The density of superposed craters is identical to that of the surrounding terrain, suggesting that groove formation has not been continuous but terminated billions of years ago (Thomas et al. 1978).

The grooves are clearly related in some way to the formation of Stickney. They are probably surface manifestations of radial fractures that formed as a result of the catastrophic disruption of Phobos during the Stickney impact (Thomas et al. 1978). Alternative explanations, such as fracturing by tidal stresses (Soter and Harris 1977) or drag forces (Pollack et al. 1978) fail to explain the relation to Stickney. Few of the grooves have the rectilinear form expected of fractures. Most have a pitted or beaded appearance, and many more resemble crater chains than graben. Some segments appear to have slightly raised rims. The beaded appearance could result from drainage of near-surface fragmental debris into depressions caused by the fractures. Alternatively the fractures may have caused some internal heating and expulsion of volatiles along the fractures (Veverka and Thomas 1979). According to this hypothesis the pits would be somewhat analogous to the sand boils produced along terrestrial faults in poorly consolidated sediments.

COMPOSITION

While no definitive information is available on the composition of either satellite, the reflectivities and densities provide good clues. Three lines of evidence suggest a composition similar to a carbonaceous chondrite—a type of meteorite that is rich in water and organics (Mason 1971). First, both satellites are dark objects, with albedos in the 5–7 percent range (Veverka 1977). Second, the spectral reflectance, in the 0.2–0.9 μm range, resembles that of carbonaceous chondrites and C-type asteroids (Pang et al. 1977). Third, both satellites have low densities; the mean values are less than 2.0 gm/cm^3, suggesting a significant volatile component. A carbonaceous chondrite composition would explain why Phobos tends to fracture under impact, since low-density, carbonaceous chondrites have low strengths. They can also contain 10–20 percent by weight of water, which could be released under impact-generated stresses and blow out fragmental debris to form the pits and raised rims that are observed along the grooves (Veverka 1978).

Lewis (1974) hypothesized that during condensation of the primitive solar nebula, carbonaceous chondrites formed only in the outer half of the asteroid belt. Phobos and Deimos may have formed in this region and been subsequently captured by Mars. Pollack and colleagues (1978) and Hunten (1978) have suggested that capture occurred during the final stages of accretion of Mars, when the planet was still surrounded by an extensive primitive atmosphere, which provided the necessary drag to capture the satellites. They are, therefore, two objects that survived the early accretion of the planet.

SUMMARY

Mars has two moons, the 21-km–diameter Phobos (figure 15.6) and the 13-km–diameter Deimos. Both are saturated with craters and retain a thick regolith. The surface of Phobos looks much like that of the Moon, with craters in various states of preservation, from fresh, bowl-shaped types to those that are barely visible. Two nonlunar characteristics are the absence of complex craters and the presence of numerous grooves, which radiate from the largest crater, Stickney, and appear to have formed by dislocation of the entire object at the time of formation of the crater. Deimos has a more muted appearance than Phobos, as though it had a thicker regolith. Craters are mostly subdued and appear to be partly filled with debris. The overall shape of Deimos is controlled by several curved facets separated by ridges. Streaks suggest movement of surface debris away from the ridges, which have a slightly higher albedo than the facets. The low densities (< 2 gm/cm³) of both bodies, their low albedos, and their spectral reflectivities suggest compositions similar to carbonaceous chondrites.

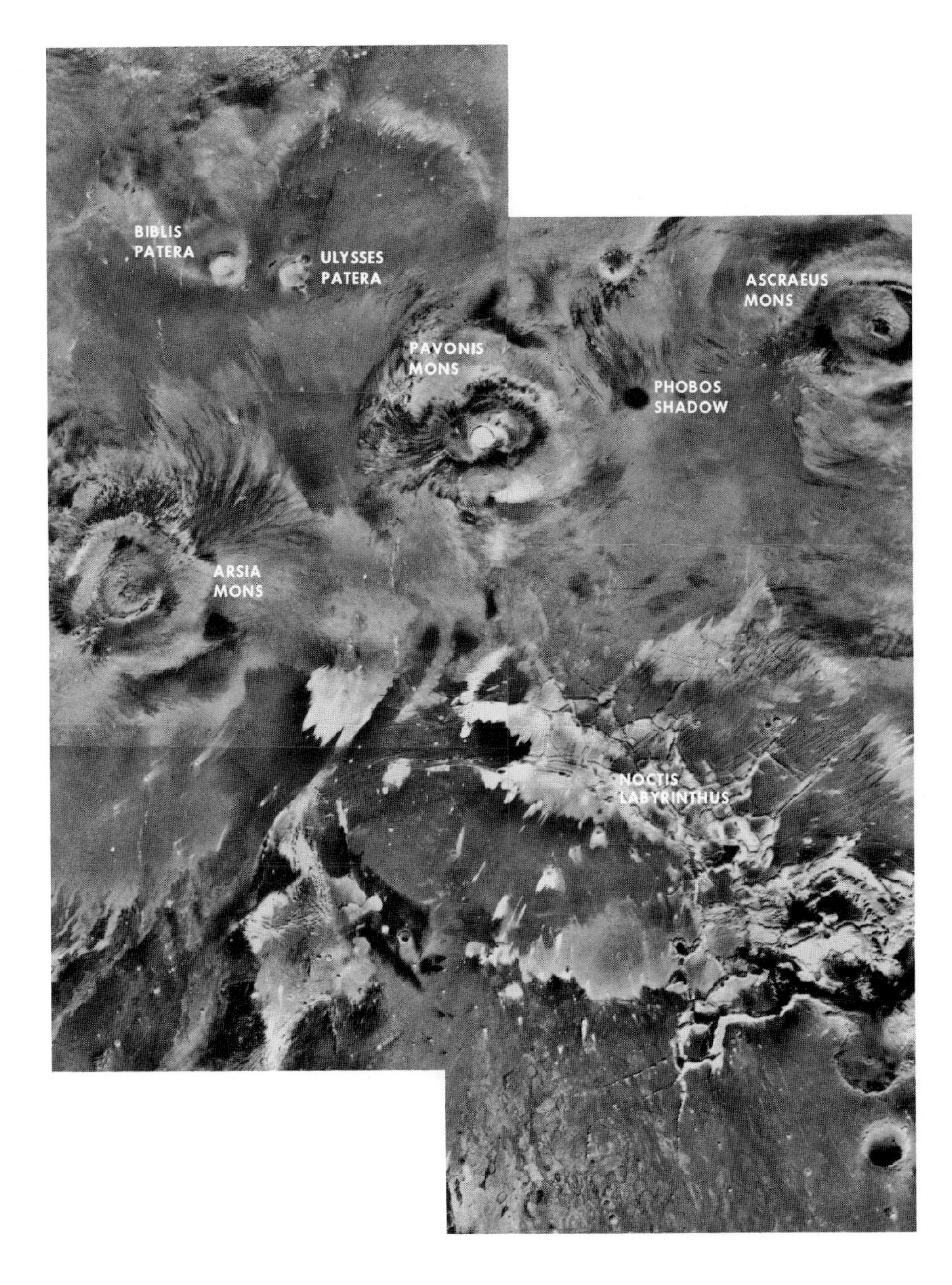

Figure 15.6. View of the shadow of Phobos over the Tharsis region. The fan-shaped arrays of flows to the northeast and southwest of the Tharsis volcanoes are particularly well-displayed. Such pictures can be used for geodetic purposes. The satellite orbit, the Sun's illumination, and the time the picture was taken are all known precisely, and therefore the position of the shadow on the ground is known. The positional information so derived can be cross-checked against positions determined from the planetary data alone.

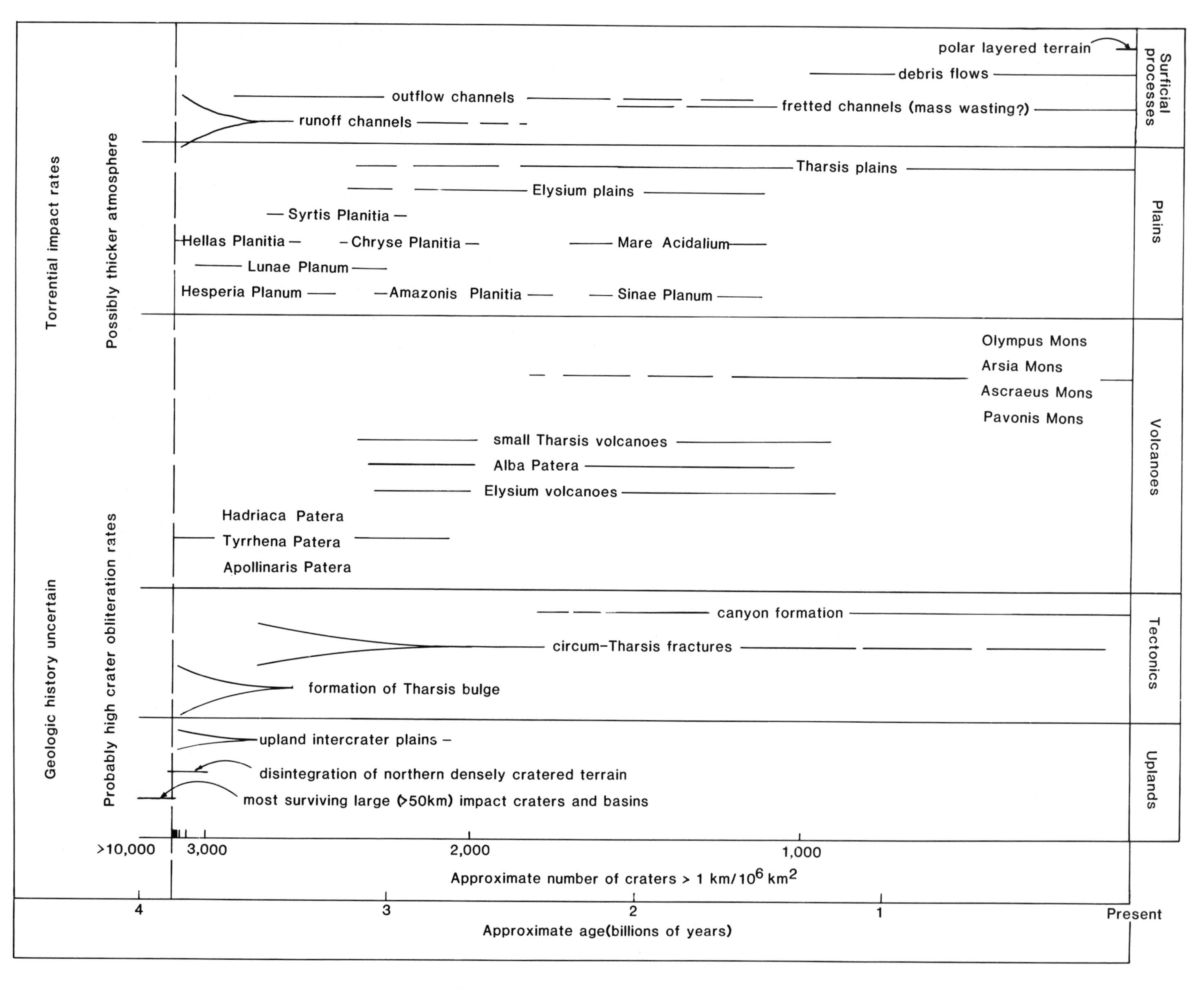

Figure 16.1. Generalized representation of the major events in martian history. The ages given at the bottom of the diagram are very tentative and could be subject to large errors.

16 SUMMARY AND CONCLUSIONS

From the discussions in the previous chapters we can now reconstruct the broad outlines of how the surface of Mars has evolved and how its development has differed from that of the Moon and Earth. Evidence from the first half billion years or so is largely destroyed as a result of the continual reshaping of the landscape by impact. On the Moon high impact rates in the final stages of accretion, around 4.6 billion years ago, are believed to have caused melting of the outer few hundred kilometers and subsequent separation of an anorthositic crust (Wood 1975; Taylor 1974). As the rate of accretion slowed, the crust began to solidify, possibly around 4.4 billion years ago, and by the time the impact rates had declined to around their present levels, around 3.9 billion years ago, the Moon's outer rind was rigid, and volcanism was relatively deep-seated. On Mars a similar sequence of events may have taken place. Extensive melting and outgassing very early (4.5 billion years ago?) were probably followed by solidification of the crust and then a rapid decline in the impact rate. As indicated in chapter 5, the time of the decline is generally believed to have coincided with that on the Moon. As the rate of impact tailed off, the landscape stabilized and began to retain geomorphologic evidence, from which the subsequent geologic history has been reconstructed.

At the time of the decline in impact rates the differences between the northern and southern hemispheres had been established, with much of the northern hemisphere at a lower elevation than the southern hemisphere. The cause of the asymmetry—whether the result of an uneven distribution of large impact basins, the sweeping up of the lower-density crust into a continentlike mass, or the collapse of sections of the crust—is not known. The Tharsis bulge was also present at this time (figure 16.1) although it may not have developed to its current size. As we saw in chapter 8, the formation of the bulge may have been triggered by the separation of the core.

Runoff channels are almost ubiquitous on surfaces that date from around 3.9 billion years ago, and if formed by running water, as appears likely, then an atmosphere that was thicker than the present one is implied. An early thick atmosphere is also consistent with the early high crater obliteration rates and the isotopic compositions of oxygen and nitrogen in the present atmosphere (see chapter 3). Much of the outgassing probably occurred during the early, widespread melting, although the abundant evidence for subsequent volcanism indicates that the planet continued to outgas throughout its history.

The almost complete lack of runoff channels on surfaces younger than about 3.5 billion years old suggests that the thick atmosphere did not long survive the decline in impact rates. A relatively thin atmosphere for most of Mars's history is also indicated by the immature drainage pattern on the old terrain, which shows that fluvial erosion was not sustained for long periods of time, and by the limited amounts of eolian erosion experienced by most equatorial plains, despite ages that in some cases are in excess of 3 billion years. Much of the carbon that was originally in the atmosphere may now be either chemically bound as carbonate in the regolith, adsorbed as carbon dioxide in the near-surface rocks, or dissolved in water or water-ice. Most of the water that outgassed—estimated at about 100 m averaged over the whole surface—probably exists as ice in the thick permafrost that is everywhere present or as groundwater at greater depths (see chapter 13). Much of the original nitrogen may have been lost by exospheric processes, but some may be chemically bound in the soil as nitrates.

We saw in chapter 7 that the planet has been volcanically active throughout its history, although seemingly at progressively declining rates. Extensive ridged plains within the cratered uplands indicate that at the time of the decline in impact rates, 3.9 billion years ago, volcanism was occurring in almost all regions of the planet. Within about a billion years of the decline many of the sparsely cratered plains formed, presumably as a result of the effusion of vast quantities of low-viscosity, iron-rich lavas, which covered most of the low-lying, heavily cratered areas of the northern hemisphere. As time passed, formation of volcanic plains became more and more restricted to areas encompassed by the Tharsis and Elysium bulges. In the last billion years or so formation of the plains appears to have been largely confined to Tharsis.

The individual volcanoes show a similar trend. The older volcanoes, such as Hadriaca, Apollinaris, and Amphitrites, are widely spaced and are not grouped together in specific provinces, whereas the younger volcanoes occur almost exclusively in Tharsis and Elysium. Older volcanoes also tend to have little vertical relief as compared to the younger ones, which may have enormous heights, a fact that has been attributed to a thickening of the lithosphere with time. Crater counts on some of the large volcanoes or on flows that originate in vents on their flanks suggest ages so young that the volcanoes may still be active. The edifices themselves may be quite old, however, having accumulated slowly, possibly for times in excess of a billion years.

The presence of the large bulge in the Tharsis region and, to a much lesser extent, of that around Elysium created stresses in the crust and, as a result, an extensive array of radial fractures (see chapter 8). Superposition relations suggest that most of the fractures are old, having formed in the first half of the planet's history, but fracturing continued into relatively recent times, as indicated by the occasional radical fracture that cuts a young lava flow. The canyons appear to have been caused largely by faulting along the radial fractures, although other processes, such as landsliding, gullying, and undermining, have contributed to their present configuration (see chapter 9). The canyons have also continued to form into the relatively recent past, as evidenced by faults that transect gullies in relatively young lava plains. They may, however, be quite ancient features that have continued to form throughout geologic time as a consequence of the stress pattern established by the Tharsis bulge.

Much of the water that cut the runoff channels early in the planet's history and some that outgassed later during subsequent volcanism may have become part of a vast artesian system trapped below the thick permafrost that developed after the atmosphere thinned. Slow percolation probably resulted in the accumulation of water in low areas, such as around the Chryse basin, leading to periodic breakouts which formed the large outflow channels (see chapter 10). Although most such channels appear to be relatively old, they are younger than the runoff channels and may have continued to form episodically until relatively late dates. The fate of the water that cut these channels is uncertain. It could not have returned to the groundwater system because of the permafrost seal. More probably the water exists as ice, possibly interbedded with volcanic and eolian debris in the high latitudes.

We saw in chapter 12 that Mars is subject to periodic changes in climate as a result of precession of the orbital and rotational axes and variations in eccentricity and obliquity. Such changes must affect the general circulation of the atmosphere and the stability of volatiles at the surface and may cause variations in atmospheric pressure as a result of adsorption and desorption of CO_2 in the regolith (see chapter 13). Evidence for such climate changes is best preserved at the poles, although the changes may also be a prime cause of the complex erosional and depositional histories experienced by all surfaces in high latitudes (see chapter 4).

At both poles are sequences of layered sediments of geologically recent age, judging from the almost total lack of superposed craters. The sediments are believed to consist largely of eolian debris, possibly mixed with ice, and the layering may be the result of variable rates of accumulation induced by climate changes. Accumulation may be currently occurring at the north pole because of the coincidence of the growth of the seasonal cap with dust-storm activity in the southern hemisphere (see chapter 12). The accumulation of dust at the southern cap may be negligible at the present time. Because of precession conditions will be reversed in 25,000 years, and the south pole will accumulate the debris raised by the dust storms. Obliquity and eccentricity variations may also affect accumulation and erosion of the deposits and may in some circumstances lead to complete erasure of the deposits, since old layered sediments are not seen.

Thus in some ways Mars resembles Earth. It has an atmosphere, albeit a much thinner one than Earth, and a surface that has been modified by the wind to produce a variety of erosional and depositional landforms. There is abundant evidence of near-surface water, either as ice or liquid; the water appears to have eroded parts of the surface and reacted with the surface materials to produce weathered products. The surface of Mars also provides abundant evidence of deformation and volcanism, and both seem to have continued into the relatively recent geologic past. Like Earth's surface, therefore, the martian surface has been subject to chemical reaction with the atmosphere, erosion by wind and water, and modification by extensive volcanism. But, despite these similarities, the differences between the two bodies are profound.

A major cause of the differences between the two planets is their contrasting tectonics. The geology of Earth's surface is dominated by the effects of plate motion. Interaction between the plates controls the distribution of continents, the formation of mountain chains, the location of volcanic and tectonic activity, and the general style of crustal deformation. Indeed there are few geologic processes that are not affected in some way by plate tectonics. New lithosphere forms continually where plates diverge at midoceanic ridges, and old lithosphere is ingested at subduction zones, where one plate dips under another. Since it is normally the oceanic lithosphere that is subducted, most of the ocean floor is relatively young, very little being older than 200 million years. Only on the continents is there any record of events that took place billions of years ago. In contrast Mars's crust is fixed, and all those features, such as linear mountain chains, transcurrent fault zones, and linear ocean deeps, that on Earth characterize interaction between the plates are absent. The stability also results in preservation of an ancient record in almost all areas of the planet.

The second reason for the differences between the surfaces of Earth and Mars is the presence of abundant liquid water on Earth. Water plays an essential role in two major processes: (1) weathering—the chemical breakdown of rock-forming minerals, mainly by hydration and carbonation, into mineral assemblages more in equilibrium with conditions at the surface—and (2) gradation—the steady reduction of surface relief by erosion and transport of material from high to low areas. Both processes are complemented by opposing processes, such as metamorphism, in which the low-temperature weathered products are reconstituted into high-temperature assemblages, and tectonic activity, which tend to accentuate surface relief. Primary igneous rock minerals, for example, are altered during weathering into clays and other minerals, which are then carried to the oceans and deposited in sedimentary basins, such as the troughs along subduction zones. Deep burial can result in melting, to form other igneous rocks, or in metamorphism and formation of new high-temperature assemblages. Uplift may then reexpose the materials to weathering and the cycle starts over. Gradation is an essential part of the cycle, in that the weathered products must be eroded and carried to the oceans if they are to be deeply buried and metamorphosed. Wind may participate, but most erosion and transport is by running water.

Earth's surface is thus a complicated system, in which material is being recycled both through the mantle, by subduction and ocean-floor spreading, and within the lithosphere and hydrosphere, by weathering, metamorphism, erosion, and deposition. There are reasons to believe that the entire system is in a crude equilibrium, because the composition of the main interfaces, the ocean, and the atmosphere appear to have remained approximately constant for the last billion years or so (Siever 1974). This is speculative, however. Particularly uncertain are the rates of tectonic activity associated with plate motions prior to 200 million years ago, since the oceanic record from before this time is almost nonexistent, and the continental record is fragmentary.

The dynamics of the martian crust are totally different. Although the planet has been volcanically and tectonically active throughout much of its history, the lithosphere appears not to have been recycled through the underlying asthenosphere. Huge volcanoes have formed, as well as vast fracture systems, but the activity is not concentrated in linear zones as on Earth, but rather affects areas of broad regional extent. Furthermore, although water has probably flowed across the martian surface at times in the past, fluvial erosion has been trivial. Where channels are present, they mostly wind between the craters or down the crater rims, but rarely has erosion been sufficiently sustained to wear away the craters themselves. Erosion of young features, such as the large volcanoes, is imperceptible.

On Earth uplift is followed by enhanced erosion and downwarping by rapid sedimentation, so that a rough equilibrium is maintained between the relief-forming processes and gradation by running water. On Mars, because of the limited fluvial erosion and deposition, no such equilibrium is achieved. If relief is created, such as by volcanic, tectonic, or impact processes, it largely remains. If any equilibrium is achieved, it is not with the surficial processes of erosion and deposition, but rather with the ability of the lithosphere to sustain the loads created by the relief. As a consequence Mars, despite its smaller size, has considerably more relief. Volcanoes have grown to almost three times the height of Mt. Everest, canyons several kilometers deep have survived for billions of years, and a broad 10-km–high upwarp of continental dimensions has apparently persisted in the region of Tharsis for much of the planet's history.

The lack of significant sediment transport implies, in addition, that weathering products are not recycled. Chemical analyses of the surface materials indicate that weathered products, such as clays and evaporites, are present. It is not clear how or where these form, but water may be abundant below the surface as ice or as liquid within artesian reservoirs, and such water is likely to be an efficient weathering agent, probably being charged with carbon dioxide. Thus, where subsurface rocks are in contact with groundwater, weathering may occur, but without erosion and removal of the weathering products alteration must be slowed, as a protective rind of weathered material accumulates around the unaltered rocks. The subsurface rocks may therefore be in a state of static equilibrium with the hydrosphere and atmosphere, in contrast to the apparent dynamic interaction that occurs on Earth. As we saw in chapter 13, this is an overgeneralization; nevertheless the rate of interaction on Mars appears to be orders of magnitude lower than that on Earth.

Thus in addition to the lack of cycling of the martian lithosphere through the upper mantle, caused by the absence of plate tectonics, the cycling of the surface materials and the interaction with the hydrosphere and atmosphere are hindered by climatic conditions that prevent flow of water across the surface. The result is an active planet with enormous surface relief, on which features with a wide range of ages are preserved in almost pristine condition.

APPENDIX A
VIKING ORBITER PICTURES

Most of the pictures in this book were taken with the cameras on the Viking orbiters. While the pictures can be used without knowledge of how the cameras worked or how the sequences were designed, an elementary knowledge is useful for understanding the data set. For a detailed description of the Viking lander pictures, see Mutch and colleagues (1978).

The cameras

Two identical cameras (Wellman et al. 1976) were mounted side by side on the scan platform of each orbiter. The cameras were slow-scan vidicons similar to those flown on the previous Mariner missions. Also on the scan platform, and bore-sighted with the cameras, were an infrared thermal mapper (IRTM) and the Mars atmospheric water detector (MAWD). The platform could be rotated on command about two axes, in order to point the instruments to specific targets. Each camera consisted of a telescope, the vidicon sensor (a vacuum tube with a light-sensitive faceplate), a filter wheel/shutter assembly, and supporting electronics. The telescope, an all-spherical Cassegrain with a 475 mm focal length, focused an image of the scene on a 14.0 × 12.5 mm area of the vidicon faceplate. The light induced on the plate an electric charge, which varied over the active area according to the scene content. Exposures could be adjusted from 3.18 msec to several seconds by means of the shutter between the telescope and the vidicon. Exposures were chosen to minimize smear and to place most of the scene intensities in the middle of the light-transfer curve of the vidicon. Different filters could also be interposed between the telescope and the vidicon.

The charge on the faceplate was read off by scanning the plate with an electron beam. The scene content was then contained in the modulation of the beam current. Many of the pictures, especially those from orbiter B, have coherent noise at the top of the frame, which was caused by microphonics during readout. After readout the faceplate was light-flooded by means of small bulbs and all traces of the previous image were erased by repeated scanning. Because 4.48 sec was blocked out for exposure and another 4.48 sec for readout and preparation time for the next exposure, each camera could take a picture every 8.96 sec, or one orbiter with two cameras could lay down a swath of pictures two frames wide, with exposures alternating every 4.48 sec.

The beam current was digitized to give 1182 elements on a line, each coded to 7 bits. The raster with which the faceplate was scanned contained 1056 lines, so that a single image contained 1182 × 1056 picture elements (pixels) and 8.74×10^6 bits. Since the cameras alternated every 4.48 sec, they produced a 2.112 Mb/sec data stream, which was recorded in parallel on seven tracks of the spacecraft tape recorders. The storage capacity of the two recorders on each spacecraft was approximately 120 frames. The data were relayed back to Earth one track at a time, at rates that ranged from 2 kb/sec to 16 kb/sec, depending on the Earth-Mars geometry. At a nominal 8 kb/sec one frame took 20 min to read back. All the orbiter frames have seven vertical lines along the right hand margin just outside the image area. These indicate by breaks in the line how complete the record is of each track. Almost all the data were received at one of the three 64-m dishes (Goldstone, Madrid, Canberra) of the Deep Space Net (DSN) and were then relayed to the Jet Propulsion Laboratory in Pasadena for reconstruction.

The angular field of view of each camera was 1.51° × 1.69° which gives a frame dimension of 39 km × 44 km for vertical viewing from 1,500 km. The bore-sight offset of the paired cameras was 1.38°, which gave a frame sidelap of 0.31°. The cameras were originally designed to optimize coverage of potential landing sites from an altitude of 1,500 km. From this altitude the spacecraft motion caused the centers of successive frames in a swath to be spaced 35 km apart along the ground track, and onlap was maintained. Usually the edges of the frames projected onto the surface were not parallel to the ground track, so that neat rectangular arrays of frames rarely resulted. More commonly the footprints were diamond-shaped, and the points of the diamonds overlapped. At altitudes other than 1,500 km, small scan platform slews had to be made between frames to optimize coverage. At higher altitudes the frames had to be spread further apart to compensate for the larger frame size and to avoid excessive overlap. At lower altitudes slews were needed to prevent gaps between successive frames. At altitudes below 1,000 km pictures were commonly taken with the scan platform in motion, to compensate for the motion of the spacecraft and to prevent blurring of the image.

Sequence planning

The pictures finally taken were the result of a complex interplay of a number of factors, such as the viewing geometry, the needs of the different scientific disciplines, the capacity of the tape recorders, DSN availability, orbit knowledge, competition from other orbiter instruments, and the need to provide relay links to the landers. Because of this complexity and the many uncertainties at the start of the planning cycles, the sequence design was interactive, and commonly the initial wishes of the experimenters could not be fully realized. Frequently also, the intent of the experimenters was frustrated by the planet itself. Sequences to photograph specific surface features, for example, may have been perfectly executed but may have turned out to have had limited usefulness because of a thick cloud cover.

The main constraint on sequencing was the viewing geometry. The orbiters were initially placed in nearly synchronous (~ 24-hr) orbits, with periapses close to 1,500 km and apoapses close to 33,000 km. The inclination of orbiter A was 39° for most of the mission; that for orbiter B was 80°, to allow for better viewing of the poles. To achieve optimum discrimination of the topography of the landing sites, the periapsis of orbiter A was initially placed 25° from the evening terminator, and the periapsis of orbiter B was at a similar angular distance from the morning terminator. As Mars moved in its orbit around the Sun, however, the illumination at periapsis changed by approximately 1/2° per day. The periapsis of orbiter A initially moved to higher sun angles; that of orbiter B initially moved toward the terminator and into the dark. Observations were ultimately made for almost two martian years, during which the illumination conditions changed constantly. In addition to the motion of Mars and the orbiters around the Sun, the normal to the orbit plane of the orbiters executed a slow conical motion, which was partly offset by slow regressive rotation of the line of apsides. The net result was that the surface below periapsis was in the dark approximately half the time and was illuminated the other half. These conditions largely controlled the use of the cameras. When the planet was illuminated near periapsis, efforts were concentrated on acquiring high-resolution coverage. When the illuminated portion of the planet could be viewed at apoapsis, synoptic coverage was obtained for color reconstruction, geodetic purposes, and the study of the atmosphere.

The first step in the sequence planning was assessing the illumination conditions and outlining the broad strategy. Latitude-time plots were prepared, on which illumination and altitude contours were plotted for several months ahead. Optimum illumination for viewing topography is with the Sun 10° to 25° above the horizon. The plots showed the times and latitudes at which these conditions were met and also indicated the spacecraft altitude at the position of optimum illumination. The broad strategy and general mix of photographs, whether predominantly high-altitude or low-altitude, could thus be specified from these plots months in advance. At this stage decisions had to be made concerning maneuvers to change the orbit period. If the period was close to the rotation period of the planet, then the ground tracks of successive passes were close together, and the orbiter moved only slowly in longitude. Such a geometry was desirable when most of the imaging was at periapsis, in order to maintain contiguity between swaths taken on successive passes. On the other hand, if most of the imaging was to be at higher altitudes, then a larger difference between the orbit period and the Mars rotation period was desired, to prevent excessive sidelap and to maximize longitudinal coverage. Orbit trims were therefore made periodically to respond to the different strategies. The orbit period that was chosen fixed the rate at which the orbiter moved in longitude; the timing of the maneuver controlled the specific longitudes covered. A further consideration in deciding upon the broad strategy was the condition of the planet itself. Imaging of the surface was avoided during dust-storm periods and where hazes and clouds were known to occur frequently at the season in question. Despite these precautions on many of the pictures the surface is still hidden from view.

The orbiters were oriented with reference to the Sun and a reference star, which had to be changed frequently. Small gas jets activated by a Sun sensor kept the solar panels facing the Sun. To fix the orientation around the Sun line, the spacecraft had a star sensor, which pointed in a direction slightly offset from the plane of the solar panels. The line of sight of the star sensor thus swept out a cone when the spacecraft was rotated around the Sun line. As Mars orbited the Sun, the orientation of the Sun line in inertial space changed, and with it the intersection of the star tracker cone with the celestial sphere. The array of stars accessible to the star sensor was thus constantly changing. A reference star was chosen by searching along the line of intersection of the star sensor cone with the celestial sphere for an appropriately bright star. Once the star had been chosen, the spacecraft was rotated through the appropriate angle about the Sun line for the sensor to lock onto that star. Stability was again achieved by means of small gas jets. Locked on the Sun and the reference star, the spacecraft was then a stable platform with its own coordinate system. Viewing from the spacecraft was defined in terms of cone and clock angles. The angle between the Sun line and the line of sight is called the cone angle; the rotation angle around the Sun line from the reference star to the line of sight is called the clock angle.

Once the timing and the nature of the maneuvers had been decided upon and the reference star had been chosen, then the sequences on specific orbits could be considered. Computer plots were generated that showed views of Mars from the perspective of the orbiters at different points in their orbits. The plots also included as overlays clock- and cone-angle grids, since pointing had to be specified in these parameters. The plots showing what was in view were compared with existing maps and other photographs, and targets were chosen. The clock-cone grids specified how the scan platform had to be oriented, and the timing was specified with respect to periapsis.

Unfortunately, poor knowledge of the orbit period commonly led to errors in pointing the cameras. Targets were specified several weeks ahead of execution. Because of the uneven gravity field, particularly around Tharsis, ground tracks could not be precisely predicted. The problem was especially severe when periapsis was lowered to 300 km. Moreover, commands were given with reference to the clock on board the spacecraft, whereas the planning charts were referenced with respect to periapsis. Commonly the time of periapsis could not be precisely predicted, again because of the variations in the orbit period. This occasionally led to timing offsets and targeting errors, despite late updates to correct for orbit perturbations. However, mission planners were sometimes able to apply "windage" through an intuitive grasp of all the motions, thereby saving sequences.

Picture processing

After the digital data were transmitted to Earth and received at the Jet Propulsion Laboratory, they were subject to a variety of processes to produce the final photoproducts. The first step was to strip out all the non-video data and produce a System Data Record (SDR). This was compiled in video format, and an Experimental Data Record (EDR), which is simply a digital archival tape of the raw data, was produced. The frames were then subject to a series of cleanup and enhancement procedures: (1) Missing pixels were filled in with the average values (DN) of the adjacent pixels. (2) Telemetry noise was removed by a program (despike) that compared DN values with those of the adjacent pixels. If the difference exceeded a specified threshold value, then the DN was replaced with the average of the adjacent pixels. This removed the "salt and pepper" texture in noisy data. (3) A blemish removal program corrected in part for known blemishes on the vidicon faceplate. However, a prominent circular blemish caused by a dust particle still remains on pictures from one camera on the A orbiter. (4) A shading correction program removed artifacts caused by the differential light response of different parts of the vidicon and produced an image with 8-bit pixels. The shading correction was not perfect, partly because the shading characteristics of the cameras changed while the mission was in progress. A slight camera shading is thus still evident in some low-contrast scenes.

At this point the processing branched. A shading-corrected version, labeled SCR-2, was produced after an appropriate stretch. Normally the range of DN values in the scene was so narrow that if the unstretched data were used to drive the film recorders, then the film produced would have virtually no contrast. The processed data were therefore "stretched" to fill the full 256-DN range of the new 8-bit data. This was effected by making a histogram of the DN in the scene, then counting pixels from the bottom of the DN range until a specified percentage (usually 2–3 percent) of the scene total had been counted. This pixel was then assigned a DN value of 1. This meant that the darkest 2–3 percent of the scene was assigned a value of zero, in other words it was made black in the final scene. Similarly pixels were counted down from the top of the histogram, and the top 2–3 percent of the scene was made white. Other pixel values were assigned by interpolation. The process is important to understand, for the stretch changed from scene to scene according to the scene content. A "busy" scene, such as one from the canyons, might have DN values in the shading-corrected version that range from 50 to 200, whereas the equivalent picture from the adjacent plains might have DN values that range from 130 to 150. The plains picture would therefore be subject to a much harsher stretch. In addition caution must be exercised in interpreting "blacks" within the scene as shadows. The blacks are mostly an artifact of the processing and are only rarely shadows. Shadows can normally be identified only in the unprocessed digital data.

A filtered version (NGF) was usually produced, in addition to the shading-corrected version. Topographic detail is generally more visible in the filtered version, particularly if there are large albedo variations or differences in illumination within the scene. The filter used was an asymmetric box filter. A moving average was calculated for a specified number of pixels along the line being processed. To this average was added a weighted value of the average used to process the pixel in the same position on the previous line. The DN for the pixel being processed was then computed by taking the difference between the actual pixel value and the weighted average, then adding 127 to place the data in the center of the DN range. The process was repeated for every pixel in the frame. The horizontal dimensions of the filter could be specified, as could the weighting of the pixel average from the previous line, which controlled the rate of decay of the filter in the vertical direction. Like all filters the NGF creates artifacts where there are strong local contrasts. The artifacts are particularly evident around the edge of the polar cap, where there is usually a dark band created by the filter. The filtered data were stretched in the same way as the shading-corrected data, to drive the film recorders.

Two kinds of filtered versions were produced: orthographic and rectilinear. Many pictures are geometrically distorted because of oblique views. The orthographic version displays the scene as though it were viewed vertically. Such transformations were only performed when the emission angle (the angle between vertical and the spacecraft, as viewed from the center of the frame) was less than 60° and the range was less than 15,000 km. Otherwise an unprojected (rectilinear) filtered version was produced. Normally the pictures were not processed until the geometry at the time the pictures were taken could be fully reconstructed. This entailed having orbit determinations before and after the event and understanding the slow rocking motion of the spacecraft as it moved between firings of the stabilizing jets. The geometric data were compiled into a Supplementary Experimental Data Record (SEDR). This was required in order to make the orthographic projections and fill out the science data block in the picture annotation. For operational reasons some pictures were processed before the SEDR was available. In these cases the science data block in the annotation is left blank.

Picture annotation

The annotations around the Viking orbiter pictures provide information

Figure A.1.

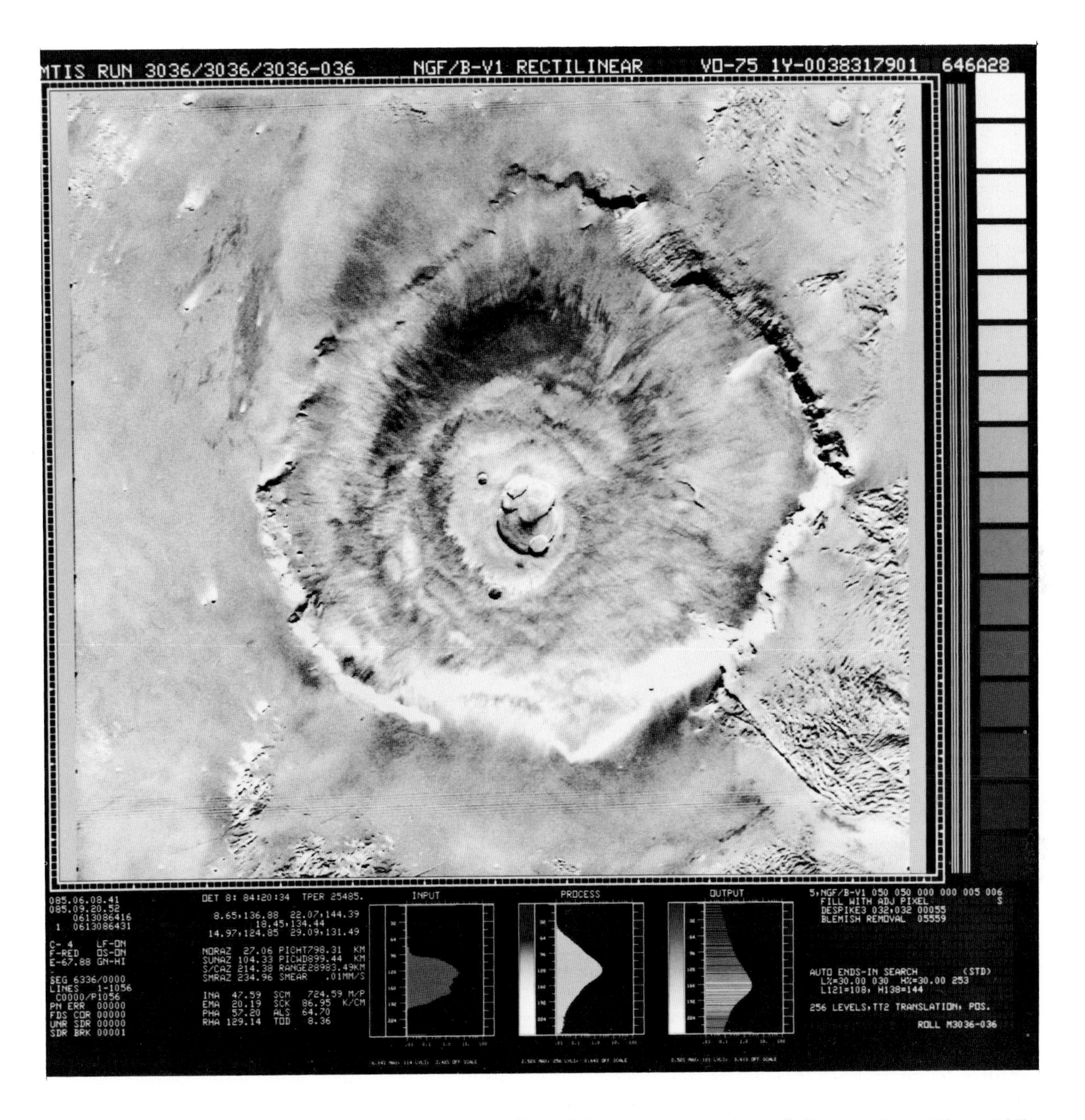

on the quality of the video data, the timing, the geometry, the state of the camera when the picture was taken, and how the picture was processed. For most purposes all that is needed is the picture number (PICNO) at the top right (figure A.1), which identifies the frame, the version in top center, and the information in the science data block, lower left center. For completeness, however, an explanation of the entire annotation is given in table A.1. The histograms at the bottom center of the pictures show the frequency of different DN values in the frames at different stages in the processing. The left histogram (input) is of the raw data. The middle histogram (process) is for the processed data that constitute the particular version (e.g. SCR-2, NGF). The histogram on the right (output) is of the stretched data used to drive the film recorders.

Qualified scientists and educators may request Viking orbiter pictures from the National Space Science Data Center, Code 601.4, Goddard Space Flight Center, Greenbelt, MD 20771.

TABLE A.1.

GENERAL

PICTURE MATRIX	1056 LINES X 1204 PIXELS/LINE = 1.27 x 10^6 PIXELS = 3168 X 3612 DISPLAY ELEMENTS
TRACK PRESENCE MASK	INDICATES PRESENCE OF SEGMENTS (DATA FRAMES) COMPRISING EACH LINE. 172 PIXELS/SEGMENT x 7 SEGMENTS/LINE = 1204 PIXELS/LINE
GRAY SCALE	16 LEVEL RECORDED GRAY SCALE (NOT TRANSLATED) DATA NUMBER (DN); 00 (BLACK), 17, 34, 51, 68, 85, 102, 119, 136, 153, 170, 187, 204, 221, 238, 255 (WHITE)

PICTURE TITLE

MTIS RUN	VRP RUN NO./VEP RUN NO./VPG RUN NO. - PRODUCT SEQUENCE NO. (relevant only for data management)	
PROCESS	RAW	
	SHADING CORRECTED - SCR 2	
	FILTERED - NGF, MF-1	
	(RECTILINEAR OR ORTHOGRAPHIC PROJECTION)	
VO 75	1X	1,2 = S/C ID X,Y = EVEN, ODD FSC
	-XXXXXXXXXX	FRAME START COUNT (FSC)
	XXXaXX	PICTURE NUMBER (PICNO)
		xxx = ORBIT REVOLUTION NO.
		A = S/C 1 Revs 1 - 1,000
		S = S/C 1 Revs > 1,000
		B = S/C 2
		XX = PICTURE SERIAL NO.

CAMERA DATA BLOCK
(lower left)

LINE		
1.	ERT (UTC) OF FIRST PICTURE SEGMENT	DDD.HH.MM.SS
2.	ERT (UTC) OF LAST PICTURE SEGMENT	DDD.HH.MM.SS
3.	30-BIT FDS COUNT (FDSC) - FIRST PICTURE SEGMENT	DECIMAL
4.	ORBITER NUMBER AND 30-BIT FDS COUNT - LAST PICTURE SEGMENT	DECIMAL
5.	--	
6.	C-CAMERA SERIAL NUMBER LF-LIGHT FLOOD (ON-OFF)	
7.	F-FILTER (BLUE, MINUS BLUE, VIOLET, CLEAR, GREEN, RED) OS-OFFSET (ON-OFF)	
8.	E-EXPOSURE TIME-MS GN-GAIN (HI-LO)	
9.	--	
10.	SEG-FULLY SYNCED/PARTIALLY SYNCED SEGMENTS (DATA FRAMES)	
11.	LINES - FIRST AND LAST LINES FOR WHICH ONE OR MORE SEGMENTS WERE RECEIVED	
12.	CXXXX/PXXXX	NO. COMPLETE LINES/NO. PARTIAL LINES
13.	PN ERROR	NO. OF SEGMENTS WITH FRAME SYNC CODE WORD ERROR
14.	FDS CORR.	NO. OF SEGMENTS WITH FDSC CORRECTIONS
15.	UNR SDR	NO. OF UNREADABLE SDR SEGMENTS
16.	SDR BRK	NO. OF LOGICAL SEQUENCE BREAKS IN SDR

SCIENCE DATA BLOCK
(lower left center)

LINE	LABEL	MEASUREMENT	UNITS
1	OET	ORBITER EVENT TIME (UTC) (MEASUREMENT TIME)	Y:DDD:HH:MM
	TPER	TIME FROM PERIAPSIS	SEC
2	--		
3	--	INTERCEPT LATITUDE (P1) UPPER LEFT	DEG
	--	INTERCEPT LONGITUDE (P1) UPPER LEFT	DEG
	--	(P3) UPPER RIGHT	DEG
	--	(P3) UPPER RIGHT	DEG
4	--	(P5) CENTER	DEG
	--	(P5) CENTER	DEG
5	--	(P7) LOWER LEFT	DEG
	--	(P7) LOWER LEFT	DEG
	--	(P9) LOWER RIGHT	DEG
	--	(P9) LOWER RIGHT	DEG
6	--		
7	NORAZ	AZIMUTH OF NORTH [a]	DEG
	PICHT	DISTANCE ON SURFACE, RETICLE 2-8	KM
8	SUNAZ	AZIMUTH OF SUN [a]	DEG
	PICWD	DISTANCE ON SURFACE, RETICLE 4-6	KM
9	S/CAZ	AZIMUTH OF SUB-SPACECRAFT POINT [a]	DEG
	RANGE	INTERCEPT RANGE (P5)	KM

10	SMRAZ	SMEAR AZIMUTH-IMAGE PLANE [a]	DEG
	SMEAR	SMEAR VELOCITY-IMAGE PLANE	MM/SEC
11	--		
12	INA	INCIDENCE ANGLE (P5)	DEG
	SCM	SCALE - FRAME CENTER	M/PIXEL
13	EMA	VIEWING ANGLE (P5)	DEG
	SCK	SCALE - FRAME CENTER (5" FILM)	KM/CM
14	PHA	PHASE ANGLE (P5)	DEG
	ALS	SUNS AREOCENTRIC LONGITUDE	DEG
15	RHA	MARS HOUR ANGLE (P5)	DEG
	TOD	TIME OF DAY ON PLANET	HRS
16	SCO	ORTHOGRAPHIC PROJECTION SCALE	KM/PIXEL

[a] Measured Clockwise from Axis defined by Reticle 5-6

PROCESSING DATA BLOCK
(lower right)

LINE			
1	PROCESS TYPE	1, RAW	
		2, SCR2	SHADING CORRECTED
		5, NGF/B-V1	NONGRADIENT HIGH-PASS FILTERED
		6, NGF/B-V2	NONGRADIENT HIGH-PASS FILTERED WITH DIFFERENT DC FEEDBACK
		3 or 4, MF-X	MTIS HIGH-PASS FILTERED-TYPE X
		XXX	FILTER WIDTH, PIXELS, OR M FACTOR MF-X
		XXX	FILTER HEIGHT, PIXELS, OR Q FACTOR MF-X
		XXX	-
		XXX	-
		XXX	FILTER STARTING LINE NO.
		XXX	DC FEEDBACK FACTOR (F)% or BINARY FACTOR, MF-X
2	PIXEL FILL-IN	FILL WITH XXX - FILLED WITH DN VALUE XXX FILL WITH ADJ PIXEL - FILLED WITH VALUE OF LEFT ADJACENT PIXEL	
	R or S	INDICATES WHETHER DATA SOURCE IS RAW OR SHADING CORRECTED DATA	
3	DESPIKE (VERS. 3)	XXX, XXX	UPPER AND LOWER TOLERANCES IN DN
	PIXEL ERROR CORRECTION	XXXXX	NO. OF ERRORS CORRECTED
4	BLEMISH REMOVAL	XXXXX	NO. OF PIXELS CORRECTED IN ALL BLEMISHES
5	--		
6	GEOMETRIC TRANSFORMATION		LABEL PRESENT IF PICTURE IS ORTHOGRAPHIC (OR POLAR STEREO) PROJECTION.
7	FILM RECORDER RESOLUTION	X1,X2,or X3	X1 = 3168 x 3168 DELS X2 = 1584 x 1584 DELS X3 = 1056 x 1056 DELS
8	--		
9	STRETCH TYPE	AUTO ENDS-IN SEARCH	(OVR) = OVERRIDE PARAMETERS
		AUTO CENTER-OUT SEARCH	(STD) = STANDARD PARAMETERS
		MANUAL SEARCH	
10	L% = XX.XX	SPECIFIED PERCENTAGE OF DATA ABOVE A SPECIFIED LOW-END START POINT IN THE PROCESSED HISTOGRAM TO BE ELIMINATED FOR ESTABLISHING THE LOW-END START POINT (IN DN) FOR THE OUTPUT HISTOGRAM.	
	YYY	SPECIFIED DN FOR THE LOW-END START POINT FOR OUTPUT DATA DISPLAY AND HISTOGRAM	
	H% = XX.XX	DEFINES THE HIGH-END PARAMETERS CORRESPONDING TO L% = XX.XX ABOVE	
	YYY	DEFINES HIGH-END PARAMETER CORRESPONDING TO YYY ABOVE	
11	STRETCH PARAMETERS (OUTPUT DATA) LXXX = YYY	XXX = COMPUTED LOW-END DN IN PROCESSED HISTOGRAM TO BE TRANSLATED TO SPECIFIED YYY DN IN OUTPUT HISTOGRAM	
	MXXX = YYY	XXX = FIXED MID-POINT DN IN PROCESSED HISTOGRAM TO BE TRANSLATED TO SPECIFIED YYY DN IN OUTPUT HISTOGRAM (3-POINT STRETCH)	
	HXXX = YYY	XXX = COMPUTED HIGH-END DN IN PROCESSED HISTOGRAM TO BE TRANSLATED TO SPECIFIED YYY DN IN OUTPUT HISTOGRAM	
12	--		
13	XXX DISPLAY LEVELS	NO. OF LEVELS SPECIFIED FOR OUTPUT DISPLAY (HISTOGRAM)	
	TT-X	OUTPUT TRANSLATION TABLE IDENTIFIER (GAMMA CORRECTION TABLE): TT-1, TT-2, TT-3	
14	--		
15	ROLL ABXXX-YYY	FILM ROLL NO. M1 = VO-1 RECTILINEAR M2 = VO-2 RECTILINEAR P1 = VO-1 ORTHOGRAPHIC P2 = VO-2 ORTHOGRAPHIC MO = DIGIFAX OR SPECIAL	

Measured Clockwise from Axis defined by Reticle 5-6

.EL

APPENDIX B **MAPS OF MARS**

The U.S. Geological Survey has published a series of maps that depict the martian surface in different ways. There are three main types:

1. Shaded relief maps, which are air-brush renditions of the surface relief.
2. Topographic maps, which are the same as the shaded relief maps except that elevation contours and albedo patterns have been superimposed.
3. Geologic maps, which are a geologist's interpretation of the configuration of the rocks at the surface.

Maps are published at a variety of scales, from 1:25 million (1:25M) of the entire planet, from which the maps following this appendix were reproduced, to 1:250,000 (1:250K) for some special site maps made at the request of the Viking project. The entire surface has been mapped in thirty separate sheets (figure B.1), designated MC charts, at a scale of 1:5,000,000 (1:5M) (table B.1). These are based almost entirely on Mariner 9 data. A new series, at a scale of 1:2,000,000 (1:2M) is being prepared from the Viking data. The 1:2M maps are referenced to the MC charts. MC-18 SE, for example, is the 1:2M map of the southeast quadrant of the 1:5M chart MC-18.

A complete description of the 1:5M map series and how the maps were made, together with a gazeteer of martian names, can be found in Batson et al. 1979. For a commentary on martian nomenclature, see Blunck 1977.

U.S. Geological Survey maps can be purchased from U.S. Geological Survey, Branch of Distribution, Denver Federal Center, Denver, CO 80225 or from U.S. Geological Survey, Branch of Distribution, 1200 South Eads Street, Arlington, VA 22202. The maps should be ordered by the I-numbers listed in table B.1. The 1:25M topographic map of the entire planet, for example, is Map I-961.

TABLE B.1. PUBLISHED MAPS OF MARS

Quad. No.	*Map* Type*	*Scale*	*Year Publ.*	*I-No.*
MC1	Geol	5M	—	—
MC2	Geol	5M	—	—
MC3	Geol	5M	1979	1154
MC4	Geol	5M	1978	1048
MC5	Geol	5M	1978	1065
MC6	Geol	5M	1978	1038
MC7	Geol	5M	1979	1140
MC8	Geol	5M	1978	1049
MC9	Geol	5M	1975	893
MC10	Geol	5M	1975	894
MC11	Geol	5M	1976	895
MC12	Geol	5M	1977	996
MC13	Geol	5M	1977	995
MC14	Geol	5M	1979	1110
MC15	Geol	5M	1976	935
MC16	Geol	5M	1979	1137
MC17	Geol	5M	1978	896
MC18	Geol	5M	1978	897
MC19	Geol	5M	1979	1144
MC20	Geol	5M	—	—
MC21	Geol	5M	1977	1020
MC22	Geol	5M	1978	1073
MC23	Geol	5M	1978	1111
MC24	Geol	5M	1979	1145
MC25	Geol	5M	1978	1077
MC26	Geol	5M	—	—
MC27	Geol	5M	1977	910
MC28	Geol	5M	1976	941
MC29	Geol	5M	1978	1008
MC30	Geol	5M	1978	1076
MC1	Shad	5M	1976	969
MC1	Topo	5M	1977	1027
MC2	Shad	5M	1976	989
MC3	Shad	5M	1975	963
MC4	Shad	5M	1975	958
MC4	Topo	5M	1976	979
MC5	Shad	5M	1978	1052
MC6	Shad	5M	1978	1121
MC6	Topo	5M	1978	1119
MC7	Shad	5M	1978	1122
MC7	Topo	5M	1978	1120
MC8	Shad	5M	1976	956
MC9	Shad	5M	1975	926
MC9	Topo	5M	1976	977
MC10	Shad	5M	1975	925

MC10	Topo	5M	1976	971
MC11	Shad	5M	1976	955
MC11	Topo	5M	1976	978
MC12	Shad	5M	1978	1079
MC13	Shad	5M	1975	929
MC13	Topo	5M	1976	967
MC14	Shad	5M	1977	1023
MC14	Topo	5M	1977	1024
MC15	Shad	5M	1978	1131
MC15	Topo	5M	1978	1135
MC16	Shad	5M	1978	1075
MC17	Shad	5M	1975	924
MC17	Topo	5M	1976	984
MC18	Shad	5M	1975	928
MC18	Topo	5M	1976	976
MC19	Shad	5M	1975	927
MC19	Topo	5M	1976	975
MC20	Shad	5M	1978	1050
MC21	Shad	5M	1978	1118
MC22	Shad	5M	1978	1123
MC23	Shad	5M	1976	1000
MC23	Topo	5M	1976	1001
MC24	Shad	5M	1979	1166
MC24	Topo	5M	1979	1167
MC25	Shad	5M	1979	1164
MC25	Topo	5M	1979	1165
MC26	Shad	5M	1975	923
MC26	Topo	5M	1976	985
MC27	Shad	5M	1979	1168
MC28	Shad	5M	1979	1169
MC29	Shad	5M	1979	1170
MC30	Shad	5M	1976	970
—	Geol	25M	1978	1083
—	Shad	25M	1975	MF67
—	Shad	25M	1975	MF64
—	Topo	25M	1976	961
—	Shad	25M	1972	810
—	Shad	25M	1975	940
—	Topo	1M	1976	986
—	Shad	1M	1975	947
—	Shad	1M	1976	957
—	Topo	1M	1977	1002
—	Shad	1M	1977	1055
—	Shad	1M	1976	939
—	Topo	1M	1976	983
—	Shad	1M	1975	946
—	Topo	1M	1976	988
—	Shad	1M	1977	1026
—	Topo	1M	1977	1046
—	Shad	5M	—	—
—	Shad	17M	1973	—
—	Shad	32M	—	—
—	CMOS	1M	1977	1069
—	CMOS	1M	1977	1068
—	CMOS	250K	1977	1059
—	CMOS	1M	1977	1061
—	CMOS	250K	1977	1060

NOTE: Geol = geologic map; Shad = shaded relief only; Topo = shaded relief, albedo and elevation contours; CMOS = corrected mosaic.

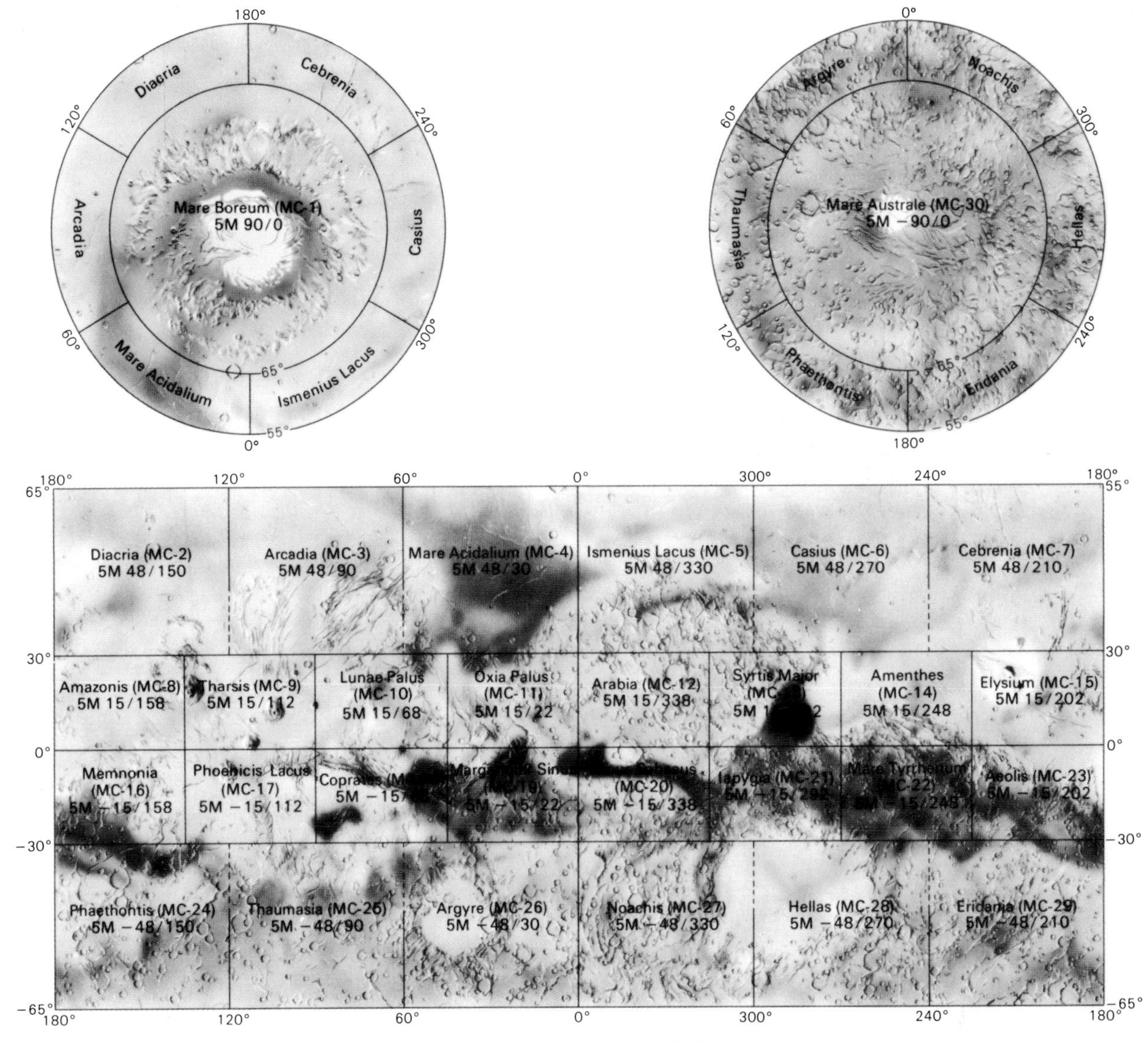

Figure B.1. Layout of the U.S. Geological Survey 1:5 million quadrangle maps of Mars. The name, the chart number in parentheses, the scale (5M in all cases), and the central latitude and longitude are given for each quadrangle.

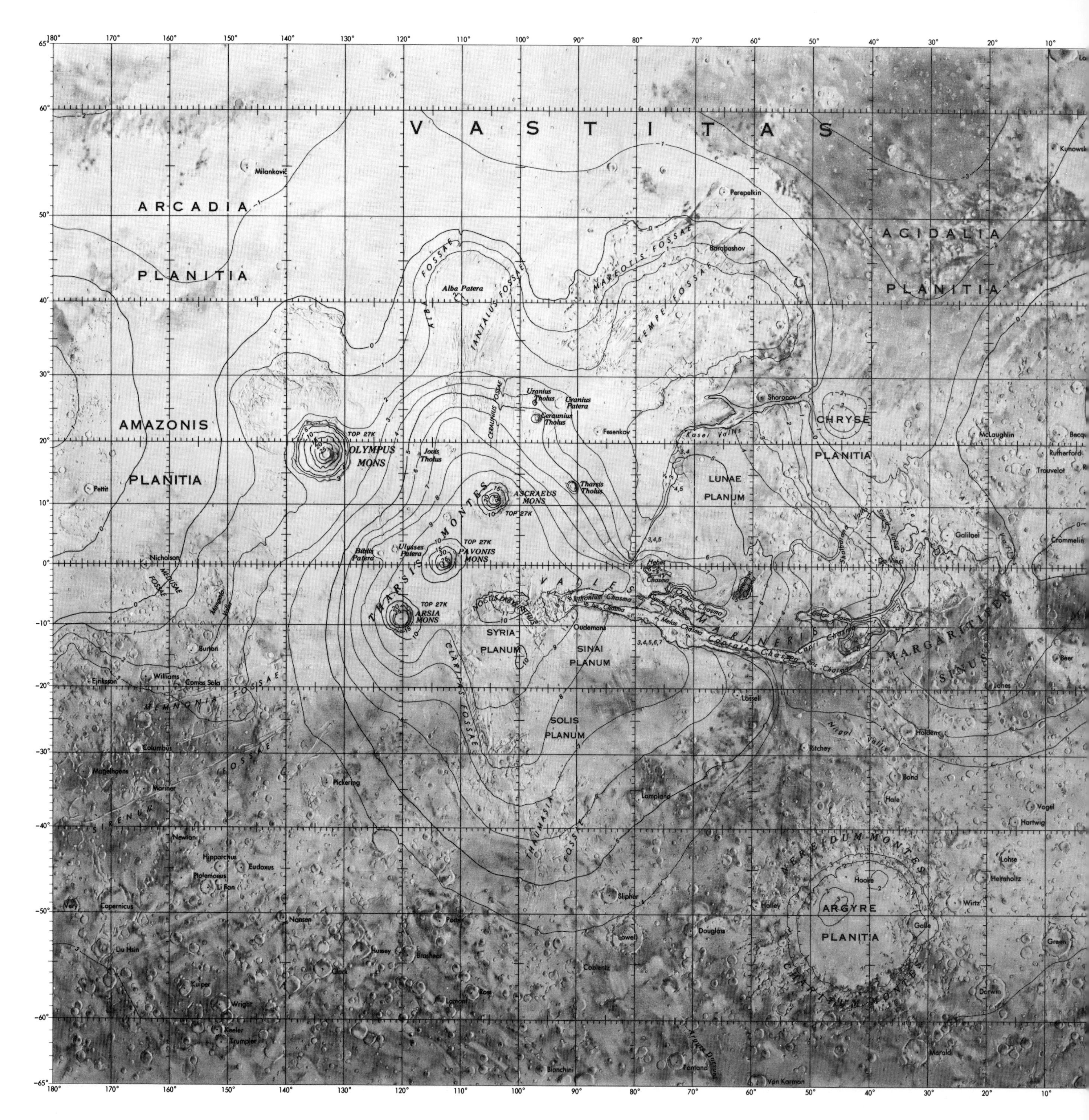

Equatorial maps of Mars (courtesy of U.S. Geological Survey).

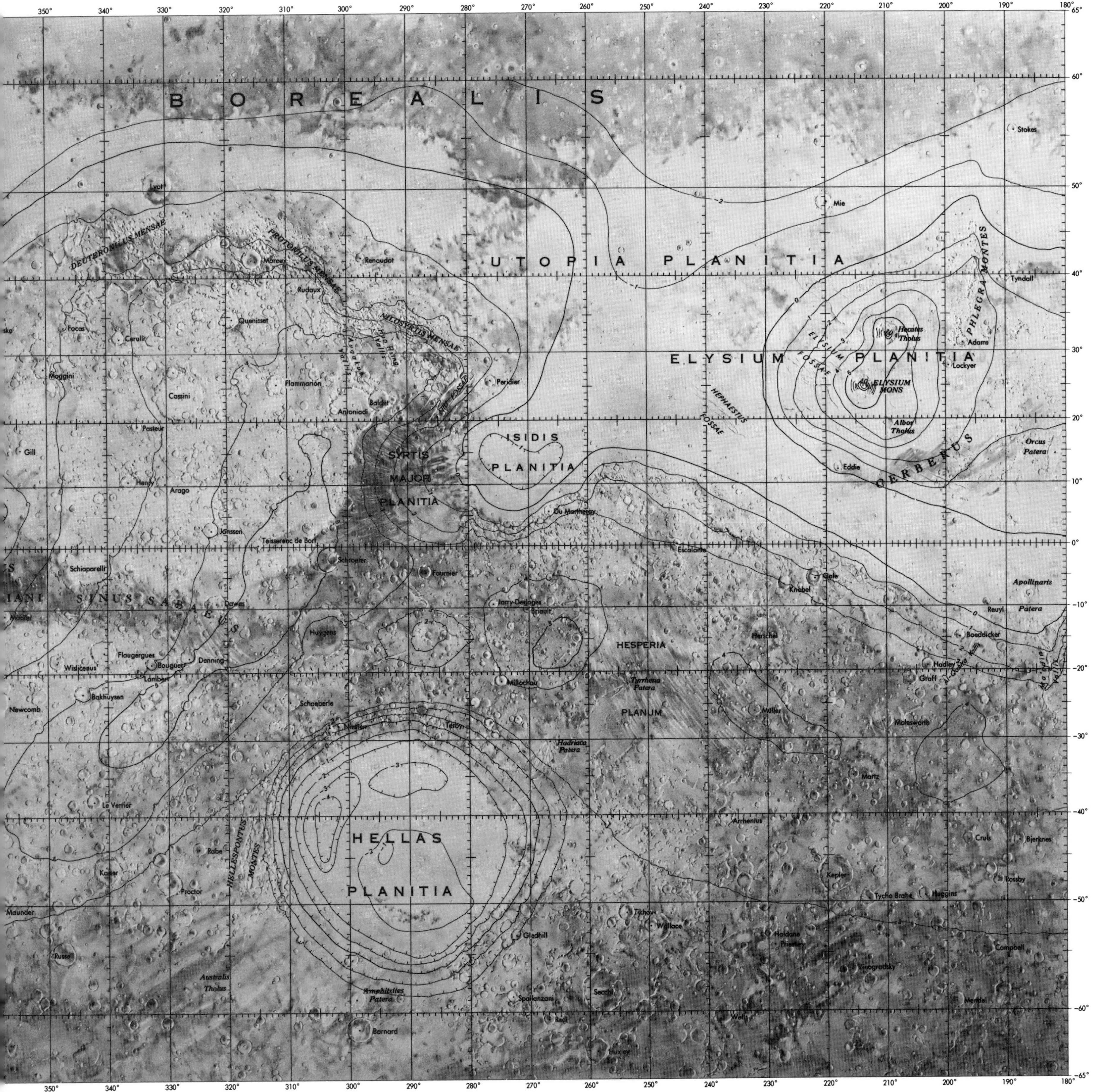

B O R E A L I S
UTOPIA PLANITIA
ELYSIUM PLANITIA
ELYSIUM MONS
Hecates Tholus
Albor Tholus
PHLEGRA MONTES
ELYSIUM FOSSAE
HEPHAESTUS FOSSAE
CERBERUS
ISIDIS PLANITIA
SYRTIS MAJOR PLANITIA
DEUTERONILUS MENSAE
PROTONILUS MENSAE
NILOSYRTIS MENSAE
SINUS SABAEUS
HESPERIA PLANUM
HELLAS PLANITIA
HELLESPONTUS MONTES
Tyrrhena Patera
Hadriaca Patera
Amphitrites Patera
Australis Tholus
Orcus Patera
Apollinaris Patera
Mie
Stokes
Tyndall
Adams
Lockyer
Eddie
Lyot
Moreux
Renaudot
Rudaux
Quenisset
Focas
Cerulli
Moggini
Cassini
Flammarion
Peridier
Baldet
Antoniadi
Pasteur
Gill
Henry
Arago
Janssen
Teisserenc de Bort
Schroeter
Schiaparelli
Fournier
Escalante
Gale
Knobel
Dawes
Huygens
Flaugergues
Bouguer
Denning
Wislicenus
Lambert
Bakhuysen
Newcomb
Schaeberle
Millochau
Herschel
Reuyl
Boeddicker
Hadley
Graff
Molesworth
Müller
Martz
Arrhenius
Cruls
Bjerknes
Kepler
Rossby
Tycho Brahe
Huggins
Campbell
Vinogradsky
Haldane
Priestley
Tikhov
Wallace
Gledhill
Secchi
Spallanzani
Redi
Wells
Mendel
Barnard
Huxley
Le Verrier
Rabe
Kaiser
Proctor
Maunder
Russell
Du Martheray
Terby
Niesten
Jarry-Desloges
Briault
Mädler

REFERENCES

Adams, J. B., and McCord, T. B. 1969. Mars: Interpretation of spectral reflectivity of light and dark regions. *J. Geophys. Res.* 74:4851–56.

Allegre, C. J., Courtillot, V. E., and Matteur, M. 1974. Evidence for lateral movements of the martian crust (Abs.). *Trans. Am. Geophys. Union (EOS)* 55:341.

Allen, C. C. 1979a. Areal distribution of martian rampart craters. *Icarus* 39:111–23.

Allen, C. C. 1979b. Volcano-ice interactions on Mars. *J. Geophys. Res.* 84:8048–59.

Anders, E., and Owen, T. 1977. Mars and Earth: Origin and abundance of volatiles. *Science* 198:453–65.

Anderson, D. M., Biemann, K., Orgel, L. E., Oro, T., Shulman, G. P., Toulmin, P., and Urey, H. C. 1972. Mars spectrometric analysis of organic compounds, water, and volatile constituents, in the atmosphere and surface of Mars. *Icarus* 16:111–38.

Anderson, D. M., Gatto, L. W., and Ugolini, F. 1973. An examination of Mariner 6 and 7 imagery for evidence of permafrost terrain on Mars. In *Permafrost: The North American contribution to the second international conference* (F. J. Sanger, ed.), pp. 449–508. Nat. Acad. Sci. U.S.A., Washington, D.C.

Anderson, D. W., Gaffney, E. S., and Low, P. F. 1967. Frost phenomena on Mars. *Science* 155:319–22.

Anderson, D. M., Schwartz, M. J., and Tice, A. R. 1979. Water-vapor adsorption by sodium montmorillonite at −5°C. *Icarus* 34:638–44.

Andresen, A., and Bjerrum, L. 1968. *Slides in subaqueous slopes in loose sand and silt.* Norwegian Geotechnical Institute Publ. no. 81.

Antoniadi, E. M. 1930. *La planète Mars, 1659–1929.* Herman et Cie, Paris.

Aramaki, S. 1956. The 1783 activity of Asama Volcano. Parts 1 and 2. *Jap. J. Geol. Geog.* 27:191–229; 28:11–32.

Aronson, J. R., and Emslie, A. G. 1975. Composition of the martian dust as derived from infrared spectroscopy from Mariner 9. *J. Geophys. Res.* 80:4925–31.

Arvidson, R. E. 1972. Aeolian processes on Mars: Erosive velocities, settling velocities, and yellow clouds. *Geol. Soc. Am. Bull.* 83:1503–08.

Arvidson, R. E. 1974a. Wind-blown streaks, splotches, and associated craters on Mars: Statistical analysis of Mariner 9 photographs. *Icarus* 21:12–27.

Arvidson, R. E. 1974b. Morphologic classification of martian craters and some implications. *Icarus* 22:264–71.

Arvidson, R. E. 1978. *Viking implications for martian aeolian dynamics.* NASA Tech. Memo. 79729, pp. 238–40.

Arvidson, R. E. 1979. *A post-Viking view of martian geologic evolution.* NASA Tech. Memo. 80339, pp. 80–81.

Arvidson, R. E., Binder, A. B., and Jones, K. L. 1978. The surface of Mars. *Sci. Am.* 238:76–89.

Arvidson, R. E., Carusi, A., Coradini, A., Coradini, M., Fulchignoni, M., Federico, C., Funicello, R., and Salomone, M. 1976. Latitudinal variation of wind erosion of crater ejecta deposits on Mars. *Icarus* 27:503–16.

Arvidson, R. E., Guiness, E. A., and Lee, S. W. 1979. Differential aeolian redistribution rates on Mars. *Nature* 278:533–35.

Bagnold, R. A. 1941. *The physics of blown sand and desert dunes.* Methuen, London.

Baird, A. K., Castro, A. J., Clark, B. C., Toulmin, P., Rose, H. J., Keil, K., and Gooding, J. L. 1977. The Viking x-ray fluorescence experiment: Sampling strategies and laboratory simulations. *J. Geophys. Res.* 82:4595–624.

Baird, A. K., Toulmin, P., Clark, B. C., Rose, H. J., Keil, K., Christian, R. P., and Gooding, J. L. 1976. Mineralogic and petrologic implications of Viking geochemical results from Mars: Interim results. *Science* 194:1288–93.

Baker, V. R. 1973. *Paleohydrology and sedimentology of Lake Missoula flooding of eastern Washington.* Geol. Soc. Am., Spec. Paper 144.

Baker, V. R. 1974. Erosional forms and processes for the catastrophic Pleistocene Missoula floods in eastern Washington. In *Fluvial geomorphology* (M. Morisawa, ed.), pp. 123–48. State Univ. of New York, Binghamton, N.Y.

Baker, V. R. 1977. *Viking: Slashing at the martian scabland problem.* NASA Tech. Memo. X-3511, pp. 169–72.

Baker, V. R. 1978. A preliminary assessment of the fluid erosional process that shaped the martian outflow channels. In *Proc. 9th Lunar and Planet. Sci. Conf.*, pp. 3205–23.

Baker, V. R., and Kochel, R. C. 1978a. Morphological mapping of martian outflow channels. In *Proc. 9th Lunar Planet. Sci. Conf.*, pp. 3181–92.

Baker, V. R., and Kochel, R. C. 1978b. Morphometry of streamlined forms in terrestrial and martian channels. In *Proc 9th Lunar Planet. Sci. Conf.*, pp. 3193–203.

Baker, V. R., and Milton, D. J. 1974. Erosion by catastrophic floods on Mars and Earth. *Icarus* 23:27–41.

Baldwin, R. B. 1963. *The measure of the Moon.* Univ. of Chicago Press, Chicago.

Ballou, E. V., Wood, P. C., Wydeven, T., Lehwalt, M. E., and Mack, R. E. 1978. Chemical interpretation of Viking 1 life detection experiment. *Nature* 271:644–45.

Banin, A., and Rishpon, J. 1979. Smectite clays in Mars soil: Evidence for their presence and role in Viking biology experimental results. *J. Mol. Evol.* 14:133–52.

Barnes, H. L. 1956. Cavitation as a geologic agent. *Am. J. Sci.* 254:493–505.

Barth, C. A. 1974. The atmosphere of Mars. *Ann. Rev. Earth Planet. Sci.* 2:333–67.

Batson, R. M., Bridges, P. M., and Inge, J. L. 1979. *Atlas of Mars.* NASA Spec. Pub. 438.

Baum, W. A. 1974. Earth-based observations of martian albedo changes. *Icarus* 22:363–70.

Belcher, D., Veverka, J., and Sagan, C. 1971. Mariner photography of Mars and aerial photography of Earth: Some analogies. *Icarus* 15:241–52.

Belton, M. J. S., Broadfoot, A. L., and Hunten, D. M. 1968. Abundance and temperature of CO_2 on Mars during the 1967 opposition. *J. Geophys. Res.* 73:4795–806.

Biemann, K. 1974. Test results on the Viking gas chromatograph–mass spectrometer experiment. *Origins of Life* 5:517–30.

Biemann, K., Owen, T., Rushneck, D. R., Lafleur, A. L., and Howarth, D. W. 1976a. The atmosphere of Mars near the surface: Isotope ratios and upper limits on the noble gases. *Science* 194:76–78.

Biemann, K., and others (11 authors). 1976b. Search for organic and volatile inorganic compounds in two surface samples from the Chryse Planitia region of Mars. *Science* 194:72–76.

Biemann, K., and others (12 authors). 1977. The search for organic substances and inorganic volatile compounds in the surface of Mars. *J. Geophys. Res.* 82:4641–62.

Bills, B. G., and Ferrari, A. J. 1978. Mars topography and geophysical implications. *J. Geophys. Res.* 83:3497–508.

Binder, A. B., Arvidson, R. E., Guiness, E. A., Jones, K. L., Morris, E. C., Mutch, T. A., Pieri, D. C., and Sagan, C. 1977. The geology of the Viking lander 1 site. *J. Geophys. Res.* 82:4439–51.

Binder, A. B., and Cruikshank, D. P. 1966. Lithologic and mineralogic investigations of the surface of Mars. *Icarus* 5:521–25.

Binder, A. B., and Jones, J. C. 1972. Spectrophotometric studies of the photometric function, composition, and distribution of the surface materials of Mars. *J. Geophys. Res.* 77:3005–19.

Blackwelder, E. 1934. Yardangs. *Geol. Soc. Am. Bull.* 45:159–66.

Blasius, K. R. 1976a. Topical studies of the geology of the Tharsis region of Mars. Ph.D. diss. Calif. Inst. Tech.

Blasius, K. R. 1976b. The record of impact cratering on the great volcanic shields of the Tharsis region of Mars. *Icarus* 29:343–61.

Blasius, K. R. 1979. Topography of six martian volcanoes from high-altitude systematic stereo imaging of Viking orbiter 1 (Abs.). *Bull. Am. Astron. Soc.* 11:573.

Blasius, K. R., and Cutts, J. A. 1976. Shield volcanism and lithospheric structure beneath the Tharsis plateau, Mars. In *Proc. 7th Lunar Sci. Conf.*, pp. 3561–73.

Blasius, K. R., Cutts, J. A., and Roberts, R. J. 1978. *Large-scale erosive flows associated with Chryse Planitia, Mars: Source and sink relationships.* NASA Tech. Memo. 79729, pp. 275–76.

Blasius, K. R., Cutts, J. A., Guest, J. E., and Masursky, H. 1977. Geology of Valles Marineris: First analysis of imaging from the Viking orbiter primary mission. *J. Geophys. Res.* 82:4067–91.

Blumsack, S. L. 1971. On the effects of topography on planetary circulation. *J. Atmos. Sci.* 28:1134–43.

Blumsack, S. L., Gierasch, P. J., and Wessel, W. R. 1973. An analytical and numerical study of the martian planetary boundary layer over slopes. *J. Atmos. Sci.* 30:66–82.

Blunck, J. 1977. *Mars and its satellites.* Exposition Press, Hicksville, New York.

Booth, M. C., and Kieffer, H. H. 1978. Carbonate formation in Mars-like environments. *J. Geophys. Res.* 83:1809–15.

Boyce, J. M. 1979. *A method for measuring heat flow in the martian crust using impact crater morphology.* NASA Tech. Memo. 80339, pp. 114–18.

Boyce, J. M., and Roddy, D. J. 1978. *Martian rampart craters: Crater processes that may affect diameter-frequency distributions.* NASA Tech. Memo. 79729, pp. 162–65.

Boyce, P. B., and Thompson, D. T. 1972. A new look at the martian "violet haze" problem. I. Syrtis Major-Arabia 1969. *Icarus* 16:291–303.

Breed, C. S., Grolier, M. J., and McCauley, J. F. 1979. Morphology and distribution of common 'sand' dunes on Mars: Comparison with the Earth. *J. Geophys. Res.* 84:8183–204.

Bretz, J. H. 1923. The channeled scablands of the Columbia plateau. *J. Geol.*, 31:617–49.

Bretz, J. H. 1932. *The Grand Coulee.* Am. Geog. Soc. Spec. Publ. 15.

Bretz, J. H. 1969. The Lake Missoula floods and the channeled scabland. *J. Geol.*, 77:505–43.

Briggs, G. A. 1974. The nature of the residual polar caps. *Icarus* 23:167–91.

Briggs, G. A., Baum, W. A., and Barnes, J. 1979. Viking orbiter observations of dust in the martian atmosphere. *J. Geophys. Res.* 84:2795–820.

Briggs, G. A., Klaasen, K., Thorpe, T., Wellman, J., and Baum, W. 1977. Martian dynamical phenomena during June–November 1976: Viking orbiter imaging results. *J. Geophys. Res.* 82:4121–49.

Brown, F. S. 1978. The biology instrument for the Viking Mars mission. *Rev. Sci. Instr.* 49:139–82.

Brown, F. S. and others (16 authors). 1978. The biology instrument for the Viking Mars mission. *Rev. Sci. Instr.* 49:139–82.

Brown, H. 1960. Density and mass distributions of meteorites. *J. Geophys. Res.* 75:1679–83.

Bruch, C. W. 1966. Instrumentation for the detection of extraterrestrial life. In *Biology and the exploration of Mars* (C. S. Pittendrigh, W. Vishniac, and J. P. Pearman, eds.), pp. 487–502. Nat. Acad. Sci. U.S.A., Nat. Res. Council Publ. 1296.

Bryan, W. B. 1973. Wrinkle ridges as deformed surface crust on ponded mare lava. In *Proc. 4th Lunar Sci. Conf.*, pp. 93–106.

Burns, J. A. 1978. The dynamical evolution and origin of the martian moons. *Vistas in Astronomy* 22:193–210.

Buvet, R., and Ponnamperuma, C. 1971. *Chemical evolution and the origin of life.* North-Holland Publ. Co., Amsterdam and London.

Capen, C. F. 1971. Yellow clouds—past and future. *Sky and Telescope* 41:2–4.

Capen, C. F. 1974. The martian yellow cloud of July 1971. *Icarus* 22:345–62.

Capen, C. F. 1976. Martian albedo feature variations with season: Data of 1971 and 1973. *Icarus* 28:213–30.

Capen, C. F., and Martin, L. J. 1972. Survey of martian yellow storms (Abs.). *Bull. Am. Astron. Soc.* 4:374.

Carr, M. H. 1973. Volcanism on Mars. *J. Geophys. Res.* 78:4049–62.

Carr, M. H. 1974a. Tectonism and volcanism of the Tharsis region of Mars. *J. Geophys. Res.* 79:3943–49.

Carr, M. H. 1974b. The role of lava erosion in the formation of lunar rilles and martian channels. *Icarus* 22:1–23.

Carr, M. H. 1975. The volcanoes of Mars. *Sci. Am.* 2344:32–43.

Carr, M. H. 1976. Changes in height of martian volcanoes with time. *Geologica Romana* 15:421–22.

Carr, M. H. 1979. Formation of martian flood features by release of water from confined aquifers. *J. Geophys. Res.* 84:2995–3007.

Carr, M. H. 1980. The morphology of the martian surface. *Space Sci. Rev.* 25:231–84.

Carr, M. H., and Schaber, G. G. 1977. Martian permafrost features. *J. Geophys. Res.* 82:4039–65.

Carr, M. H., Blasius, K. R., Greeley, R., Guest, J. E., and Murray, J. E. 1977a. Observations on some martian volcanic features as viewed from the Viking orbiters. *J. Geophys. Res.* 82:3985–4015.

Carr, M. H., Crumpler, L. S., Cutts, J. A., Greeley, R., Guest, J. E., and Masursky, H. 1977b. Martian impact craters and emplacement of ejecta by surface flow. *J. Geophys. Res.* 82:4055–65.

Carr, M. H., Masursky, H., and Saunders, R. S. 1973. A generalized geologic map of Mars. *J. Geophys. Res.* 78:4031–36.

Carr, M. H., and others (13 authors). 1976. Preliminary results from the Viking orbiter imaging experiment. *Science* 193:766–76.

Chapman, C. R. 1974. Cratering on Mars. 1. Cratering and obliteration history. *Icarus* 22:264–71.

Chapman, C. R. 1976. Asteroids and meteorite parent bodies: An astronomical perspective. *Geochim. et Cosmochim. Acta* 40:710–19.

Chapman, C. R., and Jones, K. L. 1977. Cratering and obliteration history of Mars. *Ann. Rev. Earth Planet. Sci.* 5:515–40.

Chapman, C. R., Pollack, J. B., and Sagan, C. 1969. An analysis of Mariner 4 cratering statistics. *Astron. J.* 74:1039–51.

Christensen, E. J. 1975. Martian topography derived from occultation radar, spectral, and optical measurements. *J. Geophys., Res.* 80:2909–13.

Chun, S., Pang, K. D., and Ajello, J. M. 1979. *Photocatalytic oxidation of organic compounds in Murchison meteorite under simulated martian conditions* (Abs.). NASA CP-2072, p. 15.

Chun, S., Pang, K. D., Cutts, J. A., and Ajello, J. M. 1978. Photocatalytic oxidation of organic compounds on Mars. *Nature* 274:875–76.

Cintala, M. J., and Mouginis-Mark, P. J. 1980. Martian fresh crater depths: More evidence for substrate volatiles? In *Lunar and Planetary Science XI*, pp. 143–45. Lunar and Planetary Sci. Institute, Houston.

Cintala, M. J., Head, J. W., and Mutch, T. A. 1975. Depth/diameter relationships for martian

and lunar craters (Abs.), *Trans. Am. Geophys. Union (EOS)* 56:389.
Clague, D. A., and Dalrymple, G. B. 1973. Age of Koko seamount, Emperor seamount chain. *Earth Planet. Sci. Lett.* 17:411–15.
Clark, B. C. 1978. Implications of abundant hygroscopic minerals in the martial regolith. *Icarus* 34:645–65.
Clark, B. C., Baird, A. K., Rose, H. J., Toulmin, P., Christian, R. P., Kelliher, W. C., Castro, A. J., Rowe, C. D., Keil, K., and Huss, G. R. 1977. The Viking x-ray fluorescence experiment: Analytical methods and results. *J. Geophys Res.* 82:4577–94.
Clark, B. C., Baird, A. K., Rose, H. J., Toulmin, P., Keil, K., Castro, A. J., Kelliher, W. C., Rowe, C. D., and Evans, P. H. 1976. Inorganic analyses of martian surface sample at the Viking landing sites. *Science* 194:1283–88.
Clark, B. R., and Mullin, R. P. 1976. Martian glaciation and the flow of solid CO_2. *Icarus* 27:215–28.
Collins, S. A. 1971. *The Mariner 6 and 7 pictures of Mars.* NASA Spec. Pub. 263.
Condit, C. D. 1978. Distribution and relations of 4- to 10-km diameter craters to global geologic units of Mars. *Icarus* 34:465–78.
Cooper, H. F. 1977. A summary of explosion cratering phenomena relevant to meteor impact events. In *Impact and explosion cratering* (D. J. Roddy, R. O. Pepin, and R. B. Merrill, eds.), pp. 11–44. Pergamon Press, New York.
Courtillot, V. C., Allegre, C. J., and Matteur, M. 1975. On the existence of lateral relative motions on Mars. *Earth Planet. Sci. Lett.* 25:279–85.
Crumpler, L. S., and Aubele, J. C. 1978. Structural evolution of Arsia Mons, Pavonis Mons, and Ascraeus Mons, Tharsis region of Mars. *Icarus* 34:496–511.
Cutts, J. A. 1973a. Nature and origin of layered deposits of the martian polar regions. *J. Geophys. Res.* 78:4231–49.
Cutts, J. A. 1973b. Wind erosion in the martian polar regions. *J. Geophys. Res.* 78:4211–21.
Cutts, J. A. 1980. *Simulation of stratigraphy of martian polar layered deposits.* NASA Tech. Memo. 81776, pp. 63–65.
Cutts, J. A., and Blasius, K. R. 1979. Martian outflow channels: Quantitative comparison of erosive capacities for eolian and fluvial models (Abs.). In *Lunar and Planetary Sciences X*, pp. 257–59. Lunar and Planetary Institute, Houston.
Cutts, J. A., and Smith, R. S. U. 1973. Eolian deposits and dunes on Mars. *J. Geophys. Res.* 78:4139–54.
Cutts, J. A., Blasius, K. R., and Roberts, W. J. 1979. Evolution of martian polar landscape: Interplay of long-term variation in perennial ice cover and dust storm activity. *J. Geophys. Res.* 84:2975–94.
Cutts, J. A., Blasius, K. R., Briggs, G. A., Carr, M. H., Greeley, R., and Masursky, H. 1976. North polar region of Mars: Imaging results from Viking 2. *Science* 194:1329–37.
Cutts, J. A., Roberts, W. J., and Blasius, K. R. 1978a. Martian channels formed by lava erosion (Abs.). In *Lunar and Planetary Science IX*, Pt. 1, p. 209. Lunar and Planetary Institute, Houston.
Cutts, J. A., Roberts, W. J., and Blasius, K. R. 1978b. Chaotic terrain and channels associated with Chryse Planitia, Mars (Abs.). In *Lunar and Planetary Science IX*, Pt. 1, pp. 206–08. Lunar and Planetary Institute, Houston.
Davies, D. W. 1979. The vertical distribution of Mars water vapor. *J. Geophys. Res.* 84:2875–79.
Dence, M. R. 1965. The extraterrestrial origin of Canadian craters. *Ann. New York Acad. Sci.* 123:941–69.
Dence, M. R. 1968. Shock zoning at Canadian craters: Petrography and structural implications. In *Shock metamorphism of natural minerals* (B. M. French, and N. M. Short, eds.), pp. 168–84. Mono Book Co., Baltimore.
Dence, M. R. 1971. Impact melts. *J. Geophys. Res.* 76:5552–65.
Dence, M. R. 1972. The nature and significance of terrestrial impact structures. In *Proc. 24th Internat. Geol. Congr., Sect. 15*, pp. 77–89.
Dence, M. R. 1973. Dimensional analysis of impact structures. *Meteoritics* 8:343–44.
Dence, M. R., Grieve, R. A. F., and Robertson, P. B. 1977. Terrestrial impact structures: Principal characteristics and energy considerations. In *Impact and explosion cratering* (D. J. Roddy, R. O. Pepin, and P. B. Merrill, eds.), pp. 247–75. Pergamon Press, New York.
Dial, A. L. 1978. *The Viking 1 landing site crater diameter-frequency distribution.* NASA Tech. Memo. 79729, pp. 179–81.
Dohnanyi, J. S. 1972. Interplanetary objects in review: Statistics of their masses and dynamics. *Icarus* 17:1–48.
Dollfus, A. 1961. Visual and photographic studies of the planets at the Pic du Midi. In *Planets and satellites* (G. P. Kuiper, and B. M. Middlehurst, eds.), vol. 3, ch. 15. Univ. of Chicago Press, Chicago.
Drake, F. D. 1965. The radio search for intelligent and extraterrestrial life. In *Current aspects of exobiology* (G. Mamikunian, and M. H. Briggs, eds.), pp. 323–45. Pergamon Press, New York.
Duxbury, T. C., and Veverka, J. 1978. Deimos encounter by Viking: Preliminary imaging results. *Science* 20:812–14.
Dzurisin, D., and Blasius, K. R. 1975. Topography of the polar layered deposits of Mars. *J. Geophys. Res.* 82:4225–48.
Dzurisin, D., and Ingersoll, A. P. 1975. Seasonal buffering of atmospheric pressure on Mars. *Icarus* 26:437–40.
Eaton, J. R., and Murata, K. J. 1960. How volcanoes grow. *Science* 132:925–38.
Elsasser, W. M., Olson, P., and Marsh, B. D. 1979. The depth of mantle convection. *J. Geophys. Res.* 84:141–55.
Evans, J. E., and Maunder, E. W. 1903. Experiments as to the actuality of the 'canals' observed on Mars. *Monthly Not. Roy. Astron. Soc. London* 53:488.
Fanale, F. P. 1976. Martian volatiles: Their degassing history and geochemical fate. *Icarus* 28:179–202.
Fanale, F. P., and Cannon, W. A. 1974. Exchange of adsorbed H_2O and CO_2 between regolith and atmosphere of Mars caused by changes in surface insolation. *J. Geophys. Res.* 79:3397–402.
Fanale, F. P., and Cannon, W. A. 1979. Mars: CO_2 adsorption and capillary condensation on clays—significance for volatile storage and atmospheric history. *J. Geophys. Res.* 84:8404–14.
Farmer, C. B., and Doms, P. E. 1979. Global seasonal variation of water vapor on Mars and the implications for permafrost. *J. Geophys. Res.* 84:2881–88.
Farmer, C. B., Davies, D. W., and LaPorte, D. D. 1976. Mars: Northern summer ice cap—water-vapor observations from Viking 2. *Science* 194:1339–41.
Fischbacher, G. E., Martin, L. J., and Baum, W. A. 1969. Martian polar cap boundaries. Final rept, part A, under contract 951547 to Jet Propulsion Lab. Pasadena, Calif. Lowell Observatory, Flagstaff, Ariz.
Flammarion, C. 1892. *La planète Mars et ses conditions d'habitabilité, vol.* 1. Gauthier-Villars et Fils, Paris.
Focas, J. H. 1961. Etude photometrique et polarimetrique des phénomènes saisonniers de la planète Mars. *Ann. d'Astrophys.* 24:309–25.
Focas, J. H. 1962. Seasonal evolution of the fine structure of the dark areas of Mars. *Planet. Space Sci.* 9:371–81.
Frey, H. 1979. Thaumasia: A fossilized early-forming Tharsis uplift. *J. Geophys. Res.* 84:1009–23.
Frieden, E. 1972. The chemical elements of life. *Sci. Am.* 227(1):52–60.
Friedmann, E. I. 1979. *Endolithic microbial life in the Antarctic dry valleys: A terrestrial model of martian environment?* (Abs.). NASA CP-2072, p. 28.
Fuller, A. O., and Hargraves, R. B. 1978. Some consequences of a liquid water saturated regolith in early martian history. *Icarus* 34:614–21.
Fuller, M. L. 1922. Some unusual erosion features in the loess of China. *Geog. Rev.* 12:570–84.
Gapcynski, J. P., Tolson, R. H., and Michael, W. H. 1977. Mars gravity field: Combined Viking and Mariner 9 results. *J. Geophys. Res.* 82:4325–27.
Gatley, I., Kieffer, H. H., Miner, E., and Neugebauer, G. 1974. Infrared observations of Phobos from Mariner 9. *Astrophys. J.* 190:497–503.

Gatto, L. W., and Anderson, D. M. 1975. Alaskan thermokarst terrain and possible martian analog. *Science* 188:255–57.
Gault, D. E. 1970. Saturation and equilibrium conditions for impact cratering on the lunar surface: Criteria and implications. *Radio Sci.* 5:273–91.
Gault, D. E., and Baldwin, B. S. 1970. Impact cratering on Mars—some effects of the atmosphere (Abs.). Trans. Am. Geophys. Union (EOS) 51:342.
Gault, D. E., and Greeley, R. 1978. Exploratory experiments of impact craters formed in viscous-liquid targets: Analogs for martian rampart craters? *Icarus* 34:486–95.
Gault, D. E., Guest, J. E., Murray, J. B., Dzurisin, D., and Malin, M. C. 1975. Some comparisons of impact craters on Mercury and the Moon. *J. Geophys. Res.* 80:2444–60.
Gault, D. E., Quaide, W. L., and Oberbeck, V. R. 1968. Impact cratering mechanics and structures. In *Shock metamorphism of natural materials* (B. M. French, and N. M. Short, eds.), pp. 87–99. Mono Book Corp., Baltimore.
Gierasch, P., and Goody, R. M. 1968. A study of the thermal and dynamical structure of the martian lower atmosphere. *Planet. Space Sci.* 16:615–46.
Glasstone, S. 1968. *The book of Mars.* NASA Spec. Pub. 179.
Gooding, J. L. 1978. Chemical weathering on Mars, thermodynamic stabilities of primary minerals (and their alteration products) from mafic igneous rocks. *Icarus* 33:483–513.
Gooding, J. L., and Keil, K. 1978. Alteration of glass as a possible source of clay minerals on Mars. *Geophys. Res. Lett.* 5:727–30.
Greeley, R. 1978. *Mars: A model for the formation of dunes and related structures.* NASA Tech. Memo. 79729, pp. 244–45.
Greeley, R., and Hyde, J. H. 1972. Lava tubes of the cave basalt, Mount St. Helens, Washington. *Geol. Soc. Am. Bull.* 83:2397–418.
Greeley, R., and Spudis, P. 1978. Volcanism in the cratered terrain hemisphere of Mars. *Geophys. Res. Lett.* 5:453–55.
Greeley, R., Iverson, J. D., Pollack, J. B., Udovich, N., and White, B. 1974a. Wind tunnel simulations of light and dark streaks on Mars. *Science* 183:847–49.
Greeley, R., Iverson, J. D., Pollack, J. B., Udovich, N., and White, B. 1974b. Wind tunnel studies of martian eolian processes. In *Proc. Roy. Soc. London*, ser. A, vol. 341, pp. 331–60.
Greeley, R., Leach, R., White, R., Iverson, J., and Pollack, J. 1980. Threshold wind speeds for sands on Mars: Wind tunnel simulations. *Geophys. Res. Lett.* 7:121–24.
Greeley, R., Papson, R., and Veverka, J. 1978. Crater streaks in the Chryse Planitia region of Mars: Early Viking results. *Icarus* 34:556–67.
Greeley, R., Theilig, E., Guest, J. E., Carr, M. H., Masursky, H., and Cutts, J. A. 1977. Geology of Chryse Planitia. *J. Geophys. Res.* 82:4093–109.
Greeley, R., White, B., Leach, R., Iverson, J., and Pollack, J. 1976. Mars: Wind friction speeds for particle movement. *Geophys. Res. Lett.* 3:417–20.
Green, D. H., and Ringwood, H. E. 1963. Mineral assemblages in a model mantle composition. *J. Geophys. Res.* 68:937–45.
Gregory, T. 1979. *A martian global plains system represented by Lunae Planum surface units.* NASA Tech. Memo. 80339, pp. 75–77.
Grolier, M. J., Erickson, G. E., McCauley, J. F., and Morris, E. C. 1974. *The desert landforms of Peru: A preliminary photographic atlas.* U.S. Geol. Survey, Interagency Rept., Astrogeology 57.
Guest, J. E., Butterworth, P. S., and Greeley, R. 1977. Geological observations in the Cydonia region of Mars from Viking. *J. Geophys. Res.* 82:4111–20.
Gunn, B. M., and Warren, G. 1962. Geology of Victoria Land between the Mawson and Mulock Glaciers, Antarctica. *New Zealand Geol. Survey, Bull.* 71:1–157.
Hanel, R. A., Conrath, B. J., Hovis, W. A., Lowman, P. D., Pearl, J. C., Prabhakara, C., and Schlachman, B. 1972. Infrared spectroscopy experiment of the Mariner 9 mission: Preliminary results. *Science* 175:305–08.
Hargraves, R. B., Collinson, D. W., Arvidson, R. E., and Spitzer, C. R. 1976. Viking magnetic properties investigation: Further results. *Science* 194:1303–09.
Harp, E. L. 1974. Fracture systems and tectonics on Mars. Ph.D. diss. Univ. Utah.
Harris, S. A. 1977. The aureole of Olympus Mons, Mars. *J. Geophys. Res.* 82:3099–107.
Hartmann, W. K. 1971a. Martian cratering II: Asteroid impact history. *Icarus* 15:396–409.
Hartmann, W. K. 1971b. Martian cratering III: Theory of crater obliteration. *Icarus* 15:410–28.
Hartmann, W. K. 1972a. *Moons and planets.* Bogden and Quidley, Tarrytown, New York.
Hartmann, W. K. 1972b. Interplanetary variations in scale of crater morphology—Earth, Mars, Moon. *Icarus* 17:707–13.
Hartmann, W. K. 1973a. Martian cratering 4: Mariner 9 initial analysis of cratering chronology. *J. Geophys. Res.* 78:4096–116.
Hartmann, W. K. 1973b. Martian surface and crust: Review and synthesis. *Icarus* 19:550–75.
Hartmann, W. K. 1974a. Geological observations of martian arroyos. *J. Geophys. Res.* 79:3951–57.
Hartmann, W. K. 1974b. Martian and terrestrial paleoclimatology: Relevance of solar variability. *Icarus* 22:301–11.
Hartmann, W. K. 1977a. Cratering in the solar system. *Sci. Am.* 236:84–99.
Hartmann, W. K. 1977b. Relative crater production rates on planets. *Icarus* 31:260–76.
Hartmann, W. K., and Raper, O. 1974. *The new Mars.* NASA Spec. Pub. 337.
Hartmann, W. K., and Wood, C. A. 1971. Moon: Origin and evolution of multiring basins. *Moon* 3:3–78.
Hartmann, W. K., and others (12 authors). 1981. Chronology of planetary volcanism by comparative studies of planetary cratering. In *Basaltic volcanism in the terrestrial planets.* Lunar and Planetary Institute, Houston.
Hawkins, G. S. 1960. Asteroidal fragments. *Astrophys. J.* 65:318–22.
Hawkins, G. S. 1963. Impacts on the Earth and the Moon. *Nature* 197:78.
Head, J. W., Settle, M., and Wood, C. A. 1976. Origin of Olympus Mons escarpment by erosion of prevolcanic substrate. *Nature* 263:667–68.
Hedin, S. 1903. *Central Asia and Tibet.* Scribners, New York.
Heim, A. 1932. *Bergsturz und Menschenleben.* Fretz and Wasmuth, Zurich.
Hess, S. L., Henry, R. M., and Tillman, J. E. 1979. The seasonal variation of atmospheric pressure on Mars as affected by the south polar cap. *J. Geophys. Res.* 84:2923–27.
Hess, S. L., Henry, R. M., Leovy, C. B., Ryan, J. A., and Tillman, J. E. 1977. Meteorological results from the surface of Mars: Viking 1 and 2. *J. Geophys. Res.* 82:4559–74.
Hess, S. L., Ryan, J. A., Tillman, J. E., Henry, R. M., and Leovy, C. B. 1980. The annual cycle of pressure on Mars measured by Viking 1 and 2. *Geophys. Res. Lett.* 7:197–200.
Hodges, C. A. 1978. Central pit craters on Mars. (Abs.). In *Lunar and Planetary Science IX*, pp. 521–22. Lunar and Planetary Institute, Houston.
Hodges, C. A. 1979. *Some lesser volcanic provinces on Mars.* NASA Tech. Memo. 80339, pp. 248–49.
Hodges, C. A., and Moore, H. J. 1978. Table mountains on Mars (Abs.). In *Lunar and Planetary Science IX*, pp. 523–25. Lunar and Planetary Institute, Houston.
Hodges, C. A., and Moore, H. J. 1979. The subglacial birth of Olympus Mons and its aureoles. *J. Geophys. Res.* 84:8061–74.
Hodges, C. A., and Wilhelms, D. E. 1978. Formation of lunar ring basins. *Icarus* 34:294–323.
Holland, H. D., and Blackburn, T. R. 1979. X-ray photoelectron spectrometric and gas exchange evidence for surface oxidation of martian regolith analogues after ultraviolet irradiation. In COSPAR, *Life sciences and space research*, vol. 17 (R. Holmquist, and A. C. Strickland, eds.), pp. 65–76. Pergamon Press, New York.
Hord, C. W., Barth, C. A., Stewart, A. L., and Lane, A. L. 1972. Mariner 9 ultraviolet spectrometer experiment: Photometry and topography of Mars. *Icarus* 17:443–56.
Hord, C. W., Simmons, K. E., and McLaughlin, L. K. 1974. Mariner 9 ultraviolet spectrometer experiment: Pressure altitude measurements on Mars.

Icarus 21:292–302.
Horowitz, N. H., Hubbard, J. S., and Hobby, G. L. 1972. The carbon assimilation experiment: The Viking Mars lander. *Icarus* 16:147–52.
Horowitz, N. H., Hubbard, J. S., and Hobby, G. L. 1976. The Viking carbon assimilation experiments: Interim report. *Science* 194:1321–22.
Horowitz, N. H., Hubbard, J. S., and Hobby, G. L. 1977. Viking on Mars: The carbon assimilation experiments. *J. Geophys. Res.* 82:4659–62.
Houck, J. R., Pollack, J. B., Sagan, C., Schaak, D., and Decker, J. A. 1973. High-altitude infrared spectroscopic evidence for bound water on Mars. *Icarus* 18:470–80.
Howard, A. D. 1978. Origin of the stepped topography of the martian poles. *Icarus* 34:581–99.
Howard, K. A. 1972. *Ejecta blankets of large craters exemplified by King crater: Apollo 16 preliminary science report.* NASA Spec. Pub. 315, pp. 29-70 to 29-77.
Howard, K. A. 1974. Fresh lunar impact craters. Reviews of variations in size. *Proc. 5th Lunar Sci. Conf.*, pp. 61–69.
Howard, K. A., and Muehlberger, W. R. 1973. *Lunar thrust faults in the Taurus-Littrow region: Apollo 17 preliminary science report.* NASA SP-330, pp. 31-22 to 31-25.
Howard, K. A., Wilhelms, D. E., and Scott, D. H. 1974. Lunar basin formation and highland stratigraphy. *Rev. Geophys. Space Phys.* 12:309–28.
Hubbard, J. S. 1976. The pyrolytic release experiment: Measurement of carbon assimilation. *Origins of Life* 7:281–92.
Hubbard, J. S. 1979. Laboratory simulation of the pyrolytic release experiments: An interim report. *J. Mol. Evol.* 14:211–21.
Hubbard, J. S., Hardy, J. P., and Horowitz, N. H. 1971. Photocatalytic production of organic compounds from CO_2 and H_2O in a simulated martian atmosphere. *Proc. Nat. Acad. Sci., U.S.A.* 68:574–78.
Hubbard, J. S., Hardy, J. P., Voecks, C. E., and Golub, E. E. 1973. Photocatalytic synthesis of organic compounds from CO and water. *J. Mol. Evol.* 2:149–66.
Huck, F. O., Jobson, D. J., Park, S. K., Wall, S. D., Arvidson, R. E., Patterson, W. R., and Benton, W. D. 1977. Spectrophotometric and color estimates of the Viking lander sites. *J. Geophys. Res.* 82:4401–11.
Huguenin, R. L. 1974. The formation of goethite and hydrated clay minerals on Mars. *J. Geophys. Res.* 79:3895–905.
Huguenin, R. L. 1976. Mars: Chemical weathering as a massive volatile sink. *Icarus* 28:203–12.
Huguenin, R. L., Adams, J. B., and McCord, T. B. 1978. Mars: Surface mineralogy from reflectance spectra (Abs.). In *Lunar Science VIII*, pp. 478–80. Lunar and Planetary Institute, Houston.
Huguenin, R. L., Miller, K. J., and Harwood, W. S. 1979a. Frost weathering on Mars: Experimental evidence for peroxide formation. *J. Mol. Evol.* 14:103–32.
Huguenin, R. L., Miller, K. J., and Harwood, W. S. 1979b. *Mars: Chemical reduction of the martian regolith by frost.* NASA CP-2072, pp. 40–42.
Huguenin, R. L., Prinn, R. G., and Madarazzo, M. 1977. Mars: Photodesorption from mineral surfaces and its effects on atmospheric stability. *Icarus* 32:270–98.
Hulme, G. 1974. The interpretation of lava flow morphology. *Geophys. J. Roy. Astron. Soc.* 39:361–83.
Hulme, G. 1976. The determination of the rheological properties and effusion rate of an Olympus Mons lava. *Icarus* 27:207–13.
Hunt, G. R., Logan, L. M., and Salisbury, J. W. 1973. Mars: Components of infrared spectra and the composition of the dust cloud. *Icarus* 18:459–69.
Hunten, D. M. 1974. Aeronomy of the lower atmosphere of Mars. *Rev. Geophys. Space Phys.* 12:529–35.
Hunten, D. M. 1978. Capture of Phobos and Deimos by protoatmospheric drag. *Icarus* 37:113–23.
Hunten, D. M. 1979. *Possible oxidant sources in the atmosphere and surface* (Abs.): NASA CP-2072, p. 42.
Imshenetskii, A. A., and Murzakov, B. C. 1977. On the search for life on Mars. *Mikrobiologiya* 46:1103–13.
Inge, J. L. 1974. *Mars—1973* (albedo map, 1:25,000,000). Lowell Observatory, Flagstaff, AZ.
Inge, J. L. 1976. *Mars—1975–1976* (albedo map, 1:25,000,000). Lowell Observatory, Flagstaff, AZ.
Inge, J. L. 1978. *Mars—1978* (albedo map, 1:25,000,000). Lowell Observatory, Flagstaff, AZ.
Inge, J. L., and Baum, W. A. 1973. A comparison of martian albedo features with topography. *Icarus* 19:323–28.
Inge, J. L., Capen, C. F., Martin, L. J., and Faure, B. Q. 1971a. Mars—1969: A new map of Mars from planetary patrol photographs. *Sky and Telescope* 41:336–39.
Inge, J. L., Capen, C. F., Martin, L. J., and Thompson, D. T. 1971b. *Mars—1971* (albedo map, 1:25,000,000). Lowell Observatory, Flagstaff, AZ.
Ingersoll, A. P. 1974. Mars: The case against permanent CO_2 frost caps. *J. Geophys. Res.* 79:3403–10.
Iverson, J. D., Greeley, R., White, B. R., and Pollack, J. B. 1976. The effect of vertical distortion in the modeling of sedimentation phenomena: Martian crater wake streaks. *J. Geophys. Res.* 81:4846–47.
James, P. B., Briggs, G., Barnes, J., and Spruck, A. 1979. Seasonal recession of Mars's south polar cap as seen by Viking. *J. Geophys. Res.* 84:2889–922.
Johansen, L. A. 1979. *The latitude dependence of martian splosh cratering and its relationship to water.* NASA Tech. Memo. 80339, pp. 123–25.
Johnston, D. H., and Toksoz, M. N. 1977. Internal structure and properties of Mars. *Icarus* 32:73–84.
Jones, K. L. 1974. Evidence for an episode of martian crater obliteration intermediate in martian history. *J. Geophys. Res.* 79:3917–32.
Jones, K. L., Bragg, S. L., Wall, S. D., Carlston, C. E., and Pidek, D. G. 1979. One Mars year: Viking lander imaging observations. *Science* 204:799–806.
Jordan, J. F., and Lorell, J. 1975. Mariner 9: An instrument of dynamical science. *Icarus* 25:146–65.
Kent, P. E. 1966. The transport mechanism in catastrophic rockfalls. *J. Geol.* 74:79–83.
Kieffer, H. H., and Palluconi, F. D. 1979. *The climate of the martian polar cap.* NASA CP-2072, pp. 45–46.
Kieffer, H. H., Chase, S. C., Martin, T. Z., Miner, E. D., and Palluconi, F. D. 1976. Martian north pole summer temperature: Dirty water-ice. *Science* 194:1341–44.
Kieffer, H. H., Christensen, P. R., Martin, T. Z., Miner, E. D., and Palluconi, F. D. 1976. Temperatures of the martian surface and atmosphere: Viking observations of diurnal and geometric variations. *Science* 194:1346–57.
Kieffer, H. H., Martin, T. Z., Peterfreund, A. R., and Jakosky, B. M. 1977. Thermal and albedo mapping of Mars during the Viking primary mission. *J. Geophys. Res.* 82:4249–91.
Kieffer, S. W., and Simonds, C. H. 1979. *The role of volatiles and lithology in the impact cratering process.* Johnson Space Center Pub. 16062. Houston.
King, J. S., and Riehle, J. R. 1974. A proposed origin of the Olympus Mons escarpment. *Icarus* 23:300–17.
Klein, H. P. 1976. General constraints on the Viking biology investigations. *Origins of Life* 7:273–79.
Klein, H. P. 1978. The Viking biological investigations: General aspects. *J. Geophys Res.* 82:4677–80.
Klein, H. P., and others (13 authors). 1976. The Viking biological investigation: Preliminary results. *Science* 194:99–105.
Klein, H. P., Lederberg, J., and Rich, A. 1972. Biological experiments: The Viking Mars lander. *Icarus* 16:139–46.
Kliore, A. J., Cain, D. L., Levy, G. S., Eshleman, V. R., Fjeldbo, G., and Drake, F. D. 1965. Occultation Experiment: Results of the first direct measurements of Mars atmosphere and ionosphere. *Science* 149:1243–48.
Kliore, A. J., Cain, D. L., Fjeldbo, G., and Seidel, B. L. 1972. Mariner 9 S-band martian occultation experiments: Initial results on the atmosphere and topography of Mars. *Science* 175:313–17.
Kliore, A. J., Fjeldbo, G., Seidel, B. L., and Rasool, I. 1969. Mariners 6 and 7: Occultation mea-

surements of the atmosphere of Mars. *Science* 166:1393–97.

Kliore, A. J., Fjeldbo, G., Seidel, B. L., Sykes, M. J., and Woiceshyn, P. M. 1973. S-band radio occultation measurements of the atmosphere and topography of Mars with Mariner 9: Extended mission coverage of polar and intermediate latitudes. *J. Geophys. Res.* 78:4331–51.

Kong, T. Y., and McElroy, M. B. 1977. Photochemistry of the martian atmosphere. *Icarus* 32:168–89.

Komar, P. D. 1979. Comparisons of the hydraulics of water flows in martian outflow channels with flows of similar scale on Earth. *Icarus* 34:156–81.

Krinsley, D. H., Leach, R., Greeley, R., and McKee, T. R. 1979. Simulated martian eolian abrasion and creation of aggregates. NASA Tech. Memo. 80339, pp. 313–15.

Kuckes, A. F. 1977. Strength and rigidity of the elastic lunar lithosphere and implications for present-day mantle convection in the Moon. *Phys. Earth Planet. Interiors* 14:1–12.

Kuiper, G. P. 1952. *The atmosphere of the Earth and planets.* Univ. of Chicago Press, Chicago.

Lachenbruch, A. H. 1962. *Mechanics of thermal contraction cracks and ice wedge polygons in permafrost.* Geol. Soc. Am. Spec. Paper 70.

Lambert, R., St, J., and Chamberlain, V. E. 1978. CO_2 permafrost and martian topography. *Icarus* 34:568–80.

Latham, G. V., Ewing, M., Press, F., Sutton, G., Dorman, J., Nakamura, Y., Toksoz, N., Duennebeir, F., and Lammlein, D. 1971. Passive seismic experiment. In *Apollo 14 Preliminary Science Report*, NASA Spec. Pub. 272, pp. 133–61.

Lawless, J. G., Kjos, K. M., Mednick, D., Odom, D., and Levi, N. 1977. *Metallic clays in prebiological chemistry.* Am. Chem. Soc., Pacific Conf., Anaheim, Calif.

Lederberg, J. 1965. Signs of life: The criterion system of exobiology. *Nature* 207:9–13.

Lederberg, J., and Sagan, C. 1962. Microenvironments for life on Mars. *Proc. Nat. Acad. Sci., U.S.A.* 48:1473–75.

Leighton, R. B., and Murray, B. C. 1966. Behavior of carbon dioxide and other volatiles on Mars. *Science* 153:136–44.

Leighton, R. B., Horowitz, N. H., Murray, B. C., Sharp, R. P., Herriman, A. G., Young, A. T., Smith, B. A., Davies, M. E., and Leovy, C. G. 1969a. Mariner 6 and 7 television pictures: Preliminary analysis. *Science* 166:49–67.

Leighton, R. B., Horowitz, N. H., Murray, B. C., Sharp, R. P., Herriman, A. G., Young, A. T., Smith. B. A., Davies, M. E., and Leovy, C. G. 1969b. *Television observations from Mariner 6 and 7. Mariner-Mars 1969 A Preliminary Report.* NASA Spec. Pub. 225.

Leighton, R. B., Murray, B. C., Sharp, R. P., Allen, J. D., and Sloan, R. K. 1965. Mariner IV photography of Mars: Initial results. *Science* 149:627–30.

Leovy, C. B. 1966. Mars ice caps. *Science* 154:1178–79.

Leovy, C. B. 1977. The atmosphere of Mars. *Sci. Am.* 237: 34–43.

Leovy, C. B., and Mintz, Y. 1969. Numerical simulation of the weather and climate of Mars. *J. Atmos. Sci.* 26:1167–90.

Leovy, C. B., and Zuruk, R. W. 1979. Thermal tides and martian dust storms: Direct evidence for coupling. *J. Geophys. Res.* 84:2956–68.

Leovy, C. B., Briggs, G. A., and Smith, B. A. 1973. Mars atmosphere during the Mariner 9 extended mission: Television results. *J. Geophys. Res.* 78:4252–66.

Levin, G. V., and Straat, P. A. 1976a. Labeled release—an experiment in radiospirometry. *Origins of Life* 7:293–311.

Levin, G. V., and Straat, P. A. 1976b. Viking labeled release biology experiment: Interim results. *Science* 194:1322–29.

Levin, G. V., and Straat, P. A. 1977a. Recent results from the Viking labeled release experiment on Mars. *J. Geophys. Res.* 82:4663–67.

Levin, G. V., and Straat, P. A. 1977b. Life on Mars? The Viking labeled release experiment. *Biosystems* 9:165–74.

Levin, G. V., and Straat, P. A. 1979a. *Status of interpretation of Viking labeled release experiment* (Abs.). NASA CP-2072, p. 52.

Levin, G. V., and Straat, P. A. 1979b. Completion of the Viking labeled release experiment on Mars. *J. Mol. Evol.* 14:167–83.

Levinthal, E. C., Fox, P. L., and Sagan, C. 1977. Lander imaging as a detector of life on Mars. *J. Geophys. Res.* 82:4468–78.

Lewis, J. L. 1974. The temperature gradient in the solar nebula. *Science* 186:440–43.

Lindsay, J. F. 1974. Depositional processes on the lunar surface. In *Proc. 5th Lunar Sci. Conf.* pp. 450–52.

Lingenfelter, R. E., Peale, S. J., and Schubert, G. 1968. Lunar rivers. *Science* 161:266–69.

Logan, L. M., Hunt, G. R., and Salisbury, J. J. 1975. The use of mid-infrared spectroscopy in remote sensing of space targets. In *Infrared and Raman spectra of lunar and terrestrial minerals* (C. Karr, ed.), pp. 117–42. Academic Press, New York.

Lorell, J., Born, G. H., Christensen, E. J., Jordan, J. F., Laing, P. A., Martin, W. L., Sjogren, W. L., Shapiro, I. I., Reasonberg, R. D., and Slater, G. L. 1972. Mariner 9 celestial mechanics experiment: Gravity field and pole direction of Mars. *Science* 175:317–20.

Lowell, P. 1908. *Mars and its canals.* Macmillan, New York.

Lowell, P. 1909. *Evolution of worlds.* Macmillan, New York.

Lowell, P. 1909. *Mars as the abode of life.* Macmillan, New York.

Lucchitta, B. K. 1977. Topography, structure, and mare ridges in southern Mare Imbrium and northern Oceanus Procellarum. In *Proc. 8th Lunar Sci. Conf.*, pp. 2691–703.

Lucchitta, B. K. 1978a. A large landslide on Mars. *Geol. Soc. Am. Bull.* 89:1601–09.

Lucchitta, B. K. 1978b. Morphology of Chasma walls, Mars. *J. Res. U.S. Geol. Survey* 6:651–62.

Lucchitta, B. K. 1980. Martian outflow channels sculpted by glaciers. In *Lunar and Planetary Science XI*, pp. 634–36. Lunar and Planetary Institute, Houston.

MacDonald, T. L. 1971. The origins of martian nomenclature. *Icarus* 15:232–40.

Madarazzo, M., and Huguenin, R. L. 1977. Petrologic implications of Viking XRF analysis based on reflection spectra and the photochemical weathering model. *Bull. Am. Astron. Soc.* 9:527–28.

Mainguet, M. 1972. *Le modelé des grès: Problèmes Généraux.* Institut Géographique Nationale, Paris.

Malin, M. C. 1976. Nature and origin of intercrater plains on Mars. Ph.D. diss. Calif. Inst. Tech.

Malin, M. C. 1977. Comparison of volcanic features: Elysium (Mars) and Tibetsi (Earth). *Geol. Soc., Am. Bull.* 84:908–19.

Marcus, A. H. 1970. Comparison of equilibrium size distributions for lunar craters. *J. Geophys. Res.* 75:4977–84.

Marov, M. V., and Petrov, G. I. 1973. Investigations of Mars from the Soviet automatic stations Mars 2 and 3. *Icarus* 19:163–79.

Martin, L. J. 1974. The major martian yellow storm of 1971. *Icarus* 22:175–88.

Martin, L. J. 1976. 1973 dust storm on Mars: Maps from hourly photographs. *Icarus* 29:363–80.

Martin, L. J., and Baum, W. A. 1969. A study of cloud motions on Mars. Final report B, under contract 951547, to Jet Propulsion Laboratory, Pasadena, Calif. Lowell Observatory, Flagstaff, AZ.

Mason, B. 1971. *Handbook of elemental abundances in meteorites.* Gordon and Breach, New York.

Mass, C., and Sagan, C. 1976. A numerical circulation model with topography for the martian southern hemisphere. *J. Atmos. Sci.* 33:1418–30.

Masson, P. 1977. Structural pattern analysis of the Noctis Labyrinthus-Valles Marineris regions of Mars. *Icarus* 30: 49–62.

Masursky, H. 1973. An overview of geologic results from Mariner 9. *J. Geophys. Res.* 78:4037–47.

Masursky, H., Boyce, J. M., Dial, A. L., Schaber, G. G., and Strobell, M. E. 1977. Formation of martian channels. *J. Geophys. Res.* 82:4016–38.

Masursky, H., and Crabill, N. L. 1976. Search for the Viking 2 landing site. *Science* 194:62–68.

McCauley, J. F. 1973. Mariner 9 evidence for wind erosion in the equatorial and mid-latitude regions

of Mars. *J. Geophys. Res.* 78:4123–37.
McCauley, J. F. 1977. Orientale and Caloris. *Phys. Earth Planet. Interiors* 15:220–50.
McCauley, J. F. 1978. *Geologic map of the Coprates quadrangle of Mars.* U.S. Geol. Survey, Misc. Inv. Map I-897.
McCauley, J. F., Breed, C. S., El Baz, F., Whitney, M. J., Grolier, M. J., and Ward, A. W. 1979. Pitted and fluted rocks in the western desert of Egypt: Viking comparisons. *J. Geophys. Res.* 84:8222–32.
McCauley, J. F., Carr, M. H., Cutts, J. A., Hartmann, W. K., Masursky, H., Milton, D. J., Sharp, R. P., and Wilhelms, D. E. 1972. Preliminary Mariner 9 report on the geology of Mars. *Icarus* 17:289–327.
McCauley, J. F., Grolier, M. J., and Breed, C. S. 1977. Yardangs. In *Proc. 8th annual geomorphology symposium* (D. O. Doerhing, ed.), pp. 233–69. State University of New York, Binghamton.
McCord, T. B., and Westphal, J. A. 1971. Mars: Narrow band photometry, from 0.3 to 2.5 microns, of surface regions during the 1969 apparition. *Astrophys. J.* 168:141–53.
McCord, T. B., Huguenin, R. L., and Johnson, G. L. 1977. Photometric imaging of Mars during the 1973 opposition. *Icarus* 31:293–314.
McCrosky, R. E., and Ceplecha, Z. 1968. *Photographic networks for fireballs.* Smithson. Astrophys. Pbs., Spec. Rept. No. 288.
McElroy, M. B., and Yung, Y. L. 1976. Oxygen isotopes in the martian atmosphere: Implications for the evolution of volatiles. *Planet. Space Sci.* 24:1107–13.
McElroy, M. B., Kong, T. Y., and Yung, Y. L. 1977. Photochemistry and evolution of Mars's atmosphere: A Viking perspective. *J. Geophys. Res.* 82:4379–88.
McGetchin, T. R., and Smyth, J. R. 1978. The mantle of Mars: Some possible geological implications of its high density. *Icarus* 34:512–36.
McGill, G. E. 1978. *Geologic map of the Thaumasia quadrangle of Mars.* U.S. Geol. Survey, Spec. Inv. Map I-1077.
McGill, G. E., and Wise, D. U. 1972. Regional variations in degradation and density of martian craters. *J. Geophys. Res.* 77:2433–41.
McKay, D. S., and Morrison, D. A. 1971. Lunar breccias. *J. Geophys. Res.* 76:5658–69.
Michaux, C. M., and Newburn, R. L. 1972. *Mars scientific model.* Jet Propulsion Laboratory Document 606-1.
Miller, S. L., and Orgel, L. E. 1974. *The origins of life on Earth.* Prentice-Hall, Englewood Cliffs, NJ.
Mills, A. A. 1977. Dust clouds and frictional generation of glow discharges on Mars. *Nature* 268:614.
Milton, D. J. 1973. Water and processes of degradation in the martian landscape. *J. Geophys. Res.* 78:4037–47.
Milton, D. J. 1974. Carbon dioxide hydrate and floods on Mars. *Science* 183:654–56.
Milton, D. J., and Roddy, D. J. 1972. Displacements with impact craters. *Proc. 24th Int. Geol. Congr., Sec. 15,* pp. 119–24.
Milton, D. J. et al. (10 authors). 1972. Gosses Bluff impact structure, Australia. *Science* 175:1199–207.
Moore, H. J., and Schaber, G. G. 1975. An estimate of the yield strength of Imbrium flows. In *Proc. 6th Lunar Sci. Conf.,* pp. 101–08.
Moore, H. J., Arthur, D. W. G., and Schaber, G. G. 1978. Yield strengths of flows on the earth, Mars, and Moon. In *Proc. 9th Lunar Planet. Sci. Conf.,* pp. 3351–78.
Moore, H. J., Hodges, C. A., and Scott, D. H. 1974. Multiringed basins—illustrated by Orientale and associated features. In *Proc. 5th Lunar Sci. Conf.,* pp. 71–100.
Moore, H. J., Hutton, R. E., Scott, R. F., Spitzer, C. R., and Shorthill, R. W. 1977. Surface materials of the Viking landing sites. *J. Geophys. Res.* 82:4497–523.
Morris, E. C. 1979. *A pyroclastic origin for the aureole deposits of Olympus Mons.* NASA Tech. Memo. 82385, pp. 252–54.
Morris, E. C., Jones, K. L., and Berger, J. P. 1978. Location of Viking 1 lander on the surface of Mars. *Icarus* 34:548–55.
Morris, R. V., and Lauer, H. V. 1979. *Experimental evidence against the UV photochemical oxidation of magnetite.* NASA Tech. Memo. 80339, pp. 225–27.
Morrison, R. H., and Oberbeck, V. R. 1975. Geomorphology of crater and basin deposits—emplacement of the Fra Mauro formation. In *Proc. 6th Lunar Sci. Conf.,* pp. 2503–30.
Mottoni, G. de. 1970. *Cartografia del pianete Marte basata su documentazione fotografica internazionale a partire del 1907. IV Opposizioni dal 1960 al 1967.* Publicazioni dell' Osservatorio Astron. di Milano-Merata, no. 22.
Mouginis-Mark, P. J. 1979a. *Distribution of fluidized craters on Mars.* NASA Tech. Memo. 80339, pp. 147–49.
Mouginis-Mark, P. J. 1979b. Martian fluidized crater morphology: Variations with crater size, latitude, altitude, and target material. *J. Geophys. Res.* 84:8011–22.
Müller, G., and Hickisch, R. 1972. On the adsorption behavior of bacteria and the soil. In *Proc. Symposium on soil microbiology* (J. Szegi, ed.), pp. 263–69. Akademiai, Budapest.
Murray, B. C., and Malin, M. C. 1973. Polar wandering on Mars. *Science* 179:997–1000.
Murray, B. C., Soderblom, L. A., Cutts, J. A., Sharp, R. P., Milton, D. J., and Leighton, R. B. 1972. Geologic framework of the south polar region of Mars. *Icarus* 17:328–45.
Murray, B. C., Soderblom, L. A., Sharp. R. P., and Cutts, J. A. 1971. The surface of Mars. I. Cratered terrains. *J. Geophys. Res.* 76:313–30.
Murray, B. C., Strom, R. G., Trask, N. J., and Gault, D. E. 1975. Surface history of Mercury: Implications for terrestrial planets. *J. Geophys. Res.* 80:2508–14.
Murray, B. C., Ward, W. R., and Young, S. C. 1973. Periodic insolation variations on Mars. *Science* 180:638–40.
Mutch, T. A., and Head, J. W. 1975. The geology of Mars: A brief review of some recent results. *Rev. Geophys. Space Phys.* 13:411–16.
Mutch, T. A., and Saunders, R. S. 1976. The geologic development of Mars: A review. *Space Sci. Rev.* 19:3–57.
Mutch, T. A., Arvidson, R. E., Binder, A. B., Guiness, E. A., and Morris, E. C. 1977. The geology of the Viking lander 2 site. *J. Geophys. Res.* 82:4452–67.
Mutch, T. A., Arvidson, R. E., Head, J. W., Jones, K. L., and Saunders, R. S. 1976. *The geology of Mars.* Princeton Univ. Press, Princeton.
Mutch, T. A., Binder, A. B., Huck, F. O., Levinthal, E. C., Morris, E. C., Sagan, C., and Young, A. T. 1972. Imaging experiment: The Viking lander. *Icarus* 16:92–110.
Mutch, T. A., et al. (10 authors). 1976. The surface of Mars: The view from the Viking 1 lander. *Science* 193:791–801.
Mutch, T. A., et al. (25 authors). 1976. The surface of Mars: The view from the Viking 2-lander. *Science* 194:1277–83.
Mutch, T. A., et al. (Many authors). 1978. *The martian landscape.* NASA SP-425.
National Aeronautics and Space Administration. 1974. *Mars as viewed by Mariner 9.* NASA SP-329.
Neugebauer, G., Munch, G., Kieffer, H. H., Chase, S. C., and Miner, E. 1971. Mariner 1969 infrared radiometer results: Temperatures and thermal properties of the martian surface. *Astron. J.* 76:719–28.
Neukum, G., and Hiller, K. In press. Martian ages. *J. Geophys Res.*
Neukum, G., and Wise, D. U. 1976. Mars: A standard crater curve and possible new time scale. *Science* 194:1381–87.
Neukum, G., Konig, B., and Arkani-Hamed, J. 1975. A study of lunar impact crater size-distributions. *Moon* 12:201–29.
Newsome, H. E. In press. A model for hydrothermal alteration of impact melt sheets with implication for Mars. *Icarus.*
Nier, A. O., and McElroy, M. B. 1976. Structure of the neutral upper atmosphere of Mars: Results from Viking 1 and Viking 2. *Science* 194:1298–300.
Nier, A. O., and McElroy, M. B. 1977. Composition and structure of Mars's upper atmosphere: Results from the neutral mass spectrometers on Vi-

king 1 and 2. *J. Geophys. Res.* 82:4341–49.
Nier, A. O., Hanson, W. B., McElroy, M. B., Spencer, N. W., Duckett, R. J., Knight, T. C., and Cook, W. S. 1976. Composition and structure of the martian atmosphere: Preliminary results from Viking 1. *Science* 193:786–88.
Nier, A. O., Hanson, W. B., McElroy, M. B., Seiff, A., and Spencer, N. W. 1972. Entry science experiments for Viking. *Icarus* 16:74–91.
Noland, M., and Veverka, J. 1977. The photometric functions of Phobos and Deimos. II. Surface photometry of Deimos. *Icarus* 30:200–11.
Noland, M., Veverka, J., and Pollack, J. B. 1973. Mariner 9 polarimetry of Phobos and Deimos. *Icarus* 20:490–502.
Nummedal, D. 1976. Fluvial erosion on Mars. In *Proc. Colloquium on water in planetary regoliths*, pp. 47–54. Hanover, NH.
Nummedal, D. 1978. *The role of liquefaction in channel development on Mars.* NASA Tech. Memo. 79729, pp. 257–59.
Oberbeck, V. R. 1975. The role of ballistic erosion and sedimentation in lunar stratigraphy. *Rev. Geophys. Space Phys.* 13:337–62.
Oberbeck, V. R., Hortz, F., Morrison, R. H., Quaide, W. L., and Gault, D. E. 1975a. On the origin of the lunar smooth plains. *Moon* 12:19–54.
Oberbeck, V. R., Morrison, R. H., and Hortz, F. 1975b. Transport and emplacement of crater and basin deposits. *Moon* 13:9–26.
Oberbeck, V. R., Quaide, W. L., Arvidson, R., and Aggarwal, H. 1977. Comparative studies of lunar, martian, and mercurian craters and plains. *J. Geophys. Res.* 82:1681–98.
Oort, J. H. 1950. The structure of the clouds of comets surrounding the solar system and a hypothesis concerning its origin. *Bull. Astron. Inst. Neths.* 11:91.
Oort, J. H. 1963. Empirical data on the origin of comets. In *The Moon, meteorites, and comets* (B. M. Middlehurst, and G. P. Kuiper, eds.), pp. 665–73. Univ. of Chicago Press, Chicago.
Opik, E. J. 1965. Mariner IV and craters on Mars. *Irish Astron. J.* 7:92–104.
Opik, E. J. 1966. The martian surface. *Science* 153:255–65.
Oro, J., and Holzer, G. 1979. The effects of ultraviolet light on the degradation of organic compounds: A possible explanation for the absence of organic matter on Mars. In *COSPAR: Life sciences and space research*, vol. 17, (R. Holmquist, and A. C. Strickland, eds.), pp. 77–86. Pergamon Press, New York.
Owen, T. 1966. The composition and surface pressure of the martian atmosphere: Results from the 1965 opposition. *Astrophys. J.* 146:257–70.
Owen, T., and Biemann, K. 1976. Composition of the atmosphere at the surface of Mars: Detection of argon-36 and preliminary results. *Science* 193:801–03.
Owen, T., Biemann, K., Rushneck, D. R., Biller, J. E., Howarth, D. W., and Lafleur, A. L. 1977. The composition of the atmosphere at the surface of Mars. *J. Geophys. Res.* 82:4635–39.
Oyama, V. I. 1972. The gas exchange experiment for life detection: The Viking Mars lander. *Icarus* 16:167–84.
Oyama, V. I., and Berdahl, B. J. 1977. The Viking gas exchange experiments from Chryse and Utopia surface samples. *J. Geophys. Res.* 82:4669–76.
Oyama, V. I., Berdahl, B. J., and Carle, G. C. 1977. Preliminary findings of the Viking gas exchange experiment and a model for martian surface chemistry. *Nature* 265:110–14.
Oyama, V. I., Berdahl, B. J., Carle, G. C., Lehwalt, M. E., and Ginoza, H. S. 1976. The search for life on Mars: Viking 1976, gas changes as indicators of biological activity. *Origins of Life* 7:313–33.
Oyama, V. I., Berdahl, B. J., Woeller, F., and Lehwalt, M. E. 1978. The chemical activities of the Viking biology experiments and the arguments for the presence of superoxide, peroxides, Fe_2O_3 and carbon suboxide polymer in martian soil. In *COSPAR: Life sciences and space research*, vol. 16 (R. Holmquist, and A. C. Strickland, eds.), pp. 3–8. Pergamon Press, New York.
Paecht-Horowitz, M. 1971. Polymerization amino acid-phosphate anhydrates in the presence of clay minerals. In *Chemical Evolution and the Origin of Life* (R. Buvet and C. Ponnamperuma, eds.), pp. 245–51. North Holland Publ. Co., Amsterdam.
Palluconi, F. D. 1979. Mars: The thermal inertia of the surface from −60° to +60°. *Bull. Am. Astron. Soc.* 11:575.
Pang, K. D., Rhodes, J. W., Lane, A. L., and Ajello, J. M. 1977. Spectral albedo and composition of Deimos: Inferences on the origin of the martian satellites (Abs.). *Bull. Am. Astron. Soc.* 9:518.
Pardee, J. T. 1942. Unusual currents in glacial Lake Missoula, Montana. *Geol. Soc. Am. Bull.* 53:1569–600.
Parker, G. G., Shawn, L. M., and Ratzleff, K. W. 1964. Officers Cave, a pseudokarst feature in altered tuff and volcanic ash of the John Day formation in eastern Oregon. *Geol. Soc. Am. Bull.* 75:393–402.
Pechmann, J. C. 1980. The origin of polygonal troughs on the northern plains of Mars. *Icarus* 42:185–210.
Peterfreund, A. R., and Kieffer, H. H. 1979. Thermal infrared properties of the martian atmosphere. 3. Local dust clouds. *J. Geophys. Res.* 84:2853–63.
Peterson, J. E. 1978. Volcanism in the Noachis-Hellas region of Mars. 2. In *Proc. 9th Lunar Planet. Sci. Conf.*, pp. 3411–32.
Phillips, R. J. 1978. *Report on the Tharsis Workshop.* NASA Tech. Memo. 79729, pp. 334–36.
Phillips, R. J., and Ivins, E. R. 1979. Geophysical observations pertaining to solid-state convection in the terrestrial planets. *Phys. Earth Planet. Interiors* 19:107–48.
Phillips, R. J., and Lambeck, K. 1980. Gravity fields of the terrestrial planets: Long wavelength anomalies and tectonics. *Rev. Geophys. Space Phys.* 18:27–76.
Phillips, R. J., and Saunders, R. S. 1975. The isostatic state of martian topography. *J. Geophys. Res.* 80:2893–97.
Phillips, R. J., Banerdt, W. B., Sleep, N. H., and Saunders, R. S. In press. Martian stress distributions: Arguments about finite strength support of Tharsis. *J. Geophys. Res.*
Phillips, R. J., Saunders, R. S., and Conel, J. E. 1973. Mars: Crustal structure inferred from Bouguer gravity anomalies. *J. Geophys. Res.* 78:4815–20.
Phinney, W. C., Simonds, C. H., Cochran, A., and McGee, P. E. 1978. West Clearwater, Quebec impact structure. Part II. Petrology. In *Proc. 9th Lunar Planet. Sci. Conf.*, pp. 2659–93.
Pieri, D. C., 1976. Martian channels: Distribution of small channels on the martian surface. *Icarus* 27:25–50.
Pieri, D. C. 1979. Geomorphology of martian valleys. Ph.D. diss. Cornell Univ.
Pieri, D. C., and Sagan, C. 1978. *Junction angles in martian channels.* NASA Tech. Memo. 79729, p. 268.
Pike, R. J. 1974. Depth/diameter relation of fresh craters: Revisions from spacecraft data. *Geophys. Res. Lett.* 1:291–94.
Pike, R. J. 1977. Size dependence in the shape of fresh impact craters on the Moon. In *Impact and explosion cratering* (D. J. Roddy, R. O. Pepin, and R. B. Merrill, eds.), pp. 489–509. Pergamon Press, New York.
Pike, R. J. 1978. Volcanoes on the inner planets: Some preliminary comparisons of gross topography. In *Proc. 9th Lunar Planet. Sci. Conf.*, pp. 3239–73.
Pike, R. J. 1980a. *Geometric interpretation of lunar craters.* U.S. Geol. Survey, Prof. Paper 1046C.
Pike, R. J. 1980b. Control of crater morphology by gravity and target type, Mars, Earth, Moon. In *Proc. 11th Lunar and Planetary Sci. Conf.*, pp. 2159–89.
Pike, R. J., and Arthur, D. W. G. 1979: Simple to complex impact craters: The transition on Mars. NASA Tech. Memo. 80339, pp. 132–34.
Pittendrigh, C. S., Vishniac, W., and Pearman, J. P. T. 1966. *Biology and the exploration of Mars.* Nat. Acad. Sci., Nat. Res. Council Publ. 1296, Washington, D.C.
Plafker, G., and Erickson, G. E. 1977. Nevados Huascarán avalanches. Peru. In *Rock slides and avalanches* (B. Voight, ed.), pp. 277–314. Elsevier, Amsterdam.

Plescia, J. B., 1979. *Tectonism of the Tharsis region.* NASA Tech. Memo. 80339, pp. 47–49.
Plescia, J. B., and Saunders, R. S. 1979a. *Evolution of martian volcanoes.* NASA Tech. Memo. 80339, pp. 241–43.
Plescia, J. B., and Saunders, R. S. 1979b. The chronology of martian volcanoes. In *Proc. 10th Lunar Planet. Sci. Conf.*, pp. 2841–59.
Plescia, J. B., and Saunders, R. S. In press. The tectonic chronology of the Tharsis region, Mars. *Geophys. Res. Lett.*
Plescia, J. B., Saunders, R. S., and Gregory, T. 1979. Geologic evolution of Tharsis volcanoes (Abs.). In *Lunar and Planetary Science* X, vol. 3, pp. 989–91. Lunar and Planetary Institute, Houston.
Plummer, W. T., and Carson, R. K. 1969. Mars: Is the surface colored by carbon suboxide? *Science* 166:1141–42.
Pollack, J. B. 1979. Climate change on the terrestrial planets. *Icarus* 37:479–553.
Pollack, J. B., and Black, D. C. 1979. Implications of the gas compositional measurements of Pioneer Venus for the origin of planetary atmospheres. *Science* 205:56–59.
Pollack, J. B., and Sagan, C. 1969. An analysis of martian photometry and polarimetry. *Space Sci. Rev.* 9:243–99.
Pollack, J. B., Burns, J., and Tauber, M. E. 1978. Gas drag in primordial circumplanetary envelopes: A mechanism for satellite capture. *Icarus* 37:587–611.
Pollack, J. B., Colburn, D., Kahn, R., Hunter, J., Van Camp, W., Carlston, C. E., and Wolf, M. R. 1977. Properties of aerosols in the martian atmosphere, as inferred from Viking lander imaging data. *J. Geophys. Res.* 82:4479–96.
Pollack, J. B., Colburn, D. S., Flaser, M., Kahn, R., Carlston, C. E., and Pidek, D. 1979. Properties and effects of dust particles suspended in the martian atmosphere. *J. Geophys. Res.* 84:2929–45.
Pollack, J. B., Greenberg, E. H., and Sagan, C. 1967. A statistical analysis of the martian wave of darkening and related phenomena. *Planet. Space Sci.* 15:817–24.
Pollack, J. B., Leovy, C. B., Mintz, Y. H., and Van Camp, W. 1976. Winds on Mars during the Viking season: Predictions based on a general circulation model with topography. *Geophys. Res. Lett.* 3:479–82.
Pollack, J. B., Veverka, J., Sagan, C., Duxbury, T. C., Acton, C. H., Born, G. H., Hartmann, W. K., and Smith, B. A. 1973. Mariner 9 television observations of Phobos and Deimos. *J. Geophys. Res.* 78:4313–26.
Ponnamperuma, C., Shimoyama, A., Yamada, M., Hobo, T., and Pal, R. 1977. Possible surface reactions on Mars: Implications for Viking biology results. *Science* 197:455–57.
Press, F., and Siever, R. 1978. *Earth.* Freeman, San Francisco.
Quaide, W. L., and Oberbeck, V. R. 1968. Thickness determinations of the lunar surface layer from impact craters. *J. Geophys. Res.* 73:5247–70.
Rapp, A. 1960. *Talus slopes and mountain walls at Tempelfjorden, Spitzbergen.* Norsk Polarinstitutt Skrifter, no. 119, p. 93.
Reimers, C. E., and Komar, P. D. 1979. Evidence for explosive volcanic density currents on certain martian volcanoes. *Icarus* 39:88–110.
Roddy, D. J. 1977. Large-scale impact and explosion craters: Comparisons of morphological and structural analogs. In *Impact and explosion cratering* (D. J. Roddy, R. O. Pepin, and R. B. Merrill, eds.), pp. 185–246. Pergamon Press, New York.
Roth, L. E., Downs, G. S., and Saunders, R. S. 1980. Radar altimetry of south Tharsis, Mars. *Icarus* 42:287–316.
Ryan, J. A. 1964. Notes on the martian yellow clouds. *J. Geophys. Res.* 69:3759–70.
Ryan, J. A., and Henry, R. M. 1979. Mars atmospheric phenomena during major dust storms as measured at the surface. *J. Geophys. Res.* 84:2821–29.
Ryan, J. A., Henry, R. M., Hess, S. L., Leovy, C. B., Tillman, J. E., and Walcek, C. 1978. Mars meteorology: Three seasons at the surface. *Geophys. Res. Lett.* 5:715–18.
Sagan, C. 1973. Sandstorms and eolian erosion on Mars. *J. Geophys. Res.* 78:4155–61.
Sagan, C., and Bagnold, R. A. 1975. Fluid transport on Earth and eolian transport on Mars. *Icarus* 26:209–18.
Sagan, C., and Lederberg, J. 1976. The prospects for life on Mars: A pre-Viking assessment. *Icarus* 28:291–300.
Sagan, C., and Fox, P. 1975. The canals on Mars: An assessment after Mariner 9. *Icarus* 25:602–12.
Sagan, C., Pieri, D. C., Fox, P., Arvidson, R. E., and Guiness, E. A. 1977. Particle motion on Mars inferred from Viking lander cameras. *J. Geophys. Res.* 78:4155–61.
Sagan, C., Veverka, J., Fox, P., Dubisch, R., French, R., Gierasch, P., Quam, L., Lederberg, J., Levinthal, E., Tucker, R., Eross, B., and Pollack, J. B. 1973a. Variable features on Mars. 2. Mariner 9 global results. *J. Geophys. Res.* 78:4163–96.
Sagan, C., Toon, O. B., and Gierasch, P. J. 1973b. Climate change on Mars. *Science* 181:1045–49.
Sagan, C., Veverka, J., Fox, P., Dubisch, R., Lederberg, J., Levinthal, E., Quam, L., Tucker, R., Pollack, J. B., and Smith, B. A. 1972. Variable features on Mars: Preliminary Mariner 9 results. *Icarus* 17:346–72.
Salisbury, J. W. 1966. The light and dark areas of Mars. *Icarus* 5:291–98.
Salisbury, J. W., and Hunt, G. R. 1969. Compositional implications of the spectral behavior of the martian surface. *Nature* 222:132–36.
Schaber, G. G. 1973. Lava flows in Mare Imbrium: Geologic evaluation from Apollo orbital photographs. In *Proc. 4th Lunar Sci. Conf.*, pp. 73–92.
Schaber, G. G., Horstmann, K. C., and Dial, A. L. 1978. Lava flow materials in the Tharsis region of Mars. *Proc. 9th Lunar Planet. Sci. Conf.*, pp. 3433–58.
Schiaparelli, G. V. 1878. *Osservazioni astronomiche e fisiche sull' asse di rotazione e sulla topografia del pianete Marte.* Atti della R. Academia dei Lincei, Memoria della ce. di scienze fisiche. Mem. 1, ser. 3 v. 10, pp. 308–439.
Schonfeld, E. 1977. Martian volcanism. *Lunar Science Conference VIII*, pp. 843–45. Lunar and Planetary Institute, Houston.
Schultz, P. H., Glicken, H., and McGetchin, T. R. 1979. Intrusive melting of water-ice on Mars. In *Lunar and Planetary Science* X, pp. 1075–77. Lunar and Planetary Institute, Houston.
Schumm, S. A. 1974. Structural origin of large martian channels. *Icarus* 22:371–84.
Scott, D. H., and Carr, M. H. 1978. *Geologic map of Mars.* U.S. Geol. Survey, Misc. Inv. Map I-1083.
Seiff, A., and Kirk, D. B. 1977. Structure of the atmosphere of Mars in summer at mid-latitudes. *J. Geophys. Res.* 82:4364–78.
Sengor, A. M. C., and Jones, E. C. 1975. A new interpretation of martian tectonics with special reference to the Tharsis region (Abs.). *Geol. Soc. Am., Abstract with Program* 7:1264.
Settle, M., and Head, J. W. 1979. The role of rim slumping in the modification of lunar impact craters. *J. Geophys. Res.* 84:3081–96.
Sharp, R. P. 1973a. Mars: Fretted and chaotic terrains. *J. Geophys. Res.* 78:4073–83.
Sharp, R. P. 1973b. Mars: Troughed terrain. *J. Geophys. Res.* 78:4063–72.
Sharp, R. P. 1973c. Mars: South polar pits and etched terrain. *J. Geophys. Res.* 78:4222–30.
Sharp, R. P. 1974. Ice on Mars. *J. Glaciology* 13:173–85.
Sharp, R. P., and Malin, M. C. 1975. Channels on Mars. *Geol. Soc. Am. Bull.* 86:593–609.
Sharp, R. P., Soderblom, L. A., Murray, B. C., and Cutts, J. A. 1971. The surface of Mars. 2. Uncratered terrains. *J. Geophys. Res.* 76:331–42.
Shaw, H. R. 1973. Mantle convection and volcanic periodicity in the Pacific: Evidence from Hawaii. *Geol. Soc. Am. Bull.* 84:1505–26.
Shoemaker, E. M. 1960. Penetration mechanics of high-velocity meteorites, illustrated by Meteor Crater, Arizona. In *Proc. 21st Internat. Geol. Congr., Pt. 18*, pp. 418–34.
Shoemaker, E. M. 1962. Interpretation of lunar craters. In *Physics and astronomy of the Moon* (Z. Kopal, ed.), pp. 283–359. Academic Press, New York.
Shoemaker, E. M. 1963. Impact mechanics at

Meteor Crater, Arizona. In *The Moon, meteorites, and comets* (B. M. Middlehurst, and G. P. Kuiper, eds.), pp. 301–36. Univ. of Chicago Press, Chicago.

Shoemaker, E. M. 1966. Preliminary analysis of the fine structure of the lunar surface in Mare Cognitum. In *The Nature of the Lunar Surface* (W. N. Hess, D. H. Menzel, and J. A. O'Keefe, eds.), pp. 23–77. Johns Hopkins Press, Baltimore.

Shoemaker, E. M., et al. (12 authors). 1966. *Progress in the analysis of the fine structure and geology of the lunar surface from Ranger VIII and IX photographs*. Jet Propulsion Lab., Tech. Rept., No. 32-800, pp. 249–337.

Shoemaker, E. M. et al. (12 authors). 1970. *Preliminary geologic investigation of the Apollo 12 landing site, Part A*. NASA SP-235, pp. 113–56.

Short, N. M., and Foreman, M. L. 1972. Thickness of impact crater ejecta on the lunar surface. *Mod. Geol.* 3:69–91.

Shreve, R. L. 1966. Sherman landslide, Alaska. *Science* 154:1639–43.

Siever, R. 1974. Comparison of Earth and Mars as differentiated planets. *Icarus* 22:312–24.

Singer, R. B., McCord, T. B., Clark, R. N., Adams, J. B., and Huguenin, R. L. 1979. Mars surface composition from reflectance spectroscopy: A summary. *J. Geophys. Res.* 84:8415–26.

Sinton, W. M., and Strong, J. 1960. Radiometric observations of Mars. *Astrophys. J.* 131:459–69.

Sjogren, W. L. 1979. Mars gravity: High-resolution results from Viking orbiter 2. *Science* 203:1006–09.

Sjogren, W. L., Lorell, J., Wong, L., and Downs, W. 1975. Mars gravity field based on a short arc technique. *J. Geophys. Res.* 80:2899–908.

Sleep, N. H., and Phillips, R. J. 1979. An isostatic model for the Tharsis province, Mars. *Geophys. Res. Lett.* 6:803–06.

Slipher, E. C. 1937. An outstanding atmospheric phenomenon on Mars. *Pub. Astron. Soc. Pacific* 49:137–40.

Slipher, E. C. 1962. *The photographic story of Mars.* Northland Press, Flagstaff, AZ.

Smith, B. A. 1970. Phobos: Preliminary results from Mariner 7. *Science* 168:828–30.

Smith, E. I. 1976. Comparison of the crater morphology-size relationship for Mars, Moon, and Mercury. *Icarus* 28:543–50.

Smith, E. I., and Hartnell, J. A. 1977. *The effect of nongravitational factors on the shape of martian, mercurian, and lunar craters: Target effects.* NASA Tech. Memo. X-3511, pp. 91–93.

Smith, E. I., and Sanchez, A. G. 1973. Fresh lunar craters: Morphology as a function of diameter, a possible criterion for crater origin. *Mod. Geol.* 4:51–59.

Smith, S. A., and Smith, B. A. 1972. Diurnal and seasonal behavior of discrete white clouds on Mars. *Icarus* 16:509–21.

Smoluchowski, R. 1968. Mars: Retention of ice. *Science* 159:1348–50.

Snyder, C. W. 1979. The extended mission of Viking. *J. Geophys. Res.* 84:7917–33.

Soderblom, L. 1976. Viking orbital colorimetric images of Mars: Preliminary results. *Science* 194:97–99.

Soderblom, L. A., and Wenner, D. B. 1978. Possible fossil H_2O liquid-ice interfaces in the martian crust. *Icarus* 34:622–37.

Soderblom, L. A., Condit, C. D., West, R. A., Herman, B. M., and Kriedler, T. J. 1974. Martian planetwide crater distributions: Implications for geologic history and surface processes. *Icarus* 22:239–63.

Soderblom, L. A., Edwards, K., Eliason, E. M., Sanchez, E. M., and Charette, M. P. 1978. Global color variations on the martian surface. *Icarus* 34:446–64.

Soderblom, L. A., Kriedler, T. J., and Masursky, H. 1973. Latitudinal distribution of a debris mantle on the martian surface. *J. Geophys. Res.* 78:4117–22.

Soderblom, L. A., Malin, M. C., Cutts, J. A., and Murray, B. C. 1973. Mariner 9 observations of the surface of Mars in the north polar region. *J. Geophys. Res.* 78:4197–210.

Soffen, G. A. 1977. The Viking Project. *J. Geophys. Res.* 82:3959–70.

Solomon, S. C., and Head, J. W. 1979. Vertical movement in mare basins: Relation to mare emplacement, basin tectonics, and lunar thermal history. *J. Geophys. Res.* 84:1667–82.

Solomon, S. C., Head, J. W., and Comar, R. P. 1979. *Thickness of martian lithosphere from tectonic features: Evidence for lithospheric thinning beneath volcanic provinces.* NASA Tech. Memo. 80339, pp. 60–62.

Soter, S., and Harris, A. 1977. Are striations on Phobos evidence for tidal stress? *Nature* 268:421–22.

Squyres, S. W. 1978. Martian fretted terrain: Flow of erosional debris. *Icarus* 34:600–13.

Squyres, S. W. 1979a. The evolution of dust deposits in the martian north polar regions. *Icarus* 40:244–61.

Squyres, S. W. 1979b. *Distribution of lobate debris aprons on Mars.* NASA Tech. Memo. 80339, pp. 50–52.

Strom, R. G. 1972. Lunar mare ridges, rings, and volcanic ring complexes. In *The Moon* (G. K. Runcorn, and H. C. Urey, eds.) pp. 187–215. Reidel, Dordrecht, Holland.

Strom, R. G., and Whitaker, E. A. 1976. *Populations of impacting bodies in the inner solar system.* NASA Tech. Memo. X-3364, pp. 194–96.

Swanson, D. A. 1972. Magma supply rate at Kilauea Volcano, 1952–1971. *Science* 177:169–70.

Swanson, D. A. 1973. Pahoehoe flows from the 1969–1971 Mauna Ulu eruptions, Kilauea Volcano, Hawaii. *Geol. Soc. Am. Bull.* 84:615–26.

Taylor, S. R. 1974. *Lunar Science: A post-Apollo view.* Pergamon Press, New York.

Thomas, P. 1979. Surface features of Phobos and Deimos. *Icarus* 40:223–43.

Thomas, P., and Veverka, J. 1979. Seasonal and secular variations of wind streaks on Mars: An analysis of Mariner 9 and Viking data. *J. Geophys. Res.* 84:8131–46.

Thomas, P., Veverka, J., and Campos-Marquetti, R. 1979. Frost streaks in the south polar cap of Mars. *J. Geophys. Res.* 84:4621–33.

Thomas, P., Veverka, J., and Duxbury, T. 1978. Origin of the grooves on Phobos. *Nature* 273:282–84.

Thompson, G. A., and Burke, D. B. 1974. Regional geophysics of the Basin and Range Province. *Ann. Rev. Earth Planet. Sci.* 2:213–38.

Thompson, D. T. 1972. Brief history of the martian "violet haze" problem. *Rev. Geophys. Space Sci.* 10:919–33.

Thompson, D. T. 1973. Time variations of martian regional contrasts. *Icarus* 20:42–47.

Thorarinsson, S. 1957. *The jokulhlaup from the Katla area in 1955 compared with other jökulhlaups in Iceland.* Reykjavik Mus. Natl. Hist., Misc. Paper 18, pp. 21–25.

Thurber, C. H., and Toksoz, M. N. 1978. Martian lithosphere thickness from elastic flexure theory. *Geophys. Res. Lett.* 5:977–80.

Tillman, J. E., Henry, R. M., and Hess, S. L. 1979. Frontal systems during passage of the martian north polar hood over the Viking lander 2 site prior to the first 1977 dust storm. *J. Geophys. Res.* 84:2947–55.

Toksoz, M. N., and Hsui, A. T. 1978. Thermal history and evolution of Mars. *Icarus* 34:537–47.

Tolson, R. H. et al. (19 authors). 1978. Viking first encounter of Phobos: Preliminary results. *Science* 199:61–64.

Tombaugh, C. W. 1968. A survey of long-term observational behavior of various martian features that affect some recently proposed interpretations. *Icarus* 8:227–58.

Toon, O. B., Pollack, J. B., and Sagan, C. 1977. Physical properties of particles comprising the martian dust storm, 1971–1972. *Icarus* 30:663–96.

Toulmin, P., Baird, A. K., Clark, B. C., Keil, K., and Rose, H. J. 1973. Inorganic chemical investigations by x-ray fluorescence analysis: The Viking Mars lander. *Icarus* 20:153–78.

Toulmin, P., Baird, A. K., Clark, B. C., Keil, K., Rose, H. J., Christian, R. P., Evans, P. H., and Kelliher, W. C. 1977. Geochemical and mineralogical interpretation of the Viking inorganic chemical results. *J. Geophys. Res.* 82:4625–34.

Toulmin, P., Clark, B. C., Baird, A. K., Keil, K., and Rose, H. J. 1976. Preliminary results from the Viking x-ray fluorescence experiment: The first

sample from Chryse Planitia, Mars. *Science* 194:81–84.
Turekian, K. K., and Clark, S. P. 1975. Nonhomogeneous accumulation model for terrestrial planet formation and the consequences for the atmosphere of Venus. *J. Atmos. Sci.* 32:1257–61.
U. S. Geological Survey. 1976. *Topographic Map of Mars.* U. S. Geol. Survey, Misc. Inv. Map I-961.
Van Tassel, R. A., and Salisbury, J. W. 1964. The composition of the martian surface. *Icarus* 3:264–69.
Varnes, D. J. 1958. Landslide types and process. In *Landslides and engineering practice* (E. B. Eckel, ed.), pp. 20–47. Nat. Res. Council, Highway Research Board, Spec. Rept. 29.
Vaucouleurs, G. de. 1954. *Physics of the planet Mars.* Faber and Faber, London.
Veverka, J. 1977. Photometry of satellite surfaces. In *Planetary satellites* (J. Burns, ed.), pp. 171–209. Univ. of Arizona Press, Tucson.
Veverka, J. 1978. The surfaces of Phobos and Deimos. *Vistas in astronomy* 22:163–92.
Veverka, P., and Thomas, P. 1979. Phobos and Deimos: A preview of what asteroids are like? In *Asteroids* (T. Gehrels, ed.), pp. 628–51. Univ. Arizona Press, Tucson.
Veverka, J., Thomas, P., and Bloom, A. 1978. *Classification of martian wind streaks.* NASA Tech. Memo. 78455, pp. 46–47.
Veverka, J., Thomas, P., and Greeley, R. 1977. A study of variable features on Mars during the Viking primary mission. *J. Geophys. Res.* 82:4167–87.
Wade, F. A., and de Wys, J. N. 1968. Permafrost features in the martian surface. *Icarus* 9:175–85.
Wald, G. 1962. Life in the second and third periods; or, why phosphorus and sulfur for high-energy bonds? In *Horizons in biochemistry* (M. Kasha and B. Pullman, eds.), pp. 127–42. Academic Press, New York.
Walker, G. P. L. 1973. Lengths of lava flows. *Phil. Trans. Roy. Soc. London*, ser. A., vol. 274, pp. 107–18.
Wallace, D., and Sagan, C. 1979. Evaporation of ice in planetary atmospheres: Ice-covered rivers on Mars. *Icarus* 39:385–400.
Ward, A. W. 1979. Yardangs on Mars: Evidence of recent wind erosion. *J. Geophys. Res.* 84:8147–66.
Ward, W. R. 1973. Large-scale variations in the obliquity of Mars. *Science* 181:260–62.
Ward, W. R. 1974. Climatic variations on Mars. 1. Astronomical theory of insolation. *J. Geophys. Res.* 79:3375–86.
Ward, W. R., Burns, J. A., and Toon, O. B. 1979. Past obliquity oscillations of Mars: The role of the Tharsis uplift. *J. Geophys. Res.* 84:243–59.
Ward, W. R., Murray, B. C., and Malin, M. C. 1974. Climatic variations on Mars. 2. Evolution of carbon dioxide atmosphere and polar caps. *J. Geophys. Res.* 79:3387–95.
Webster, P. J. 1977. The low-altitude circulation of Mars. *Icarus* 30:626–49.
Wellman, J. B., Landauer, F. P., Norris, D. D., and Thorpe, R. E. 1976. *The Viking orbiter visual imaging subsystem.* Am. Inst. of Aeronautics and Astronautics, Paper 76–124.
Wells, E., and Veverka, J. 1978. Mars: Simulation of the spectrophotometric effects of martian dust storms (Abs.). *Bull. Am. Astron. Soc.* 10:569.
Wentworth, C. K. 1931. *The mechanical composition of sediments in graphic form.* Iowa State Univ. Studies in Nat. History, vol. 14, no. 3.
West, M. 1974. Martian volcanism: Additional observations and evidence of pryoclastic activity. *Icarus* 21:1–11.
Wetherill, G. W. 1975. Late heavy bombardment of the Moon and terrestrial planets. In *Proc. 6th Lunar Sci. Conf.*, pp. 1539–61.
White, B. R., Greeley, R., and Iverson, J. D. 1979. *Numerical solutions to particle flow on Mars.* NASA Tech. Memo. 80339, pp. 322–24.
Whitney, M. 1978. The role of vorticity in developing lineation by wind erosion. *Geol. Soc. Am. Bull.* 89:1–18.
Whitney, M., and Dietrich, R. V. 1973. Ventifact sculpture by windblown dust. *Geol. Soc. Am. Bull.* 84:2561–82.
Wilhelms, D. E. 1973. Comparison of martian and lunar multiringed circular basins. *J. Geophys. Res.* 78:4084–95.
Williams, R. S. 1978. *Geomorphic processes in Iceland and on Mars: A comparative appraisal from orbital images.* Geol. Soc. Am., 91st Annual Mtg., Abstracts with Programs, p. 517.
Wilshire, H. G., Offield, T. W., Howard, K. A., and Cummings, D. 1972. *Geology of the Sierra Madera cryptovolcanic explosion structure, Pecos County, Texas.* U.S. Geol. Survey, Prof. Paper, 549-H, pp. H1-H42.
Wilson, I. G. 1971. Desert sand flow basins and a model for the development of ergs. *Geog. J.* 137:180–99.
Wilson, I. G. 1973. Ergs. *Sedimentary Geol.* 10:77–106.
Wilson, L. 1976. Explosive volcanic eruptions. III. Plinian eruption columns. *Geophys. J. Roy. Astron. Soc.* 45:543–56.
Wise, D. U. 1974. Martian fault patterns and time sequence in relation to volcanism, northern Tharsis ridge area. (Abs.). *Trans. Am. Geophys. Union (EOS)* 55:341.
Wise, D. U. 1975. Faulting and stress trajectories near Alba Volcano northern Tharsis ridge of Mars. (Abs.). In *Proc. Int. Colloquium of Planet. Geol., Rome*, pp. 430–33.
Wise, D. U. 1977. *Timing of deformational events in the northern Tharsis bulge of Mars.* NASA Tech. Memo. X-2511, pp. 59–60.
Wise, D. U., Golombek, M. P., and McGill, G. E. 1979. Tharsis province of Mars: Geologic sequence, geometry, and a deformation mechanism. *Icarus* 38:456–72.
Wood, C. A., and Head, J. W. 1976. Comparisons of impact basins on Mercury, Mars, and the Moon. In *Proc. 7th Lunar Sci. Conf.*, pp. 3629–51.
Wood, C. A., Head, J. W., and Cintala, M. J. 1978. Interior morphology of fresh martian craters: The effects of target characteristics. In *Proc. 9th Lunar Planet. Sci. Conf.*, pp. 3691–709.
Wood, J. A. 1975. The Moon. *Sci. Am.* 233:92–102.
Woodruff, N. R., and Siddoway, F. H. 1965. A wind erosion equation. *Proc. Soil Sci. Soc. Am.* 29:602–08.
Woronow, A. 1978. A general cratering history model and its implications for the lunar highlands. *Icarus* 34:76–88.
Wu, S. S. C. 1978. Mars synthetic topographic mapping. *Icarus* 33:417–40.
Wu, S. S. C., Schafer, F. J., Nakata, G. M., and Jordan, R. 1973. Photogrammetric evaluation of Mariner 9 photography. *J. Geophys. Res.* 78:4405–10.
Young, R. S., Painter, R. B., and Johnson, R. D. 1965. *An analysis of the extraterrestrial life detection problem.* NASA SP-75.
Yung, Y. L., and Pinto, J. P. 1978. Primitive atmosphere and implications for the formation of channels on Mars. *Nature* 273:730–32.
Zill, L. P., Mack, R., and DeVincenzi, D. L. In press. Mars ultraviolet simulation facility. *J. Mol. Evol.*

INDEX